3D Flash Memories

Rino Micheloni
Editor

3D Flash Memories

Editor
Rino Micheloni
Performance Storage BU
Microsemi Corporation
Vimercate
Italy

ISBN 978-94-017-7510-6 ISBN 978-94-017-7512-0 (eBook)
DOI 10.1007/978-94-017-7512-0

Library of Congress Control Number: 2016936991

Printed on acid-free paper

This Springer imprint is published by Springer Nature
The registered company is Springer Science+Business Media B.V. Dordrecht

Once again to my wife Sabrina
and my daughters Laura and Greta

—Rino Micheloni

Foreword

In 1903, the Wright brothers made the first powered flight into the third dimension. After the fulfillment of this ancient dream, predictions of longer, faster, and safer flights were easy to make. Fundamental, practical breakthroughs, such as this first powered flight, are momentous when achieved. But even more striking are the unforeseen achievements that these breakthroughs enable and inspire. Few would have predicted that humans would leave footprints on the surface of the moon only 66 years after the Wright brothers' first flight.

The semiconductor industry touches nearly all parts of our lives and quietly scales precision manufacturing to volumes unimaginable with other methods. Wafers of nonvolatile memory (NVM) routinely are fabricated with trillions of individual cells. Although challenging, this density of components was predictable from the early days of Gordon Moore's observations that the number of components on a 2D integrated circuit would double every year, and later every 2 years ("Moore's Law"[1]). Harder to predict, in 1965, were the inventions of the internet, smart phones, and autonomous vehicles. These are economically possible largely because of Moore's Law.

The doubling of components is slowing to every 2.5 years.[2] But we have also arrived at an unprecedented inflection point for Moore's Law. Component density can continue to increase by building into the third dimension. Other technologies have already added an "extra" dimension. These include pressure-sensitive touchscreens, video gesture capture, autonomous vehicle navigation, 3D printing, and drones that maneuver throughout the sky.

Traditional 2D semiconductor processing obviously includes the third dimension. Material thicknesses are varied; devices are formed by multiple layers of

[1]G. Moore, "Cramming more components onto integrated circuits", *Electronics Magazine* 19 pp. 24–27, (1965); Additional information at: http://www.computerhistory.org/semiconductor/timeline/1965-Moore.html.

[2]Recently, Intel has stated that since the 22 nm process node two and a half years is the pace of doubling of components per integrated circuit: http://www.cnet.com/news/keeping-up-with-moores-law-proves-difficult-for-intel/.

materials and shapes; and interlayer interconnects enable complex routing. Furthermore, "FinFET"[3] 3D transistors have been shipping in microprocessors for several years. However, the processing that is currently utilized for 3D (or "vertical") NAND flash is fundamentally different. 2D processes fabricate devices that can be fully enumerated by viewing the *x–y* plane of the wafer from above. However, 3D flash also fabricates devices that are enumerated in the *z* direction, perpendicular to the plane of the wafer. Just as in 1965 when Gordon Moore published his projections for the future of semiconductor fabrication, we cannot foresee all of the benefits that scaling in this additional dimension will provide.

It would be shortsighted to dismiss 3D semiconductor processing as just an incremental step in the progression of micro/nano-electronic fabrication. This new dimension multiplies the number of elements that can be created with the same number of photolithographic steps. As experience in manufacturing is gained, this multiplier grows from dozens, to hundreds, and possibly beyond. Fabricating devices in the third dimension provides a new degree of freedom for creativity in designs, architectures, and layouts. It offers challenges to create breakthroughs in testing, integration, power density, heat dissipation, and systems-on-chip. 3D semiconductors can become a profitable platform for experimenting with interactions between these fields to yield new economies of scale that do not yet exist.

3D upon 3D systems are being imagined in which a stack of varied devices can replace a computer, and eventually a portion of a data center. What is not yet seen is the new field of view these advances will clear so that we can envision what is beyond them. Becoming efficient at building 3D flash is likely to be a critical step on the path to creating what has not yet been imagined. It is often said that what used to take a roomful of electronics now easily fits in our pockets. 3D processing may allow us to make the same statement about a roomful of *today's* technology. It may also allow us to say that what once took a pocketful of components now fits in a blood cell.

To accelerate the implementation and adoption of 3D systems, clear, practical information is needed. Rino Micheloni's new book provides a wealth of hard-to-find information that can help push the industry forward. It can expand the number of people who confidently take the first step into new manufacturing possibilities. Just as the practical skills needed for creating powered flight were learned in a bicycle shop, the practical skills for creating the next breakthroughs of nano-electrical systems are being learned in 3D flash fabs. This large industry will serve as an invaluable incubator from which future jobs, employees, and inventors will grow.

University professors are encouraged to build on the information in this book and make 3D complexity clear and understandable for the next generations of graduates. 3D processing and design will likely move from the "future topics" section of classes to core fundamentals of the engineering curriculum. Students will

[3]C.H. Lee et al., "Novel body tied FinFET cell array transistor DRAM with negative word line operation for sub 60 nm technology and beyond", *VLSI Technical Digest*, pp. 130–131, (2004).

learn how to efficiently transport charge at the 3D nanoscale and perhaps have routine undergraduate lab exercises on the manipulation of electron spin from one layer to the next.

The important lessons learned by fabricating 3D flash in high volume will provide insights for forming nanoscale "vertical" structures that can act as scaffolding for future manufacturing techniques. These scaffolds might support and direct self-assembling structures or even biological growth. Such structures and know-how may revolutionize power efficiency, health care, and system miniaturization.

As cost and power efficiencies are realized, products with tremendous processing, sensing, and data densities can be expected. These products may make it possible to automate and optimize tasks from the critical to the mundane to the joyous. Affordable weather forecasts for your precise location may become available. Autonomous transport will free up many human hours lost while driving, enabling an increase in personal productivity. Ever more capable personal assistants will provide custom coaching for your athletic performance, artistic expression, or even social interactions.

This high density of processing, sensing, and data will enable early-adopters to differentiate their products and services in new ways, creating profitable market opportunities. As 3D processing becomes more commonplace, its techniques may be applied to existing products to improve their utility, cost, and power consumption. Optimal design, across all electronics, might need to be redefined. We are likely to see old industries disrupted and new ones created.

Thanks to Rino and his many authors, contributors, supporters, editors, and staff for creating an accessible source of difficult-to-obtain knowledge on 3D memory. I believe your work will serve as a catalyst for accelerating advances in 3D wafer fabrication. This, in turn, will accelerate advances in the many fields that depend on semiconductor technology.

Plano, TX, USA
2016

Charles H. Sobey
Chief Scientist at ChannelScience

Preface

It is difficult to understate the impact of NAND flash memory in the last decade. Although flash memory has existed for a few decades, the most recent generations of NAND memory have enabled wholesale change in many aspects of our daily lives. The digital music player revolution was greatly accelerated with a conversion from miniature Hard Disk Drives (HDDs) to NAND flash memory, providing the ability for customers to carry more than just the initial ground-breaking (at the time) “1000 songs in your pocket”. Now we’re carrying movie collections, video podcasts, albums, video games, and home videos in that same pocket, due in large part to advances made in NAND flash technology. Who would have guessed that a major feature for today’s smart phone buyer would be the amount of local storage (i.e., NAND flash) maintained in the phone?

In addition to the huge advances enabled by NAND flash in consumer devices, a similar dynamic is emerging in the systems that power the internet. The introduction of NAND flash (and SSDs) to storage architectures has completely disrupted the existing storage giants, causing many of them to acquire start-ups versus building their own systems that can capitalize on the performance benefits of NAND-based SSDs over traditional HDDs. Entire racks of HDDs are being replaced by a single instance of an all-flash-array, and end customers are finding themselves paying less in total cost of ownership. The presence of NAND flash, as well as other next-generation nonvolatile memories, in future storage architectures is also driving an entirely new cycle of innovation in software and hardware design, creating a Storage Renaissance of sorts.

To put it simply, NAND flash continues to disrupt our world, and it doesn’t appear to be slowing down. 3D NAND is the next enabler of continued advancement and disruption, and this book provides an excellent foundation for anyone interested in the technology, where the technology is heading next, and its impact on the industry.

Derek D. Dicker
Vice President and Performance
Storage BU Manager, Microsemi Corporation

Acknowledgments

Writing a technical book is always a challenge as there are so many details, diagrams, graphs, and numbers that it is very easy to have "bugs" (mistakes, errors), like in all engineering projects. This is why I am especially grateful to all the people who volunteered to review the chapters.

This book would have never reached a conclusion without all the contributed chapters. I know that authors spent weekends and nights to put together the best possible material. Thank You All!

I also want to thank Cindy Zitter from Springer: this is the sixth book she is helping me with. I do really appreciate her continuous support.

Last but not least, let me thank Luca Crippa for his tireless dedication to this project: he contributed with all the amazing bird's-eye views and cross sections of Chaps. 3–6. I am sure that the reader will be impressed by these 3D views.

Rino Micheloni

Contents

About the Editor

Dr. Rino Micheloni is Fellow at Microsemi Corporation where he currently runs the Non-Volatile Memory Lab in Milan, with special focus on NAND flash. Prior to joining Microsemi, he was Fellow at PMC-Sierra, working on NAND flash characterization, LDPC, and NAND signal Processing as part of the team developing flash controllers for PCIe SSDs. Before that, he was with IDT (Integrated Device Technology) as Lead Flash Technologist, driving the architecture and design of the BCH engine in the world's 1st PCIe NVMe SSD controller. Early in his career, he led flash design teams at STMicroelectronics, Hynix, Infineon, and Qimonda; during this time, he developed the industry's first MLC NOR device with embedded ECC technology and the industry's first MLC NAND with embedded BCH.

Rino is IEEE Senior Member, he has co-authored more than 50 publications, and he holds 241 patents worldwide (including 119 US patents). He received the STMicroelectronics Exceptional Patent Award in 2003 and 2004, and the Qimonda IP Award in 2007.

Rino has published the following books with Springer: *Inside Solid State Drives* (2013), *Inside NAND Flash Memories* (2010), *Error Correction Codes for Non-Volatile Memories* (2008), *Memories in Wireless Systems* (2008), and *VLSI-Design of Non-Volatile Memories* (2005).

Introduction

NAND Flash Memories: 3D or 5D?

Flash memory has been a disruptive technology from its industrial inception in the early '90s. Believe it or not, innovation is still ongoing after more than 25 years.

At the beginning it was NOR flash and Intel was the key player. NOR didn't ask for Error Correction Code (ECC), but programming parallelism was pretty low because programming was based on *Channel Hot Electron* (CHE), which is known to be extremely power consuming. Later Toshiba introduced a new type of flash, called NAND, where cells are organized like n-MOS transistors in a NAND logic gate. In the NOR architecture, two memory cells share a bitline contact; in a NAND matrix there is only one bitline contact every NAND string, which is usually made of 64 or 128 cells. Therefore, the NAND array is natively smaller than NOR and, because of the lower number of contacts in the array, technology shrink is also easier. Indeed, NAND has replaced DRAM in driving the process technology race and today we are talking about feature sizes down to 14–15 nm.[4] The other major difference with respect to NOR is the usage of Fowler–Nordheim tunneling for programming; because of its negligible power consumption, tunneling allows programming 16, 32, or 64 kB at the same time. Think about the importance of programming speed when you are taking a picture with your digital camera!

Thanks to its storage density, NAND has changed our lives. USB keys have replaced floppy disks and flash cards (SD, eMMC) record our pictures and movies instead of analog films.

In both applications, storage density is the enablement factor. Flash vendors have spent billions and billions of dollars to shrink the technology in order to make NAND a consumer product. In the first years of the twenty-first century, a third NAND dimension was introduced to foster this memory density race even further:

[4]S. Lee et al., "A 128 Gb 2b/cell NAND Flash Memory in 14 nm Technology with t_{prog} = 640 μs and 800 Mb/s I/O Rate", 2016 IEEE International Solid-State Circuits Conference (ISSCC), Dig. Tech. Papers, pp. 138–139, San Francisco, USA, Feb. 2016.

multilevel storage. In practice, each single physical cell can contain more than one bit of information. SLC is the acronym used for 1 bit per cell, MLC is used for 2 bits per cell, while TLC refers to cells able to store 3 digital bits. QLC, i.e., 4 bits in the same cell, might also become a reality relatively soon.[5]

But again, this was not enough for a market increasingly hungry for storage density. To be more specific, cellular phones, and later on smart phones, came into the game with very severe space constraints; their unbelievable selling volumes forced the market to find another storage dimension (the fourth one): multi-die stacking. In other words, several pieces of silicon are stacked inside a single physical package, one on top of each other. Nowadays, after years of development, assembly technologies and design solutions allow mass production of 8-die stacks. The combination of a 8-die stack with a sub-20 nm 128 Gb (MLC) puts 1 Tb of NAND memory in a single 14 mm × 18 mm footprint. Not happy enough, most of the flash vendors are also engineering a 16-die stacking, allowing 2 Tb NAND in a single package!

NAND R&D and NAND fabs are extremely expensive and suppliers are always looking for new applications. In the last 4–5 years Solid State Drives (SSDs) have emerged as the new killer applications for flash: first in the consumer space (mainly driven by Apple products) but now expanding to enterprise applications as well (more details in Chap. 1). SSDs fueled a new wave of innovations. In 2013, at the Flash Memory Summit conference in the Silicon Valley, Samsung announced the first 3D-based commercial product, after 10 years of research and development. What is 3D in this case? Basically, multiple layers (up to 48, as we speak[6]) of memory cells are grown within the same piece of silicon. This is actually the fifth storage dimension in the history of flash.

3D is a brand new technology, not only because of its multi-layer architecture, but also because it is based on a new type of NAND memory cell. So far, NAND has always been based on "floating gate", where the information is stored by injecting electrons in a piece of polysilicon completely surrounded by oxide (this is why it is called floating gate). On the contrary, most of 3D memories are based on "charge trap" cells. Actually, this is not a brand new type of cell: several memory vendors developed this technology in the past, because they thought it could be more scalable than floating gate, even with planar layout. History tells us that they never succeeded because charge trap cells had poor reliability. Now, with 3D NAND, charge trap has a new chance. Certainly, a lot of work has been done on the materials side, but there are still a lot of uncertainties and question marks.

---

[5]S. Ohshima, "Advances in 3D Memory: High Density Storage for Hyperscale, Cloud Applications, and Beyond", Keynote 5, Flash Memory Summit, Santa Clara (CA), USA, Aug. 2015.

[6]D. Kang et al., "256 Gb 3b/cell V-NAND Flash Memory with 48 Stacked WL Layers", 2016 IEEE International Solid-State Circuits Conference (ISSCC), Dig. Tech. Papers, pp. 130–131, San Francisco, USA, Feb. 2016.

Don't rush. Floating gate is not dead. At least one flash supplier, Micron, is working on it.[7] At the end of the day, floating gate has been developed and manufactured for decades; at least for the first few 3D generations, all the know-how developed over time might be of help, especially considering that, initially, most of the effort needs to be spent on the vertical integration itself.

Chapter 2 covers both charge trap and floating gate technologies, including a comparison between the two technologies in terms of both reliability and scalability.

There is a plethora of different materials and vertical architectures out there. This book walks the reader through this new 3D journey. New cells, new materials, new vertical architectures; basically, each vendor has its own unique solution. Chapter 3 (3D Stacked NAND), Chap. 4 (BiCS and P-BiCS), Chap. 5 (3D Floating Gate), Chap. 6 (3D VG NAND), and Chap. 7 (3D Advanced Architectures) offer a wide overview of how 3D can materialize. Visualizing 3D structures can be a challenge for the human brain. This is why all these chapters contain a lot of bird's-eye views and cross sections along the three axes.

During the opening keynote of the Intel Developer Forum (IDF) 2015 in San Francisco, Intel announced that it will be soon shipping SSDs based on its new 3D Xpoint memory, co-developed with Micron. Intel has claimed that 3D Xpoint is much faster and more reliable than the existing flash storage. As we speak, there are no public details about this memory cell, but Intel disclosed that it is based on a crosspoint architecture. Chapter 8 presents an overview of Resistive Random Access Memory (RRAM) crosspoint arrays, from memory cells to selectors, from reliability to the challenges of 3D integration.

Not to forget, 3D can be combined with the other storage dimensions mentioned in the first part of this introduction, i.e., multilevel storage and die stacking. Chapter 9 deals with all the most recent innovations in the packaging technology, including Through Silicon Vias (TSV).

From the early days, NAND flash has strongly leveraged ECC techniques to improve reliability. Starting from BCH, Chap. 10 describes the evolution towards the most recent commercial solutions, which are based on Low Density Parity Check (LDPC) codes. Indeed, among the codes that can be effectively integrated on a small piece of silicon, LDPC is definitely the one that gets closer to the Shannon limit.

Looking forward, the combination of TLC/QLC with several 3D layers might require higher correction capabilities. Chapter 11 provides an insightful overview on some of the most recent advancements in the field: asymmetric algebraic codes and non-binary LDPC codes.

Last but not least, Chap. 12 is centered on the implications of 3D flash memories from a system perspective. In this chapter, 3D NAND is combined with other

[7]T. Tanaka et al., "A 768 Gb 3b/cell 3D-Floating-Gate NAND Flash Memory", 2016 IEEE International Solid-State Circuits Conference (ISSCC), Dig. Tech. Papers, pp. 142–143, San Francisco, USA, Feb. 2016.

memories for next generation hybrid SSDs. In addition, workload optimized techniques to improve SSD's write performances are introduced.

Is 14 nm the last step for planar cells? How many generations of 3D are possible? Can 3D go down to cell's feature sizes below 20 nm? Can 100 layers be integrated within the same piece of silicon? Is QLC possible with 3D? Will 3D be reliable enough for enterprise and datacenter applications? These are some of the questions that this book tries to answer. Enjoy!

Rino Micheloni

Chapter 1
The Business of NAND

Rahul N. Advani

1.1 Memory Industry Transformation

The evolution of the memory industry in the last decade has been nothing short of transformational arguably some of the largest changes the industry has seen in its entire history. Since the focus of this book is NAND Flash, we will examine the dramatic changes in (1) the vendor landscape, (2) the fundamental technology used to create the NAND memory cell, and (3) changes in usages in different segments which have made SSDs the critical growth product in NAND. Let us examine all these changes in a little more detail.

1.1.1 NAND and Memory Vendor Landscape Consolidation

There are many factors underlying the dramatic consolidation in the memory vendor landscape. Let us first examine the extent of the change and some secular patterns in that regard. Figure 1.1 shows that in a short span of seven years we went from 95 % of the memory industry bits coming from nine vendors in 2008 to only five vendors today. And the consolidation is likely to continue amongst NAND and *Hard Disk Drive* (HDD) vendors who feel the pressure of declining markets to collaborate and/or acquire NAND technology and supply. So, the question to be asked is: what are the factors causing these significant changes in the industry?

1. The scale of memory fabrication facilities and the investment required in equipment has now made the cost of entry in the several billions of US dollars for a new state-of-the-art fabrication facility needed to compete in volume

R.N. Advani (✉)
Performance Storage BU, Microsemi Corporation, Aliso Viejo, USA
e-mail: rahul.advani@microsemi.com

R. Micheloni (ed.), *3D Flash Memories*, DOI 10.1007/978-94-017-7512-0_1

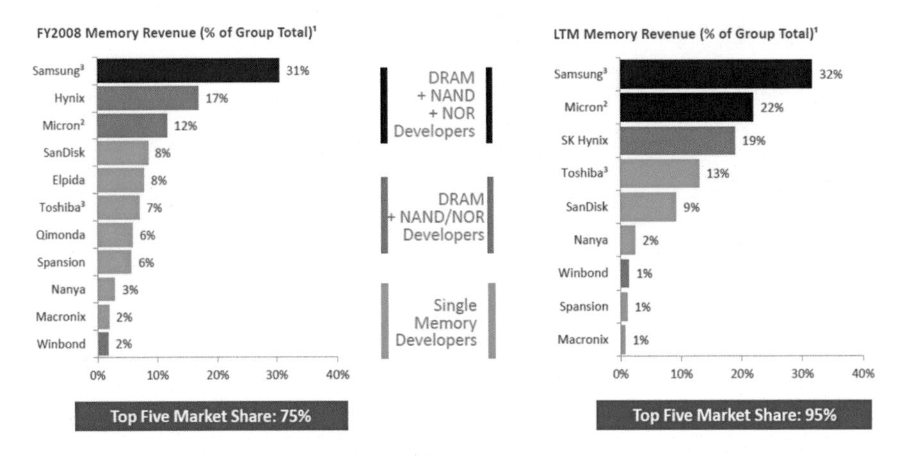

Fig. 1.1 Consolidation of the memory vendor industry. In a short 6 year span the memory vendor landscape has consolidated to 5 large vendors who provide over 95 % of the memory bits. "Group total" is defined as only those companies on the page, although others may exist. Micron includes NAND sold to Intel from IM Flash and Elpida revenue prior to merger included. Samsung and Toshiba include total memory revenue as reported. *Source* Micron, micron.com, 2014 winter analyst conference

memories. Moreover, this investment would need to be repeated every generation to remain competitive. Thus, a limited number of vendors are capable of this scale of continual investment to upgrade fabrication facilities and successfully hire and retain the top memory technical and business experts.

2. Market dynamics of down cycles for a commodity memory market have also played their part in the demise of large vendors like Qimonda, Elpida, Numonyx and a number of smaller vendors. This is an almost unavoidable part of an industry where investments in technology and fabrication facilities have to be made three to four years before volumes are expected. This causes significant challenges in predicting demand and supply correctly and thus ushers frequent short-term excesses and shortages. In the downturns, the vendors without the financial strength or the exposure to higher margin segments find it hard to sustain profitability, and often end up selling their assets to larger well-funded players.
3. Supporting controller and firmware technology has increasingly become required for a NAND vendor to be competitive. Thus, not only is leadership a requirement for the raw NAND products, but also in controller and firmware technology for today's SSD's with SATA, SAS, and PCIe interfaces, and in mobile handsets and tablets with eMMC and eMMC-like products. Removable cards and USB's use comparatively simpler controllers and firmware and are also becoming a smaller part of the overall NAND bits consumed.
4. Most industry experts now predict that hard drives will be increasingly replaced by NAND-based products, as NAND costs drop every generation. One can see examples of applications that inherently require moderate densities converting to NAND, e.g. MP3 players with 8–64 GB which have almost completely

transitioned to NAND (note that besides cost, NAND provides the added benefits of increased reliability, ruggedness and battery life). Mobile handsets have gone in the same direction and are entirely NAND based. Going forward, applications requiring lower capacities (256 GB and lower in 2016, and increasing at approximately 50 % every 18 months for the decade after) will be likely replaced by NAND-based solutions. This recognition is going to cause the hard drive and SSD industry players to collaborate and converge through acquisitions, mergers, joint ventures and technology sharing agreements. Relationships like this would, in principle, bring drive manufacturing and test expertise from the HDD vendors and access to low cost/bit memory supply from the NAND vendors. Thus, one can predict increasing collaborations and alliances between hard drive and NAND industry in the near future.

1.1.2 NAND Technology Transitions

The significant memory technology transition of this decade can be argued to be the pervasive movement in volume NAND from two dimensional (2D—scaling in x and y dimension every node using advanced photolithography equipment) to three dimensional (3D—scaling in the z dimension with the NAND cell becoming vertical and numbers of layers being added with advanced etching equipment). The 3D vertical memory requires the scaling of the layers from the 32 and 48 layers today to 64 layers and higher over the next few years. Process technology scaling in 2D was already getting prohibitively expensive and technically challenging every successive node. Yet, vendors continued to feel the pressure to provide further cost/bit reductions to compete effectively with HDD's and thus have transitioned to 3D NAND products which helps them achieve lower cost structures longer term.

The ongoing transition to 3D NAND requires significant monetary investment in large-scale fabrication facilities (fabs) with a different equipment set. Figure 1.2 shows the one time significant increase in capital expenditure (CapEx) for the first generation of 3D, which is expected to be in the billions of dollars (note: the same level of investment is not expected for future 3D NAND generations as some of the equipment would be re-used, like it had been for years in the 2D scaling era). Additionally, there is also the expected learning curve costs for NAND manufacturers who will be working with this new technology in production volumes. This is likely to depress yields initially and increase cost/bit for the first generation 3D NAND products. Finally, initial transition costs will also be higher for downstream makers of SSD's, embedded products like eMMC, and other products, as new technical challenges, tradeoff's and failure modes will have to be accommodated increasing development costs. *The promise of 3D NAND is though that these are one time transition costs and that by the second generation of 3D NAND the cost/bit will be lower than what 2D NAND could have provided.* In terms of vendor dynamics, Samsung has introduced production 3D NAND generations since 2014 and it can be

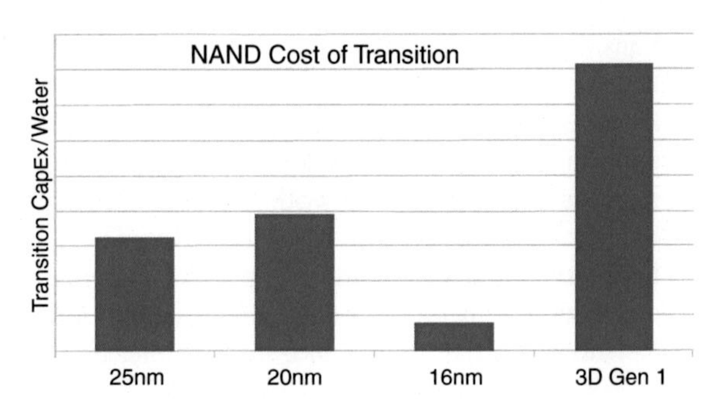

Fig. 1.2 One time significant increase in capital expenditures for the first generation of 3D NAND which would require more and complex etchers over the leading photo-lithographic equipment required for the 2D NAND nodes. *Source* Micron, micron.com, 2014 winter analyst conference

expected to be slightly ahead in the learning curve than other NAND vendors currently introducing production 3D NAND products. This advantage will have a near term impact, but is expected to be short lived.

1.1.3 NAND Usage Model Changes

The last element of the significant changes in the NAND ecosystem is in the segments themselves that use NAND. The applications that have driven NAND bit usage have been changing over time in terms of units and capacity/unit growth. Figure 1.3 shows some of the market usage changes over the last few years.

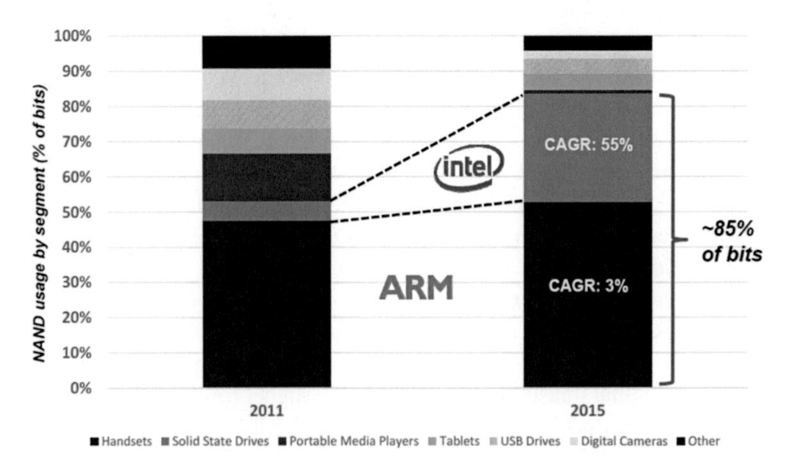

Fig. 1.3 The dramatic rise of SSD in the last few years as NAND costs decline. The change though has strengthened the Intel based x86 ecosystem in driving NAND roadmap directions as much as mobile handset/tablet ARM ecosystem had in the past (Intel eco-system as defined by Intel, laptop and server vendors downstream using their processors, and the software vendors aligning with x86 architecture)

- Growth of Client SSD's. The growth of SSD's in laptop PC's over the last few years has been both in the percentage of laptops adopting SSD's (attach rate) and the higher capacity SSD's used as the prices of NAND decline. The impact of growth of client SSD's is larger than just which application NAND is used in. The controller makers, SSD makers and customer base using client SSD's (mostly Intel X-86 based) are often significantly different from the mobile ecosystem players (ARM-based) whose influence on roadmaps has been impacted. Some have argued that client SSD's have now become the critical outlet market for NAND in oversupply situations (rather than the raw NAND market which used to play a similar role when smaller 3rd party card and USB makers consumed a higher % of the NAND bits in comparison to a few years ago).
- Growing Enterprise SSD's. This is a complex space where a number of interfaces like SAS, SATA and PCIe are being used today. We will explore this segment in detail in the next section due to its (1) technical and business complexity and its (2) potential to be a leading user of NAND bits within the next five years, and thus a critical driver for NAND component roadmaps.
- Decline of removable form factors like Flash cards and USB, in terms of the percentage of industry NAND bits utilized. Improving network access, higher capacities available in clients and the growth of cloud storage is further reducing the need for memory cards and USB's.

As segment specific as these changes seem, they have and will continue to have profound impacts on NAND component strategy and roadmaps (capacities, segment focus, and the organizations to support). At the product level, players with controller and firmware competencies applicable to end products like SSD's and eMMC will be well positioned. The key change in usage models that impact NAND is SSD's and the complex Enterprise SSD space. Let us investigate that in more detail in Sect. 1.2.

1.2 Solid State Drives

1.2.1 Enterprise SSD's

The Enterprise SSD market (Fig. 1.4) is one of the most interesting segments in the NAND landscape for multiple reasons. The previous section showed that the traditional NAND markets in cards, USB, and mobile are only declining or growing modestly, while the enterprise SSD and client SSD segment have grown rapidly and are expected to continue to do so in the next few years. The reason for this is the inherent growing need for storage and performance in enterprise applications and storage, the reduction of cost/GB of SSD's, and the increasing movement of client data to the cloud. Secondly, from a NAND component perspective, enterprise SSD capacity requirements are well aligned with the higher monolithic capacities of 3D NAND, already at 256 Gb and growing every generation. This means that 4 TB and

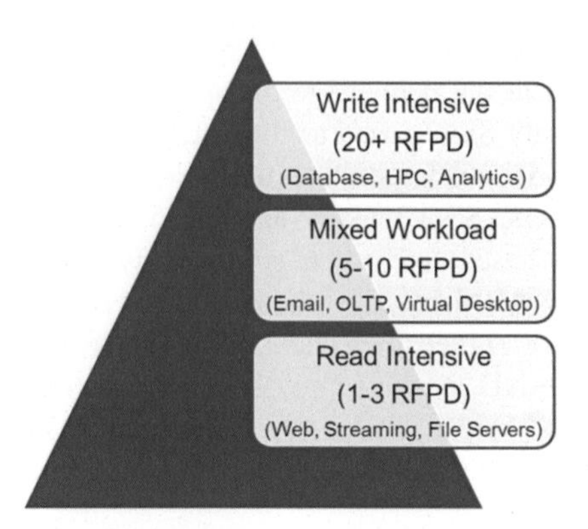

Fig. 1.4 Enterprise SSD usually classified in terms of an endurance metric and application space they target (RFPD—Random Fills Per Day for 3 or 5 years for cloud and enterprise usages is typically warrantied). The traditional classification for enterprise SSD's have yielded three distinct SSD drive specifications: "Read Intensive" at 1–3 RFPD, "Mixed Workload" at 5–10 RFPD, and "Write Intensive" at 20+ RFPD

8 TB SSD's will be possible and available soon in standard 2.5″ form factors. Lastly, the enterprise SSD market drives requirements that have the most to gain from the increased performance and features available in 3D NAND; for example, 3D NAND is capable of achieving 533 MT/s and above with greater than 8 die stacks due to lower parasitic load capacitance, as opposed to 266 and 333 MT/s being the norm for 8 high stacks with 2D NAND. Additional NAND features like program suspend, higher endurance, multi-plane operation and others which are suitable in enterprise SSD applications.

In enterprise, it is also important to understand the memory hierarchy tiers to see how the use of SSD's are better meeting the overall application requirements at the lowest cost. Figure 1.5 shows a classic memory hierarchy where different technologies play today—volatile DRAM which is the fastest external memory is the first tier after the processor, followed by the emergent persistent tier of NVRAM and NVDIMM solutions (discussed in Sect. 1.4). The SSD tiers after that offer performance and power advantages, over the lowest cost/bit hard drives.

For SSD's, even with its higher initial purchase cost, adoption has grown rapidly in enterprise for the last decade due to (1) higher performance of the order of 100–1000× better than HDD's, (2) significantly better reliability due to no moving parts (HDD's have been one of the highest failure rate elements in datacenter) and (3) significantly lower power, which is a major component of operational costs in datacenters. All these factors leading to SSD's having a lower overall total cost of ownership for a growing number of applications and explaining the high adoption rates. Today, enterprise SSD usage has even reached a level of maturity where there is active tiering of data between different types of SSD's (the term 'hot' and 'cold' data is often used). The highest performance tier today is performance-optimized PCIe SSD's, followed by cost-optimized tiers using SAS and SATA SSD's, and finally followed by the ubiquitous and lowest cost/bit HDD's for greater capacity.

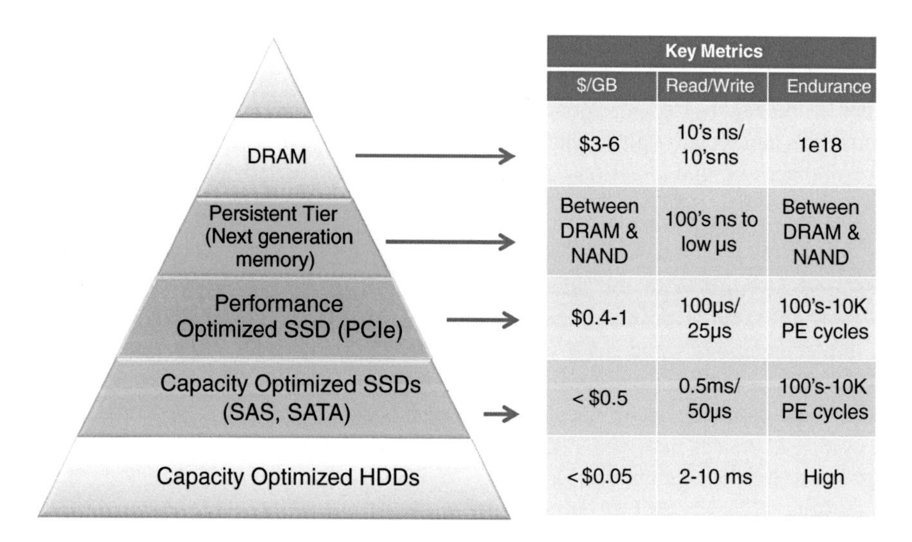

Fig. 1.5 Traditional memory hierarchy shows the multiple levels of memory being used between DRAM and hard drives (HDD). Multiple tiers of SSD's with persistent memory and NAND technologies have changed datacenter architectures and economics

The balance today is towards greater spending for SSD's at the expense of HDD's. This is not a strict distinction and can be most often seen in larger scale cloud storage implementations with the resources to invest in data and software optimizations, and enterprise by vertically optimizing hardware to specific application software (e.g. database applications, server virtualization).

The next level in enterprise SSD's worth noting is the changes in the segmentation between the three most common storage interfaces—SAS, SATA, and PCIe. The SAS interface has been the most established in enterprise with continued support and strong adoption coming from the enterprise storage space and scale up storage SAN implementations. SAS hard drives have enterprise features (e.g. dual port, end-end data protection, scalability) as well as an established ecosystem of host-based and interconnect technologies that are considered to have enterprise-grade reliability, making the adoption of SAS SSD's easier and thus historically the interface of choice for larger storage implementations. PCIe SSD's are ramping the fastest due to the following reasons:

- They provide the highest performance and lowest latency;
- The growth of server based storage implementations where cloud vendors prefer the direct connect PCIe interface already available in servers;
- Manufacturing efficiencies and lower costs by adopting client form factors like 2.5 in. drives or M.2 form factor;
- The growing PCIe enterprise ecosystem of technologies required for scale up enterprise implementations.

SATA based SSD's have most often been used because of cost and have been leveraging the high volume and manufacturing efficiency of client SSD's, sometimes at the expense of enterprise reliability features due to the use of client SSD controllers and technologies. This space is expected to decline as shown in Fig. 1.6, where lower end and client based PCIe offer the same cost advantages and higher performance than SATA.

Cloud Storage With the advent of cloud storage, server based storage is growing rapidly at the expense of traditional large centralized storage implementations using SAN (*Storage Area Network*), NAS (*Network Area Storage*) or DAS (*Direct Attached external Storage*) boxes. This trend is also being adopted by enterprise IT and server attached storage is expected to comprise greater than 80 % of all storage within a decade (Source: Wikibon Server SAN Research Project 2015). With the architectural parallelism to address reliability and the strong need for lower costs due to cloud business models, cloud vendors initially led the ramp of lower cost SATA based SSD's inside cloud datacenters, and are now taking the lead in the growth of PCIe SSD's the same way (with quicker infrastructure aligned qualifications, high volumes, and focus on cost effectiveness using standard form factors for manufacturing efficiencies). The larger cloud based vendors are thus driving not only datacenter architectures, but also component level products such as SSD's.

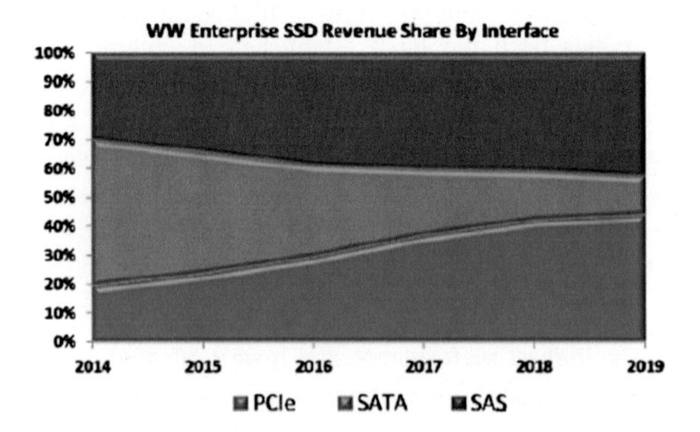

Fig. 1.6 Enterprise SSD growth by interface. One can see the dramatic rise in PCIe as the interface of choice, over the next few years. Most of this would be initially enterprise server attached and growing in enterprise storage applications as PCIe switches and enterprise features mature. *Source* IDC, worldwide solid state drive forecast, 2015–2019, May 2015: IDC #256038 and SSD market perspectives, Sept 2015

With that historical context of enterprise SSD's and the cloud storage trend in mind, one can now see the growing attractiveness of standards based *NVM Express* (NVMe) PCIe SSD's, which are available from multiple vendors ensuring second sources for production systems. PCIe SSD's are now widely deployed in servers that leverage the PCIe bus which was originally a peripheral bus, but now is also being used to provide direct storage connectivity. In fact, the SSD use case has driven the evolution of the current NVMe protocol, which focused on optimizing access to NAND while minimizing historical storage protocol overhead. There is now strong support for NVMe PCIe SSD's from Intel architecture, component providers, established SSD providers, NAND vendors, and test equipment providers. The full benefits and attraction of PCIe based storage SSD's can now be realized, including:

- Highest performance and lowest latency over both SATA and SAS. High-end PCIe SSD's can provide greater than one million I/O's per second read and latencies in the ~10's μs.
- Leveraging client volume economics, since PCIe is popular for client SSD's too. This is reflected even in the form factor preference where enterprise is gravitating towards the form factors used in client like 2.5 in. drives and M.2 for manufacturing efficiencies.
- Growing ecosystem support for PCIe storage (i.e. components like PCIe switches) and investments in *NVMf* (*NVMe over fabric*) and technologies allowing larger-scale deployments.

The performance advantages of PCIe SSD's combined with the trend of the pendulum swinging towards server-based storage from traditional DAS, SAN and NAS storage implementations, indicates the trend towards standards-based PCIe NVMe emerging as the leading SSD interface in enterprise over the next few years. The following Table 1.1 provides a real world performance comparison of SSD's with different interfaces used in enterprise.

Recent announcements from the Intel-Micron joint venture about a new material based 3D crosspoint technology is worth a mention. It is a technology expected to have higher performance and endurance than even a PCIe NAND-based SSD, and capable of replacing part of DRAM usage too. But, with the very limited information that has been publically divulged—the price points, usage models and long term success of this technology is not clear. The reader can find more details about crosspoint memory arrays in Chap. 8.

Table 1.1 Comparison of performance between different interfaces for enterprise SSD shows that PCIe SSD's significantly outperforms on both performance and latency

	SATA SSD (6 GB/s)	SAS SSD (12G)	PCIe GEN3 (X8)
Sequential read (MB/s)	~500	~1000	>4000
Random write (4k IOPS)	30–90k	~100k	200–300k
Latency	Good	Good	Best
Endurance	Same (dictated by NAND choice and controller ECC)		

1.2.2 Build-Your-Own (BYO) and Custom SSD's

As the title depicts, *Build-Your-Own* (BYO) SSD is a model where a large SSD consumer decides to make the SSD versus buying an off the shelf standard SSD. This model is relatively recent in the enterprise market, but is gaining popularity at the largest consumers of SSD's. While this model is emerging in enterprise, one can see that it has existed in a different form for many years in the client segment where multiple large ('Tier 1') PC laptop manufacturers used this model to optimize SATA SSD's. This resulted in the significant growth of client SSD manufacturers in Taiwan and China in particular using firmware, NAND type, and form factor differentiation. The motivations for creating one's own BYO SSD are (Fig. 1.7):

- NAND cost advantages: leveraging the arbitrage between raw NAND and SSD prices. This difference was larger in the past, and even with the narrowing of this gap today to 30–50 %, there is yet a significant cost savings for BYO SSD's. This advantage can be further enhanced with tailoring over provisioning to the application need, discussed in Sect. 1.2.2.1.
- Operational advantages: this is achieved by streamlining the SSD development and deployment cycles to align more closely with other platform transitions and deployments, and by optimizing the SSD's feature set and behavior to what the operator cares about the most. In addition, since the SSD operator is more intimate with the implementation and architecture of the SSD, BYO gives them an operational advantage when it comes to supporting and improving the product post deployment.
- Time-To-Market (TTM) advantages: this is achieved in two ways. (1) Joint qualifications can start earlier with the custom BYO SSD's reducing the final

Fig. 1.7 Pictorial summary of the areas where customizations and optimizations are made in *BYO SSD's* to give it an advantage over *Standard SSD's*

Applications	Platform "Knobs"				
	Power Used	DRAM Density	Flash Density	Flash Type/Cost	Host I/F BW
Cold Storage	Low	Low	High	Low	Medium
2.5" SSD (SAS/PCIe)	Low	Medium	Medium	Medium	Medium
Low/Mid PCIe SSD	Medium	Medium	Medium	Medium	High
Enterprise PCIe SSD	High	High	High	High	High
Caching Adapter	High	High	Low	High	High

Low Medium High

Fig. 1.8 Number of platform tradeoff's (Knobs) that one can optimize to get SSD's with different characteristics leveraging the same controller. Examples of knobs are changes in firmware, power envelope, DRAM density, flash density and configuration (greater number of die's allows higher parallel performance in principle), flash type (SLC, MLC, TLC), and host interface that aligns to the architecture

qualification timeline and allowing a faster volume production ramp. (2) Accelerating development and qualification by accepting a limited feature set acceptable. End user, who is well aware of the workload needs and failure modes, are in a good position to make the best tradeoff's.

- Technical advantages: this is arguably one of the largest areas of advantages of the BYO model. As seen in Fig. 1.8, there are many 'platform knobs' one can tweak, like picking the power/performance envelope, leveraging low cost NAND, optimizing DRAM density and host interface to get the lowest cost SSD tuned to the specific application needs. Having granular control of the firmware also allows for close integration with the higher levels of host software. For the largest cloud implementations, one can also see the *Flash Translation Layer* (FTL) migrating from the SSD to being integrated with the host itself (Host-based FTL) to give the host and application granular control of the NAND (including data wear leveling, the timing of garbage collection, etc.). One could argue that BYO SSD's are no different an approach than has been used for over a decade by the largest cloud vendors for custom servers to reduce server costs by architecting a minimum viable option from themselves, rather than buying a full-featured branded server.

By using a flexible controller architecture and firmware modifications, vendors can support disparate requirements spanning from the lowest cost/performance cold storage to the high-performance SSD's for database applications. Let us examine these BYO modifications in the context of the three most common SSD metrics.

1.2.2.1 Endurance

The program-erase (PE) cycles is the main metric for endurance and varies between different NAND types: industry norm is

- SLC (i.e. 1 bit/cell) at 10k Program/Erase (PE) cycles and higher;
- MLC (i.e. 2 bit/cell) at 3k PE cycles, and;
- TLC (i.e. 3 bit/cell) typically with less than 1K PE cycles.

For more details about SLC/MLC/TLC storage, please refer to Chap. 3.

Using these different types of NAND has thus been the basis of higher Random Fills Per Day (RFPD) implementations of SSD's, as was defined for enterprise in Fig. 1.4. Many vendors today use the same controller and SLC NAND for write-intensive SSD's, enterprise-grade MLC NAND for mixed read/write workloads, and consumer-grade MLC or even TLC for read-intensive SSD's. There is a considerable gap between NAND types of SLC, MLC and TLC in pricing, sometimes greater than 5× in cost/bit between SLC and TLC, which makes the choice of NAND one of the key ones for an SSD in terms of cost, performance, and endurance required.

Over-Provisioning (OP) is an extra portion of Flash storage added to the SSD and made available to the Flash controller to perform various memory management functions. In essence, OP helps during garbage collection and spreads out Flash Program and Erase operations to different NAND blocks to increase SSD's endurance. The most common OP in the industry is 28 %+ in enterprise SSD's and 7 % in client SSD's which are less performance and endurance sensitive. The OP numbers are designed to provide the required storage capacity to the user (e.g. in enterprise applications, 1024 GB of NAND with 28 % over provisioning result in 800 GB of user data space, while in the client space 512 GB of NAND with 7 % overprovisioning is equivalent to an SSD of 480 GB user data). Yet, the 28 and 7 % are arbitrary choices, and in a BYO implementation you can vary towards greater or lower overprovisioning by knowing the workload needs, e.g. endurance, sequential versus random performance, read to write ratios, etc.

1.2.2.2 Performance/Power Envelope

The characteristics of NAND performance vary significantly between SLC, MLC, and TLC. Once the application performance level is determined, a BYO or Custom SSD can arguably pick lower performance cost effective MLC or TLC NAND, and compensate with a controller with a greater number of Flash channels (please refer to Chap. 10 for more details) to obtain similar performance levels. The tradeoff of greater Flash channels sometimes requires a larger controller and thus increased power, which would then have to be managed to meet the standard power envelopes of SSD's (e.g. 9, 12, 25 W). Thus, an example of a BYO SSD would use (1) a cheaper NAND (MLC instead of SLC, TLC instead of

MLC), (2) greater number of channels for performance, (3) a strong ECC engine for endurance extension and (4) the lowest OP to meet the application performance and endurance requirements.

Power: total SSD power is dominated by the NAND itself, controller and sometimes DRAM (if used extensively in high-performance enterprise SSD's). NAND power is a function of the speed of the NAND being used (say 333 MT/s vs. 400 MT/s), how many Flash channels are used (1, 4, 8, 16, 32 channel controllers are available today), type of NAND and its programming model, and how many die are active at any time. Firmware is capable of automatically throttling in times when the system or SSD temperature rises beyond system temperature specifications, although this causes uneven performance/user experience and is thus not preferred. The better approach is to understand the power/performance envelope and system-level temperature constraints when designing a BYO SSD (optimized to the application needs with NAND type, speed, die that are active, power islanding, eliminating un-needed features, program and erase times, and use of controller and NAND optimizations).

1.2.2.3 Cost of NAND Flash

Total cost of multi terabyte enterprise BYO or Custom SSD's are yet dominated by NAND costs (which can be 80 % or higher of the total SSD cost). Thus, one can see that picking the right NAND is probably the most important decision from a cost perspective (e.g. using MLC instead of SLC, or using TLC instead of MLC). From the business perspective, having the right vendor relationships, supply chain and manufacturing efficiencies is critical to the success of the BYO model. There is an argument that supporting various NAND types from different vendors helps create competition and reduce cost (similar to using multiple vendor sources when buying any type of memory), and that involves additional development costs, but is viable only when volumes are significant.

The last factor to consider is technology leadership in introducing new nodes with lower cost structures. Amongst NAND vendors there are differences between focus on introducing nodes early versus focus on the most aggressive die sizes, even if introduced a quarter or two later. There are also differences in yield when ramping a new NAND technology node which often are dependent on choices in technology like using a floating gate or charge trap based Flash memory cell. Thus, understanding the differences between the vendors in terms of technology preferences, historical ramp rates and motivations is key to aligning NAND vendors with end customers.

The example of Fig. 1.9 assumes a 2 TB Enterprise SSD, controller with a higher number of Flash channels and stronger ECC that allows the use of lower

Standard SSD	***BYO SSD***
MLC NAND	TLC NAND
28% over provisioning	15% over provisioning
8 Flash channel controller	16 Flash channel controller
Standard ECC (40-60 bit/1K)	Strong ECC (100b+/1K, LDPC)

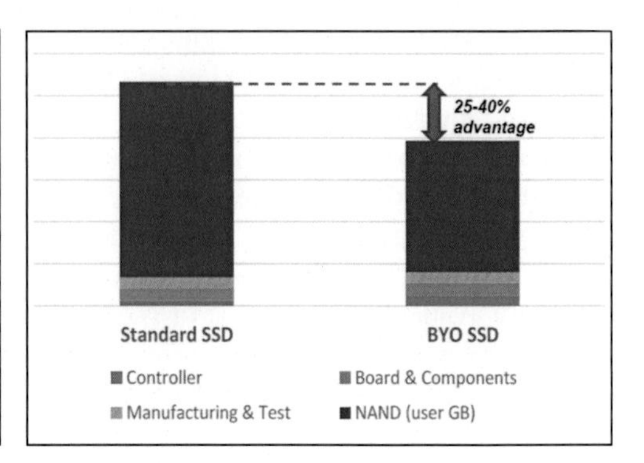

Fig. 1.9 Practical example comparing a 2 TB BYO SSD with a standard SSD

cost NAND. Savings can be as much as 25–40 % for just the initial hardware cost (not including the additional operational and technical integration advantages).

1.2.3 Economics of SSD Controllers

Throughout the previous section, we have alluded to the importance of using controllers with a flexible architecture to create customized BYO SSD's or the spectrum of read to write intensive SSD products (product lines) using the same controller. Let us first examine the economics behind SSD controllers and firmware stack.

1. Controller development costs in the latest PCI Gen 3 and SAS 12G generations have risen substantially and are now typically in the range of several tens of millions of dollars for the initial silicon development alone, after which several revisions are often required to fix bugs adding cost, risk and schedule delays. Schedule delays in this industry are very costly since a particular SSD product line usually aligns with the lowest cost NAND available for a targeted timeframe, and that in turn governs architectural choices like ECC strength, DRAM interface, etc. Thus, if a controller has a revision, it would take an additional 6–9 months and cause the SSD to perhaps be misaligned to the lowest cost NAND node that lasts only 12–15 months. This exquisite and narrow planning windows can be avoided with a flexible controller architecture capable of supporting multiple NAND process technology nodes and DRAM interfaces.
2. The personnel skillset required for high-speed design and specific protocol optimizations (PCIe, SAS or NVMe) are differentiated and not easy to hire and retain. In fact, firmware is not just used for algorithms and to address bugs found

in the silicon, but increasingly to optimize solutions for different usage models in the industry.
3. Firmware and configuration optimizations often cost a fraction of the silicon development. Besides, there are often customer specific optimizations which encourage architectural consistency in order to re-use the firmware work for many products over multiple generations; in fact, if one writes customer specific firmware to align to the higher management layers in a customer's infrastructure, there is a natural proclivity to keep that consistent across the product line and for new generations of SSD's being developed.
4. Product validation costs can also be substantial and cycle times long for enterprise SSDs. Thus, time-to-market solutions prefer to leverage previous work like test scripts which are often written and built upon over the years with a certain architecture in mind. Thus, a consistent controller architecture and firmware stack helps re-use test and manufacturing infrastructure reducing cost and improving product introduction timelines.

The key flexible controller silicon (high level) design points include (1) number of Flash channels, (2) DRAM interface, (3) enterprise features supported, (4) power islanding to turn off features and parts not required for a specific implementation, (5) strong ECC that allows to support multiple Flash generations, (6) flexibility in supporting protocol and special commands in NAND. Besides the higher level design points mentioned above, some granular examples of what SSD controller designers have to comprehend in their designs.

- Protocol communication between Flash devices: not only does NAND from different vendors differ (ONFI and Toggle protocols), but sometimes even between products of an individual NAND vendors offerings. Examples are changing from five to six bytes of addressing, or adding prefix commands to normal commands. Having the architecture that allows the protocol to be done by firmware allows the flexibility to adapt to these changes. Additionally, having a firmware-defined protocol allows Flash vendors to design-in special access and command abilities.
- Flash has different rules for order of programming and reading: a flexible controller and firmware optimizations can adapt to variable rules and use different variations of newer Flash generations that might not have been available while developing the controller silicon. Having both the low-level protocol handling as well as control of the programming and reading in firmware, allows for flexible solutions able to use many types of Flash.
- Fine-tuning algorithms/product differentiation: moving up to the higher level algorithms, like garbage collection and wear leveling, there are many intricacies in Flash. Being able to control a wide range of low level functions up to the algorithms in firmware, allows for fine-tuning of higher level algorithms to work best with the different types of Flash and application requirements. This takes

advantage of the differences Flash vendors put into their product so they can be best leveraged for diverse applications with a single set of controllers.

1.2.4 Client SSD's

Client SSD's have a different market dynamic, which is governed more by cost and alignment with using the lowest cost NAND components. The controller technology involved, while not trivial, is more readily and cost effectively available from multiple vendors. The endurance and performance requirements being modest in comparison to enterprise SSD's, some of the key to success and differentiators in the client SSD area are summarized below.

- NAND component technology cost effectiveness, especially with respect to the use of the latest generation TLC, which is being increasingly adopted for client SSD's (and possibly QLC, i.e. 4 bit/cell in the future). Related to that is how quickly a vendor can introduce qualified client SSD products based on the latest NAND node, after the raw NAND components become available.
- Controllers and firmware, which have been one of the key factors governing the success of vendors with significant market share. Although some client SSD controller suppliers provide production worthy firmware along with the controller (called the 'turnkey firmware' model), which allows SSD vendors to be able to introduce products without investing in extensive firmware teams.
- Flexibility on features and hardware form factors, which is arguably the main reason why smaller vendors, who do not manufacture NAND, have succeeded in the market. Many of the tier 2 players in the industry offer this flexibility as a value as tier 1's often don't like to fragment their offerings.

The client SSD space at this point is dominated by products from the NAND vendors themselves since they have access to NAND at fabrication cost, and one can expect that to continue with the 3D NAND transition using 3D NAND.

1.3 NAND Component Technology Evolution: The 3D NAND Transition

With this context and background on memory vendors and growth of SSD's, let us now discuss the main topic of this book: 3D NAND Flash memories. The discussion about NAND component technology scaling can be understood in terms of two important criteria, (1) the need to lower cost/bit to continue growing the NAND percentage of total storage (often at the expense of HDD's), and (2) limitations of

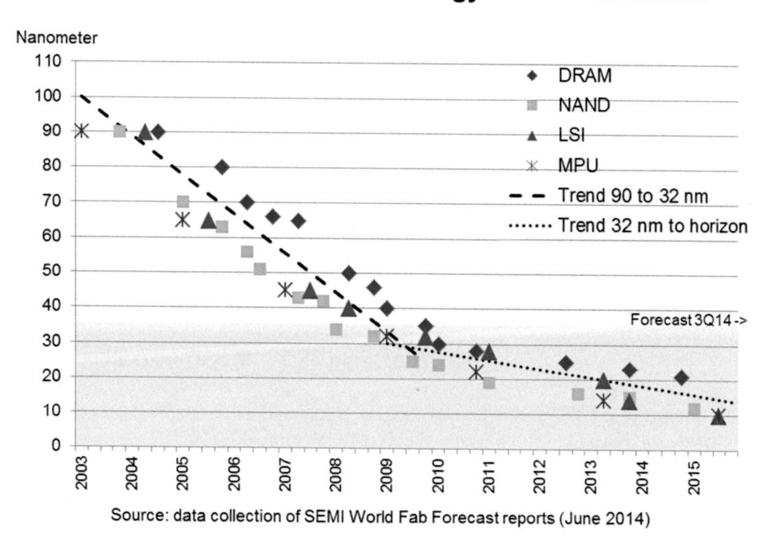

Fig. 1.10 Long term semiconductor node shrinks have clearly slowed down below 30 nm (in 2008). Most industry experts believe that the 3D scaling approach will continue to allow NAND to scale for the next decade and will be an approach increasingly used in other semiconductor memories

2D technology in endurance and performance that led to the exploration of 3D NAND as the future for this technology. Figure 1.10 shows the evolution of scaling for various CMOS semiconductor technologies. It's clear that after 30 nm, scaling has slowed down due to the technical complexity of scaling, and the equipment sets getting prohibitively expensive every generation. For NAND, in 2D NAND, we were getting to the point that the sheer number of electrons that define a state (a '1' or a '0') had become in single digits, especially for TLC which needs 8 states. This was impacting endurance, performance and increasing ECC requirements every generation.

The other question that can reasonably be asked is regarding the pace of scaling as a vendor: why not be satisfied with the slower pace of scaling and focus on profitability? The reason lies in the reality that NAND is yet a small a percentage of the storage bits worldwide, whereas HDD's dominate the total number of bits shipped every year (Fig. 1.11). There have been many articles written on the demise of HDD's with the coming of increasingly cost-effective NAND solutions. The evidence shows the narrowing cost/bit difference between NAND and HDD's over the last few years, but is that enough? One could argue that this trend should examined segment by segment.

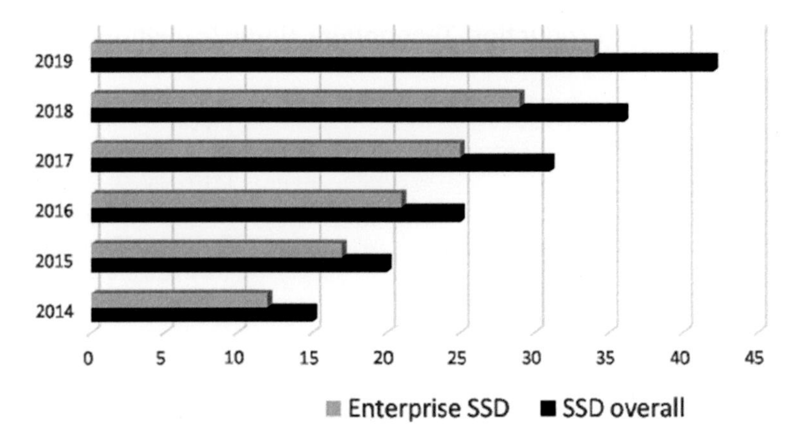

Fig. 1.11 Estimate of overall and enterprise SSD's versus HDD's. Graph shows the significant room for growth for NAND. *Source* Mark Webb, MK Ventures, 2015

1. Client and consumer applications: increasingly have reached the tipping point where the useful/typical capacities for the device are within reach of NAND-based solutions having attractive comparative costs to existing HDD solutions.
   a. MP3 players are the earliest example. Starting in 2007–2008, the MP3 segment switched almost entirely to NAND in a short period of time when the typical capacities of 30–60 GB could be cost effectively served using NAND (there are additional advantages of power, reliability, ruggedness, size and weight which cause NAND to be preferred).
   b. Digital still camera and other consumer applications using Flash cards. Here the combination of the performance, replacement ease of cards to increase capacity, size and weight have moved even the highest end professional camera's towards NAND based solutions.
   c. PC's (laptops and desktops). This is the largest part of the client market and have traditionally used hard drives. SSD adoption in laptops has grown over the last few years and consumer SSD's today routinely use TLC NAND and are often sold for less than $100. Current estimates show greater than 20 % of laptops already use SSD's and one can see that as NAND cost approaches $0.1/GB in 2018–19, almost half of laptops will have SSD's rather than HDD's (as 512 GB SSD's are often sufficient within a laptop, with backup's over the network or cloud). Some would argue that the non-NAND BOM costs should be added to the cost of SSD's when comparing to HDD's, but there are also aspects such as power, performance and reliability which are pushing PC's towards the adoption of SSD's.

TLC is the new MLC

IN THE 2D TO 3D TRANSITION, IN ORDER TO ADDRESS BOTH COST/BIT AND TO IMPROVE TECHNICAL SPECIFICATIONS, NAND VENDORS HAVE ROLLED BACK THE PROCESS NODE SEVERAL GENERATIONS AND SCALED VERTICALLY. WAFER COSTS BEING HIGHER FOR 3D NAND, A KEY GOAL IS TO MAKE THE 3D TLC PRODUCTS TECHNICALLY EQUIVALENT TO 2D MLC ENDURANCE AND PERFORMANCE SPECIFICATIONS AND POSITION THEM AS SUCH FOR THE INITIAL RAMP.

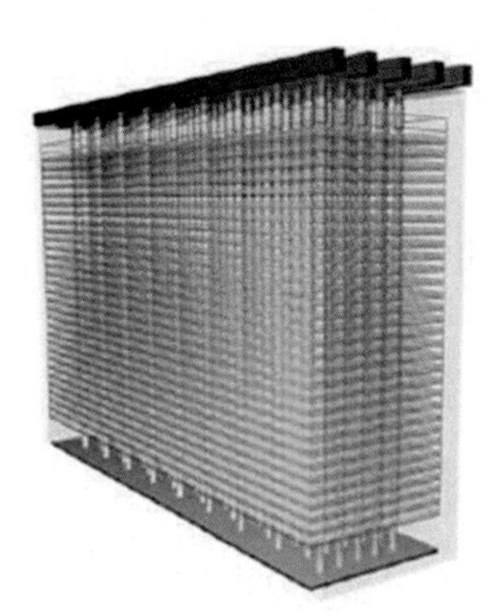

IMAGE: MICRON.COM

2. Mobile handsets: is the largest segments in terms of NAND bits used. Typical densities of 8–128 GB are easily met with NAND cost effectively. Size, weight,

and battery life have also encouraged vendors towards NAND usage typically in the form of eMMC and custom embedded products. Flash cards are sometimes used in emerging markets to increase capacities. This segment is entirely NAND based for storage already.

3. Enterprise and cloud datacenters: is where the largest growth rates of NAND usage is predicted in the next few years. The memory hierarchy displayed in Fig. 1.5 shows that SSD's, which have come in at a performance tier in enterprise applications, are now beginning to expand to the capacity tier. Hard dives will continue be driven towards higher capacities as larger bulk storage. As higher monolithic densities are introduced in successive 3D NAND generations, one can predict there will be a split in the requirements for NAND component technologies for lower performance and higher capacity applications (in comparison to typical NAND specs, even lower performance NAND will be 100's of times faster than HDD's in performance) versus lower density but performance oriented applications. More on this in the next section.

While this author does not believe that HDD will be completely replaced, the technical superiority of NAND-based solutions in power, performance, and reliability will have significant impact on reducing the number of HDD's in many application segments and continue to force HDD's towards higher capacities.

1.3.1 3D NAND Component Technology

The previous section showed scaling slowing for 2D NAND nodes and being unable to meet the inexorable push for cost reductions required to grow the NAND market. The promise of 3D scaling is that 32–48–64 layer (and beyond) products are expected to continue scaling for several generations at the same rate as past 2D nodes did in cost per bit. The other reason is technical specifications degradation in the final 2D nodes had become significant enough to encourage the 3D NAND transition which inherently starts with better endurance and performance characteristics.

3D cost and price positioning. In 3D NAND, the Gb/mm^2 is significantly higher, but the comparison is more complex than that. Since in 3D NAND the focus is on scaling the NAND cell vertically, the vendors have rolled back the process technology node several generations to reduce the risk. The meaning of that, from a specification (and price) perspective, is that vendors are positioning TLC in 3D NAND as equal to MLC in 2D NAND. This is very fortuitous as it is almost necessary to sell 3D TLC at 2D MLC prices, to overcome the lower yields expected in the initial volume ramp of a new vertical 3D NAND based technology. There are also likely differences between the cost of 32 layer floating gate (ONFI vendors) and 48 layer charge trap (Toggle vendors) products announced by different vendors in terms of initial manufacturing efficiency. An individual vendor's pace of transition to 3D is dependent on the following aspects.

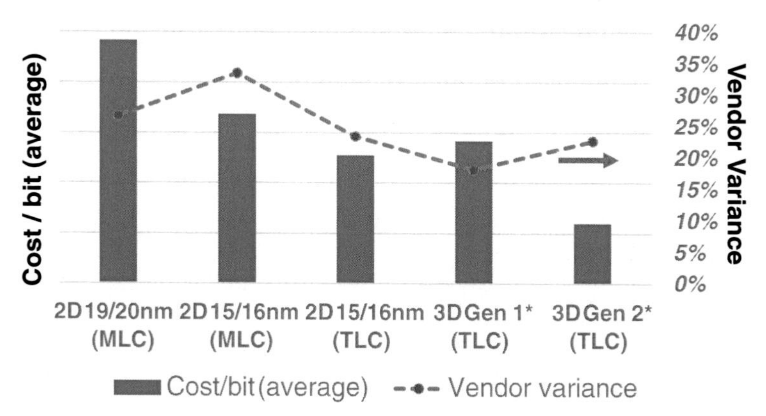

- For the 1st gen 3D products:
  - Toshiba and Hynix have announced 48 layer using *Charge trap* memory cell
  - Micron/Intel has announced 32 layers using *Floating gate* memory cell
- Calculations of Cost/bit and Vendor Variance for cost/bit are based on assumptions on die size, yield, and manufacturing costs for increased layers in 3D

Fig. 1.12 The bars show the cost/bit declining in 2D NAND. Moving to 3D NAND the expectation is a temporary increase in the first generation in cost/bit as vendors address the issues of the new technology and get yields up to similar levels as 2D NAND. Variance between vendors in cost/bit is in the 15–30 % range demonstrating differences between the leading and lagging vendors (assuming constant yield)

- *Comparison with the cost structure of the last 2D NAND products*. If a vendor's die sizes and cost structures were higher than their competitors in the last 2D node products, they would feel greater pressure to ramp their 3D products quickly.
- How quickly the *yield curve* comes up for 3D NAND products versus the mature yields of the previous 2D products (the yield curve norm in the industry over the years has been 80–85 % yield for mass production parts). This means that some vendors will have to wait for their 2nd generation 3D products to achieve cost effective 3D products (see Fig. 1.12).
- The emerging 3D TLC being able to reach the technical specifications of the last 2D MLC nodes—thus achieving around 2–3k PE cycles for 3D TLC and 5–10k PE cycles for the enterprise variant in high volume will be important.
- *Wafer cost* between the 32 Layer (ONFI vendors) and 48 Layer (Toggle vendors) products announced. Manufacturing costs are expected to scale with higher number of layers.
- *Floating gate versus charge trap*, some industry experts expect there will be manufacturing yield differences initially between different 3D NAND products.

Technical specifications. There has been natural degradation of performance and endurance specifications in 2D NAND over the generations due to smaller memory cell geometries and fewer number of electrons that can be stored in those memory cells. MLC nodes, which used to be capable of 10k PE (Program/Erase) cycles are typically specified at 3k PE cycles or lower currently. Most recent 2D TLC products are specified at just a few hundred PE cycles. Thus the key metrics that declined over the generations are:

- Reliability (PE cycles) decreasing;
- BER (bit error rate) increasing;
- ECC requirements increasing (discussed in greater depth in Chap. 10).

As pointed out earlier, for wafer cost/bit equivalency reasons, NAND vendors have to make the characteristics of 3D TLC comparable to those of the last 2D MLC (e.g. specifications like 3k program endurance cycles, higher speeds than 400 MT/s, program and erase times). While this may seem like a simple goal, getting there with high yields, using new equipment sets, and an entirely new 3D technology is likely to prove challenging. Thus, one can also expect uneven production ramps between the different vendors taking multiple years to be consistent across the industry.

Quad level cell (QLC, 4 bit/cell). If 3D TLC will be similar to 2D TLC, there is the question of what replaces 2D TLC. After the initial hesitation, 2D TLC has successfully created market acceptance for a few 100 PE cycles, especially in mobile and client SSD products. There are multiple announced efforts in the industry by NAND vendors to create QLC products (quad level cell, 4 bit/cell with 16 states) in 3D NAND. With 3D NAND TLC and QLC coming, it is not hard to imagine monolithic die densities 4–8 times greater than the 128 Gb monolithic die densities today. Additionally, one needs to keep in mind that SSD controllers have strong built-in ECC, because of the stringent requirements from the last few generations of 2D NAND. State of the art controllers today support ECC levels of 40–100 bit/1 kB (BCH) and LDPC (*Low Density Parity Check*), which has increasingly been built into newer controllers. A number of these controllers will exist for a long time and with the unknowns of failure modes to be expected in 3D NAND, there is continued momentum towards keeping ECC in controllers strong. This will help create a number of controllers with ECC capable to support 3D QLC products and the power/performance to create products with cost structures that continue to effectively compete with hard drives.

1.3.2 3D NAND Output Dominated by TLC

The net effect of the 3D NAND ramp is that TLC, which is a smaller part of the industry bits today, will be the dominant product for all the vendors over the next few years. Figure 1.13 shows how TLC is expected to become more than 50 % of the industry bits by 2017 and continue to grow. One can expect that it will take until 2017 before all vendors have cost-effective 3D NAND TLC products in high volume production (the industry crossover to 3D NAND is formally defined by greater than 50 % of the shipping bits having transitioned to it). The notable

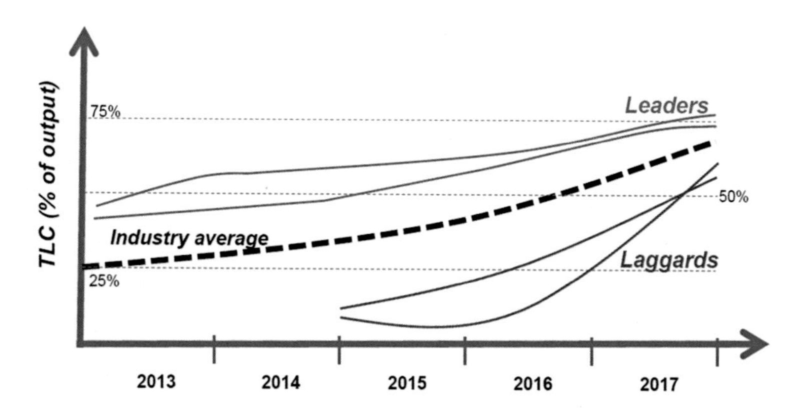

Fig. 1.13 The industry average of the TLC bits is expected to rise with the 3D transition. Individual vendors like Sandisk and Samsung have been the leaders, Toshiba around industry average, and Micron and Hynix are expected to be ramping TLC output in the near future

exception to this is Samsung, who has already introduced multiple generations of 3D NAND products in volume production.

There are some issues that this 3D NAND transition creates:

- There are a number of lower end markets who do not need 256 Gb (32 GB) or greater as the minimum density for their applications (embedded, low end mobile phones, cost sensitive removable NAND products);
- There is also a class of applications (DRAM backup, caching) that boosts performances by using multiple lower individual monolithic dice in parallel rather than few large capacity parts.

These factors will either force the NAND vendors to continue supporting low density 2D NAND products longer or introduce less cost effective and lower monolithic density 3D products.

1.3.3 Floating Gate Versus Charge Trap

Finally, in terms of NAND components, an interesting technology battle to watch is the use of either *Floating Gate* (FG) or *Charge Trap* (CT) technology for the fundamental NAND memory cell (Fig. 1.14). While most of the NAND vendors have chosen *Charge trap*, a notable exception is Micron/Intel joint venture, whose choice is *Floating gate* based NAND cell technology in their 3D NAND products. There is one set of industry process technology experts who argue that *Floating Gate* is a well-known technology with well understood failure modes that have been known and addressed for multiple generations. They further argue that floating gate will have initial manufacturing and yield advantages. The other set of experts argue that *Charge Trap* is fundamentally more scalable as one moves to subsequent

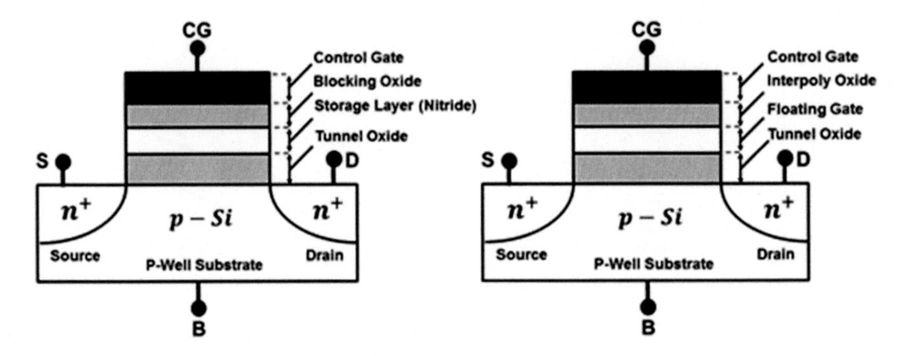

Fig. 1.14 Shows the two different memory cell types being used by different vendors. Samsung, Toshiba/Sandisk, and Hynix are using charge trap (*left*), while Intel/Micron has chosen to use floating gate (*right*). *Note* This figure is the same as Fig. 2.7 where there is a deeper technology analysis of both FG and CT

generations of 3D NAND products. While a clear answer may not be clear today in terms of which technology has longer term potential, one expects it will be evident by the end of the decade.

1.3.4 Packaging Innovation: TSV NAND

While most of this section has been about NAND components transitioning to 3D NAND, there is another area of innovation that often goes overlooked: packaging technologies. There are several technologies in this area, which are summarized below.

Interface chips that separate NAND generational requirements and provide a simpler high-level consistent interface like eMMC products, SmartNAND from Toshiba, ClearNAND from Micron, or other custom interface products in leading high end mobile products. There are some indicators from vendors at technical conferences that these may also be brought to raw NAND products to improve signal integrity for higher performance products. These products would be constructed with standard BGA and MCP packages, but have an additional logic chip which might be used (1) to reduce IO power and capacitance, (2) perform ECC, (3) handle higher level functions like garbage collection and other Flash NAND management routines, thus providing a simpler higher level interface for the main application processor.

TSV (*Through-Silicon* Via) for higher density applications. Traditional multi-die packages have individual die's attached at the edge to the substrate, making the total connection lengths several millimeters. TSVs are like steel girders inside a tall building as shown in Fig. 1.15: this reduces the length of the connection by about 10×. The advantages of the lower connection lengths are:

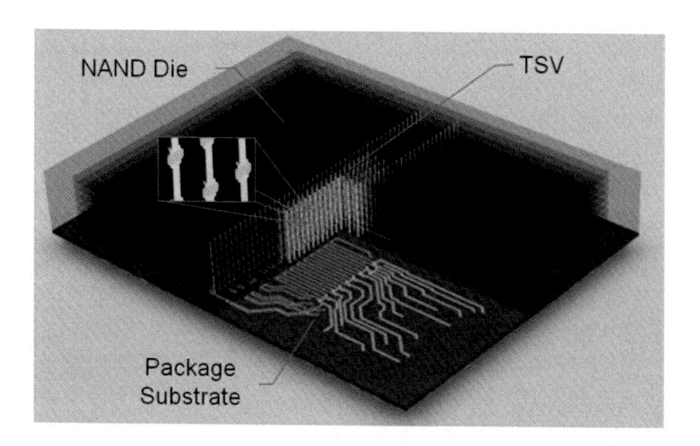

Fig. 1.15 Toshiba through silicon via (TSV) based die. This was introduced and demonstrated along with PMC Sierra at flash memory summit in August 2015. Source of diagram: Toshiba, http://toshiba.semicon-storage.com/ap-en/company/news/news-topics/2015/08/memory-20150806-1.html

- Performance—higher speeds can be supported since capacitance is reduced, thus improving *Signal Integrity* (SI). A real world example is the support of higher than 533 MT/s with 16 die stack packages and trace lengths of greater than 3 inches, which is hard to support with traditional packaging.
- Lower power—I/O voltage supply (VCCQ) can be reduced from 1.8 to 1.2 V, while VCC of the NAND core can move from 3.3 to 1.8 V. Toshiba and PMC-Sierra showcased TSV NAND at Flash Memory Summit 2015, demonstrating >50 % reduction in NAND power. The downside of TSV NAND packaging is the additional die size and cost associated with this technology. Yet, in higher die stack packages (with 4, 8, or 16 dies in a package) the advantages in power, signal integrity and performance make it an attractive option.

Since many SSD vendors use 1, 2, 4, 8, 16 die stacked packages to create different capacity SSD's, the successful introduction of these packaging technologies will require consistency between standard packaging and TSV based products from a controller point of view, so that SSD manufacturers can interchangeably use Interface chip/TSV or regularly packaged raw NAND.

1.4 New Memory Technologies on the Horizon

If one takes a broader view of the challenges in the memory area, arguably one of the most important one is the limited scaling of DRAM in comparison to NAND (note that DRAM and NAND continue to dominate the landscape of memories, making up more than 90 % of memory revenue). Most believe that all the near term issues of 3D NAND will eventually be resolved or absorbed by Flash controllers adjusting to the nuances of the new 3D NAND. Industry experts widely believe that we are entering a decade where NAND has a clear direction with 3D scaling. There is not as clear a direction for DRAM scaling. Figure 1.16 shows the emerging differences between current and projected NAND and DRAM scaling.

3D scaling is harder to do in higher speed and performance DRAM component technology. DRAM is thus looking at options that include relaxing the specification to get to higher monolithic densities with current CMOS technologies to an earlier introduction of DDR5. The latter seems to emerging as the industry direction.

In systems that use DRAM DIMM's (*Dual In-Line Memory*) on a separate board, there are other approaches like *Load Reduced DIMM* (LR-DIMM), which architecturally are similar to the interface chip described in Sect. 1.3.4. The interface buffer chip here provides the high speed interface to the host processor on one side and connects to DRAM components on the other. Thus, one does get higher bandwidth performance, but pays in terms of the additional latency and cost of using the LR buffer chip. But DIMM's only address a part of DRAM usage, and there are many cases (like SSD's) that use DRAM components and thus need the actual component die to scale.

Looking at new memory material technologies, there are two lenses to view them from (Fig. 1.17). There is the industry focus on investigating *DRAM replacement* with opportunities to invest in materials that offer the promise of being non-volatile with either (1) a lower cost/bit than DRAM or (2) significantly better performances (read/write/erase performances and features like byte level addressing) than NAND. The second area of investment is popularly referred to as *Storage Class Memories* (32 Gb and higher monolithic densities with performance between DRAM and NAND). Let's take a look at the three leading contenders.

MRAM (*Magnetoresistive Random-Access Memory*). This has been one of the promising material technologies that offers read/write speeds that are close enough to DRAM to be interesting as *DRAM replacement*. MRAM production parts until recently have been at a significantly lower density than DRAM (less than 1 Gb) and

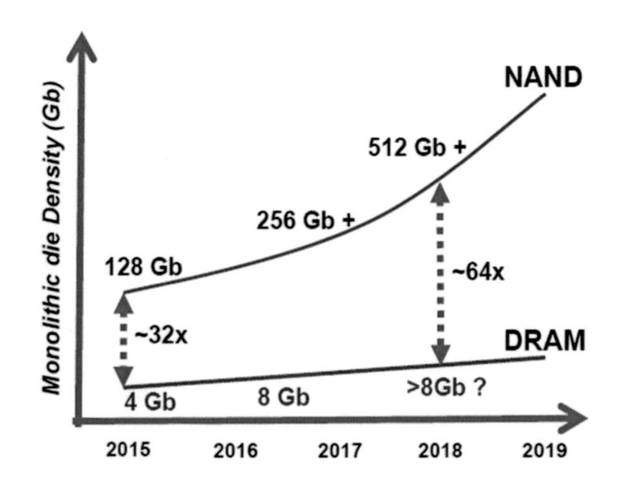

Fig. 1.16 A significant gap in scaling is emerging between mainstream densities of DRAM and NAND. There are some DRAM products based on TSV technology, but they tend to be niche high end products

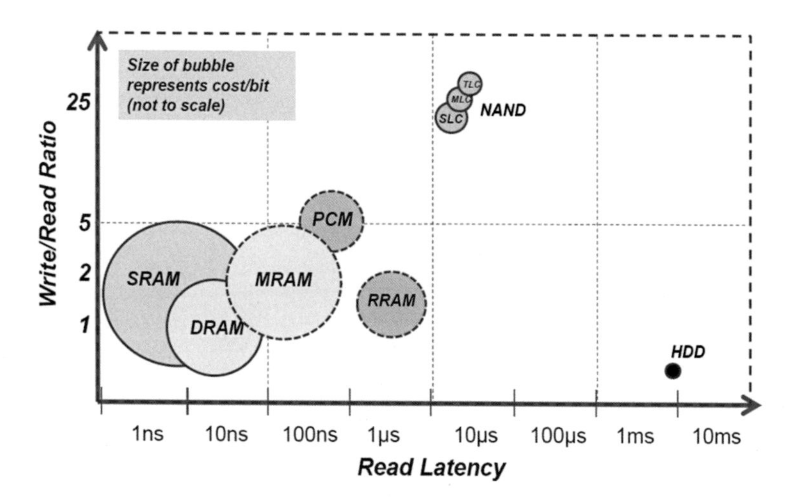

Fig. 1.17 Next generation memory technologies classified in terms of read/write timings ratios and latency. One can see MRAM as close to DRAM and PCM in a range that could drive both DRAM replacement and storage class memory usages

have also used as an *embedded DRAM* in ASICs. The next major milestone in this space are (1) monolithic densities in the multi Gb range using process technology nodes similar to DRAM, and (2) involvement of the Tier 1 memory vendors who arguably would bring the lower cost economics required for mass adoption.

PCRAM (*Phase Change RAM*). Some of the PCRAM products can be introduced both as a *DRAM replacement* and as a *Storage class memory* within high performance SSD products. The products are slated for broader introduction in the coming years and the technology is expected to have a cost structure between DRAM and 3D NAND. The aspects to monitor for PCRAM based products will be (1) how the memory vendors price and position this technology, (2) how many large Tier 1 vendors will be selling these solutions by the end of the decade, and (3) how deeply some of the leading enterprise server and storage architecture leaders integrate the technology into their systems.

ReRAM (*Resistive RAM*). There are a number of efforts in the industry investigating resistive memories including smaller startups and larger vendors, but this technology is mostly at the test chips stage for larger Gb monolithic densities. Mass production may take a few years to arrive. The draw of using this effect for memories is that as charge based technologies become more challenging in counting fewer and fewer electrons which make a '1' or a '0'; the resistive effect may be scalable longer term as the industry continues to feel pressure to reduce cost/bit. In this space, besides monitoring the progress of test chips, one should also keep an eye on progress of *1TnR* (*n* meaning multiple) where multiple memory cells require only one transistor, which will be required to scale cost effectively to Gb level monolithic die densities required to compete with mainstream DRAM and NAND.

There are also a number of other related and different technologies vying to solve the problems or take advantage of the opportunities in the industry. The author's belief is that there is no single technology that will meet all the needs and it will take a combination of new materials technologies, packaging advancements, and system and architectural innovations that will be the long-term answer.

1.5 What Do We Look for in the Next 5 Years?

The technology industry changes too fast for analysis like this to be more than a good snapshot of current status. Thus, as the industry transitions to 3D NAND over the next few years, some of the key markers to watch are reported below.

- *Memory vendor landscape.* Question: is there further consolidation in memory vendors in the next few years as laggards continue to get acquired by leaders of the memory or hard drive industry?
- *3D NAND leadership.* Question: which vendors will take the lead in the 3D NAND generation in ramping production volumes and what will be the response from the other NAND vendors?
- *Percentage of industry bits that will be 3D in 2017.* Question: will the industry have transitioned to 3D NAND (as defined by 50 % of the industry bits being 3D NAND) in 2017 or 2018? When will 50 % of the wafers be 3D NAND based?
- *Memory cell technology.* Question: will there be convergence to either Charge Trap or Floating Gate technology by the second or third generation of 3D NAND products?
- *NAND cell economics.* Question: when will QLC (4 bits/cell) be introduced to the market, and what will be the specifications of that in comparison to 2D TLC today?
- *Enterprise SSD's.* Question: who will be the leaders in PCIe SSD's (client and enterprise) longer term? Will the BYO SSD model become pervasive with large cloud operators?
- *3D X-point memories.* Question: how broadly accepted will 3D X-point from Intel/Micron be in a few years? How deeply will Intel integrate the use of this memory in their server architecture?
- *New Materials for memory.* Question: which next generation memory material will be successful longer term? Will this success be broad based or in niche applications? Will MRAM be able to ramp multi Gb products cost effectively?

Chapter 2
Reliability of 3D NAND Flash Memories

A. Grossi, C. Zambelli and P. Olivo

2.1 Introduction

Reliability represents one of the major antagonist towards the unstoppable technological evolution of hyperscaled NAND memories, since the correct operations must be assured throughout the entire lifetime. In particular, the ability of keeping unaltered the stored information even after a consistent number of write operations and for long times must be guaranteed.

A growth of the memory devices storage capacity without increasing the area occupation is constantly requested by the market: in order to satisfy such requirements, an increase of the memory density and of cell shrinking is mandatory. Nowadays, the transition from planar to three-dimensional architectures appears as the most viable solution for the integration of non-volatile memory cells in Tera-bit arrays. *Charge Trap* (CT) NAND memory cells are considered as one of the most promising technology for 3D integration because of a better scalability than *Floating Gate* (FG) NAND. Despite the high theoretical potentialities demonstrated by CT memories, several reliability issues affect such technology. Moreover, the transition from 2D to 3D changed the impact of the previously known reliability issues and generated new problems. Recently, in order to overcome such problems, new 3D vertical FG type NAND cell arrays have been proposed with promising performances.

A. Grossi (✉) · C. Zambelli · P. Olivo
Dipartimento di Ingegneria, Università degli Studi di Ferrara, Ferrara, Italy
e-mail: alessandro.grossi@unife.it

C. Zambelli
e-mail: cristian.zambelli@unife.it

P. Olivo
e-mail: piero.olivo@unife.it

R. Micheloni (ed.), *3D Flash Memories*, DOI 10.1007/978-94-017-7512-0_2

In this chapter the main reliability mechanisms affecting 3D NAND memories will be addressed, providing a comparison between 3D FG and 3D CT devices in terms of reliability and expected performances. Starting from an analysis of basic reliability issues related to both physical and architectural aspects affecting NAND memories, the specific physical mechanisms impacting the reliability of 2D CT NAND will be addressed. Then, a review of the main problems experimentally observed in different 3D CT cell concepts is reported. Finally, 3D FG memory concept is briefly introduced in order to understand the related reliability implications, and a comparison between 3D CT and 3D FG arrays is provided in terms of reliability and expected performances.

2.2 NAND Flash Reliability

During its lifetime a NAND Flash module undergoes a large number of Program/Erase (P/E) cycles. Every cycle involves very high electric fields applied to the tunnel oxide. The reliability of the entire memory requires that the tunnel oxide is able to correctly operate under stress conditions. It is obvious that huge efforts are to be spent to determine the right process for the tunnel oxide creation (in terms of thickness, material, growth, defectivity, interface, …) and the most effective algorithms in order to achieve a successful and reliable NAND technology.

In this section we will analyze the basic physical mechanisms related to the tunnel oxide, which affect both memory endurance and data retention. "Endurance" of a memory module is defined as the minimum number of P/E cycles that the module can withstand before leading to a failure. "Retention" is the ability of storing the information over time even when the external power supply is not applied. The tunnel oxide, which is a thin oxide, may be also responsible for other effects, such as erratic bits and over-programming, which might induce read errors.

2.2.1 Endurance

In NAND flash cells, program and erase operations rely on charge transport through thin oxides; this is accomplished via Fowler-Nordheim (FN) tunneling into/from a storage layer, which can be either a polysilicon FG [1] or an interfacial trapping layer in CT technology [2, 3]. Electron tunneling is responsible for a slow, but continuous, oxide wear out because of traps creation and interfacial damages; as a result, there might be charge trapping/detrapping into the tunneling oxide or undesired charge flowing into/from the storage layer.

As the number of P/E cycles increases, the above mentioned effects strongly impact writing operations. For instance, electron trapping reduces the tunneling efficiency so that, under constant voltage and time conditions, the charge injected into/from the storage layer decreases cycle after cycle.

To counteract "*endurance*" effects, all writing algorithms are based on a sequence of program/erase pulses, each one followed by a verify operation. This sequence proceeds until the expected amount of charge is correctly transferred into/from the storage layer. As the number of P/E cycles increases, the programming time is expected to reduce, whereas the erase time is expected to grow.

Without these write and verify algorithms (Chap. 3) it would be impossible to control the actual amount of charge transferred into/from the storage layer and Multi Level Cell (MLC) architectures would not exist [4].

Even if endurance is controlled by sophisticated (but slow and power consuming) algorithms, traps creation, charge trapping/detrapping, and interface damages still degrade the tunnel oxide. As a result, it gets really problematic to retain the stored information for extremely long times, which, at the end of the day, is a basic requirement of the non-volatile paradigm.

2.2.2 *Data Retention*

As mentioned in the previous section, the ability of keeping the stored information unaltered for a long time, i.e. the charge trapped into the storage layer, is mandatory for non-volatile memories. However, even with no bias applied, electron after electron, charge loss can lead to a read failure: a programmed cell can be read as erased if its threshold voltage (V_T) shifts below 0 V in case of Single-Level-Cell (SLC), or towards a lower threshold level with respect to the initial threshold voltage in case of MLC programming [5].

The higher the number of P/E cycles the worse the retention is, as it can be appreciated in Fig. 2.1, which shows how the cumulative V_T distributions of MLC programmed cells changes over time. Charge loss from the storage layer moves the V_T distributions towards lower values: the rigid shift of the cumulative V_T distributions is related to the oxide degradation and traps generation at the interface between storage layer and tunnel layer. These traps may be responsible for charge loss from the storage layer towards the silicon substrate. In fact, an empty trap, suitably positioned within the oxide, can activate *Trap Assisted Tunneling* (TAT) mechanisms characterized by a significantly higher tunnel probability with

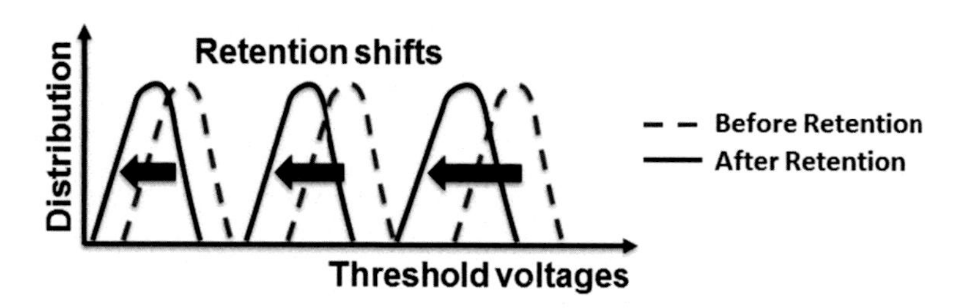

Fig. 2.1 Threshold voltage shifts induced by retention

respect to a triangular barrier unmodified by the trap presence. Moreover, an electron trapped inside the oxide during writing operations may be detrapped later on, when the cell is read or even when the cell is not addressed. As a result, the empty trap may enhance the TAT phenomenon (assuming a positive charged trap) and, in addition, it can increase the electron field at the storage layer-tunnel oxide interface, thus raising the probability of electron tunneling. It is clear that these mechanisms are strongly related to the oxide degradation and, therefore, data retention gets shorter with the number of applied writing pulses. In the MLC case, the cells programmed at higher V_T are more prone to data retention issues.

2.2.3 *Erratic Bits and Over-Programming*

The *Fowler-Nordheim* (FN) tunneling mechanism for writing and erasing data in NAND Flash has been used for several decades, demonstrating a sufficient level of reliability.

Nevertheless, it has been found that anomalous FN tunneling currents can occur in random periods of time, thus leading to significant variations of the threshold voltage after the writing operation [6] (see Fig. 2.2). This phenomenon is known as *erratic bits*.

In a NAND array, the presence of this phenomenon is detrimental for the performances of the memory as the unpredictable increase of the cell's threshold voltage may eventually induce the *over-programming* issue. As shown in Fig. 2.3, conductive cells featuring relatively large threshold voltage are erroneously read as

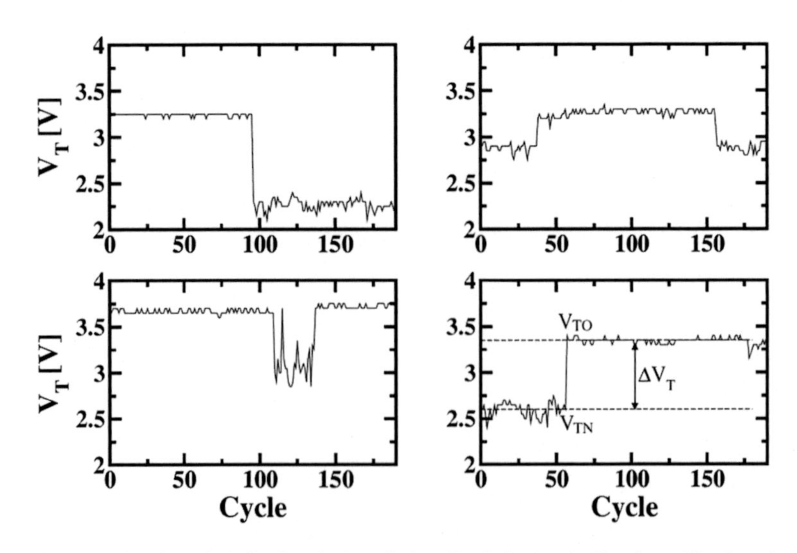

Fig. 2.2 Example of erratic behaviors in four flash cells. Cells threshold voltage V_T plotted versus the number of cycles exhibits RTN [6]

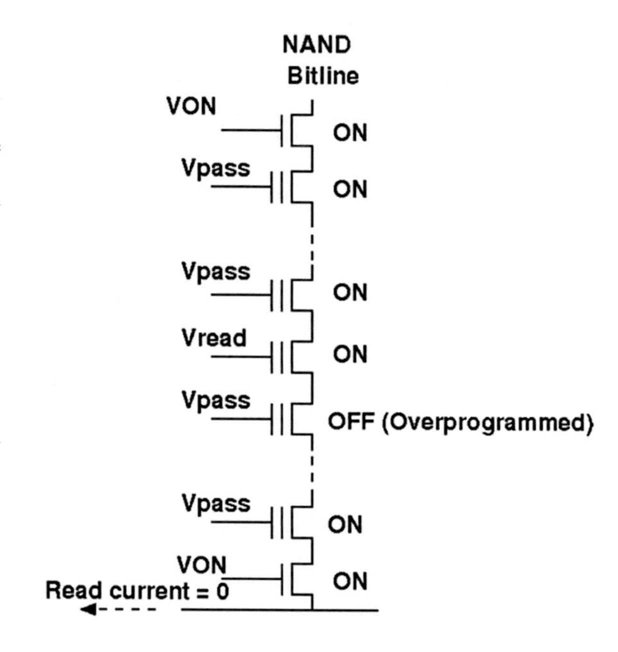

Fig. 2.3 Effect of an over-programmed cell in a NAND flash string. In normal conditions the status of the cell to be read (supposed to be ON) is correctly detected, since all other cells are driven by a V_{PASS} so that they behave as ON pass transistors. In the presence of an over-programmed cell ($V_T > V_{PASS}$), the current flow through the string is inhibited and the absence of current is attributed to a programmed status of the cell to be read, thus producing a read error [6]

OFF if over-programmed, and they can electrically isolate the NAND string. Such behavior generates read errors and consequent read throughput loss due to the additional work done by the *Error Correcting Codes* (ECC) trying to repair the failed bits.

Since erratic behaviors are intimately related to the electron tunneling mechanism, they can potentially affect all the cells of an array [6].

Anomalous tunneling has been related to the presence/absence of a cluster of positive charges in the tunnel oxide that strongly affects the FN tunneling operation. As a first approximation, erratic behaviors can, therefore, be described in terms of a two level *Random Telegraph Noise* (RTN) affecting the threshold voltage during cycling, in which the normal and the anomalous threshold voltage levels are the result of the presence of a cluster of more than 2, or less than 3, positive charges in the tunnel oxide, respectively [7, 8].

2.3 Architecture Dependent Reliability Issues

Architectural solutions for memory operations may also affect the overall reliability, by inducing errors and even cell failures [1]. The most common effects are the so called "disturbs", that can be interpreted as the influence of an operation performed on a cell (Read or Write) on the charge content of a different cell.

Read disturbs are the most frequent source of disturbs in NAND architectures. This kind of disturb may occur when reading many times the same cell without any

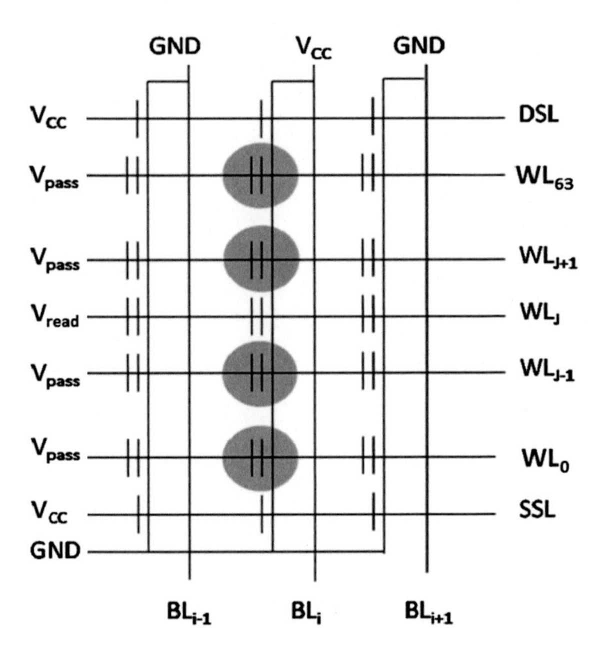

Fig. 2.4 Representation of read disturb in a NAND flash array. The cells potentially affected by read disturb are marked in *gray*

erase operation. All the cells belonging to the same string of the cell to be read must be ON, independently of their stored charge. The relatively high V_{PASS} applied to the control gate may induce a charge increase, especially if the read operation is repeated many times. These cells suffer a positive shift of their threshold voltage, which may lead to read errors. Figure 2.4 shows the typical read disturb configuration. As a matter of example, a 64 cell string is considered in the following, however the reported considerations can be extended to longer strings as well.

The probability of suffering from read disturb increases with the number of P/E cycles (i.e. towards the end of the memory useful lifetime) and it is higher in damaged cells. Read disturb does not cause permanent oxide damages: it can be reset by a simple erase operation.

In case of MLC programming, cells with lower V_T are slightly more vulnerable to shift than cells with higher V_T (see Fig. 2.5). In fact, the lower the threshold voltage the higher the voltage difference ($V_{PASS} - V_T$) across the tunnel oxide, which translates in a higher tunneling current. For cells in the erased state (ER) we observe a systematic shift of the cell's V_{TH} to the right (i.e. to higher values). The shift for P1 and P2 is much lower, since the read disturb effect becomes less prominent as V_T increases. For cells in P3, on the contrary, the average V_T shifts to the left. This is mainly due to charge loss (retention), which outweighs read disturb.

Two other important types of disturbs arise during the write operation: *Pass disturb* and *Program disturb*, which are shown in Fig. 2.6 (left and right), respectively. The former is similar to the read disturb and affects cells belonging to the same string of a cell to be programmed. With respect to the read disturb, the Pass disturb is characterized by a higher V_{PASS} voltage applied to cells that are not

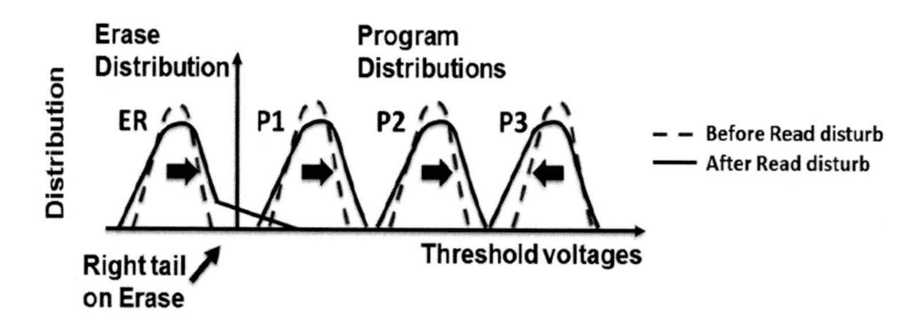

Fig. 2.5 Threshold voltage shift induced by read disturb

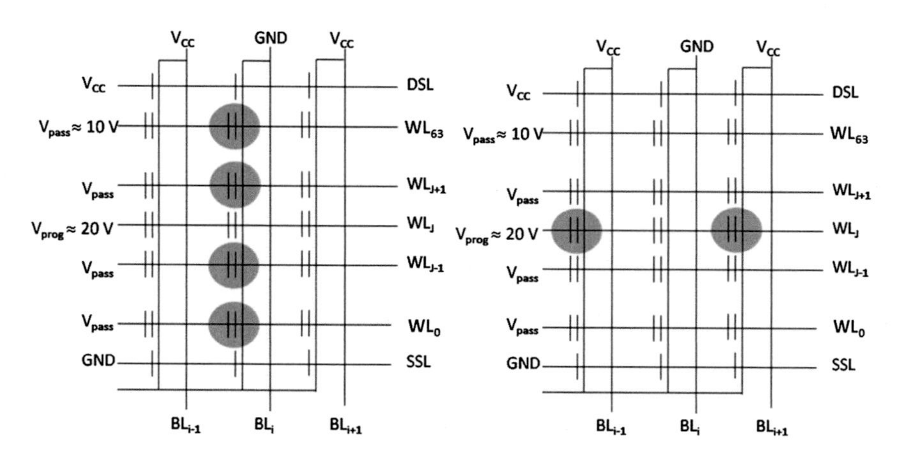

Fig. 2.6 Representation of pass disturb (*left*) and program disturb (*right*) in a NAND flash array. The cells potentially affected by disturbs are marked in *gray*

to be programmed (thus enhancing the electric field applied to the tunnel oxides and the probability of undesired charge transfer). On the other hand, the pass disturb can be repeated for a limited number of times (i.e. the number of cells in the string minus 1). In fact, when a string (block) has been fully programmed, an erase operation must be necessarily performed before any other reprogram.

The Program disturb, on the contrary, affects cells that are not to be programmed and belong to the same wordline of those that are to be programmed. In this case, again, no cumulative effects are present.

Edge Wordline disturbs affect the cells belonging to the first and last wordline, which connect the cell strings to the string selectors [9]. This disturb is due to a difference between the V_T of the cells belonging to WL_0 and WL_{63} with respect to the average V_T of all other cells. Such difference can be ascribed to three effects:

- different potentials at their terminals with respect to the other cells depending on the specific WL selected for programming;
- a different cell geometry due to the fact that these cells are located between a cell and a transistor (differently from cells belonging to WL_1–WL_{62}), therefore with a different field underneath their channels and a modified programming dynamics;

- a different cell lithography, especially when extremely scaled technology nodes are considered;
- the presence of a large *Gate Induced Drain Leakage* (GIDL) [10] current generated at the drain edge of DSL/SSL transistors due to their drain potential raised by channel boosting: such a field can efficiently trigger electron-hole pair generations followed by an acceleration of the electrons toward the channel of WL_0 and WL_{63} cells. These electrons can be injected into the floating gate of these cells, thus provoking an undesired increase of their threshold voltages.

In order to overcome such problem, the most common solution is to introduce two or more dummy WL before WL_0 and after WL_{63}, shielding the Edge Wordline Disturb: in such a way, the difference between the edge cells and all other cells in terms of potential at their terminals and cell geometry is minimized.

2.4 2D Charge Trap: Basics

The basic concept of a CT NAND memory cell consists of a metal oxide semiconductor device where the FG is replaced by an insulating charge trapping layer. Such storage layer, typically made of silicon nitride, is isolated by means of a tunnel oxide and a blocking oxide as sketched in Fig. 2.7 where the FG cell structure is reported for comparison. The tunnel oxide plays a basic role for the control of the device threshold voltage, whose value represents, from a physical point of view, the stored information. The blocking oxide prevents electrons from passing to/from the control gate. Electrons transferred into the storage layer give a threshold voltage variation. In quiescent conditions, thanks to the two oxides, the stored charge is supposed not to leak away, thus granting the nonvolatile paradigm fulfillment. Oxides are available in different materials depending on the *Back-End-Of-Line* (BEOL) process. The most common materials are: pure silicon dioxide (SiO_2) for blocking oxides, and either SiO_2 or a barrier engineered stack of Oxide-Nitride-Oxide (SiO_2-Si_3N_4-SiO_2) for tunnel oxides. A 2D planar Silicon-Oxide-Nitride-Oxide-Silicon (SONOS) cell is used as an example in this section [3].

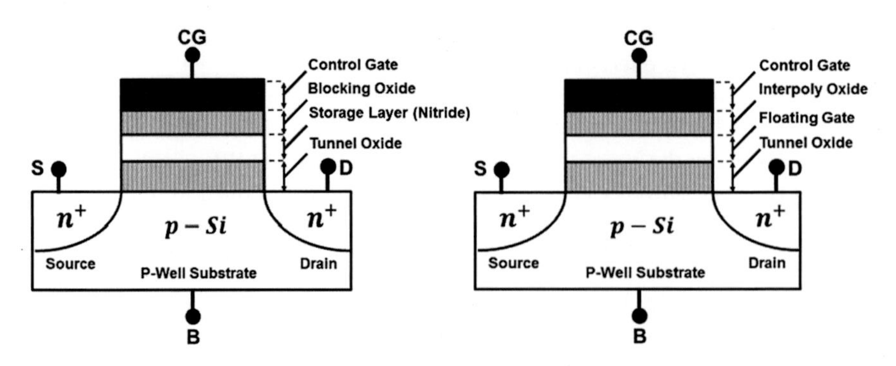

Fig. 2.7 *Left* Example of a charge trap device. *Right* Example of floating gate device

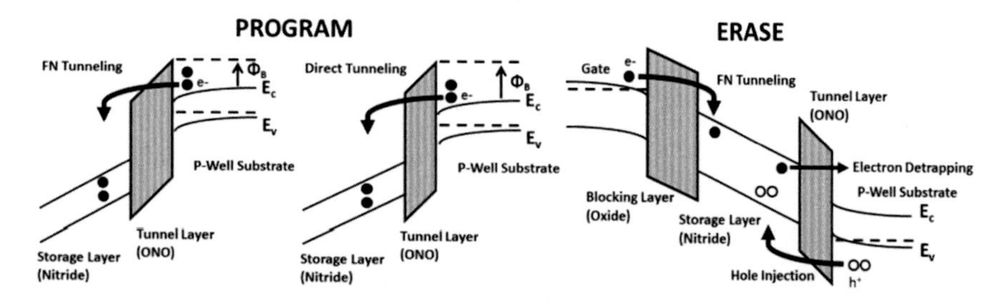

Fig. 2.8 Band diagrams of tunneling mechanisms in planar SONOS CT cell during programming (*left*) and erase (*right*). The two different conditions triggering FN or DT are sketched for programming [3]

High electric fields applied to the tunnel oxide allow electron transfer across the thin insulator to the storage layer. The physical mechanism used for injecting electrons into the storage layer depends on the applied electric field and oxide barrier thickness. In case of high electric fields and large oxide barriers, injection mainly occurs through FN tunneling, whereas in case of low electric field and thin oxide barrier, electrons mainly transfer through *Direct Tunneling* (DT): in this case there is a higher read margin window but retention is worse [3]. In CT cells electron tunneling involves the MOS channel/substrate and it requires appropriate biasing of control gate and bulk terminals (see Fig. 2.8), while drain and source are left floating. Erase operation occurs either through electron detrapping from the storage layer or hole injection from the substrate into the storage layer; at the same time, such operation causes an electron injection from the control gate to the storage layer through FN tunneling, and this is the reason for the well-known "erase saturation" problem [11]. The results of charge separation experiments [12] demonstrate that both electron detrapping and holes injection mechanisms contribute to the erase of a previously programmed CT device: electron detrapping dominates the first part of the transient, whereas hole injection prevails after the removal of the trapped electron charge due to electron emission.

2.5 2D Charge Trap: Reliability Issues

Despite the huge potential, several reliability issues affect CT memories, especially endurance and retention.

2.5.1 Endurance Degradation

The band diagram depicted in Fig. 2.9 describes oxide degradation mechanisms for blocking and tunnel layers. During programming operations (left), electron

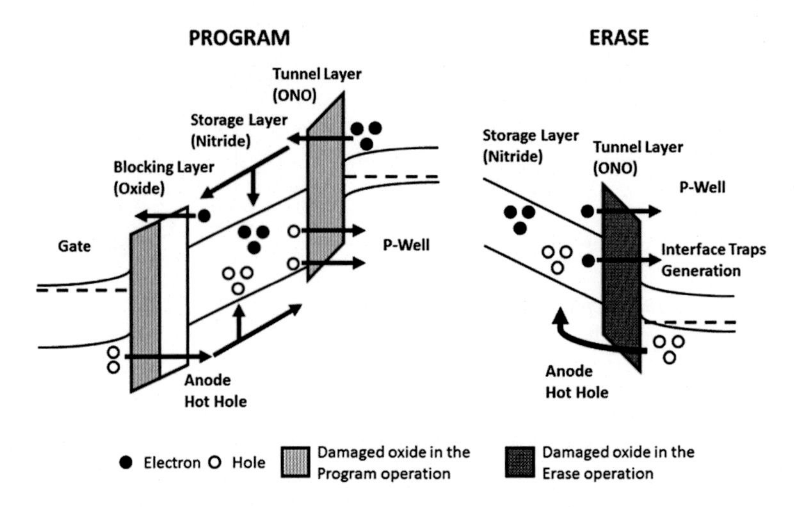

Fig. 2.9 Band diagram sketch of charge transport and trapping/detrapping during program (*left*) and erase (*right*) in planar SONOS CT cell [3]

injection occurs through either FN or DT, damaging the Tunnel Layer; damages to the Blocking Layer are caused by *Anode Hot Hole Injection* (AHHI). Moreover, electrons and holes going through blocking layer and tunnel layer from the storage layer contribute in a marginal, but not negligible, way to oxide degradation. During erasing (right), the hot hole injection from the substrate generates interface traps at the oxide/nitride interface, causing several damages to both storage and tunnel layers, as well as electrons transfer through the tunnel layer [11]. The generation of such interface traps between oxide and nitride interface is the main cause of endurance degradation: in programmed cells, electrons sitting in shallow traps can easily escape via oxide damages induced by cycling, resulting in a charge reduction that may cause read errors.

2.5.2 *Data Retention*

Data retention is one of the major issues of CT cells, especially at high temperature. Charge loss mechanisms of CT cells has been deeply investigated [13], identifying two main discharging paths: the first is related to thermal excitation of trapped carriers, the second one is due to direct tunneling through the thin tunnel oxide.

The charge loss processes are schematically depicted in Fig. 2.10. For each electron trapped inside the silicon nitride, two discharge mechanisms have to be considered. The first one is the direct *Trap-to-Band* (TB) tunneling from the storage layer traps to the conduction band of the substrate or of the gate; the second one is the thermal emission from traps to the conduction band of the storage layer.

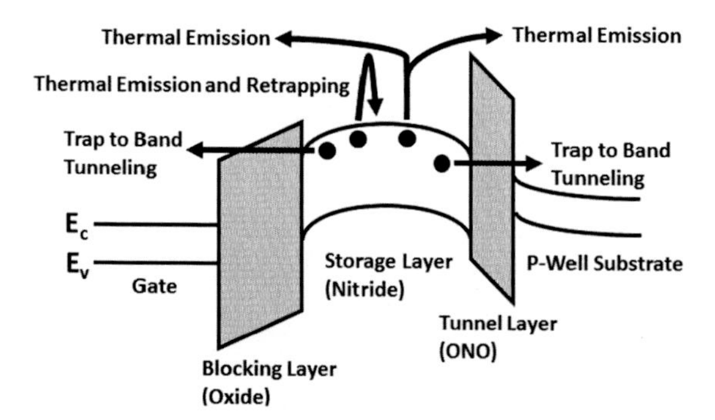

Fig. 2.10 Mechanisms involved in the discharging of programmed planar SONOS CT cell: trap-to-band tunneling through the tunnel layer, trap-to-band tunneling through the blocking layer, thermal emission above the oxide barriers, thermal emission and subsequent retrapping [13]

When thermal emission is considered, the charge loss is the result of two subsequent steps: the emission process and the escape of the electrons towards the bulk and gate electrodes. After emission, retrapping is also possible: the tunneling rate through the oxide barrier of the electrons emitted in the storage layer conduction band could be comparable with the emission and the recapture rates. Here we consider a simplified model where electrons leave the ONO layer only if their energy is higher than the lowest between the tunnel oxide and the top oxide barriers. Consequently, the tunneling of thermally excited carriers towards the bulk and the control gate at energies lower than the oxide conduction band are neglected, assuming that carriers with such an energy are recaptured in the same traps.

In additions, a fast initial charge loss has been observed on a small percentage of cells [14] (see Fig. 2.11). This V_T transient phenomenon has been attributed to the dielectric relaxation effect in the high-k layer, to charge trapping/detrapping, or to mobile charges in the blocking layer [14]. Such mechanism, denoted as *fast detrapping*, is mainly related to electrons trapped in shallow traps which have lower stability than electrons in deep traps; they can easily escape via oxide damages within 1 s after programming.

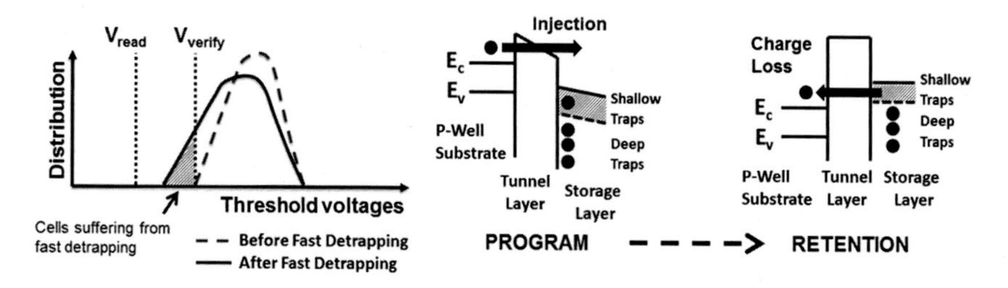

Fig. 2.11 Threshold voltage shift induced by fast detrapping (*left*). Band diagram sketch of fast detrapping effect (*right*) [14]

The same effect is observed after erase too: since the threshold voltage after program/erase does not immediately settle to the final value, there is a wrong estimation of the error bits during the verification step; of course, there is a dependency from the time interval between program/erase and read operations. Waiting for the final V_{TH} would significantly increase the total program/erase time and, of course, this is not acceptable. The transient threshold voltage shift after erase is due to hole re-distribution in the charge trap layer [15].

2.5.3 Threshold Voltage Shift During Sensing

Sensing the cell's threshold voltage during retention has been identified as one of the main reliability issues of CT cells [16]. This V_T decrease can be understood within the process of the temperature-activated charge transport through the blocking layer. The charge loss can be minimized when V_T sensing time is decreased down to microseconds. Moreover, blocking oxides engineered by adding a thin SiO_2 layer at the trapping layer/blocking oxide interface exhibit significant suppression of charge loss. Experimental data show that, for identically programmed devices, charge loss rate significantly increases when the V_T sensing operation is repeated more frequently. Furthermore, a similar amount of charge loss is observed when the cumulative sensing time is the same (same numbers of sensing measurements—dashed line in Fig. 2.12). These results indicate that charge loss might be strongly affected by the V_T sensing operation as well as retention time. The charge loss dependency from the V_T sensing time was evaluated by varying the sensing time from microseconds to few seconds. Shorter V_T sensing time is seen to minimize the initial charge loss and significantly reduce charge loss rate.

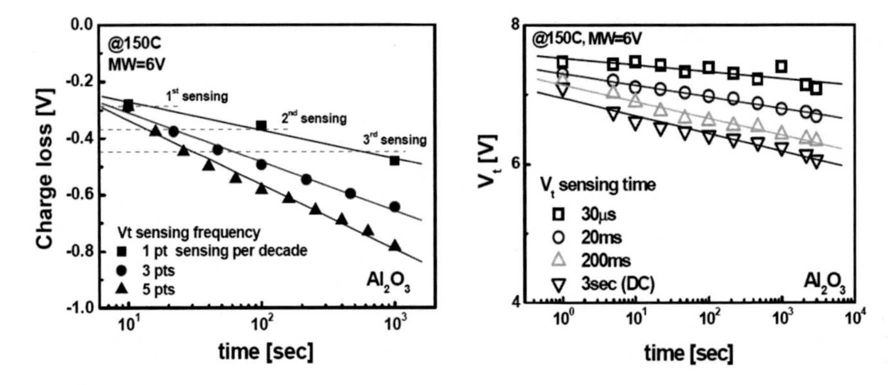

Fig. 2.12 *Left* Retention charge loss as measured by the V_T shift from the programmed V_T value (read window MW = 6 V) for different DC V_T sensing frequencies (V_T sensing time = 3 s). *Right* Programmed V_T dependency on retention time (retention charge loss) for various V_T sensing times [16]

2.6 From 2D to 3D Charge Trap NAND

Three-dimensional architectures appear today as the most viable solutions for the integration of non-volatile memory cells in Tera-bit arrays [17–19]. Two different approaches are possible in order to obtain 3D NAND devices (see Fig. 2.13): the first and simplest is to build the cell on a thin polysilicon substrate as usually done in 2D planar arrays, and stack more levels (Chap. 3) [20]. The second and most interesting approach, defined as vertical channel, is to build a CT cell with a cylindrical channel [17]. Both architectures have a physically large cell size (indicated with the feature process size $F_{process}$), mainly due to a channel width wider than planar devices, although offering a smaller equivalent area occupation due to the stacking of multiple tiers [21, 22]. The former solution does not offer any advantage over conventional planar CT cells in terms of P/E and retention, whereas the latter allows improving the cells programming performance, compared to planar devices, thanks to the shape of the CT cell, also known as *Gate-All-Around* (GAA) [23, 24]. Nevertheless, 3D NAND memories face new reliability issues because of the cylindrical shape and the multi-layer stacking. To understand these new reliability problems, basic concepts of 3D NAND cells are briefly introduced in this section; for more details about each single 3D architecture please refer to Chaps. 4, 5, 6, and 7. In the following sections, the reliability issues affecting 3D devices will be discussed by reviewing the main problems experimentally observed in different 3D NAND arrays [20, 21].

The cross section diagram along a single WL plane of the 3D vertical channel NAND reported in Fig. 2.13 (left) is shown in Fig. 2.14. It is worth highlighting that a *String-Select-Line* (SSL) group is equivalent to a 2D planar flash memory cell array. The 3D vertical flash memory has a nitride layer inside an *Oxide-Nitride-Oxide* (ONO) stack, which acts as a CT layer along the circumference of the thin poly-silicon vertical channel. Please note that each CT cell in this 3D vertical NAND memory is surrounded by the metal gates [24].

The GAA-CT cell is considered one of the most promising solution for 3D integration [25]. This is due to the curvature effect that relaxes the erase saturation problem: the electric field in the blocking oxide is lower than the one in the tunnel

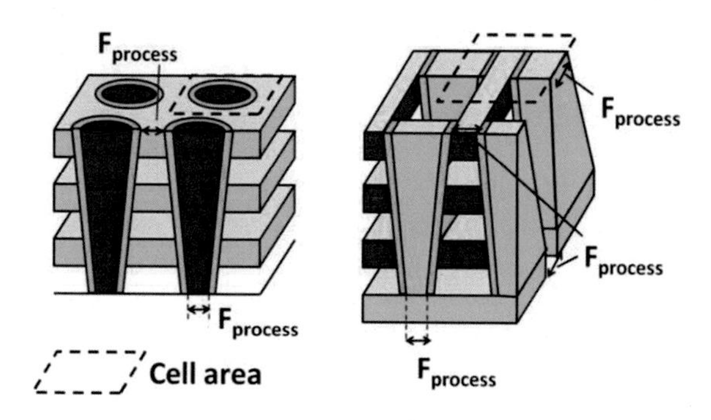

Fig. 2.13 3D NAND vertical (*left*) and horizontal channel (*right*) architectures with the corresponding feature process size $F_{process}$ [21]

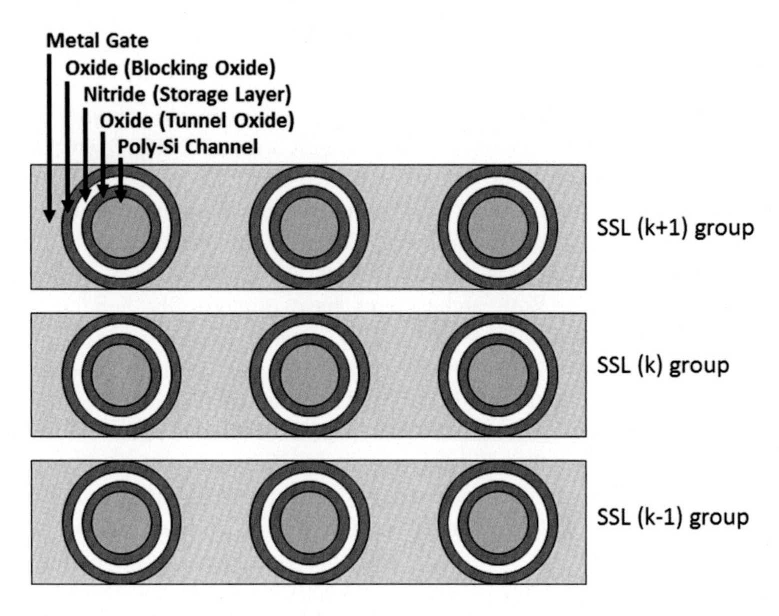

Fig. 2.14 Horizontal section of 3D vertical flash memory cell array [24]

oxide, allowing to increase the electrons detrapping from nitride traps towards the substrate. Moreover, thanks to the reduction of corner and fringing field effects during program, erase and read, GAA-CT cells allow more uniform trapped charge distributions in the storage layer and provide, in turn, steeper incremental step pulse programming (ISPP) transients than planar cells [26].

Figure 2.15 (left) shows a comparison between the GAA-CT cell (solid) and the planar CT cell (dashed), given the same thickness of gate dielectrics, in case of programming operation at $V_G = 12$ V with neutral nitride. The energy band profile

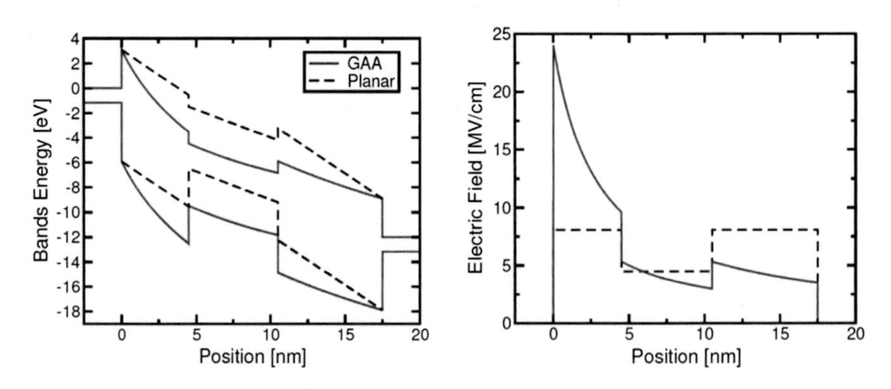

Fig. 2.15 Comparison between a GAA-CT cell and a planar CT cell having the same thickness of the gate dielectrics in terms of energy-band profile (*left*) and electric field (*right*) during program, for $V_G = 12$ V and neutral nitride [27]

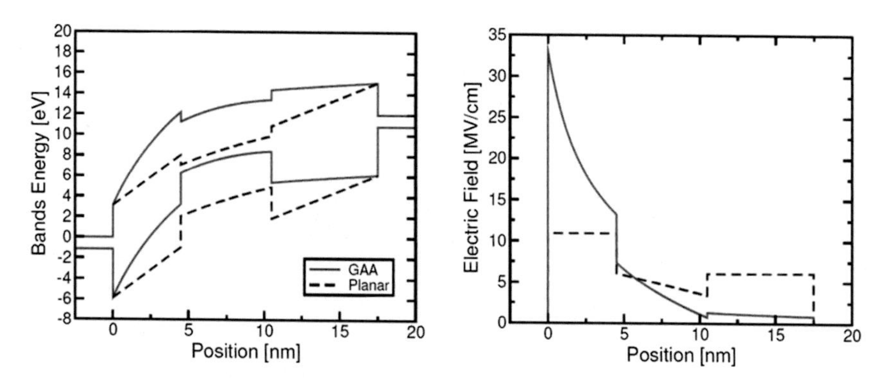

Fig. 2.16 Comparison between a GAA-CT cell and a planar CT cell having the same thickness of the gate dielectrics in terms of energy-band profile (*left*) and electric field (*right*) during erase, for $V_G = -12$ V and neutral nitride [27]

clearly shows a reduction of the thickness of the energy barrier which prevents electron tunneling from the substrate to the nitride. Compared to the planar case, the electric field inside GAA-CT dielectrics is not constant, showing a maximum value at the substrate/tunnel oxide interface. This maximum value is about three times larger than the electric field in the tunnel oxide of the planar device: of course, this is a strong improvement for the programming dynamics [27]. In addition to that, the electric field in the GAA-CT blocking oxide is lower than in the planar case, thus resulting in a reduced electron leakage from the nitride to the gate during programming (see Fig. 2.15 right).

A comparison between GAA-CT and planar cell during erase at $V_G = -12$ V is shown in Fig. 2.16. As for positive V_G, in the case of GAA-CT the electric field reaches a maximum at the substrate/tunnel oxide interface, and it is quite larger than the electric field present in the planar device (right). This behavior enhances the hole tunneling current from the substrate to the nitride during erase. In addition, the lower electric field at the gate/blocking oxide interface prevents electron injection from the gate, thus relieving the erase saturation issues [27].

2.7 3D Charge Trap: Reliability Issues

Even if the transition from 2D to 3D devices can leverage the advantages of the ring shape stack of the memory cell, all the reliability issues affecting planar devices (i.e. endurance, retention and read disturbs) are still there. On top of that, there are new reliability challenges due to vertical charge loss (i.e. through Top/Bottom oxides) and lateral charge migration (i.e. towards spacers). In case of 3D CT arrays the causes are ascribed to either uneven electrical field distribution in *Bottom Oxide* (BTO), equivalent to the blocking oxide in 2D CT, and *Top Oxide* (TPO), equivalent to the tunnel oxide in 2D CT, due to the cylindrical geometry (see Fig. 2.17).

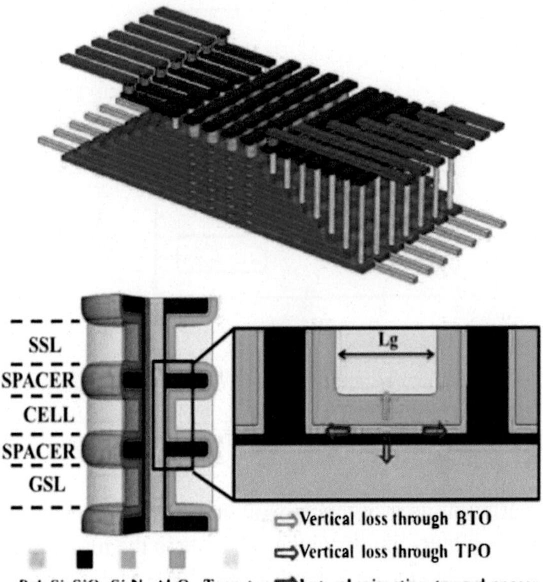

Fig. 2.17 *Top* bird's eye view of 3D CT memory. *Bottom* 3D CT-NAND string with one cell and two select transistors and schematic diagram of the charge loss paths [28]

As a result, charge loss in 3D memories is worse than what we have seen with planar devices, and this is considered the most critical reliability issue for high density and high reliability 3D integration [28]. In this section, physical mechanisms related to the vertical structure of the memory devices, such as vertical charge loss and lateral charge migration, will be described.

2.7.1 Vertical Charge Loss Through Top and Bottom Oxides

A constraint for the 3D vertical arrays, not present in planar devices, is that the charge-trap layer cannot be easily interrupted between layers. This fact creates additional leakage paths for the charge, from each cell's active area towards other cells on the same string, as schematically illustrated in Fig. 2.18. In addition to vertical charge loss through Top and Bottom oxides (along Y axis), lateral charge leakage (along X axis) represents an extra source of retention loss for cells in 3D vertical arrays, which should be carefully considered for the reliability assessment of the technology [29].

The involved physical mechanisms accounting for charge distribution evolution during time, along X and Y axes, are illustrated in Fig. 2.19. Charge transport in the conduction band of the charge trapping layer is described based on the drift-diffusion transport scheme. The interaction between free carriers and trapped carriers is governed by the carrier capture phenomenon calculated by *Shockley-Read-Hall*

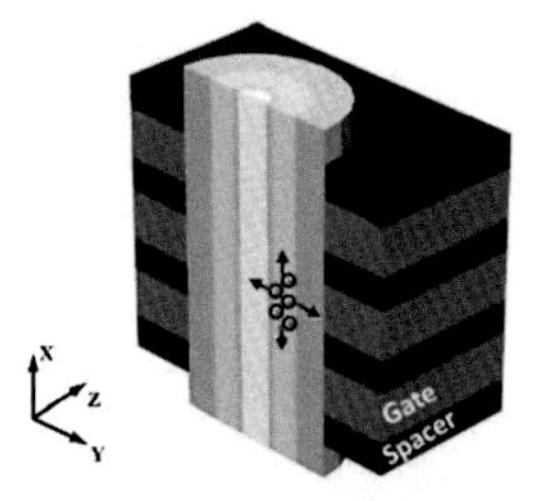

Fig. 2.18 3D CT-NAND structure and charge loss along X and Y axes [29]

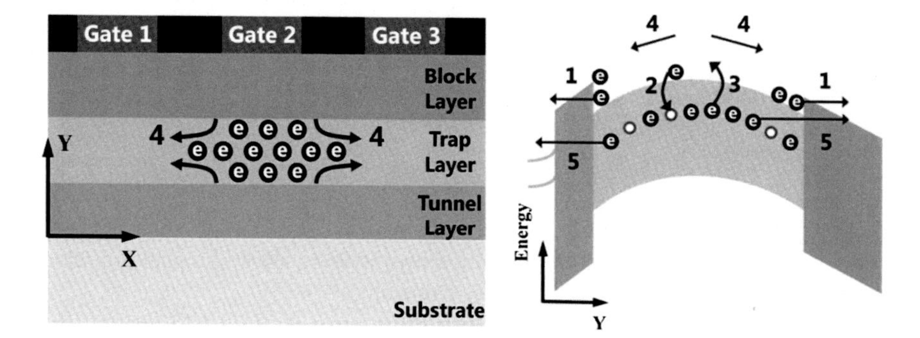

Fig. 2.19 Dominant physical mechanisms along X (*left*) and Y (*right*) axes: 1-DT/FN Tunneling, 2 and 3-carrier capture and emission, 4-drift and diffusion transport, 5-TB tunneling [27]

(SRH) theory, and carrier emission contributed by thermal and Poole-Frenkel effects. Besides, Band-to-Trap (BT) tunneling and Trap-to-Band emission should be taken into account as additional charge capture and loss mechanisms.

Figure 2.20 shows the simulated *Remaining Charge Percentage* (RCP), defined as the percentage of the initial charge which remains inside the storage layer, during

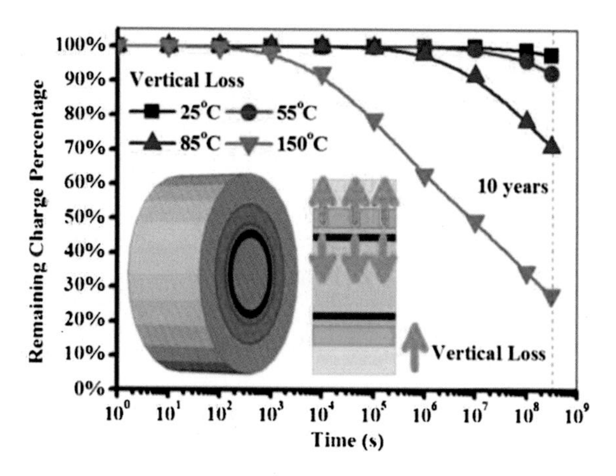

Fig. 2.20 Simulated vertical loss transients. The inset shows the schematic diagram of device structure and the cross-sectional schematic of charge loss paths [28]

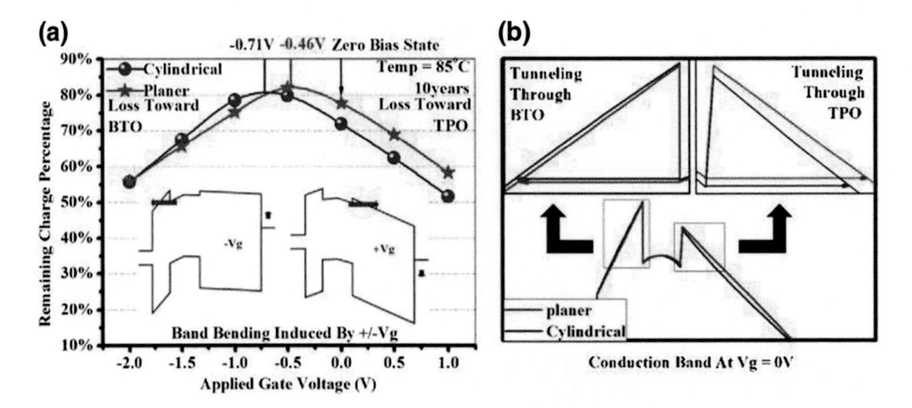

Fig. 2.21 **a** RCP of cylindrical and planar devices, with insets showing the schematic band diagram under positive and negative bias. **b** Conduction band diagram of cylindrical and planar devices at $V_G = 0$ V [28]

vertical charge loss transients [28]. The charge loss characteristics strongly depends on the temperature, which can be explained by enhanced Poole Frenkel emission from trap to conduction band at elevated temperature.

In order to distinguish the role of charge loss through BTO and charge loss through TPO, the RCP under different gate bias is given in Fig. 2.21 (left); in addition, the RCP under different stresses of planar device with identical structure parameters is also plotted as reference. As shown in the energy band diagram, the charge loss through BTO occurs under negative bias while charge loss through TPO dominates with positive bias. Compared to the planar device, charge loss towards TPO is higher and charge loss from BTO is lower in cylindrical devices. This can be explained by the conduction band diagram of Fig. 2.21 (right). Due to the uneven distribution of the electric field in BTO and TPO, the conduction band graphs of the cylindrical device are not straight any longer, but convex in BTO and concave in TPO. As a result, charge loss through BTO is slightly reduced [28].

2.7.2 *Lateral Migration Towards Spacers*

Lateral charge migration towards the spacer (i.e. the region between each layer) region (along X axis, referring to Fig. 2.18) is another key path for the charge loss due to the difficulty of cutting the CT layer between memory cells. Different shapes of the CT layer exhibit different lateral migration performances. To analyze this shape dependency, *Bit Cost Scalable* (BiCS)-type structures and *Terabit Cell Array Transistor* (TCAT)-type structures have been studied [28]; more details about both arrays can be found in Chap. 4. In order to focus on lateral migration performance, the cells were programmed to the same threshold voltage (6 V). Lateral migration is accelerated by temperature (Fig. 2.22). The considerable threshold voltage loss

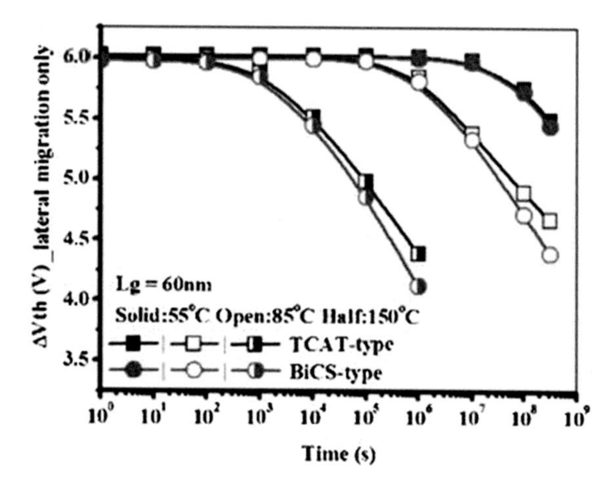

Fig. 2.22 Comparison between TCAT-type and BiCS-type devices [28]

caused by the lateral migration indicates that carriers in the nitride significantly migrate laterally. Figure 2.23a shows the distribution of trapped carriers versus elapsed time in TCAT-type and BiCS-type devices: lateral migration of trapped charges can clearly be observed as time proceeds. TCAT-type devices show better retention characteristics. Shape dependency of lateral migration can be explained by the lateral charge profile evolution shown in Fig. 2.23b: the corner of TCAT-type devices suppresses the migration of trapped charges, which can be described in device-level simulations not only by the higher charge density along the channel direction, but also by the sharp peak at the corner of the CT layer.

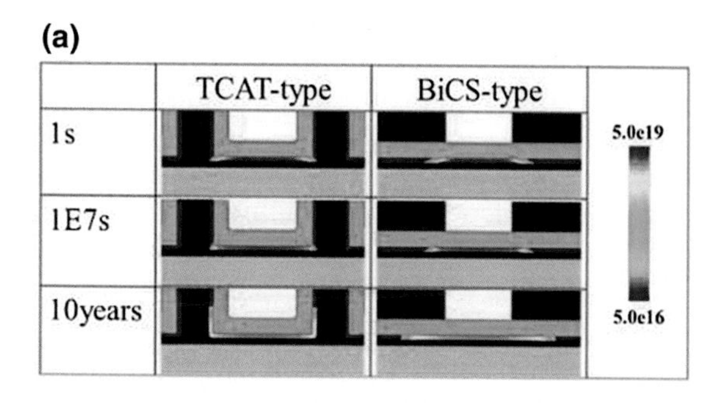

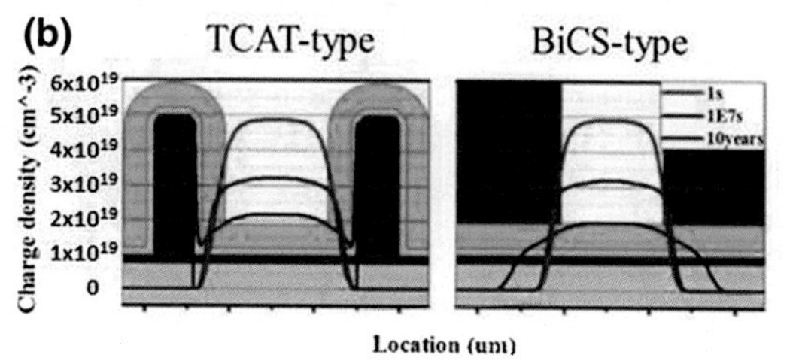

Fig. 2.23 **a** Simulated trapped charge distribution at different elapsed time and T = 85 °C. **b** Simulated lateral charge profile evolution (cutline at *middle* of CT layer along the channel) [28]

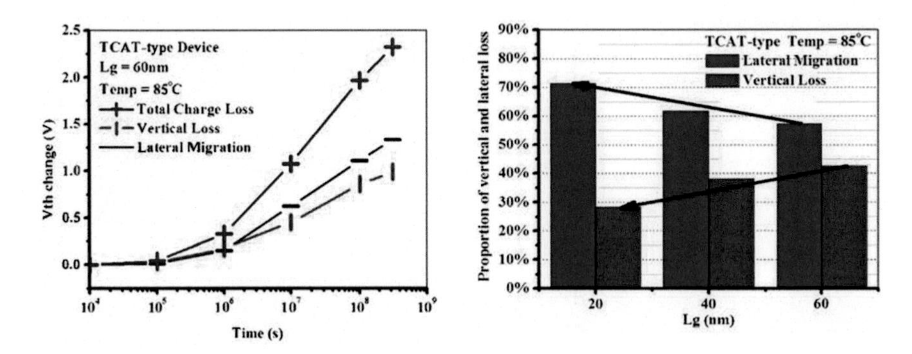

Fig. 2.24 Split of threshold voltage change by vertical loss and lateral migration versus time (*left*) and versus channel length (right) for TCAT devices [28]

Based on the above discussion, the impact of vertical loss and lateral migration in retention has been compared for TCAT and summarized in Fig. 2.24 (left). It can be seen that lateral migration is the dominant charge loss mechanism. The charge loss behavior with different channel length is shown in Fig. 2.24 (right). As the channel length reduces, lateral migration accounts for a larger percentage of charge loss, which indicates that lateral migration should be a more critical issue than vertical loss in high-density and high-reliability design of 3D GAA-CT memories.

2.7.3 Transient V_T Shift

Transient V_T shift after erase, previously described for 2D cells (Sect. 2.5.2), is observed on 3D devices too. In the GAA-CT cell, a smaller diameter of silicon nanowire shows a better erase efficiency due to the electric field concentration effect on the tunnel oxide (Fig. 2.25, left). However, the GAA-CT device also shows a transient V_T shift after erase: the amount of V_T shift is related to the amount of read window (defined as the voltage difference between the 'programmed' and 'erased' states in case of SLC architectures, or between two adjacent levels in case of MLC architectures) and the V_T shift in GAA-CT is well correlated to that in planar CT devices (Fig. 2.25, right), thus implying the same mechanism. It is worthwhile noting that the transient V_T shift can be reduced by scaling the channel length (L_G) and the diameter of the silicon nanowire (W_{NW}) in GAA-CT, probably due to a compensation effect by charge crowding and lateral charge spreading. As shown in Fig. 2.26, when the diameter of nanowire goes below 6 nm, the tendency of the drain current I_D to increase with time disappears. Therefore, 3D GAA-CT device with small nanowire diameter shows advantages in fast erase operation [15].

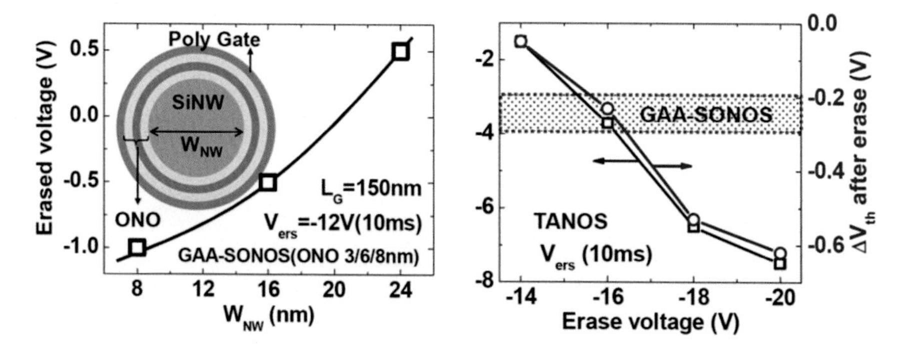

Fig. 2.25 *Left* Erased saturation voltage versus W_{NW} in GAA-CT device. A smaller W_{NW} shows a better erase efficiency due to the electric field concentration effect on the tunnel oxide. *Right* Correlation of V_T window and V_T shift after erase operation. The V_T shift in GAA-CT is well correlated to that in planar CT devices, implying the same mechanism for the transient V_T shift [15]

2.7.4 *Program and Pass Disturbs*

All kinds of 3D NAND suffer from two program disturbs. Besides the traditional program and pass disturbs affecting 2D architectures (see Fig. 2.6), in 3D NAND also the disturbs related to the vertical structure are to be taken into account.

2.7.5 *Vertical Hole Design Limitations*

Since scaling and design of 3D NAND are completely different from planar NAND, and with different implications on the memory reliability, new methodologies are

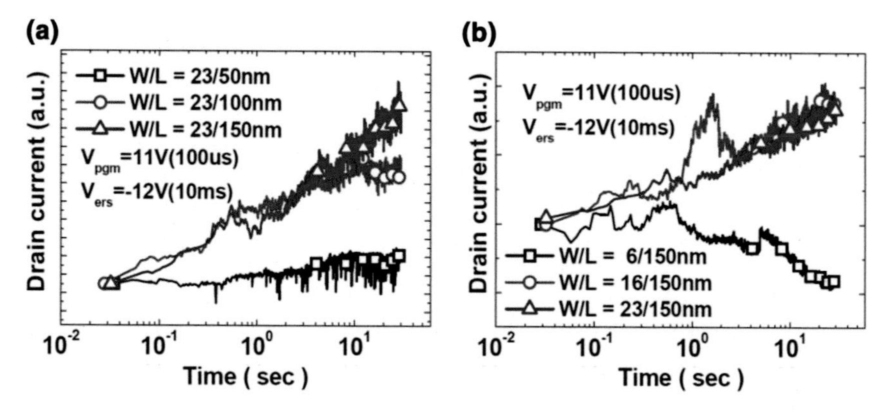

Fig. 2.26 Transient I_D of GAA-CT devices with different L_G (**a**) and W_{NW} (**b**). Transient V_T shift becomes smaller when L_G and the W_{NW} are scaled. I_D fluctuations in GAA-CT may be due to single electron effects or random telegraph noise [15]

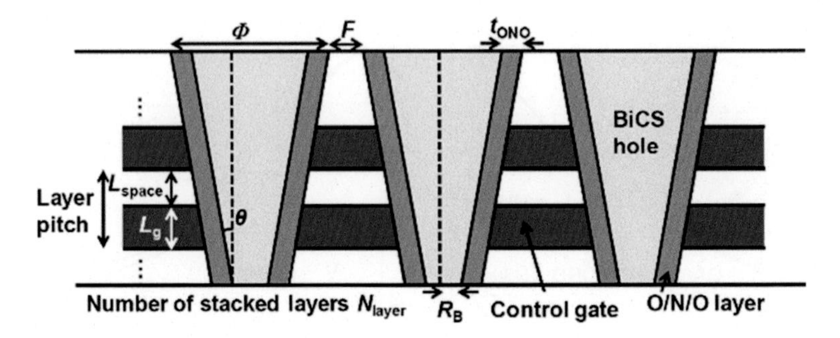

Fig. 2.27 Cross sectional view of 3D NAND [30]

required. One of the problems of 3D NAND is the reduced cells density per single memory layer. As a matter of example, Fig. 2.27 shows the simplified cross sectional view of the Bit-Cost Scalable (BiCS) type 3D NAND. The BiCS hole must be filled with ONO film ($\sim$20 nm) and silicon channel. Since it is not possible to aggressively scale the ONO film because of read window and reliability, the diameter of the BiCS hole is not so scalable. Therefore, the number of stacked layers (N_{layer}) should be increased to compensate this drawback. Moreover, as shown in Fig. 2.27, there is an additional limitation due to the finite taper angle θ in the BiCS hole: the memory cell at the top of the stack is always larger than the cell at the bottom. In other words, once the bottom of the stack reaches the minimum cell's size of a specific technology, then the area can't be shrunk anymore. On the other hand, since the minimal line and space lithography pattern are not required for the control gate (CG) formation in 3D NAND, CG length L_g and spacing L_{space} can be independently chosen. This design flexibility is allowed for 3D NAND. Therefore, suitable device design for 3D NAND can be explored in terms of L_g and L_{space}. Programming and disturbance characteristics are evaluated for the memory operation, as shown in next sections [30].

2.7.5.1 V_T Shift Induced by Stored Electrons During Programming

Given a specific electron density in the nitride layer, 3D and 2D cells can be compared in terms of the resulting V_T shift of the programmed cell, as shown in Fig. 2.28 [30]. L_g and L_{space} should not be too small and too large, respectively. The smallest shift is observed at L_g = 10 nm and L_{space} = 50 nm. When L_{space} is large, the spacing region determines the V_T of the cell. Therefore, the effect of the stored electrons on V_T relatively decreases for large L_{space} and the corresponding read window decreases. Large L_g shows high V_T shift because the potential of the center region of the channel in the target cell is mainly controlled by the stored electrons.

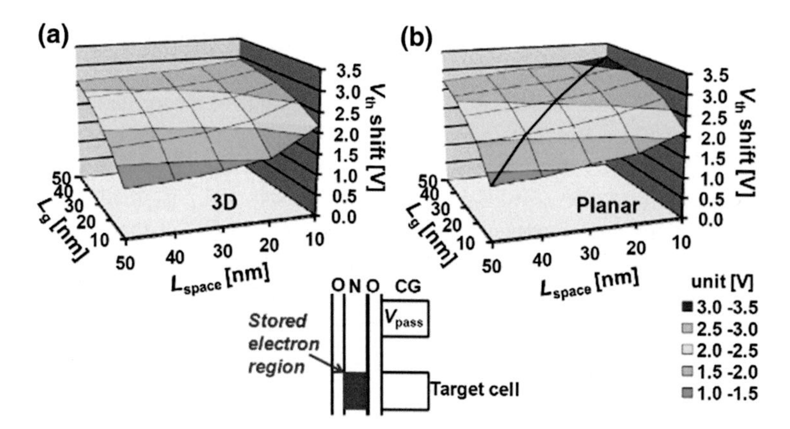

Fig. 2.28 V_T shift of the programmed cell (corresponding to read window) as a function of L_g and L_{space} when electrons (electron density: 1×10^{19} cm^{-3}) are stored in 3D NAND (**a**) and planar NAND (**b**) [30]

2.7.5.2 V_T Shift Induced by Neighboring Cells

V_T shift induced by the electrons stored in the neighboring cell was investigated as shown in Fig. 2.29 [30]. As L_g and L_{space} decrease below 20–30 nm, V_T shift due to the neighboring cell drastically increases in both 3D and planar NAND. When L_{space} is small, stored electrons couple with the channel of the target cell. This is severe when L_g is small because stored electrons potential can affect the entire channel region of the target cell. Moreover, at small L_g and L_{space}, the V_T shift degradation in 3D NAND is higher than in planar NAND. The substrate-channel coupling in planar NAND reduces the V_T shift by because of the coupling between the channel and stored electrons.

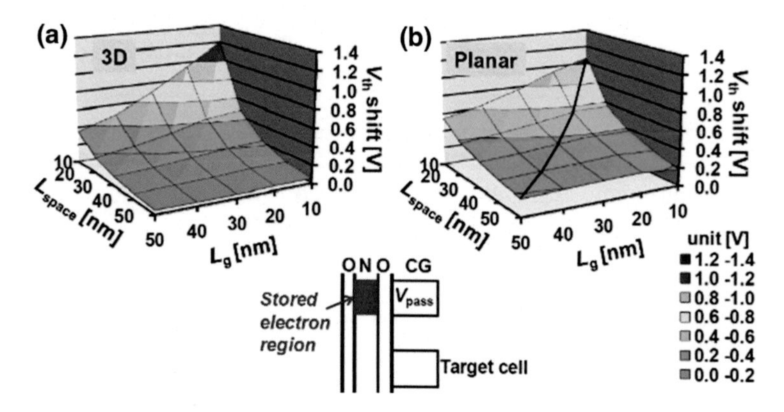

Fig. 2.29 V_{th} shift as a function of L_g and L_{space} when electrons (electron density: 1×10^{19} cm^{-3}) are stored in the neighboring cell in 3D NAND (**a**) and planar NAND (**b**) [30]

2.7.5.3 Electric Field in the Tunnel Oxide During Programming

Evaluations of the electric field in the tunnel oxide during the programming has been performed [30]: Fig. 2.30 plots the electric field of the tunnel oxide in the channel (NAND string) direction. The figure shows that the electric field spreads out in the lateral direction (fringing electric field exists). If L_g is small, the tunnel oxide electric field cannot concentrate at the center of the CG of the programmed cell. Therefore, the electric field in the tunnel oxide of the programmed cell (E_{ox_pgm}) decreases at small L_g. If L_{space} is small, then the electric field penetrates into the neighboring cells; however, if L_g is large, the penetration is only at the edge of the cell. Therefore, the electric field at the center of the CG remains low for the neighboring cells (low E_{ox_ngb}).

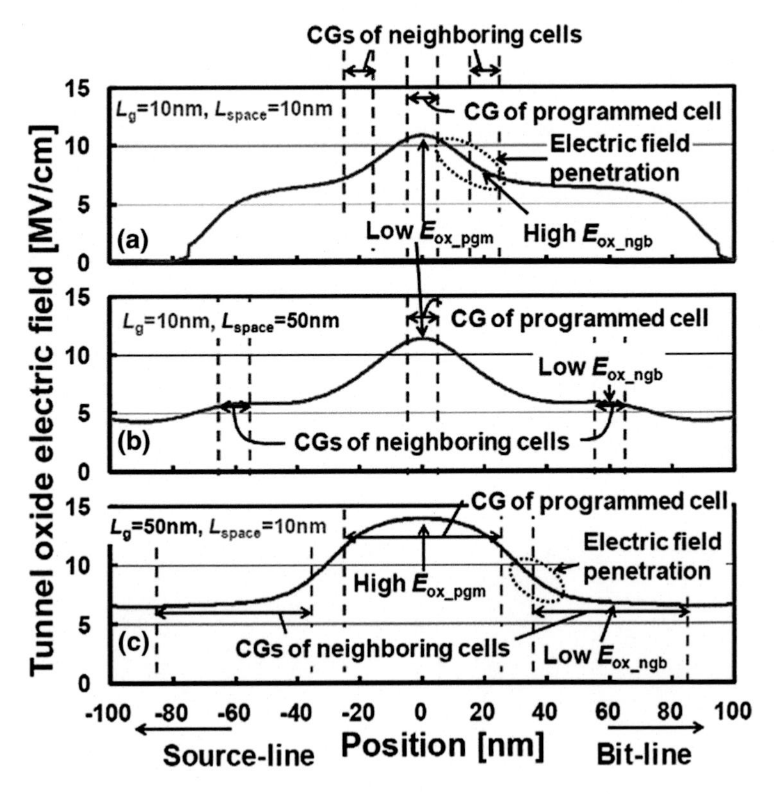

Fig. 2.30 Electric field of the tunnel oxide in the channel direction. 3D NAND string with: L_g = 10 nm and L_{space} = 10 nm (**a**), L_g = 10 nm and L_{space} = 50 nm (**b**), L_g = 50 nm and L_{space} = 10 nm (**c**) [30]

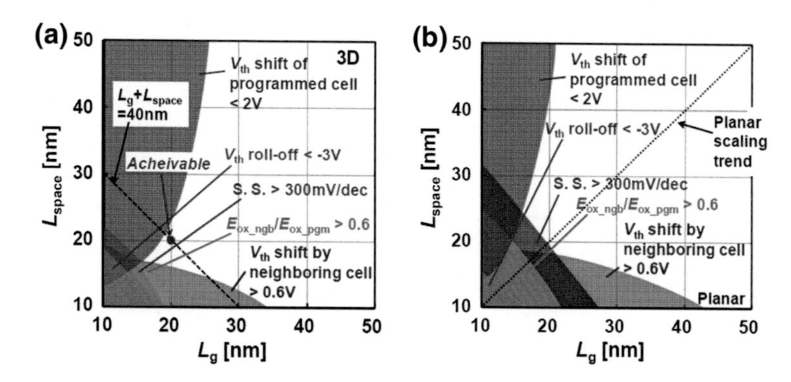

Fig. 2.31 L_g and L_{space} design window for 3D NAND (**a**) and planar NAND (**b**) [30]

2.7.5.4 Design Window of L_G and L_{SPACE}

Figure 2.31 shows the design window of L_g and L_{space} for 3D (a) and planar (b) NAND [30]. The criteria for unacceptable regions (shaded regions in Fig. 2.31) are assumed as follows; V_T roll-off < −3 V, Subthreshold Slope (S. S.) > 300 mV/dec, V_T shift < 2 V in the programmed cell, V_T shift > 0.6 V by the neighboring cell and $E_{ox\_ngb}/E_{ox\_pgm} > 0.6$. $L_g = L_{space}$ = 20 nm (layer pitch of 40 nm) is achievable in 3D NAND in terms of the electrical characteristics. Same L_g and L_{space} are preferable to cope with the tradeoff between the large V_T shift for the programmed cell and the small V_T shift induced by the neighboring cell. For further improvements, the diameter of the BiCS hole should be decreased. Table 2.1 summarizes the comparison between 3D and 2D: 3D NAND achieves very good on-current (I_{on}), S.S. and low program voltage (V_{pgm}) compared to planar NAND. Slight degradations in V_T roll-off and V_T shift caused by the stored electrons in the neighboring cell are observed only at the small L_g and L_{space} region.

2.8 3D CT Versus State-of-the-Art 2D FG

In this section, a comparison between a particular 3D CT NAND (denoted as *Stacked Memory Array Transistor*—SMArT [31], depicted if Fig. 2.32) and state-of-the-art 2D FG is reported, in terms of both performances and reliability. A detailed description of 2D FG reliability problems is provided in [1].

V_T distribution widths of MLC 3D CT cells are ~30 % smaller compared to the 2y-nm FG, because of their interference free nature (Fig. 2.33, left). The widening of cells V_T distributions during cycling is compared in Fig. 2.33 (right), where

Table 2.1 Summary of 3D NAND cells [30]

	I_{on}	V_{th} roll-off	S.S.	V_{th} shift (programmed cell)	V_{th} shift (neighboring cell)	Tunnel oxide electric field (E_{ox_ngb}/E_{ox_pgm})	V_{pgm}
Planar NAND	Poor	Fair	Poor	Fair	Fair	Fair	20 V
3D NAND	Very good	Poor at small L_g, L_{space}	Very good	Fair	Good at large L_g and L_{space}, poor at small L_g and L_{space}	Fair	17 V
Preferable scaling parameter	L_{space}	–	–	L_{space}	L_g	–	L_{space}

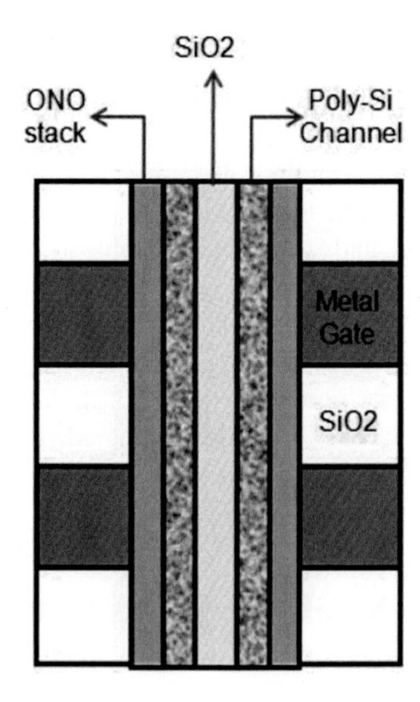

Fig. 2.32 SMArT cell schematic [31]

SMArT cells show no widening up to 5k cycles, whereas FG cells start broadening from 3k [31].

On the other hand, 3D CT have worst retention performances as shown in Fig. 2.34, where the post cycling V_T shifts at high temperature are compared. In 3D CT cells the V_T shifts are so large that the distributions are no longer separated [31].

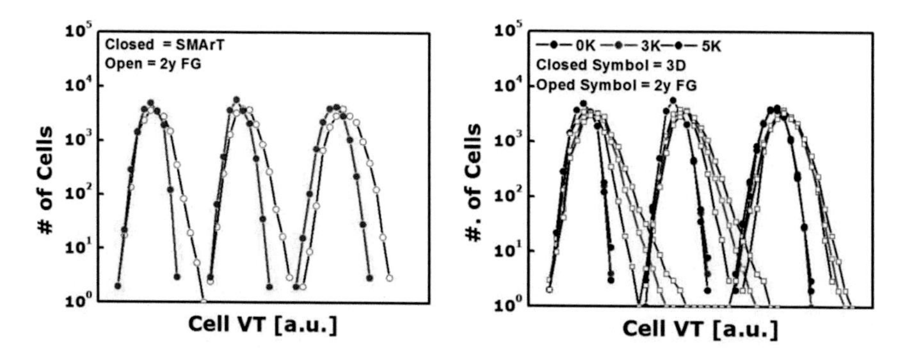

Fig. 2.33 *Left* Comparison of cells V_T distributions of 2y node FG and SMArT cells. *Right* Comparison of V_T widening during program-erase cycling [31]

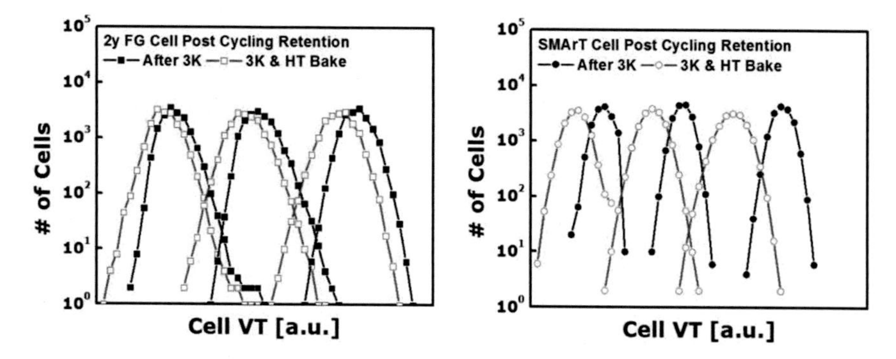

Fig. 2.34 Cells V_T distributions of HT retention after cycling of 2y node FG (*left*) and SMArT cells (*right*) [31]

2.9 3D-FG Nand

Recently, 3D vertical FG type NAND cell arrays have been proposed to overcome the retention and overall reliability issues of 3D CT NAND cell arrays [32–36]. In this section, an overview of the proposed 3D FG cells and their main reliability problems is reported. Figure 2.35 shows the bird's-eye view of published 3D vertical FG type NAND cell's structures: *Extended Sidewall Control Gate* (ESCG) [32], *Dual Control-gate with Surrounding Floating-gate* (DC-SF) [33, 34], and *Separated-Sidewall Control Gate* (S-SCG) cells [35].

ESCG and DC-SF cells suffer from interference and disturbance problems when integrated into an array due to the direct coupling effect of neighboring cells in the same string. S-SCG overcomes such a problem, strongly reducing interference and disturbance effects. In the S-SCG structure, the Source/Drain (S/D) region can be implemented by electrically inverting the pillar surface, and high CG coupling capacitance can be achieved. S-SCG structure allows obtaining highly reliable

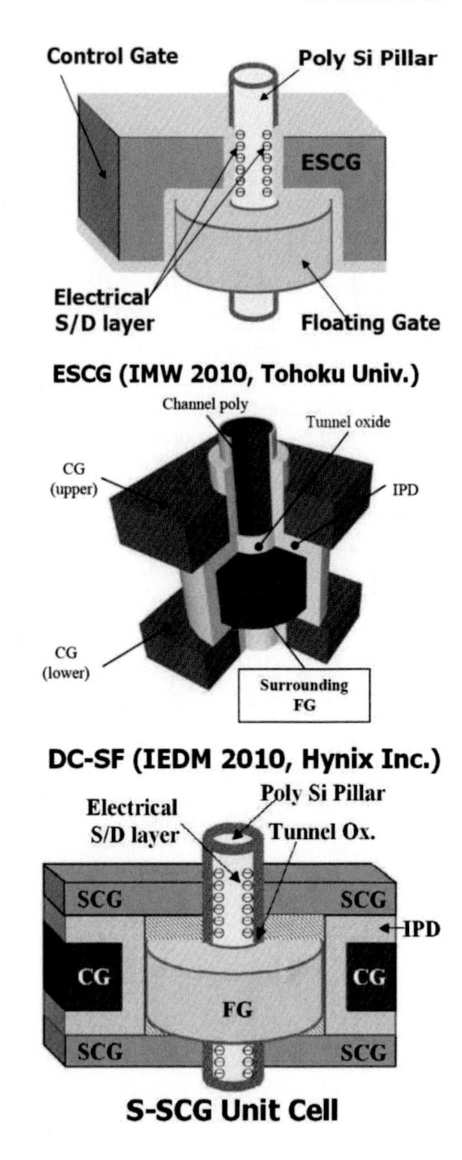

Fig. 2.35 Bird's eye views of the most recent 3-D vertical FG type NAND cell schemes: ESCG (*top*), DC-SF (*middle*) and S-SCG (*bottom*) [35]

MLC operation, high speed P/E operation and good read current margin. More details about these architectures can be found in Chap. 5.

2.9.1 DC-SF Interference and Retention Results

The interference between a programmed cell and an adjacent cell in case of 3D-FG NAND array with DC-SF cells was studied as shown in Fig. 2.36 (left) [33].

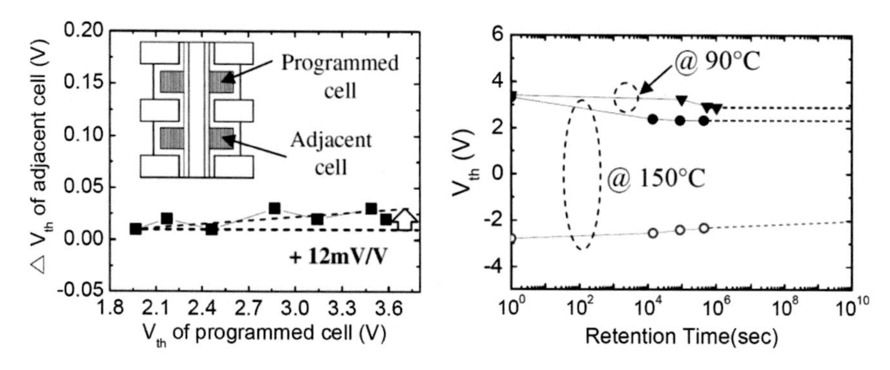

Fig. 2.36 *Left* FG-FG interference characteristics (V_T variation in the adjacent cell as a function of V_T in the programmed cell). Very small FG-FG coupling value of 12 mV is obtained. *Right* Data retention characteristics of the DC-SF NAND flash cells [33]

Negligible V_T shift of adjacent cell (12 mV) was obtained as the V_T of programmed cell is increased by 3.6 V, suggesting that each CG acts as shielding layer in a string. The data retention characteristics of DC-SF cells at two different temperatures (90 and 150 °C) are reported in Fig. 2.36 (right), showing the increase of charge loss with temperature. In case of high temperature condition, a program and erase V_T shift due to charge-loss of 0.9 and 0.2 V after a retention time of 126 h is shown, respectively.

2.9.2 S-SCG Interference Results

In 3D FG NAND arrays with S-SCG cells two critical interference coupling paths exist: the former is the indirect coupling path while the latter is direct. Figure 2.37 shows the interference effects as a function of the S-SCG initial V_T. The S-SCG structure can sufficiently suppress the indirect interference effect; however, the direct coupling from neighboring FG to the channel of S-SCG remains a very serious problem. ESCG and DC-SF cells have remarkable interference problems by this direct coupling effect as well, which directly influences the parasitic transistor below the S-SCG. To suppress this direct coupling effect, S-SCG cell applies the SCG voltage to control the parasitic transistor [35].

2.9.3 S-SCG Performance and Reliability Advantages

The S-SCG cell strongly reduces both the interference effect and the disturbance problem with good performance, and it has good potentials of highly reliable MLC operation. Moreover, lower operation voltages than conventional 3D CT is

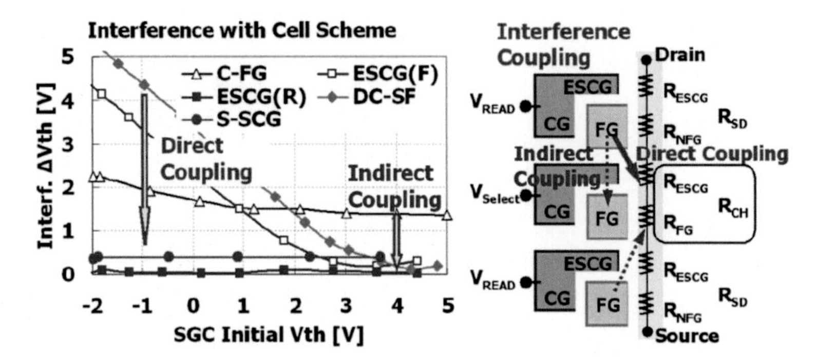

Fig. 2.37 Interference effect in 3-D vertical FG NAND cells (*left*) and cross sectional view of interference coupling paths of the conventional ESCG cell array (*right*) [35]

Fig. 2.38 The effective cell size of S-SCG cell in comparison with other 3-D vertical NAND cells at 20 nm technology [35]

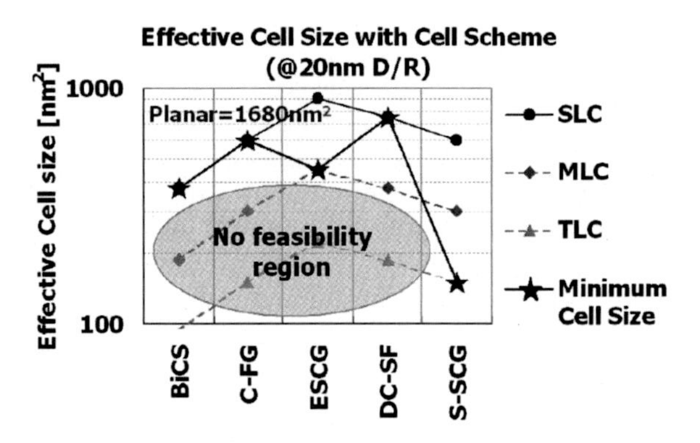

required: this implies that cell operation is more effective because of higher coupling ratio. The vertical cell height is decreased compared to that of the conventional FG cell by using a predeposited SCG layer. Figure 2.38 shows the effective cell size with 3D vertical NAND cell schemes at 20 nm technology. Although the

Table 2.2 3D NAND cells comparison summary [35]

D/R=20nm Technology (Aspect Ratio=32)			CT type	FG type			
			BiCS	C-FG	ESCG	DC-SF	S-SCG
Cell Perfromances	CG C/R		-	NG	G	VG	VG
	Reliability		NG	G	G	G	G
	Disturbance		G	G	NG	NG	VG
	Interference		G	NG	G	NG	VG
MLC feasibility (bit/cell)			Not Easy (1bit)	Not Easy (1bit)	Normal (2bit)	Not Easy (1bit)	Easy (3bit)
Minimum IPD		nm	-	-	12	12	7
Minimum Gate Elctrode		nm	-	-	6	6	6
Cell height		nm	40	40	60	50	40
# of stacked cell		ea	16	16	11	13	16

G=Good, NG=Not Good, VG=Very Good

cell size of the proposed S-SCG cell is larger than that of CT type NAND cell by about 60 %, the possibility of implementing MLC operation allows obtaining lower bit costs. Less than half of the bit cost can be achieved by implementing TLC operation to the proposed S-SCG cell. Finally, we show the MLC feasibility and the number of stacked cells in comparison with conventional 3D cells in Table 2.2 [35].

2.10 3D-CT Versus 3D-FG

In this section a final comparison between 3D CT and 3D FG in terms of performance and reliability is reported. For the comparison of the structures, vertical schematic of the conventional 3D CT structure and 3D FG cell are shown in Fig. 2.39 [34]. Unlike planar CT device, the CT nitride layer in a string of the conventional 3D CT is continuously connected from top to bottom CGs along the channel side, and it acts as a charge spreading path, which is an unavoidable problem of 3D CT cell. As a result, this causes degradation of data retention characteristics and poor distribution of cell state. In 3D FG cells, on the contrary, the FG is completely isolated by the tunnel oxide and the Inter Poly Dielectric (IPD). This approach allows obtaining a significantly reliable structure, able to contain charges without any problem related to leakage paths [34].

In order to compare the cell size, DC-SF as 3D FG and BiCS/TCAT as 3D CT are considered. The effective cell size is estimated and plotted in Fig. 2.40 as a function of the number of stacked cells. Even if the physical size of DC-SF cell is assumed to be 54 % larger than that of conventional BiCS/TCAT, DC-SF allows fabricating 1 Tb arrays with 3 bit/cell and 64 stacked cells or 2 Tb arrays with 3 bit/cell and 128 stacked cells, thanks to the small FG-FG interference.

Moreover, 3D FG ensure reliable retention characteristics and lower operation voltage than that of conventional 3D CT [34]. As a result, the 3D FG structure allows highly enhanced device performance for 3D NAND flash memory compared to 3D CT. In the following, a summary of 3D FG advantages and disadvantages compared to 3D CT is reported [37]:

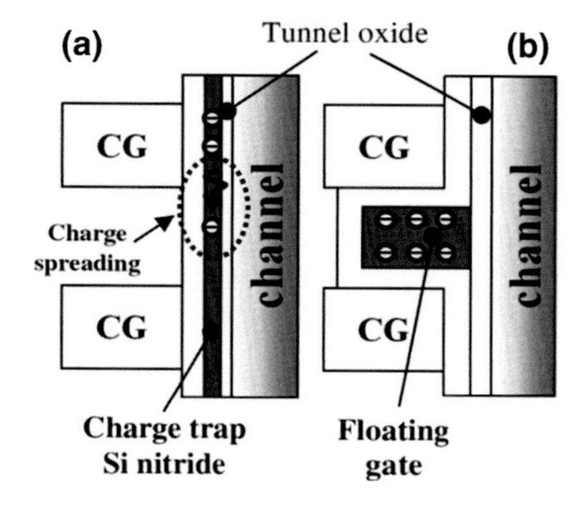

Fig. 2.39 Comparison of 3D NAND flash cell structures (**a**) CT cell (BiCS) (**b**) 3D-FG cell [34]

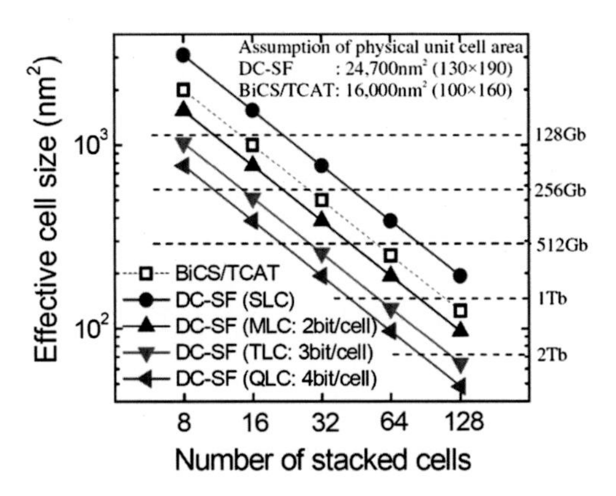

Fig. 2.40 Effective cell sizes for various DC-SF NAND flash structures [34]

Advantages of 3D FG:

- Lower charge spreading resulting in less read errors, and consequently less ECC intervention, in particular in multilevel architectures;
- better data retention because of a more stable charge into the storage layer;
- direct connection between channel poly and p-well, allowing bulk erase (no GIDL).

Disadvantages of 3D FG:

- Larger cell size;
- larger 3D pillar due to the presence of the floating gate, and hence lower scalability;
- floating gate coupling effect: even if the S-SCG structure can sufficiently suppress the indirect interference effect, direct coupling remains a very serious problem that strongly reduces the programming speed.

Even if 3D FG shows a relevant number of reliability advantages compared to 3D CT, the higher scalability still makes 3D CT the most attractive solution for the integration in hyper-scaled arrays. Moreover, it must be pointed out that enhanced programming algorithms and error correction techniques allows mitigating the previously described reliability issues.

References

1. R. Micheloni et al., *Inside NAND Flash Memories* (Springer, 2010)
2. C. Lee et al., Multi-level NAND flash memory with 63 nm-node TANOS (Si-Oxide-SiN-Al_2O_3-TaN) cell structure, in *VLSI Symposium Technical Digest* (2006), pp. 21–22
3. G. Van Den Bosch, Physics and reliability of 2D and 3D SONOS devices, in *IEEE International Memory Workshop (IMW)*, Tutorial, 18–21 May 2014

4. A. Grossi et al., Bit error rate analysis in charge trapping memories for SSD applications, in *IEEE International Reliability Physics Symposium (IRPS)*, June 2014, pp. MY.7.1–MY.7.5
5. Y. Cai et al., Threshold voltage distribution in MLC NAND flash memory: characterization, analysis, and modeling, in *Design, Automation Test in Europe Conference Exhibition (DATE)*, Mar 2013, pp. 1285–1290
6. A. Chimenton et al., A statistical model of erratic behaviors in NAND flash memory arrays. IEEE Trans. Electron Devices **58**, 3707–3711 (2011)
7. T. Ong et al., Erratic erase in ETOX TM flash memory array, in *Proceedings of VLSI Symposium Technical*, pp. 83–84 (1993)
8. C. Dunn et al., Flash EPROM disturb mechanisms, in *Proceedings of IEEE International Reliability Physics Symposium (IRPS)*, Apr 1994, pp. 299–308
9. C. Zambelli et al., Analysis of edge wordline disturb in multimegabit charge trapping flash NAND arrays, in *IEEE International Reliability Physics Symposium (IRPS)*, 10–14 Apr 2011, pp. MY.4.1–MY.4.5
10. J. Lee, C. Lee, M. Lee, H. Kim, K. Park, W. Lee, A new programming disturbance phenomenon in NAND flash memory by source/drain hot electrons generated by GIDL current, in *Proceedings of the NVSM Workshop* (2006), pp. 31–33
11. A. Arreghini et al., Experimental extraction of the charge centroid and of the charge type in the P/E operations of the SONOS memory cells, in *IEDM 2006 Technical Digest* (2006), pp. 499–502
12. L. Vandelli et al., Role of holes and electrons during erase of TANOS memories: evidences for dipole formation and its impact on reliability, in *IEEE International Reliability Physics Symposium (IRPS)*, 2–6 May 2010, pp. 731–737
13. A. Arreghini et al., Characterization and modeling of long term retention in SONOS non volatile memories, in *Solid State Device Research Conference, 37th European ESSDERC*, 11–13 Sept 2007, pp. 406–409
14. C.-P. Chen et al., Study of fast initial charge loss and it's impact on the programmed states VT distribution of charge-trapping NAND flash, in *IEEE International Electron Devices Meeting (IEDM)* (2010), pp. 5.6.1–5.6.4
15. J.K. Park et al., Origin of transient Vth shift after erase and its impact on 2D/3D structure charge trap flash memory cell operations, in *IEEE International Electron Devices Meeting (IEDM)*, 10–13 Dec 2012, pp. 2.4.1–2.4.4
16. H. Park et al., Charge loss in TANOS devices caused by Vt sensing measurements during retention, in *IEEE International Memory Workshop (IMW)*, 16–19 May 2010, pp. 1–2
17. H. Tanaka et al., Bit cost scalable technology with punch and plug process for ultra high density flash memory, in *IEEE Symposium on VLSI Technology*, 12–14 June 2007, pp. 14–15
18. J. Jang et al., Vertical cell array using TCAT (terabit cell array transistor) technology for ultra high density NAND flash memory, in *IEEE Symposium on VLSI Technology*, 16–18 June 2009, pp. 192–193
19. S.J. Whang et al., Novel 3-dimensional dual control-gate with surrounding floating-gate (DC-SF) NAND flash cell for 1 Tb file storage application, in *IEEE International Electron Devices Meeting (IEDM)*, 6–8 Dec 2010, pp. 29.7.1–29.7.4
20. W. Kim et al., Multi-layered vertical gate NAND flash overcoming stacking limit for terabit density storage, in *IEEE Symposium on VLSI Technology*, 16–18 June 2009, pp. 188–189
21. A. Goda et al., Scaling directions for 2D and 3D NAND cells, in *IEEE International Electron Devices Meeting (IEDM)*, 10–13 Dec 2012, pp. 2.1.1–2.1.4
22. H.T. Lue et al., 3D vertical gate NAND device and architecture, in *IEEE International Memory Workshop (IMW)*, Tutorial, 18–21 May 2014
23. Y. Fukuzumi et al., Optimal integration and characteristics of vertical array devices for ultra-high density, bit-cost scalable flash memory, in *IEDM Technical Digest* (2007), pp. 449–452
24. Y. Kim et al., Coding scheme for 3D vertical flash memory, in *IEEE International Conference on Communications (ICC)*, 8–12 June 2015

25. E. Nowak et al., In-depth analysis of 3D silicon nanowire SONOS memory characteristics by TCAD simulations, in *IEEE International Memory Workshop (IMW)*, 16–19 May 2010, pp. 1–4
26. H.-T. Lue et al., Understanding STI edge fringing field effect on the scaling of charge-trapping (CT) NAND flash and modeling of incremental step pulse programming (ISPP), in *IEDM Technical Digest* (2009), pp. 839–842
27. S.M. Amoroso et al., Semi-analytical model for the transient operation of gate-all-around charge-trap memories. IEEE Trans. Electron Devices **58**(9), 3116–3123 (2011)
28. X. Li et al., Investigation of charge loss mechanisms in 3D TANOS cylindrical junction-less charge trapping memory, in *IEEE International Conference on Solid-State and Integrated Circuit Technology (ICSICT)*, 28–31 Oct 2014, pp. 1–3
29. Z. Lun et al., Investigation of retention behavior for 3D charge trapping NAND flash memory by 2D self-consistent simulation, in *International Conference on Simulation of Semiconductor Processes and Devices (SISPAD)*, 9–11 Sept 2014, pp. 141–144
30. Y. Yanagihara et al., Control gate length, spacing and stacked layer number design for 3D-stackable NAND flash memory, in *IEEE International Memory Workshop (IMW)*, 20–23 May 2012, pp. 1–4
31. E.-S. Choi et al., Device considerations for high density and highly reliable 3D NAND flash cell in near future, in *IEEE International Electron Devices Meeting (IEDM)*, 10–13 Dec 2012, pp. 9.4.1–9.4.4
32. M.K. Seo et al., The 3-dimensional vertical FG NAND flash memory cell arrays with the novel electrical S/D technique using the extended sidewall control gate (ESCG), in *IEEE International Memory Workshop (IMW)*, May 2010, pp. 146–149
33. S. Aritome et al., Advanced DC-SF cell technology for 3-D NAND flash. IEEE Trans. Electron Devices **60**(4), 1327–1333 (2013)
34. S.J. Whang et al., Novel 3-dimensional dual control-gate with surrounding floating-gate (DC-SF) NAND flash cell for 1Tb file storage application, in *IEEE International Electron Devices Meeting (IEDM)*, 6–8 Dec 2010, pp. 29.7.1–29.7.4
35. M.K. Seo et al., A novel 3-D vertical FG NAND flash memory cell arrays using the separated sidewall control gate (S-SCG) for highly reliable MLC operation, in *IEEE International Memory Workshop (IMW)*, 22–25 May 2011, pp. 1–4
36. K. Parat et al., A floating gate based 3D NAND technology with CMOS under array in *IEEE International Electron Devices Meeting (IEDM)*, 7–9 Dec 2015
37. B. Prince, 3D vertical NAND flash revolutionary or evolutionary, in *IEEE International Memory Workshop*, Tutorial, 17–20 May 2015

Chapter 3
3D Stacked NAND Flash Memories

Rino Micheloni and Luca Crippa

3.1 Introduction

In planar Flash memories the most popular memory cell is based on the *Floating Gate* (FG) technology, whose cross section is shown in Fig. 3.1. Cross section and the associated floating gate model are sketched in Fig. 3.2. Basically, a MOS transistor is built with two overlapping gates: the first one (FG) is completely surrounded by oxide, while the second one forms the *Control Gate* (CG) terminal. The isolated gate constitutes an excellent "trap" for electrons, enabling long charge retention. The operations used to inject and remove electrons from the isolated gate are called Program and Erase, respectively. These operations modify the threshold voltage V_{TH} of the memory cell, which is, at the end of the day, a MOS transistor. By applying a fixed voltage to the cell's terminals, it is then possible to discriminate two storage levels: when the gate voltage is higher than V_{TH}, the cell is ON ("1"), vice versa it is OFF ("0").

To minimize the space on silicon, memory cells are packed together to form a matrix. Depending on how the cells are organized inside the matrix, it is possible to distinguish between NAND and NOR Flash memories. NAND memories are the most widespread in the storage systems; NOR architecture is described in great details in [1].

In the NAND string, memory cells are connected in series, in groups of 32, 64, 128, or even 150 [2], as shown in Fig. 3.3. Two select transistors are placed at the edges of the string, to ensure the connection to the source line (through M_{SL}) and to

R. Micheloni (✉) · L. Crippa
Performance Storage BU, Microsemi Corporation, Vimercate, Italy
e-mail: rino.micheloni@ieee.org

L. Crippa
e-mail: luca.crippa@ieee.org

R. Micheloni (ed.), *3D Flash Memories*, DOI 10.1007/978-94-017-7512-0_3

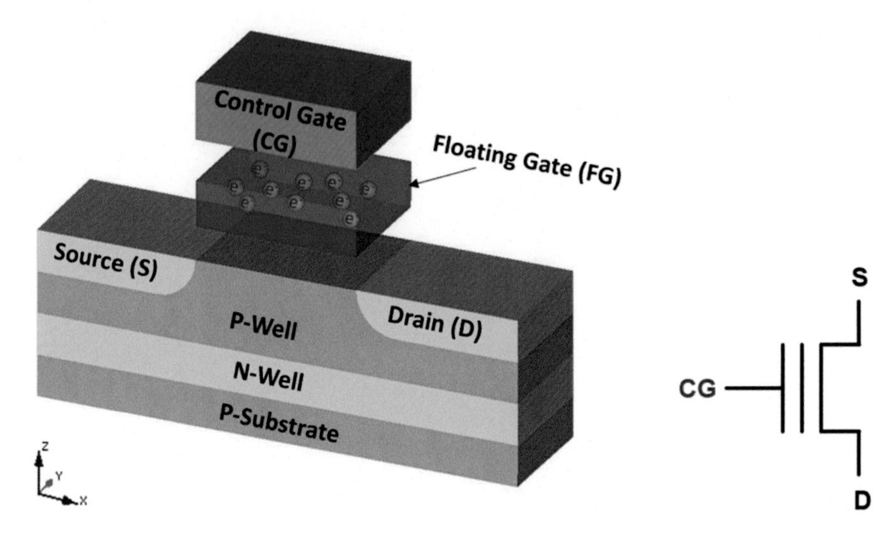

Fig. 3.1 Floating gate memory cell and its schematic symbol

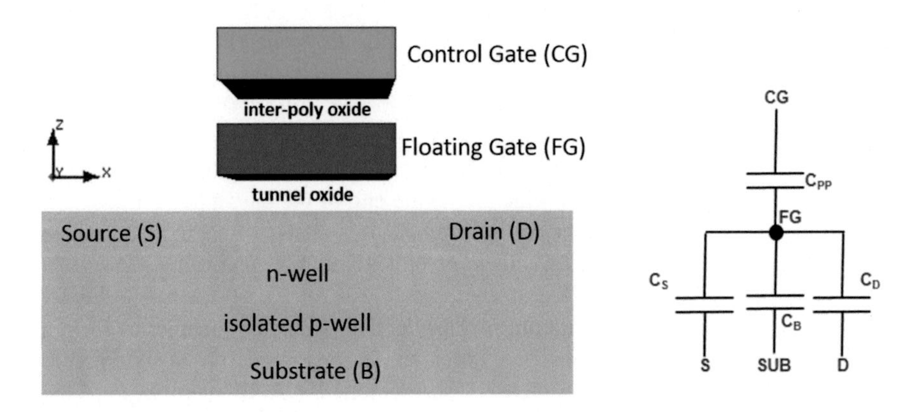

Fig. 3.2 Floating gate memory cell and the corresponding capacitive model

the bitline (through M_{DL}). Each NAND string shares the bitline contact with another string. Control gates are connected through wordlines (WLs).

Logical pages are made of cells belonging to the same wordline. The number of pages per wordline is related to the storage capability of the memory cell. Depending on the number of storage levels, Flash memories are referred to in different ways: SLC memories store 1 bit per cell, MLC memories store 2 bits per cell, TLC memories store 3 bits per cell, and QLC memories store 4 bits per cell (Fig. 3.4).

All the NAND strings sharing the same group of wordlines are erased together, thus forming a so called Flash Block. In Fig. 3.3 two blocks are shown: by using a

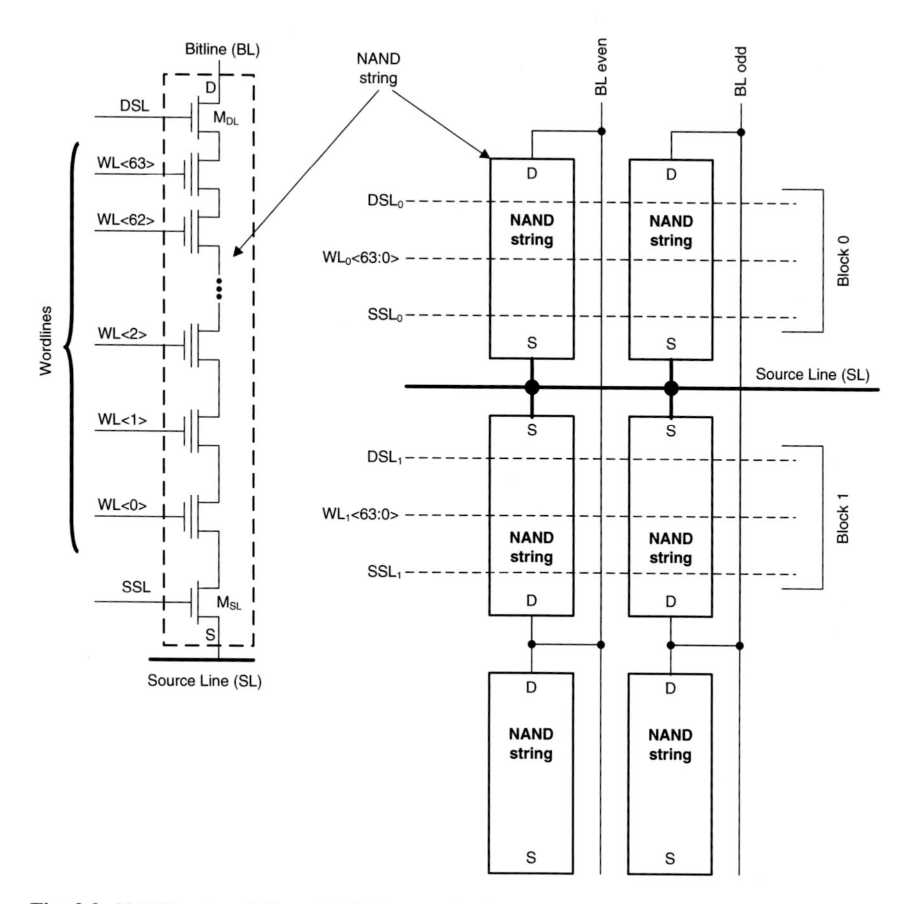

Fig. 3.3 NAND string (*left*) and NAND array (*right*)

bus representation, one block is made of WL_0<63:0>, while the other one includes WL_1<63:0>.

In terms of silicon area, a NAND Flash device is mainly a memory array but, in order to perform read, program, and erase, there is a need for additional circuits. In Fig. 3.5 a block diagram of a NAND device is sketched. The Memory Array can be split into 2 or more areas (or planes). A wordline is highlighted along the horizontal direction, while a bitline is shown in the vertical direction. The Row Decoder is located between the planes: this circuit has the task of properly biasing all the wordlines belonging to the selected NAND string. All the bitlines are connected to Page Buffers or Sense Amplifiers (Sense Amp). The purpose of sense amplifiers is to convert the current sunk by the memory cell into a digital value. In the peripheral area there are charge pumps and voltage regulators, logic circuits, and redundancy structures. I/O PADs are used to communicate with the external world.

NAND memories contain information organized in pages and blocks (Fig. 3.6). As already mentioned, a block is the smallest erasable unit and it contains multiple

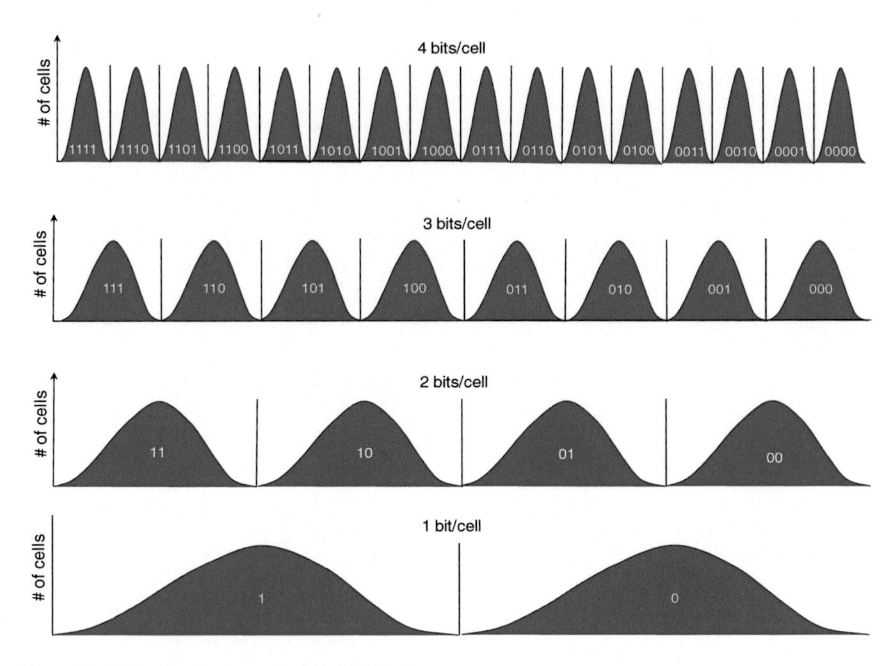

Fig. 3.4 Multi-level storage in NAND flash memory

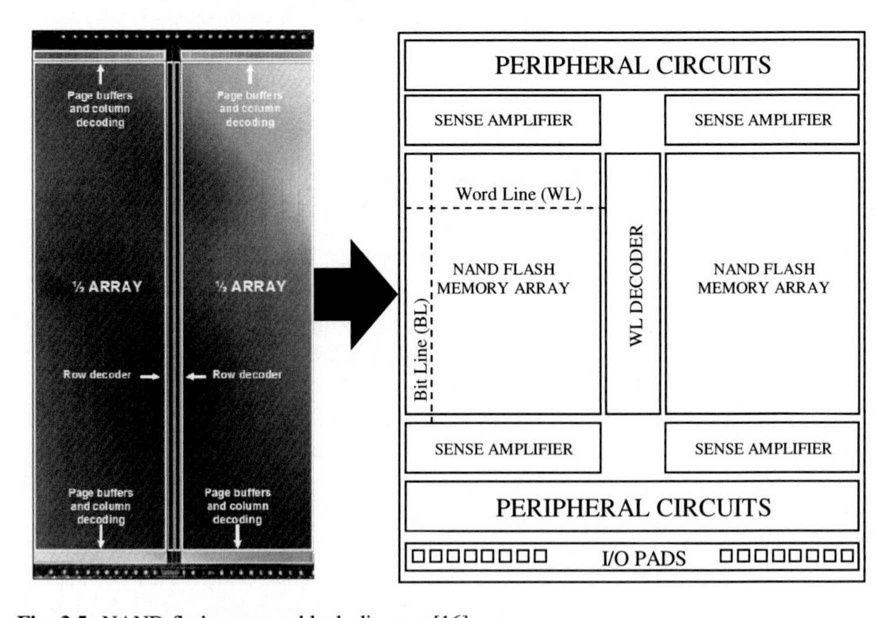

Fig. 3.5 NAND flash memory block diagram [16]

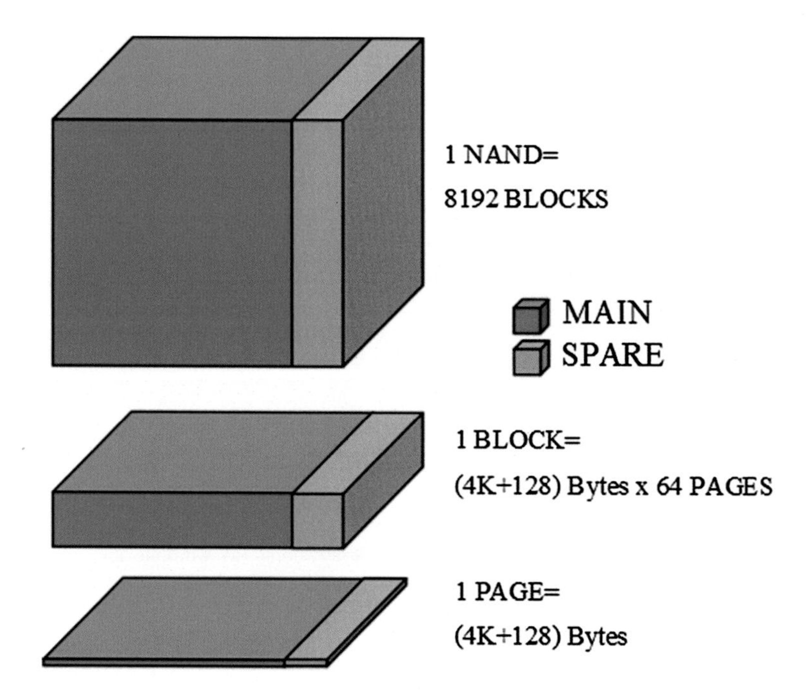

Fig. 3.6 NAND memory logic organization

logical pages. A page is the smallest addressable unit for reading and writing. Each page is composed by main area and spare area. Spare area is used for *Error Correction Code* (ECC) and firmware (FW) meta-data. The NAND logical address is built around the Row address and Column address concept. The Row address identifies the addressed page, while the Column address is used to identify a single byte inside the page.

3.2 Floating Gate Cell

The schematic structure of a 48 nm floating gate NAND cells is shown in Fig. 3.7 [3]; FG and CG are typically made of polysilicon. For all operations, the control gate electrode is capacitive coupled to the floating gate. The dielectric between FG and CG is referred to as *Inter-Poly Dielectric* (IPD) and it is typically made of a silicon Oxide/silicon Nitride/silicon Oxide triple layer (ONO). The alterable threshold voltage V_{TH} of a floating gate cell, which represents the bit of information, depends on the coupling strength between FG and CG and the amount of charge stored in the FG.

A cross section of a FG NAND array along the wordline direction is shown in Fig. 3.7a. The CG is wrapped around the FG to improve the capacitive coupling between CG and FG; this coupling reduces the operation voltages of the floating

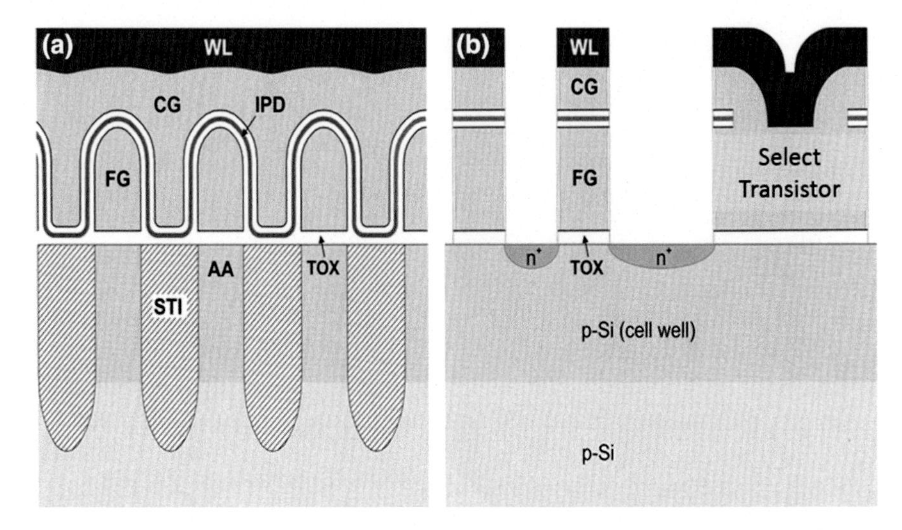

Fig. 3.7 Cross sections of a floating gate NAND array along (**a**) wordline and (**b**) bitline directions

gate cells and ensures a reliable operation, as described in Chap. 2. The active areas (AA) of two neighboring NAND strings are separated by shallow trench insulation (STI). The memory cell transistor gate oxide is denoted as *Tunnel Oxide* (TOX) because the charge (used to store a bit of information) is transferred trough this SiO_2 dielectric by quantum mechanical tunneling. It is a very crucial point for cell's reliability that the charge is transferred only through the TOX during program and erase operations. Every charge transferred through IPD (i.e. oxide between FG and CG) needs to be absolutely avoided to prevent severe reliability issues.

A cross section of a NAND string along the BL direction is shown in Fig. 3.7b. The floating gate cells are patterned by a vertical WL etch step. In the etched spaces between the floating gate cells shallow n^+-junctions are implanted in order to define memory cell transistors and reduce the string resistance. To improve charge retention, the sidewall of the floating gate is passivated through a thermal oxidation process.

The generated high quality thermal *Sidewall Oxide* (SWOX) forms an effective tunneling barrier against charge loss from the FG (Fig. 3.8). At this point, the space between FG cells is filled with silicon oxide (*Inter Wordline Dielectric*, IWD) which, usually, has a lower electrical quality compared to the tunnel oxide. The select transistors (M_{DL} and M_{SL}) are processed together with the floating gate cells and, as a consequence, they use TOX as gate dielectric. The select transistor gate length is typically in the range of 150–200 nm. To build a real transistor, the wordline layer is contacted to the floating gate layer. This contact is done by removing the ONO IPD in the middle of the select transistors prior to the CG poly-Si deposition.

The whole fabrication process of a floating gate NAND technology is typically based on 30–40 lithographic mask steps and it includes 2 poly-Si and 3 metal levels. To obtain the highest memory density, typically 3 masks are designed by

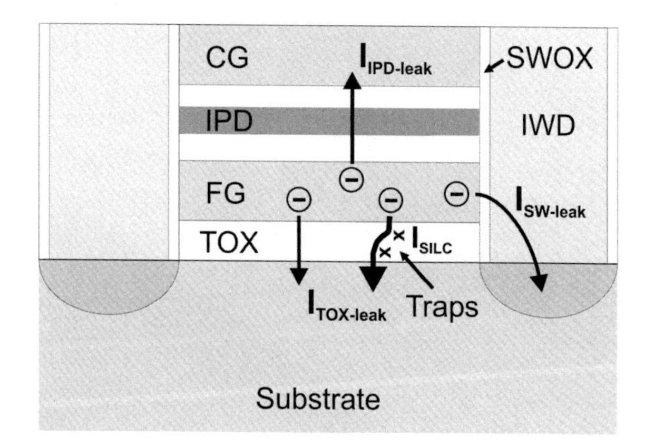

Fig. 3.8 SWOX, IWD, and possible leakage paths

using the smallest size of that particular process node: active area/STI, wordline and bitline. There are some other process steps with stringent lithographic requirements such as contacts to the bitline, source contacts, and CG to FG contacts in the string select transistors.

Before moving to 3D architectures, it is worth taking a look at the bird's-eye view of a planar array, as sketched in Fig. 3.9; in fact, with 3D architectures,

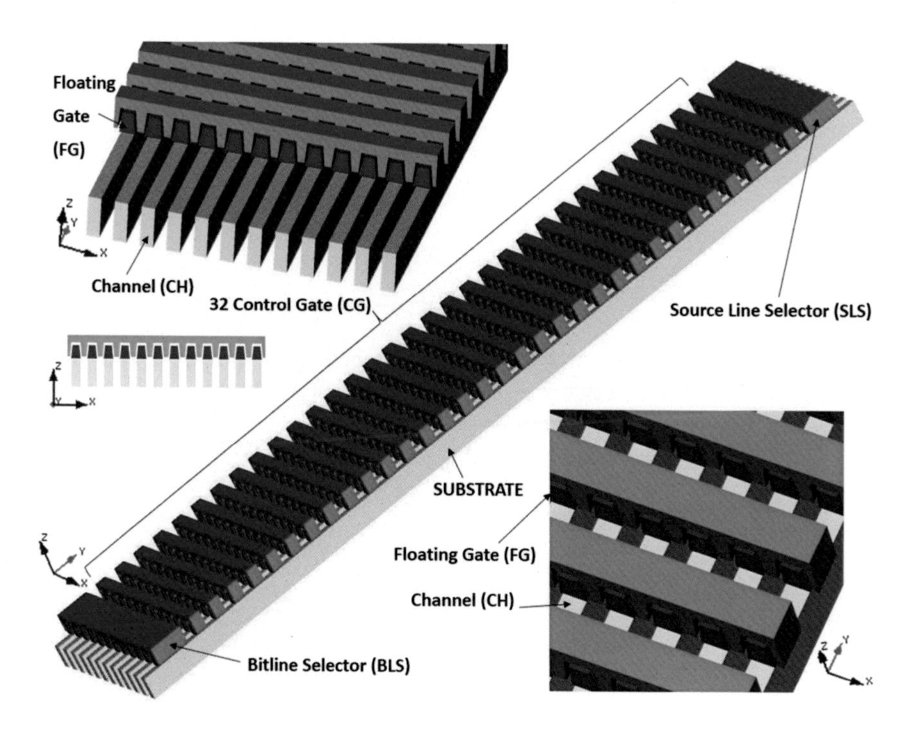

Fig. 3.9 Bird's-eye view of a planar NAND string

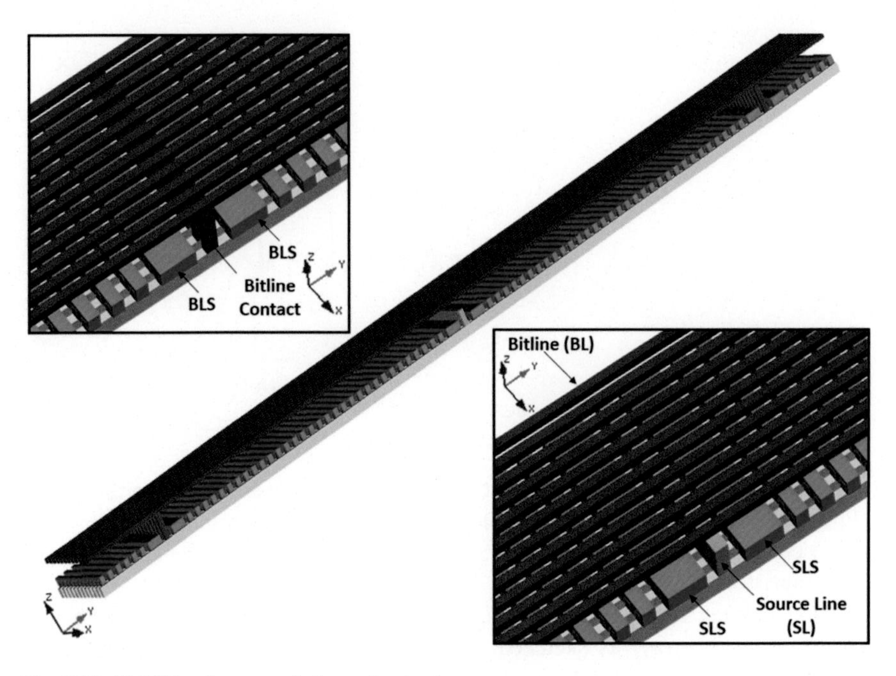

Fig. 3.10 NAND strings coupled together in the array

bird's-eye views become an essential tool. In Fig. 3.10 NAND strings are coupled together to form a memory array; in order to save space, 2 NAND strings can share either the Source Line (SL) or the Bitline contact, as shown in the insets of Fig. 3.10.

3.3 NAND Basic Operations

In this Section we provide a brief summary of how read, program, and erase operations are performed inside a NAND Flash memory; all these operations are managed by the internal microcontroller [4].

3.3.1 Read

The reading operation is designed to address specific memory cells within the array and measure the information stored therein. With reference to Fig. 3.11, when we read a NAND memory cell, its gate is driven at V_{READ} (0 V), while the other cells are biased at $V_{PASS,R}$ (usually 4–5 V), so that they can act as pass-transistors, independently from the value of their threshold voltage. In fact, an erased Flash cell

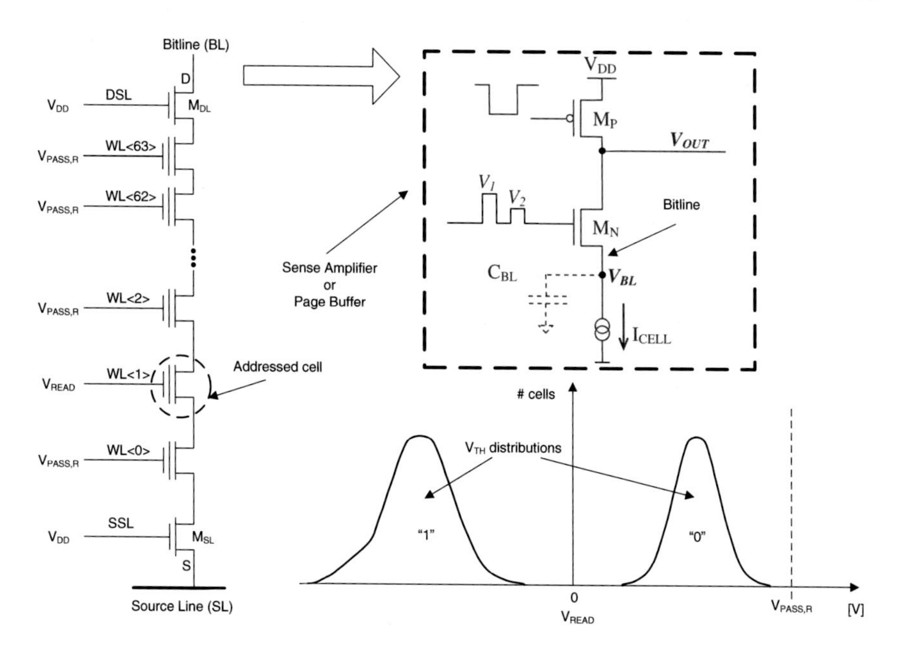

Fig. 3.11 NAND string biasing during read and SLC V_{TH} distributions

has a V_{TH} smaller than 0 V; vice versa, a written cell has a positive V_{TH} but, however, smaller than 4 V. In practice, by biasing the gate of the selected cell with a voltage equal to 0 V, the series of all the cells will conduct current only if the addressed cell is erased.

There are different reading techniques, starting from the one using the bitline parasitic capacitor, and ending with the one that integrates the current on a little dedicated capacitor. The above mentioned techniques can be used for SLC/MLC/TLC/QLC NAND memories. When there are more than 2 V_{TH} distributions, multiple basic reading operations are performed at different gate voltages. Historically, the first reading technique used the parasitic capacitor of the bitline to integrate cell's current [5–7].

This capacitor is precharged at a fixed value (usually 1–1.2 V): only if the cell is erased and sinks current, then the capacitor is discharged. Several circuits exist to detect the bitline parasitic capacitor state: the structure depicted in the inset of Fig. 3.11 is present in almost all solutions. The bitline parasitic capacitor is indicated with C_{BL} and the NAND string is equivalent to a current generator. During the charge of the bitline, the gate of the PMOS transistor M_P is kept grounded, while the gate of the NMOS transistor M_N is kept at a fixed value V_1. Typical value for V_1 is around 2 V. At the end of the charge transient the bitline has a voltage V_{BL} equal to

$$V_{BL} = V_1 - V_{THN} \tag{3.1}$$

where V_{THN} indicates the threshold voltage value of transistor M_N. At this point, transistors M_N and M_P are switched off. C_{BL} is free to discharge. After a time T_{VAL}, the gate of M_N is biased at $V_2 < V_1$, usually 1.6–1.4 V. When T_{VAL} time is long enough to discharge the bitline voltage under the value:

$$V_{BL} < V_2 - V_{THN} \tag{3.2}$$

M_N turns on and the voltage of node OUT (V_{OUT}) becomes equal to the one of the bitline. Finally, the analog voltage V_{OUT} is converted into a digital format by using simple latches.

3.3.2 Program

V_{TH} is modified by means of the *Incremental Step Programming Pulse* (ISPP) algorithm (Fig. 3.12): a voltage step (whose amplitude and duration are predefined) is applied to the gate of the cell. Afterwards, as shown in Fig. 3.13 [8], a verify operation is performed, in order to check if cell's V_{TH} has exceeded a predefined voltage value (V_{VFY}). If the verify operation is successful, the cell has reached the desired state and it is excluded from the following program pulses. Otherwise, another cycle of ISPP is applied to the cell; this time the program voltage is incremented by ΔV_{pp} (or ΔISPP).

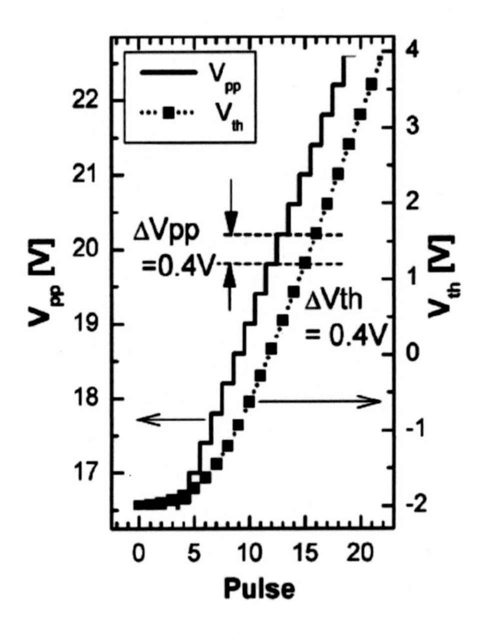

Fig. 3.12 Incremental step pulse programming (ISPP)

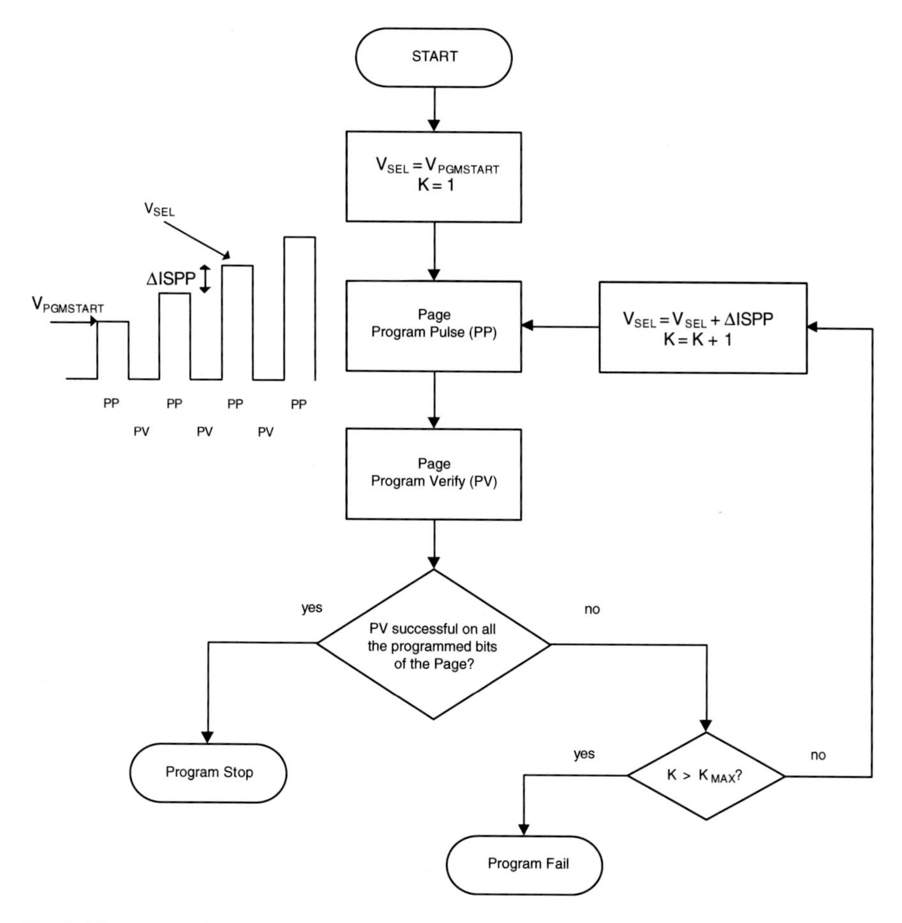

Fig. 3.13 Program flow based on incremental step programming pulse (ISPP) algorithm

During the programming pulse a high voltage is applied to the selected wordline, but the program operation has to be bit-selective. In other words, we need the ability of selecting/deselecting each single memory cell within a wordline. Therefore, on all memory cells where the program operations has to be inhibited, a high channel potential is needed to reduce the voltage drop across the tunneling dielectric and to prevent the electron tunneling from the channel to the floating gate, as indicated by Fig. 3.14a. In the first NAND devices the channel was charged by applying 8 V to the bitlines of the program inhibited NAND strings. This method suffers from several disadvantages [5], especially power consumption and high stress for the oxide between adjacent bitlines.

Less power consuming is the self-boost program inhibit scheme. By charging the string select lines and the bitlines connected to inhibited cells to V_{cc}, the select transistors are diode connected (Fig. 3.14b). When the wordline potentials rise (selected wordline to V_{pp} and unselected wordlines to V_{ppass}), the channel potential

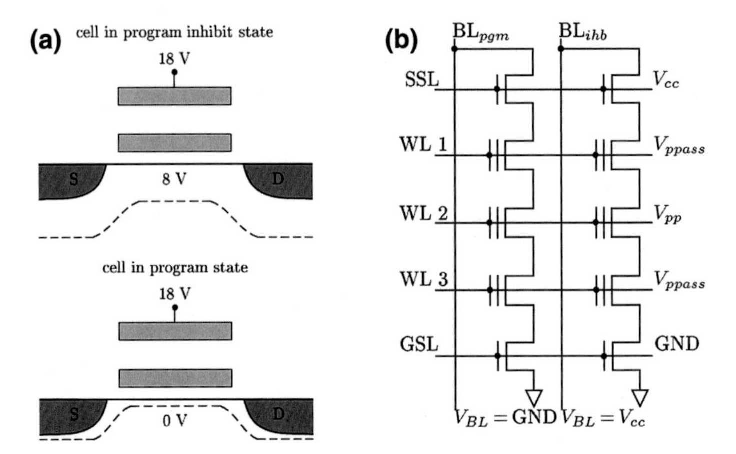

Fig. 3.14 Self boosted program inhibit scheme. **a** Conditions of cells selected for program and program inhibit. **b** Bias conditions during the self boosted program inhibit

is boosted by the parasitic series capacitors of control gate, floating gate, channel and bulk. When the voltage of the channel exceeds $V_{cc} - V_{TH,SSL}$, then SSL transistors are reverse biased and the channel of the NAND string becomes a floating node.

As the memory cells are organized in a matrix, all the cells along the wordline are biased at the same voltage even if they are not intended to be programmed, i.e. they are "disturbed". Two important typologies of disturbs are related to the program operation: *Pass disturb* and *Program disturb*: their impact on reliability is described in Chap. 2.

3.3.3 Erase

The NAND array is placed in a triple-well structure, as shown in Fig. 3.15a. Usually, each plane has its own triple-well. The source terminal is shared by all the blocks: in this way the matrix is smaller and the number of circuits for biasing the iP-well is drastically reduced.

The electrical erase is achieved by biasing the iP-well with a high voltage and keeping the wordlines of the block to be erased at ground (Fig. 3.15c).

As it is for the Programming operation, Erase exploits the physical mechanism known as Fowler-Nordheim tunneling (Chap. 2). Because the iP-well is common to all the blocks, erase of unselected blocks is prevented by leaving their wordlines floating. In this way, when the iP-well is charged, the potential of the floating wordlines raises thanks to the capacitive coupling between the control gates and the iP-well. Of course, the voltage difference between wordlines and iP-well should be low enough to avoid Fowler-Nordheim tunneling.

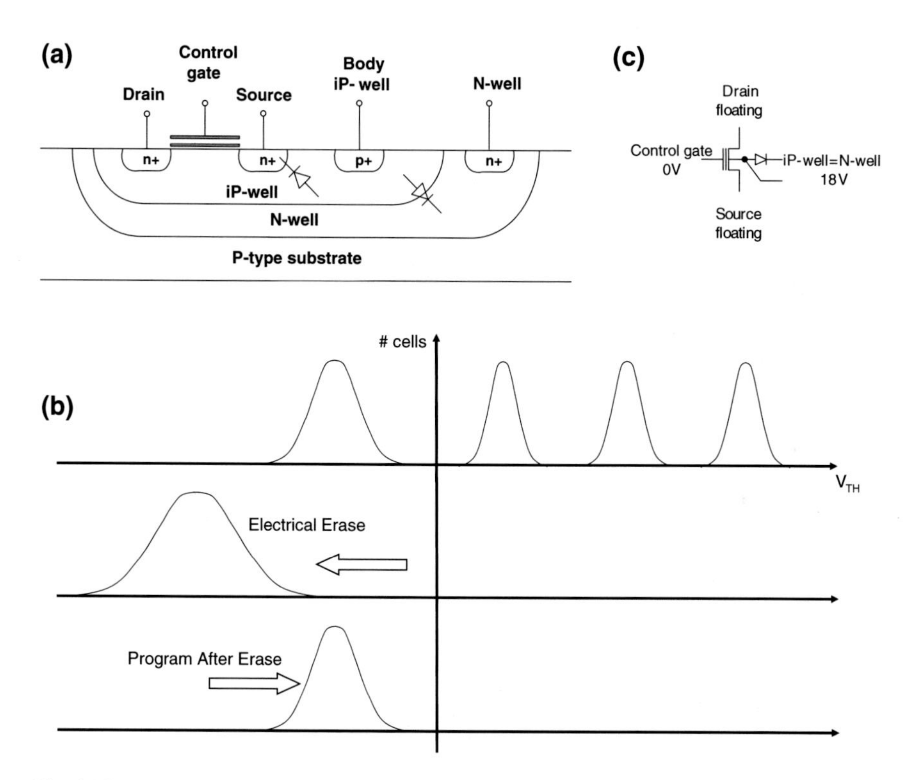

Fig. 3.15 **a** NAND matrix in triple-well. **b** Erase algorithm. **c** Erase biasing on the selected block

Figure 3.15b sketches the erase algorithm phases. NAND specifications are quite aggressive in terms of erase time. Therefore, Flash vendors try to erase the block content in few erase steps. As a consequence, a very high electric field is applied to the matrix during the Electrical Erase phase. As a matter of fact, erased distribution is deeply shifted towards negative V_{TH} values. In order to minimize floating gate coupling (Chap. 2), a *Program After Erase* (PAE) phase is introduced, with the goal of placing the distribution close to the 0 V limit (of course, with the appropriate read margin).

Each erase pulse is followed by an *Erase Verify* (EV) operation. During this phase all the wordlines are kept at ground. Goal is to verify the presence of memory cells with a V_{TH} higher than 0 V. If EV is not successful, it means that there are some columns which are still programmed. If the maximum number of erase pulses is reached, than the erase operation fails. Otherwise, the voltage applied to the matrix iP-well is incremented by ΔV_E and another erase pulse follows.

Tables 3.1 and 3.2 summarize the erase voltages.

Table 3.1 Voltages applied to the selected block during the electrical erase pulse

	T_0	T_1	T_2	T_3	T_4
BL_{even}	Float	Float	Float	Float	Float
BL_{odd}	Float	Float	Float	Float	Float
DSL	Float	Float	Float	Float	Float
WLs	0 V	0 V	0 V	0 V	0 V
SSL	Float	Float	Float	Float	Float
SL	Float	Float	Float	Float	Float
iP-well	0 V	V_{ERASE}	V_{ERASE}	0 V	0 V

Table 3.2 Voltages applied to the un-selected block during the electrical erase pulse

	T_0	T_1	T_2	T_3	T_4
BL_{even}	Float	Float	Float	Float	Float
BL_{odd}	Float	Float	Float	Float	Float
DSL	Float	Float	Float	Float	Float
WLs	Float	Float	Float	Float	Float
SSL	Float	Float	Float	Float	Float
SL	Float	Float	Float	Float	Float
iP-well	0 V	V_{ERASE}	V_{ERASE}	0 V	0 V

3.4 3D Stacked Architecture

Market request for bigger and cheaper NAND Flash memories triggers continuous research activity for cell size shrinkage. For many years, workarounds for all the scalability issues of planar Flash memories have been found. Some examples are the improved programming algorithms for controlling electrostatic interference between adjacent cells [9], and the double patterning techniques to overcome lithography restrictions.

Unfortunately, other physical phenomena prevent further size reduction of planar memory cell. Nowadays, the number of electrons stored inside the floating gate is extremely low: only tens of electrons differentiate two V_{TH} distributions levels in a process technology node below 20 nm. As reported in literature, channel doping spread [10] and random telegraph noise [11] can induce large native threshold distributions, and electron injection statistics [12] can cause additional variability after program, thus impacting both cell endurance and retention. Shrink of NAND string dimensions increases the electric field between wordlines, resulting in an increased failure rate during cycling.

3D arrays represent a promising opportunity for overcoming the bounds of planar devices. In the last decade all the top Flash vendors have spent hundreds of millions of dollars in R&D to identify a post floating gate technology with the following characteristics: large scale manufacturability, cheap process technology, reliable cell compatible with multi-level cell approach, high bit density, and compliance with current NAND device specification.

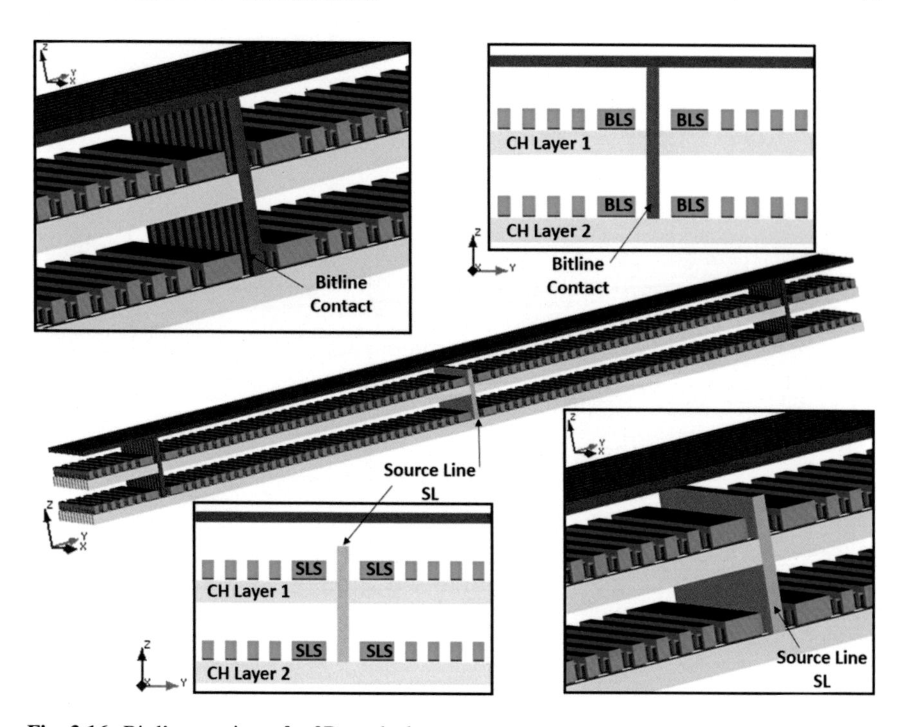

Fig. 3.16 Bird's-eye view of a 3D stacked memory

As described in the previous chapter, a fundamental process change related to the evolution from 2D to 3D memories is the transition from conventional *Floating Gate* (FG) to *Charge Trap* (CT) cells [9]. Almost all planar NAND technologies in production today use polysilicon floating gate as storage element. On the contrary, most of the 3D architectures presented so far use charge trap technology, which requires a simpler fabrication process because of the thinner cell's layer stack. There are exceptions to this rule: 3D NAND architectures based on floating gate are addressed in Chap. 5.

In this chapter we focus on the most straightforward 3D architecture, the Stacked one, which is built by using arrays with horizontal channels and horizontal gates. As shown in Figs. 3.16 and 3.17, this array is a simple stack of planar memories. Drain contacts and bitlines are common to NAND strings of different layers, whereas all the other terminals (source, source selector, wordlines and drain selector) can be decoded separately, layer by layer. Since this 3D organization is the natural evolution of the conventional planar array, this architecture was developed in the early days of the 3D exploration; of course, many considerations about cost, process technology and electrical performances can simply be derived from planar memories.

From a process technology perspective, the biggest issue is represented by the thermal budget required to grow additional silicon layers: it must be limited as

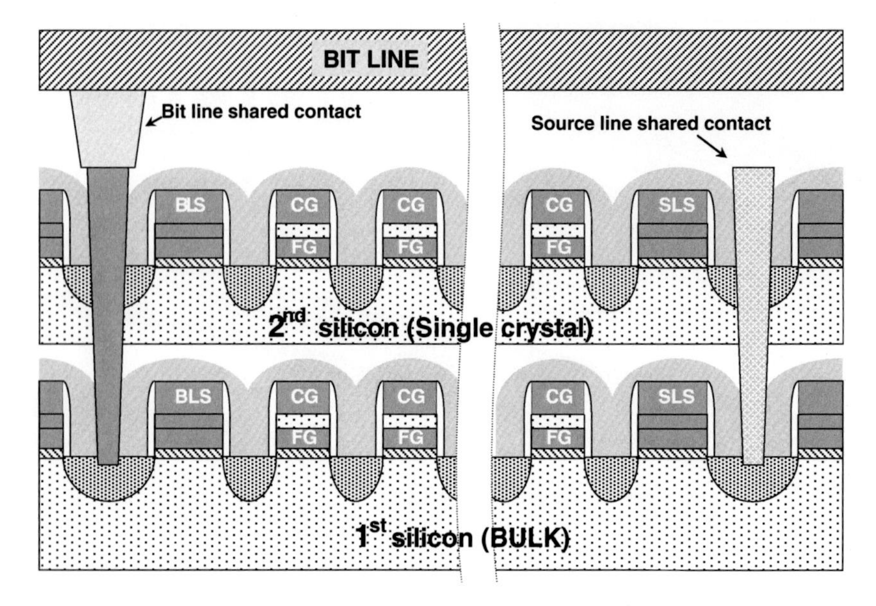

Fig. 3.17 Cross-section of a 3D stacked memory

much as possible to avoid degradation of bottom layers and to guarantee a uniform behavior among cells. Big advantage of horizontal channel/gate architecture is the flexibility: each level is manufactured separately, thus removing many issues that are present with other approaches (e.g. channel and junction doping). Process technology can be easily changed to let the cells work either in enhancement or in depletion mode, but typically enhanced mode is preferred because it allows reusing the know-how developed for planar memories. From an electrical point of view, the biggest difference compared to conventional memories is the floating substrate. In fact, Fig. 3.16 shows that this architecture doesn't allow a direct contact the body, and this constraint impacts device operations, especially the erase one.

From an economic point of view, this approach is not very effective since it multiplies the costs to realize a planar array by the number of layers; in fact, making one layer of such 3D Stacked NAND Flash requires at least 3 critical process modules (bitline/wordline/contact). The only improvement compared to conventional planar memories is represented by circuitry and metal interconnections, because they are shared. In order to limit wafer cost, number of vertical layers must be as low as possible and, to compensate this limitation, it is fundamental to use small single cells. Many publications using this array organization have been presented [13, 14]. Flexibility and reuse of know-how developed with planar CT cells are probably the reasons to explain the remarkable activity in this area.

Figure 3.18 shows a schematic description of 2 NAND matrix layers [15]: on the first one both matrix (MAT1) and peripheral circuits are formed; on the second layer only matrix MAT2 is present. Main peripheral circuits are: sense amplifiers,

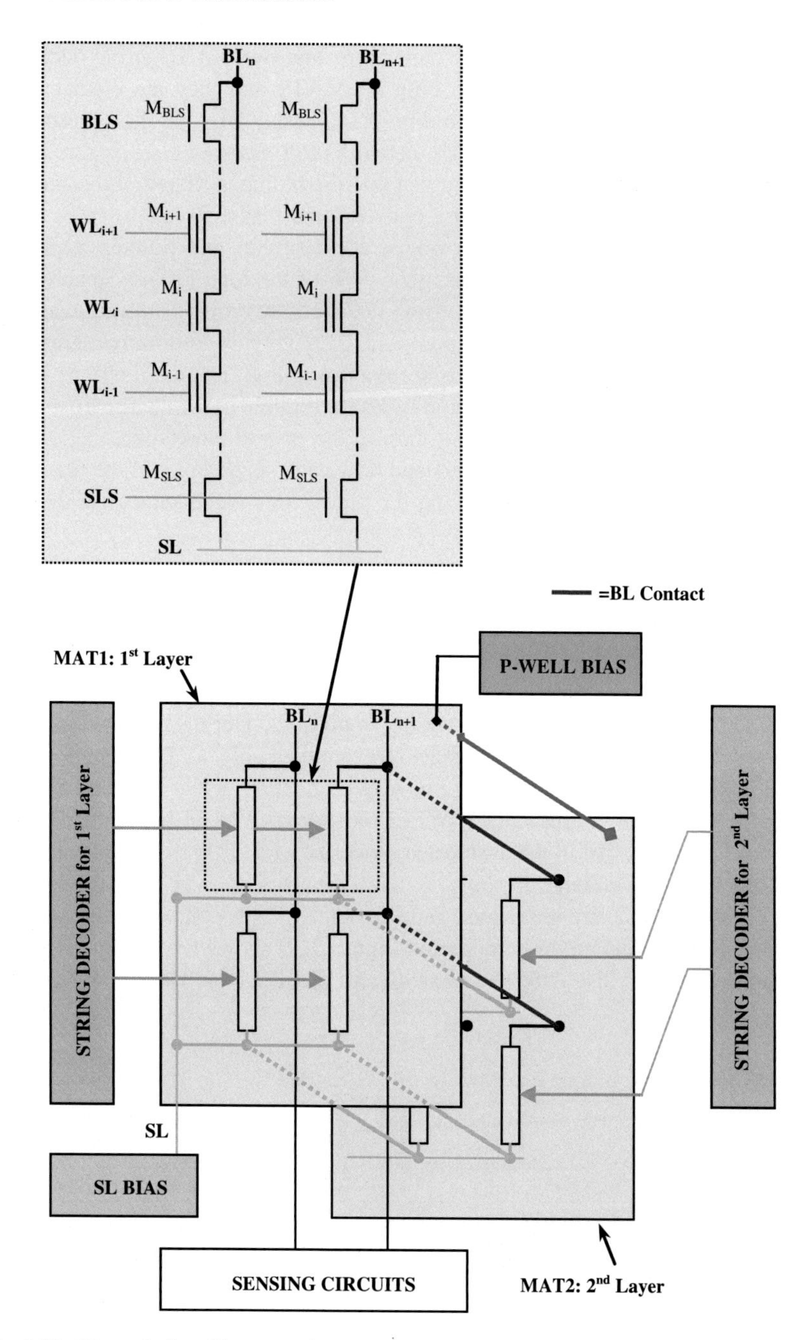

Fig. 3.18 3D stacked architecture scheme

source line (SL) and P-WELL voltage generators, and two NAND string decoders, one for each layer. Bitlines (BL) are only on MAT1 and they are connected to MAT2 trough the contacts shown in Fig. 3.18. Metal BLs are not present on MAT2; this is true for SL and P-WELL networks too. Sensing circuits can access both MAT1 and MAT2 and, due to the fact that the bitline is shared, the capacitive load of a BL is comparable to that of a conventional planar device.

Therefore, there is no penalty on power consumption and timings. Only the parasitic load of the vias is added (less than 5 % of the total bitline capacitance). Thanks to the two independent string (row) decoders, wordline parasitic load is in the same range of a planar device. Furthermore, there are no additional program and read disturbs, because only one layer is accessed at a time. Just the doubling of the P-WELL parasitic load is a penalty, but it can be negligible as the charge time of the P-WELL capacitor is not the dominant term in the overall erase time.

At this point it is important to understand how each single layer in the Stack can be properly addressed without disturbing the others: this is the subject of the next section.

3.5 Biasing of 3D Stacked Layers

Table 3.3 summarizes NAND string biasing conditions. During read and program operations, the required biasing voltages are applied only to the strings of the selected layer MAT1. Strings of MAT2 have WLs floating, while BSL and SLS are biased at 0 V. During Erase, since P-WELL is shared, WLs of unselected MAT2 layer are left floating, like in the unselected blocks of MAT1. In this way, erasing of MAT2 blocks is prevented.

MAT1 and MAT2 are fabricated independently; as a consequence, memory cell's V_{TH} distributions might be different. Figure 3.19 shows what happens after a single programming pulse ΔISPP1: two different distributions, D_{MAT1} and D_{MAT2},

Table 3.3 NAND string biasing conditions for MAT1 and MAT2

BL_n		Read	Program	Erase
		V_{PRE} (0.5–1 V)	0/V_{DD}	Floating
Selected layer 1	BLS1	V_{PASS}	V_{DD}	Floating
	Selected WL1	V_{READ}	V_{PROG}	0 V
	Unselected WL1	V_{PASS}	$V_{PASSPGM}$	0 V
	SLS1	V_{PASS}	0 V	Floating
Unselected layer 2	BLS2	0 V	0 V	Floating
	WL2	Floating	Floating	Floating
	SLS2	0 V	0 V	Floating
SL		0 V	V_{DD}	Floating
P-WELL		0 V	0 V	18–20 V

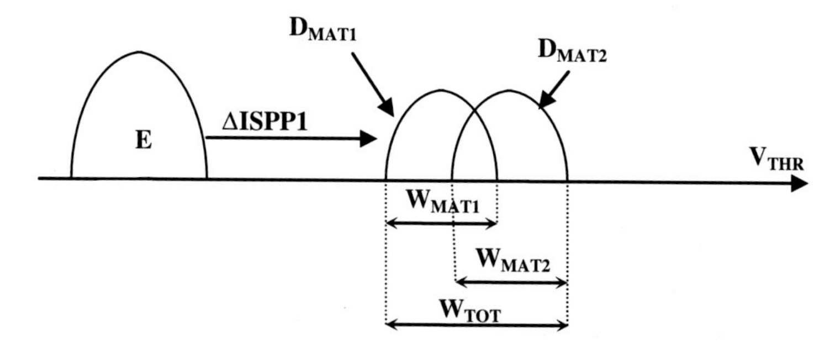

Fig. 3.19 MAT1 and MAT2 distributions after a single ISPP step

are formed. In such a situation, a conventional ISPP would degrade program speed. In fact, the starting voltage $V_{STARTPGM}$ of the ISPP algorithm is determined by the fastest cell, which is located at rightmost side of the V_{TH} distribution. Therefore, due to the enlargement of W_{TOT}, a higher number of program steps is required.

Increasing the number of ISPP steps means reducing the program speed. The proposed solution in [15] is a dedicated programming scheme for each MAT layer. Depending on the addressed MAT layer, program parameters such as $V_{PGMSTART}$, ΔISPP, and maximum number of ISPP steps are properly chosen (Fig. 3.20).

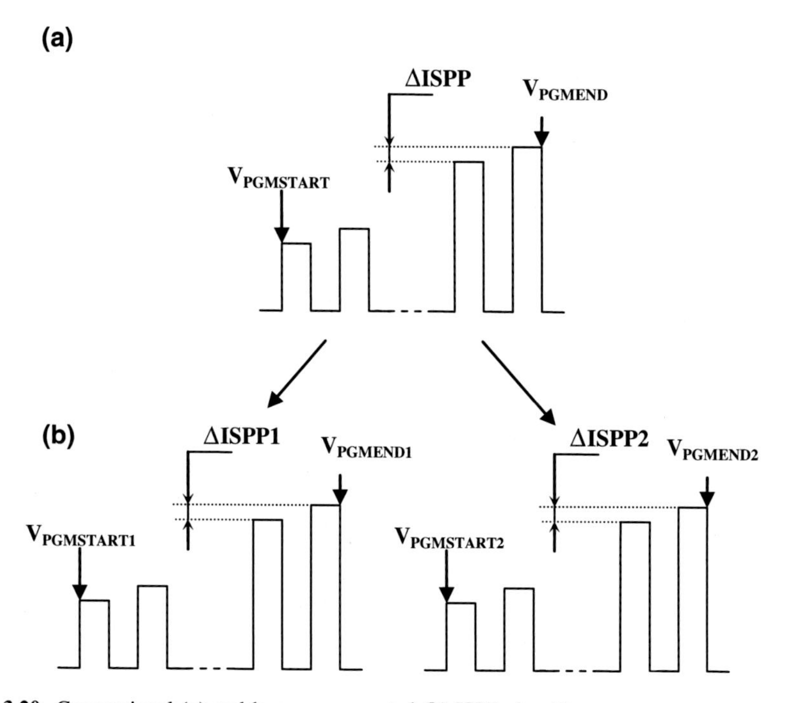

Fig. 3.20 Conventional (**a**) and layer compensated (**b**) ISPP algorithm

A dedicated control scheme can be used for erasing too: with both P-WELLs at the same erase voltage, slightly different voltages are applied to the wordlines of the different layers. It is worth highlighting that, because of the two row decoders, it is possible to randomly erase 2 blocks, one for each MAT layer.

In the next four chapters we'll take a close look at the architectural details of the following 3D options:

- 3D Charge Trap
- 3D Floating Gate
- 3D Advanced Architectures
- 3D VG.

In each chapter, a lot of bird's-eye views and cross sections will help the reader to understand the three dimensional implications of these new flash technologies.

References

1. G. Campardo, R. Micheloni, D. Novosel, *VLSI-Design of Non-Volatile Memories* (Springer, 2005)
2. S. Lee et al., A 128 Gb 2b/cell NAND flash memory in 14 nm technology with t_{prog} = 640 μs and 800 MB/s I/O Rate, in *2016 IEEE International Solid-State Circuits Conference (ISSCC)*, Digest of Technical Papers, San Francisco, USA, Feb 2016, pp. 138–139
3. N. Chan, M.F. Beug, R. Knoefler, T. Mueller, T. Melde, M. Ackermann, S. Riedel, M. Specht, C. Ludwig, A.T. Tilke, Metal control gate for sub-30 nm floating gate NAND memory, in *Proceedings of 9th NVMTS*, Nov 2008, pp. 82–85
4. R. Micheloni, L. Crippa, A. Marelli, *Inside NAND Flash Memories*, Chap. 6 (Springer, 2010)
5. K.-D. Suh et al., A 3.3 V 32 Mb NAND flash memory with incremental step pulse programming scheme. IEEE J. Solid-State Circ. **30**(11), 1149–1156 (1995)
6. Y. Iwata et al., A 35 ns cycle time 3.3 V only 32 Mb NAND flash EEPROM. IEEE J. Solid-State Circ. **30**(11), 1157–1164 (1995)
7. J.-K. Kim et al., A 120-mm 64-Mb NAND flash memory achieving 180 ns/Byte effective program speed. IEEE J. Solid-State Circ. **32**(5), 670–680 (1997)
8. R. Micheloni, L. Crippa, A. Marelli, *Inside NAND Flash Memories*, Chap. 12 (Springer, 2010)
9. R. Micheloni, L. Crippa, A. Marelli, *Inside NAND Flash Memories* (Springer, 2010)
10. T. Mizuno et al., Experimental study of threshold voltage fluctuation due to statistical variation of channel dopant number in MOSFET's. IEEE Trans. Electron Devices **41**(11), 2216–2221 (1994)
11. H. Kurata et al., The impact of random telegraph signals on the scaling of multilevel flash memories, in *Symposium on VLSI Technology* (2006)
12. C.M. Compagnoni et al., Ultimate accuracy for the NAND flash program algorithm due to the electron injection statistics. IEEE Trans. Electron Devices **55**(10), 2695–2702 (2008)
13. S.M. Jung et al., Three dimensionally stacked NAND flash memory technology using stacking single crystal Si layers on ILD and TANOS structure for beyond 30 nm node, in *IEDM Technical Digest* (2006)
14. E.K. Lai et al., A multi-layer stackable thin-film transistor (TFT) NAND-type flash memory, in *IEDM Technical Digest* (2006)

15. K.-T. Park et al., A fully performance compatible 45 nm 4-Gigabit three dimensional double-stacked multi-level NAND flash memory with shared bit-line structure. IEEE J. Solid-State Circ. **44**(1), 208–216 (2009)
16. R. Micheloni et al., A 4 Gb 2b/cell NAND flash memory with embedded 5b BCH ECC for 36 MB/s system read throughput, in *Solid-State Circuits Conference, ISSCC*, Digest of Technical Papers (2006), pp. 497–506

Chapter 4
3D Charge Trap NAND Flash Memories

Luca Crippa and Rino Micheloni

4.1 Introduction

Different criteria can be adopted to sort 3D architectures but, probably, the classification based on topological characteristics is the most effective since the choice of a specific topological organization has a direct impact on cost, electrical performances and process technology integration. The following cases can be considered:

- horizontal channel and gate (Chap. 3);
- vertical channel and horizontal gate;
- horizontal channel and vertical gate (mainly Chap. 7).

The array with horizontal channel and gate is the 3D Stacked option discussed in the previous chapter. In the 3D approach with horizontal gate and vertical channel, the planar (2D) NAND Flash string of Fig. 4.1a is rotated by 90°, as shown in Fig. 4.1b. In order to improve electrical performances, a channel fully wrapped around by gate is adopted (Fig. 4.1c and d) [1]. With this specific configuration, thanks to the curvature effect, it is possible to enhance the electric field across the tunnel oxide and to relax the electric field across the blocking oxide [2, 3], thus improving power and reliability (Chap. 2).

To simplify descriptions, in this book "vertical channel with horizontal gate" is shortened to "vertical channel".

This chapter starts off with 2 vertical channel architectures named BiCS (*Bit Cost Scalable*) and P-BiCS (*Pipe-Shaped* BiCS), respectively. BiCS was proposed for the

L. Crippa (✉) · R. Micheloni
Performance Storage BU, Microsemi Corporation, Vimercate, Italy
e-mail: luca.crippa@ieee.org

R. Micheloni
e-mail: rino.micheloni@ieee.org

R. Micheloni (ed.), *3D Flash Memories*, DOI 10.1007/978-94-017-7512-0_4

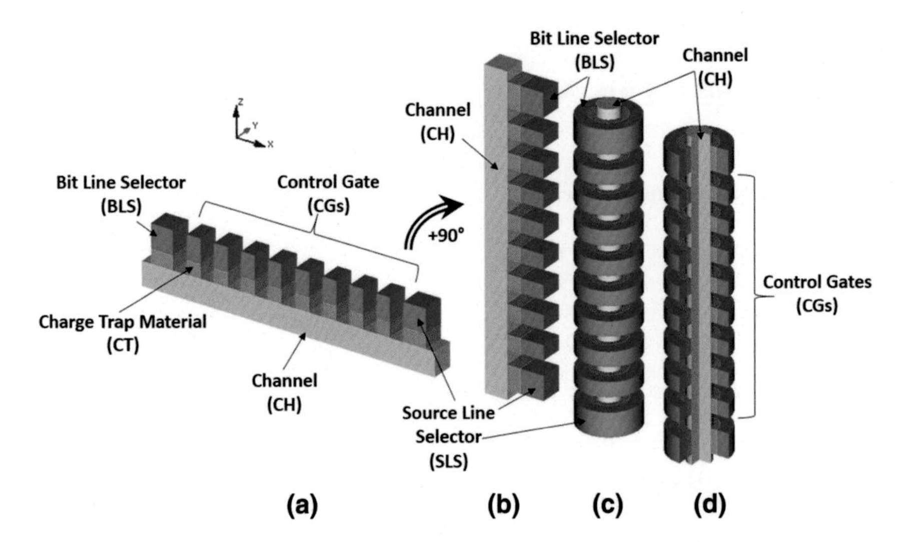

Fig. 4.1 NAND Flash string with horizontal gate and vertical channel: **a** planar, **b** planar rotated by 90°, **c** vertical channel with cylindrical shape and **d** its cross section

first time by Toshiba in 2007 [4, 5], and another version called P-BiCS was presented in 2009 [6–8] to improve retention, source selector performances and source line resistance. Both options are based on a Charge Trap memory cell (Chap. 2). BiCS can definitely be considered as a major milestone in the history of 3D Flash.

In the next 2 sections we will dig into the details of BiCS and P-BiCS, while the second part of the chapter is devoted to the evolutionary path of V-NAND, which is the first 3D architecture that reached volume production.

4.2 BiCS

Bird's-eye views of the BiCS Flash memory are sketched in Fig. 4.2, while Figs. 4.3 and 4.4 include diagrams of the equivalent circuit [5].

First of all, there is a stack of *Control Gate* (CG) plates (green rectangles). The lowest gate plate corresponds to the gate of the *Source Line Selector* (SLS) of the NAND string. Holes are punched through the entire stack and plugged with poly-silicon, thus forming a series of vertical memory cells connected in a NAND architecture. At the top of the structure we have *Bit Line Selectors* (BLS's) and *Bitlines* (BLs) [9].

Memory cells work in depletion-mode [2, 3] with the body poly-silicon being un-doped or lightly uniformly n-doped, to avoid the process complexity of forming p-n junctions within the vertical polysilicon plug (or pillar). The intersection of a control gate plate with a pillar identifies a single memory cell; each string is connected to a bitline through an upper select transistor (BLS). The bottom of the

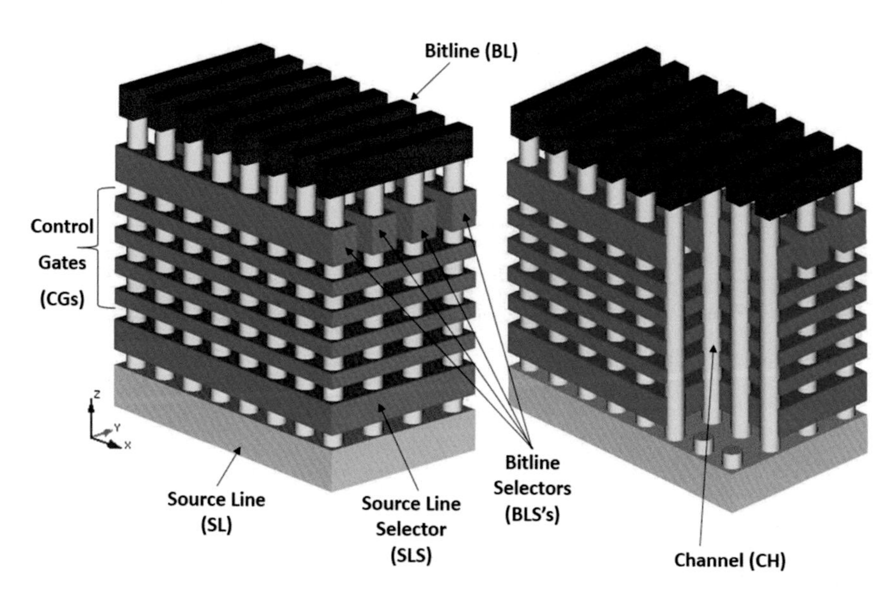

Fig. 4.2 BiCS architecture

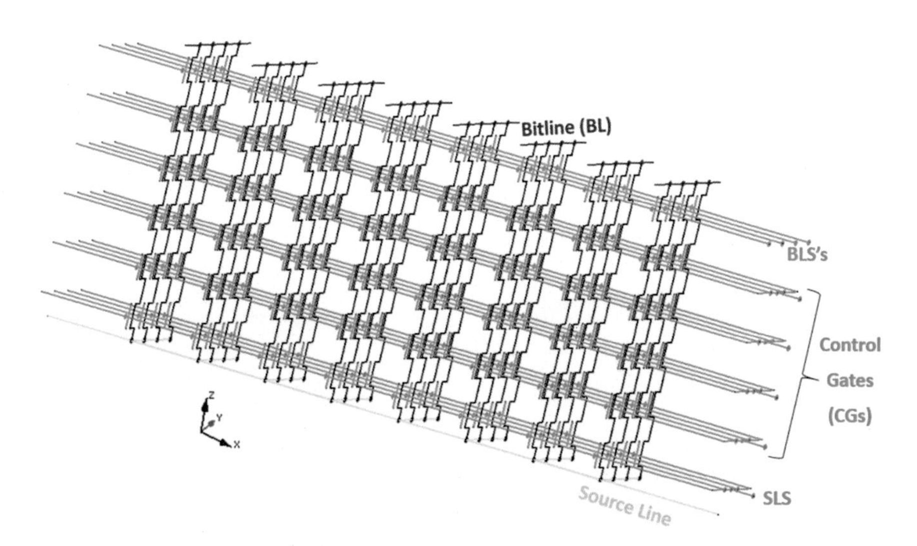

Fig. 4.3 Schematic circuit of a BiCS array

memory string is connected to the common source diffusion, which is formed on the silicon substrate. Figure 4.5 highlights a single NAND Page in the SLC case (i.e. 1 bit/cell); basically, there is one selected memory cell per bitline.

Bit density can be increased by adding more control gate plates [10, 11], while the number of the critical lithography steps remains constant because the whole stack of control gates is completely punched through with one lithography step only.

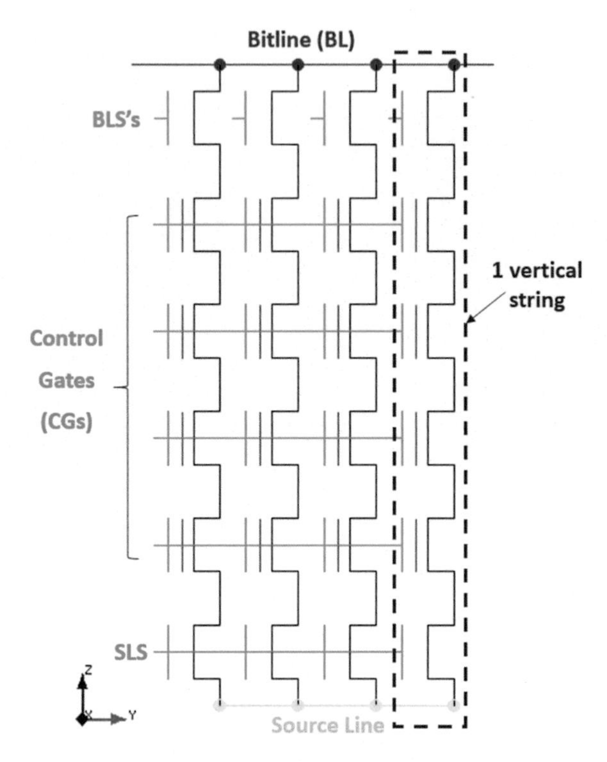

Fig. 4.4 Y-Z view of BiCS equivalent circuit schematic

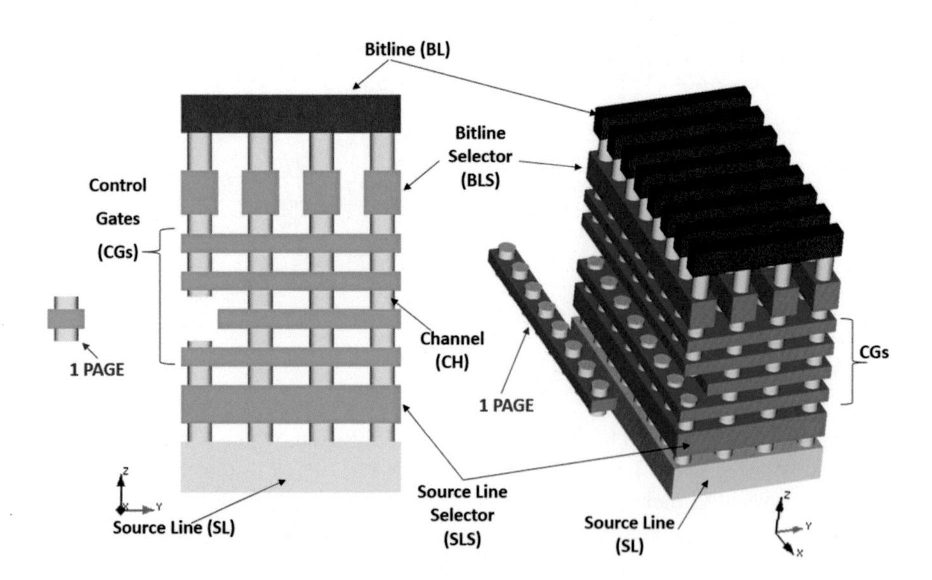

Fig. 4.5 SLC NAND flash page in the BiCS architecture

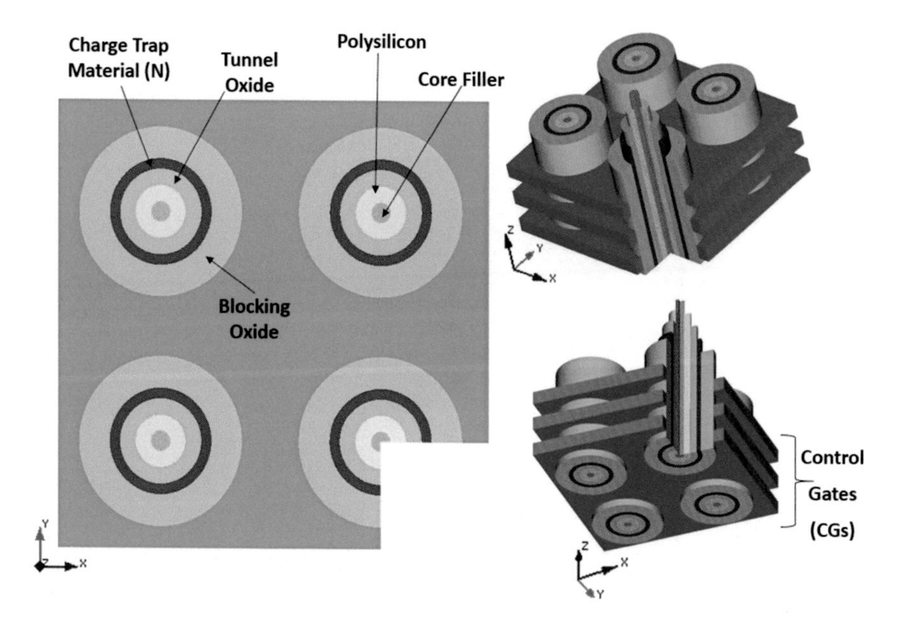

Fig. 4.6 BiCS memory cells

As discussed in Chap. 2, in the BiCS design the floating gate is replaced by a charge trap material. Multiple views of BiCS cell's structure are reported in Fig. 4.6, where tunnel oxide, field oxide, polysilicon channel and core filler are highlighted.

It is worth highlighting the fact that the body of the vertical transistor is completely made of polysilicon; in fact, the aspect ratio of the punched hole is so high that it becomes almost impossible to achieve good production yield without non-selective deposition process steps.

Unfortunately, is such a configuration, it is very difficult to control the trap density at grain boundary, and this causes a large variation of the sub-threshold characteristics of the vertical transistor. In order to reduce the trap density fluctuation, the body polysilicon needs to be much thinner than the depletion width. Basically, the concept is to reduce the volume of the polysilicon and, therefore, the total number of traps, as sketched in Fig. 4.7. Given its shape, the body of this vertical transistor is referred to as *Macaroni Body* [5]. The center of the macaroni shape is filled with dielectric film (*filler* in the following) to make the 3D process integration easier.

In Figs. 4.8 and 4.9 the basic building block of BiCS memory is split in different sections in order to provide a greater level of details for the different elements [6, 9]. Cell's structure is clearly visible in the Middle section where the reader can find all the layers seen in Fig. 4.6. In the Upper section it is worth highlighting that the charge trap material stops before the bitline selector, which is a standard transistor built with field oxide only. In other words, both the tunnel oxide and the charge trap

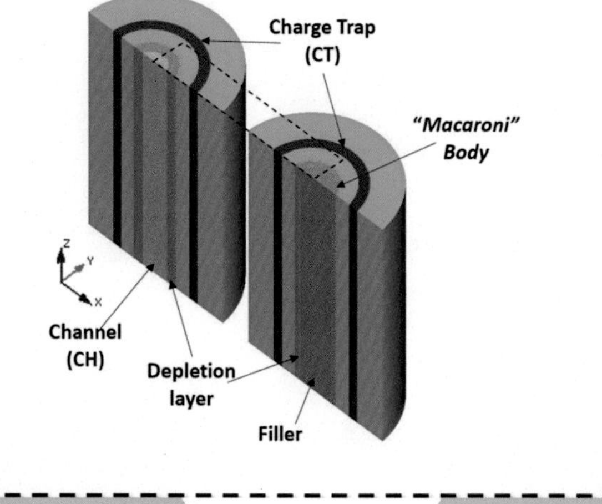

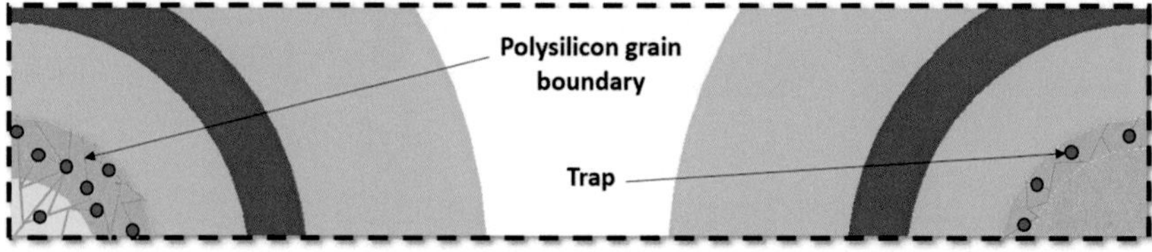

Fig. 4.7 Vertical transistor with (*right*) and without (*left*) *macaroni* body

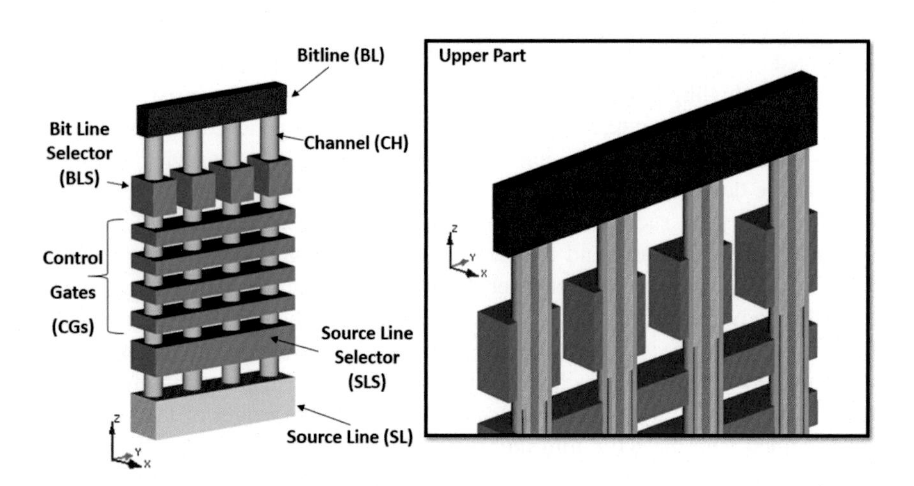

Fig. 4.8 BiCS vertical cross-section: *upper part*

material are replaced by the polysilicon. The same applies to the Source Line Selector which becomes a standard nMOS transistor, as shown in Fig. 4.9.

Figures 4.10 and 4.11 show a simplified fabrication sequence of a BiCS array [12]. In the first step all the plates (green parallelepipeds) for SLS, control gates, and bitline selectors are manufactured. Then each single stripe of BLS's is defined.

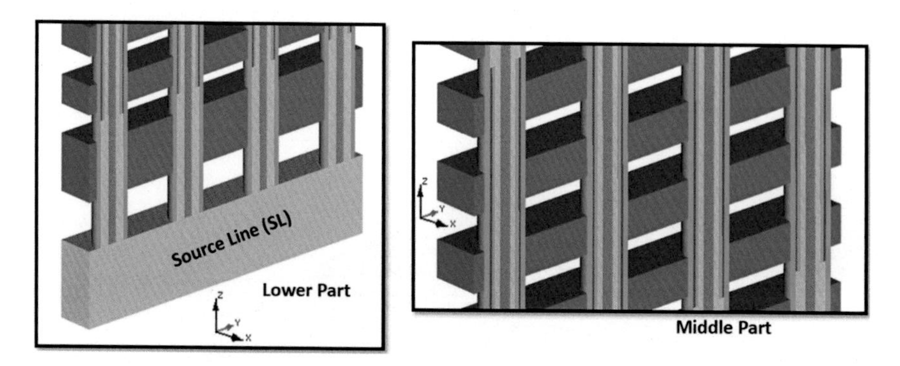

Fig. 4.9 BiCS vertical cross-section: *middle* and *lower parts*

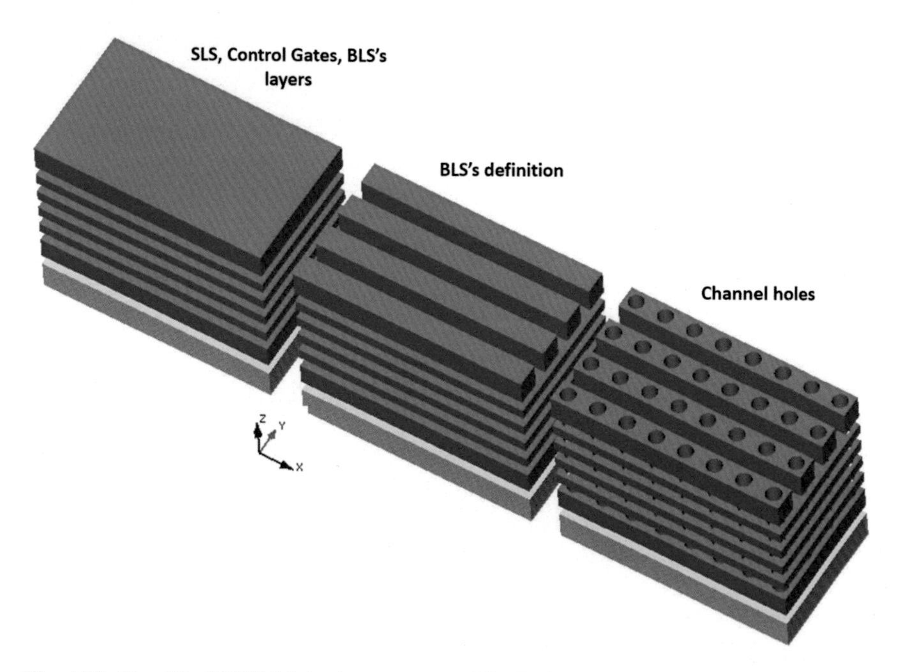

Fig. 4.10 Simplified BiCS fabrication sequence: CGs, BLS definition, and channel holes

At this point the stack is ready to be drilled: in Fig. 4.10 channel holes are clearly visible. Figure 4.11 shows the next step: each single channel hole is filled with polysilicon (yellow cylinder), thus forming a pillar. Bitlines are defined during the back-end phase. To sum up, the BiCS cell array consists of multi-stacked control gate plates and polysilicon pillars fabricated through control gate plates. Memory cells are placed at the intersections of control gate plates and polysilicon pillars.

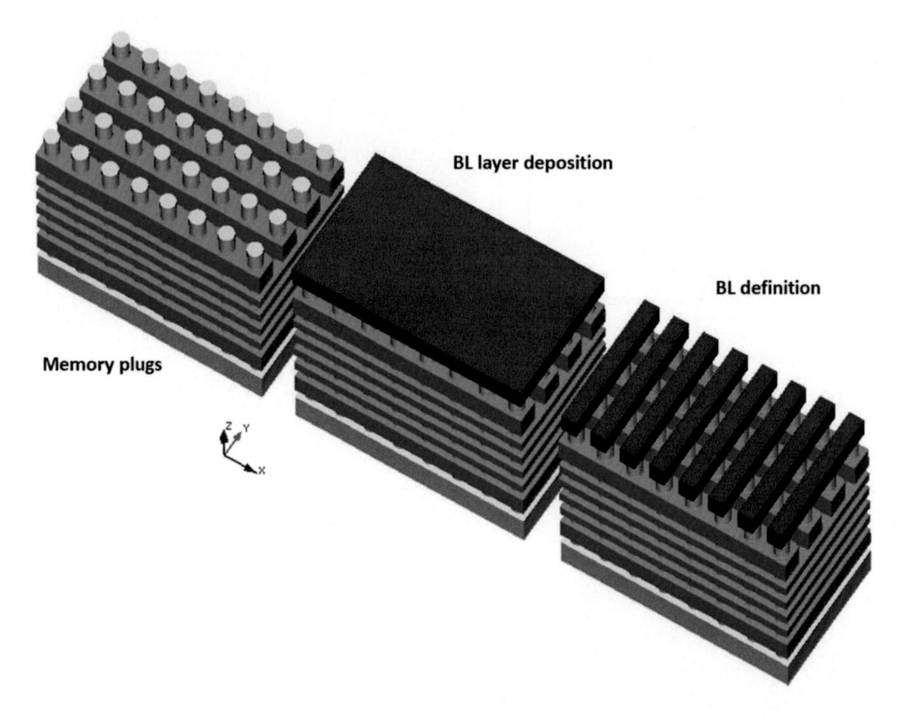

Fig. 4.11 Simplified BiCS fabrication sequence: memory plugs, BL layer deposition, and BL definition

In the following, in order to simplify the drawings, memory cell's structure is defined by the Channel layer (yellow) only.

Edges of control gate plates form a double-sided staircase structure as sketched in Fig. 4.12 [4, 5, 12, 13]. All the plates (CGs, SLS, SL) are contacted on the same side (right hand side in this example), because the other one (left hand side) is used for connecting bitline selectors. Orange rectangles represent metal connections. Figures 4.13 and 4.14 are the top and bottom bird's-eye views of Fig. 4.12.

As we have seen in Fig. 4.5, multiple Flash Pages live on the same CG plate, thus amplifying the impact of disturbs, especially during the programming phase. For minimizing disturbs, the whole stack of control gates, SLS, and SL is etched to form a slit, which, as a matter of fact, creates blocks of memory plugs. A vertical cross section of 3 adjacent blocks is shown in Fig. 4.15, while Fig. 4.16 is the corresponding bird's-eye view [12].

Historically, BiCS evolved in a different architecture called P-BiCS, which is the subject of the following section.

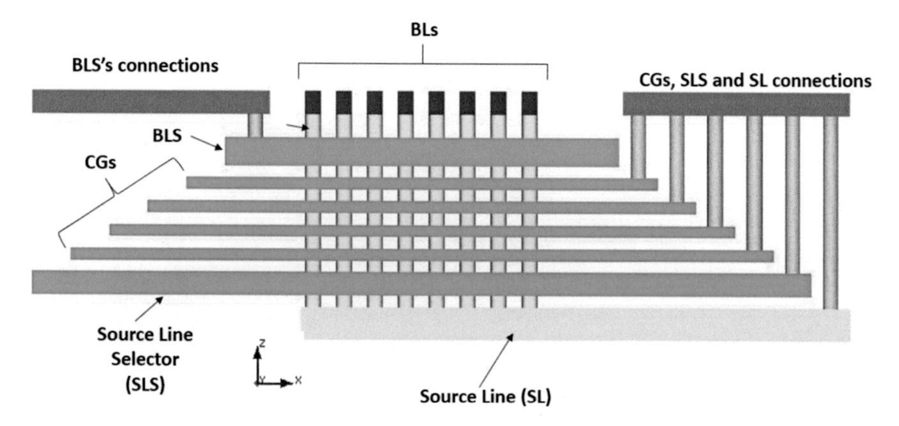

Fig. 4.12 Vertical cross section of BiCS with gate connections

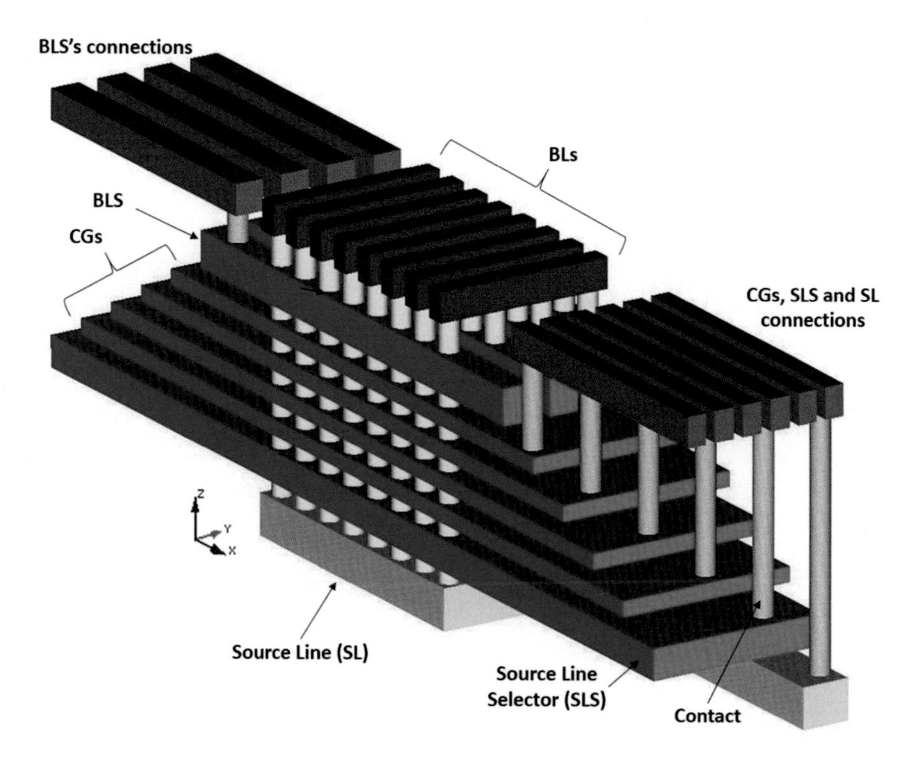

Fig. 4.13 Top bird's eye view of BiCS with gate connections

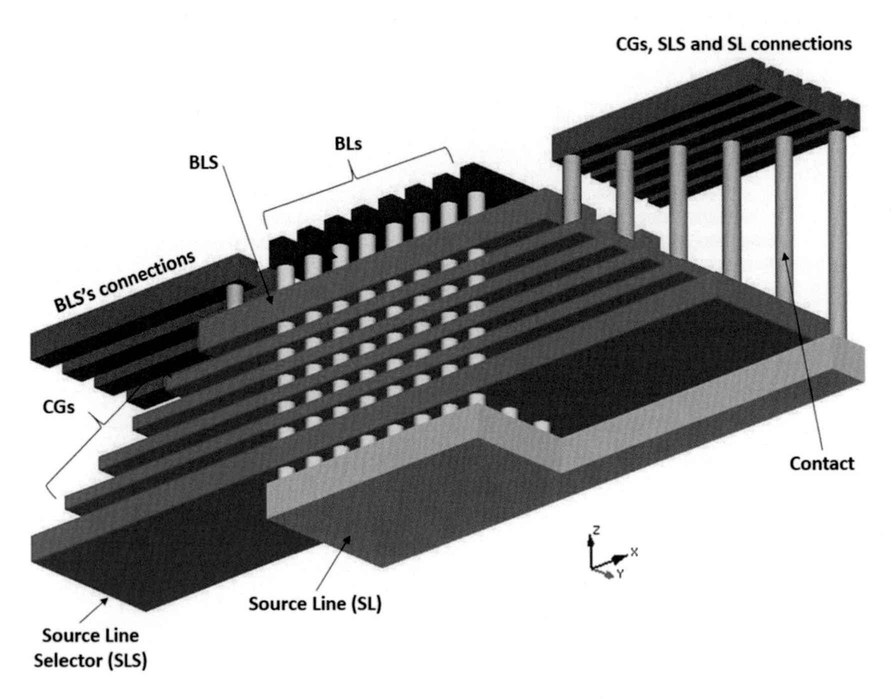

Fig. 4.14 Bottom bird's-eye view of BiCS with gate connections

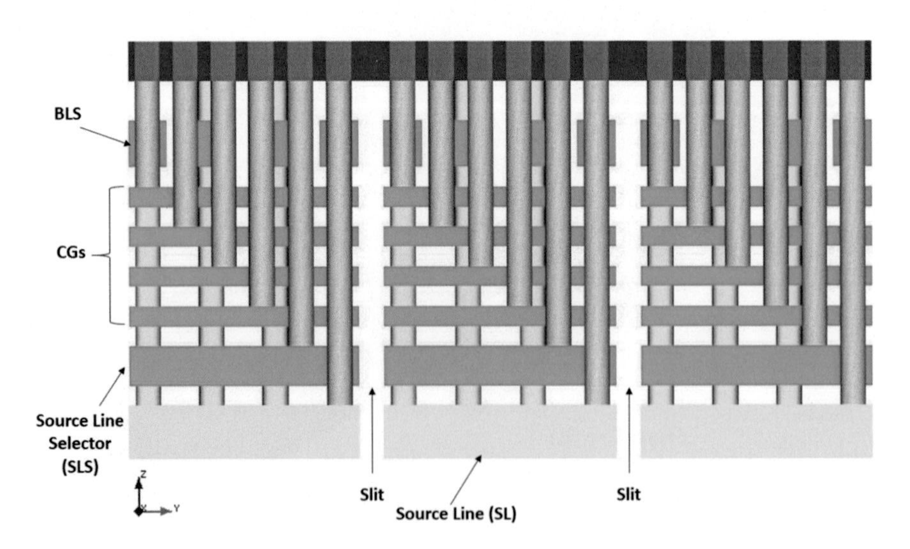

Fig. 4.15 Front view of 3 adjacent blocks to highlight the presence of slits

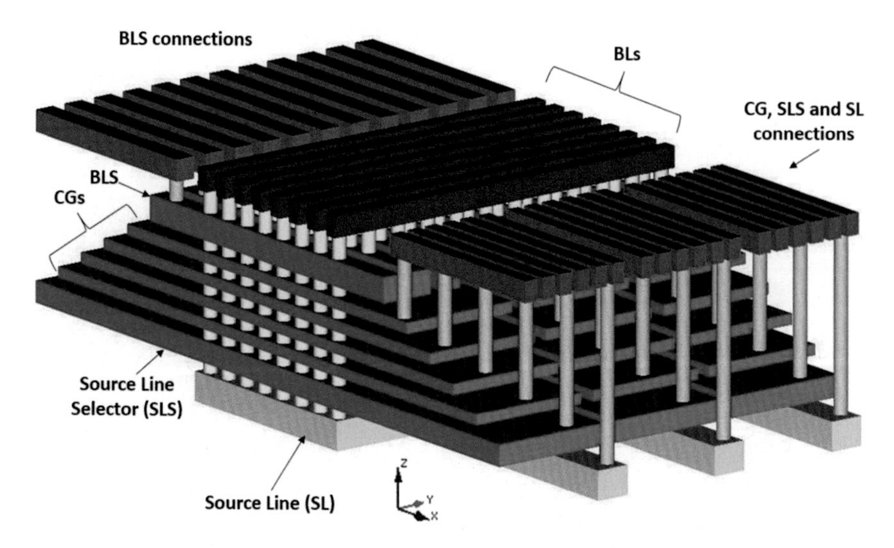

Fig. 4.16 Top bird's-eye view of 3 adjacent blocks

4.3 P-BiCS

P-BiCS (*Pipe-Shaped* BiCS) Flash was developed to solve some critical problems of BiCS, such as poor cell's reliability, poor cut-off characteristics of the Source Line Selector (placed at the bottom of BiCS), and high resistance of the Source Line [6, 7]. By adopting a U-shape vertical NAND string, as shown in Fig. 4.17, P-BiCS solves all the above mentioned problems. In particular, P-BiCS shows three main advantages compared to a straight-shaped BiCS:

- P-BiCS has better data retention and a wider V_{TH} window because the fabrication process is less stressful for the tunnel oxide;
- Because the Source Line is placed at the top of the stack, it is easier to connect it to a metal mesh, which results in a lower parasitic resistance;
- The Source Line Selector is at the same height of the BLS and, therefore, its cut-off characteristics can be tightly controlled, thus improving the functionality of the memory array itself.

A P-BICS array is built starting from the basic structure shown in Fig. 4.17 [7]. Two NAND strings are connected via the so-called Pipe Connection at the bottom of the structure, thus forming a U-shaped string (highlighted in cyan): one of the terminals of this "U" is connected to the bitline, while the other one is tied to the Source Line. It is worth pointing out that the 2 NAND strings are mirrored such that they can share the same source line plate (blue parallelepiped).

The basic building block of Fig. 4.17 can be replicated multiple times along the X direction to form a memory array as shown in Figs. 4.18, 4.19 and 4.20.

Figure 4.21 is the schematic of the equivalent circuit [8].

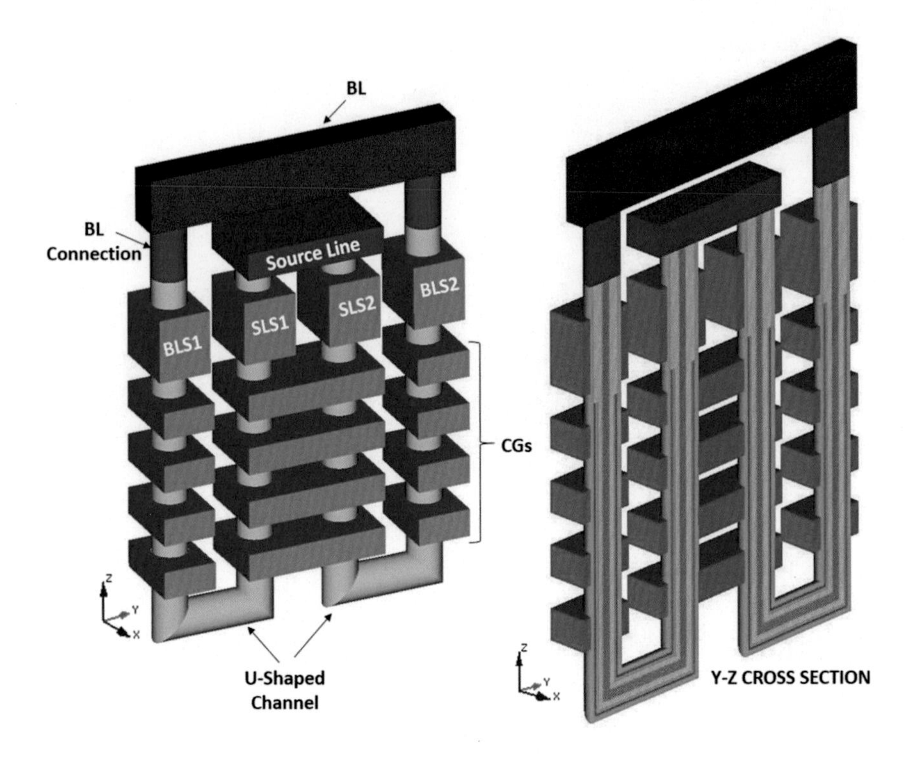

Fig. 4.17 Basic building of a P-BICS NAND flash and its vertical cross-section

For the sake of clarity, Blocking Oxide and Tunnel Oxide layers have been removed from Fig. 4.18 onward; only Charge Trap (pink) and Channel (yellow) are visible.

At this point we are ready to build a bigger array by replicating the basic building block of Fig. 4.17 along both X and Y directions, as sketched in Figs. 4.22 and 4.23.

For a better understanding of Fig. 4.22, it is worth removing some of the backend structures, as shown in Fig. 4.24. Starting from the right hand side, the reader can see how the array would look like by removing bitlines, and bitlines and Source Line at the same time (far left in Fig. 4.24).

As already pointed out, the Source Line plate is shared among all the NAND strings which are adjacent along both the X and the Y directions. This mirrored architecture minimizes the parasitic resistance of the source connection.

Next part of the manufacturing process is designed to build all the necessary connections for biasing CGs, BLs, BLS's, SLS's, and SL [8]. As we have seen in the previous section, in the BiCS architecture each control gate plate is shared by several neighboring rows of NAND strings, in order to reduce the silicon area. In P-BiCS this is not possible as 2 different control gates of the same NAND string belong to the same layer within the stack. As a result, a branched control gate

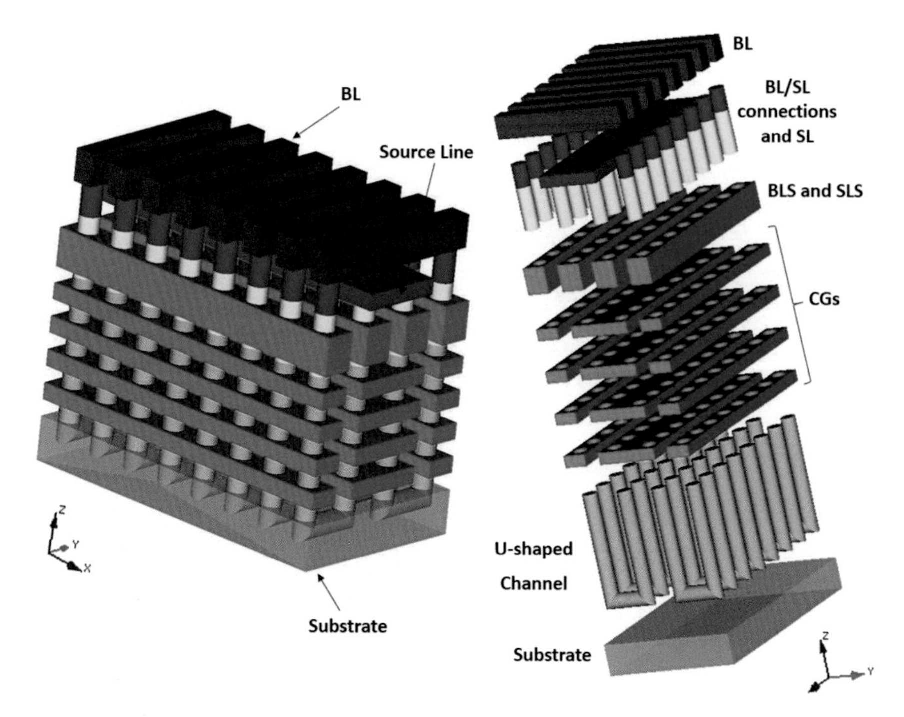

Fig. 4.18 Top bird's-eye view of a P-BiCS NAND array

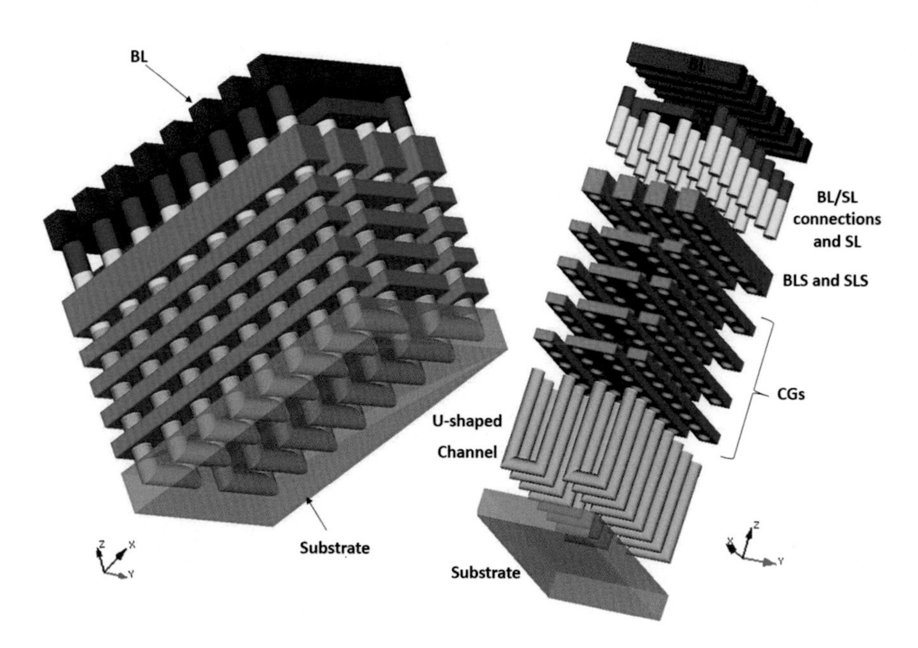

Fig. 4.19 Bottom bird's-eye view of a P-BiCS NAND array

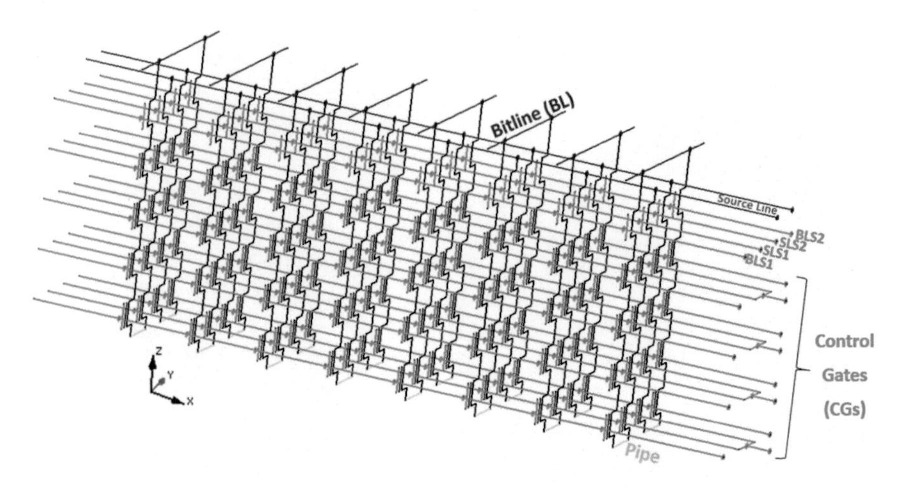

Fig. 4.20 P-BiCS circuit schematic

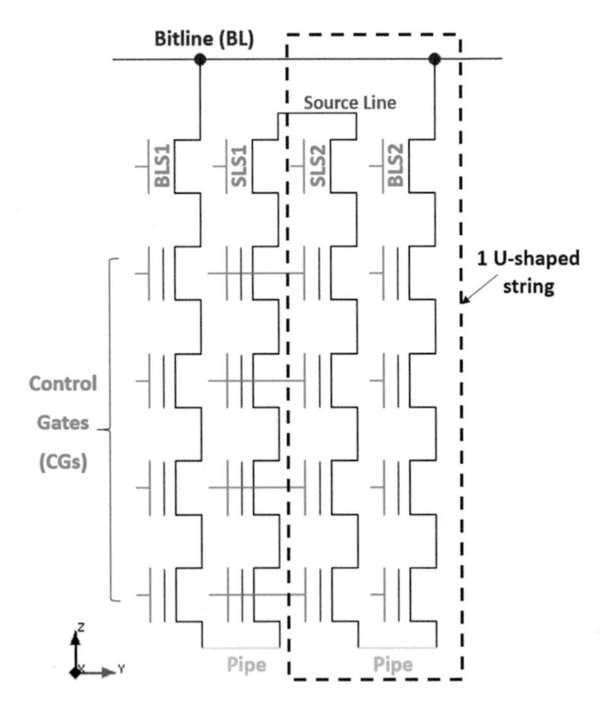

Fig. 4.21 Y-Z view of P-BiCS circuit schematic

configuration is adopted: Figs. 4.25 and 4.26 are bird's-eye views of P-BiCS with all the array connections, while the corresponding X-Z cross-section is drawn in Fig. 4.27. In this example, a NAND string of 8 Flash cells is considered.

To better appreciate the construction details of Figs. 4.25, 4.28 and 4.29 show a zoom of the right hand side from different angles.

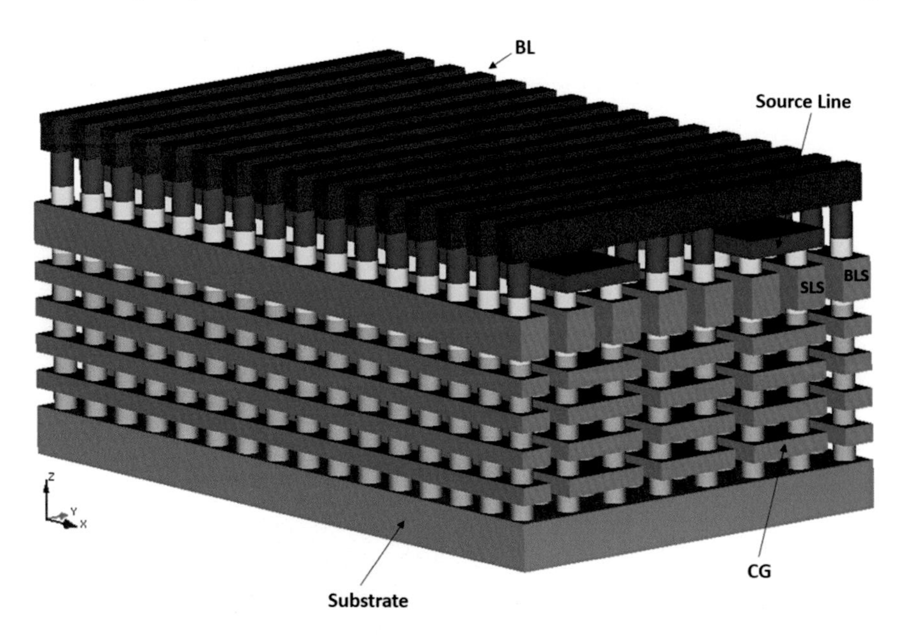

Fig. 4.22 P-BiCS NAND array

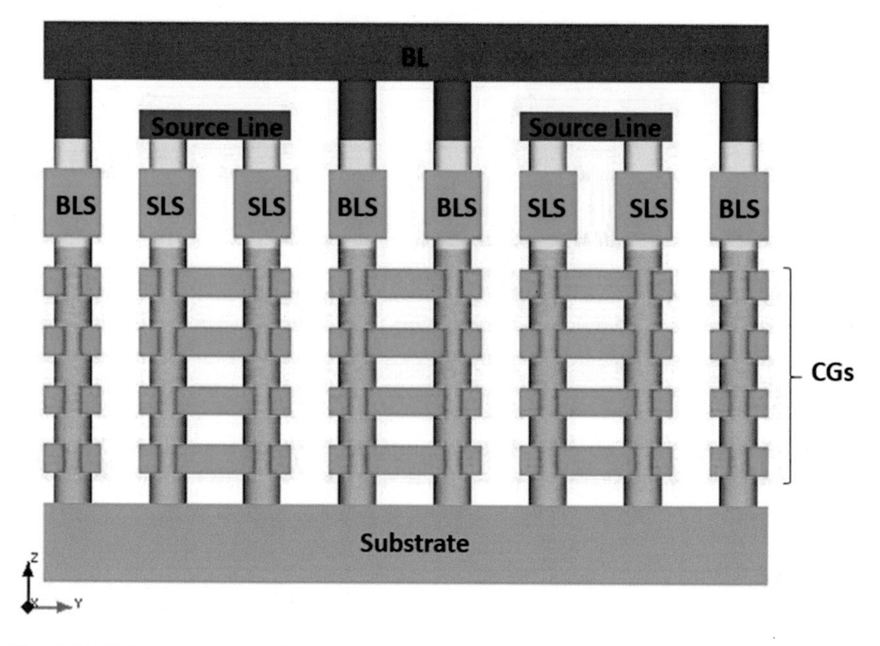

Fig. 4.23 Y-Z cross section of Fig. 4.22

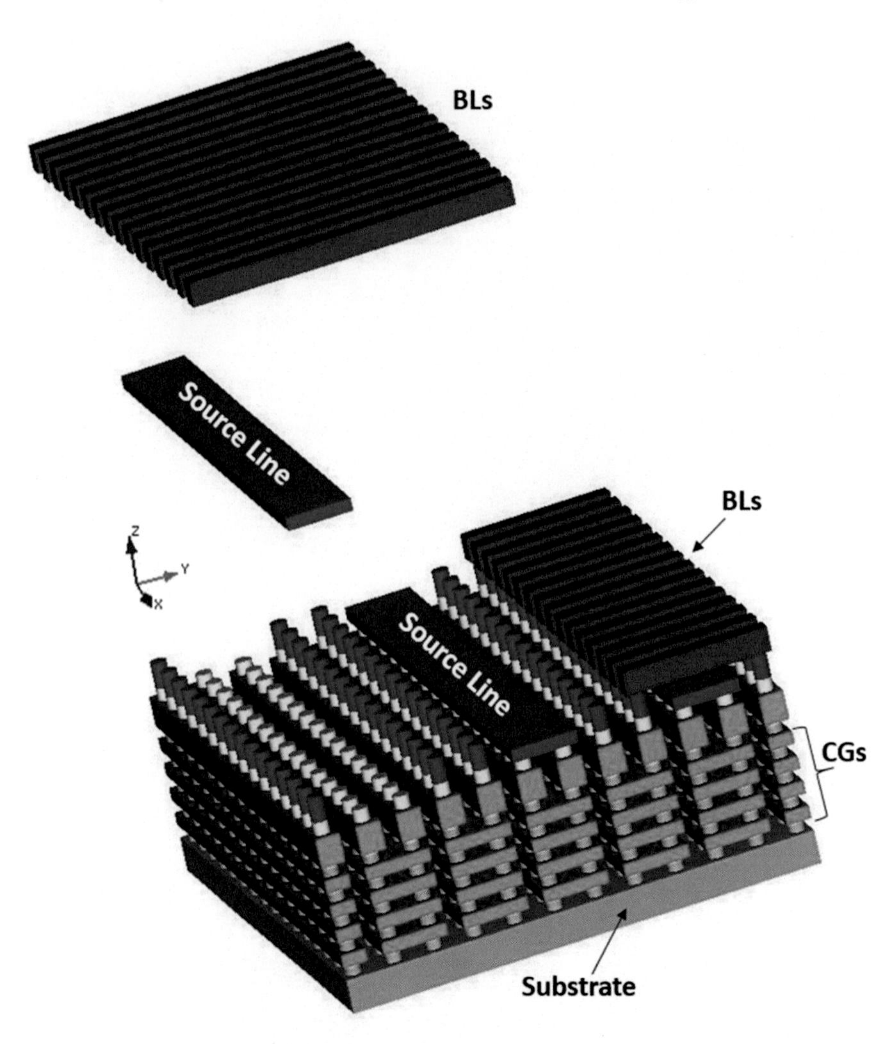

Fig. 4.24 P-BiCS array with partial removal of source line and bitlines

Let's now have a closer look to the control gates, which are realized by using fork-shaped plates [8]. Please refer to Figs. 4.30 and 4.31. Each branch of the fork controls cells of two adjacent pages. Indeed, a P-BiCS NAND string of 8 cells is formed by vertically stacking 4 pairs of control gates. Each pair of control gate plates is arranged in a staggered layout (please refer to CG0 and CG7 in Fig. 4.25 as an example). Of course, this architecture can be scaled to include more cells in the NAND string. For example, with 16/24 layers we can build strings of 32/48 cells.

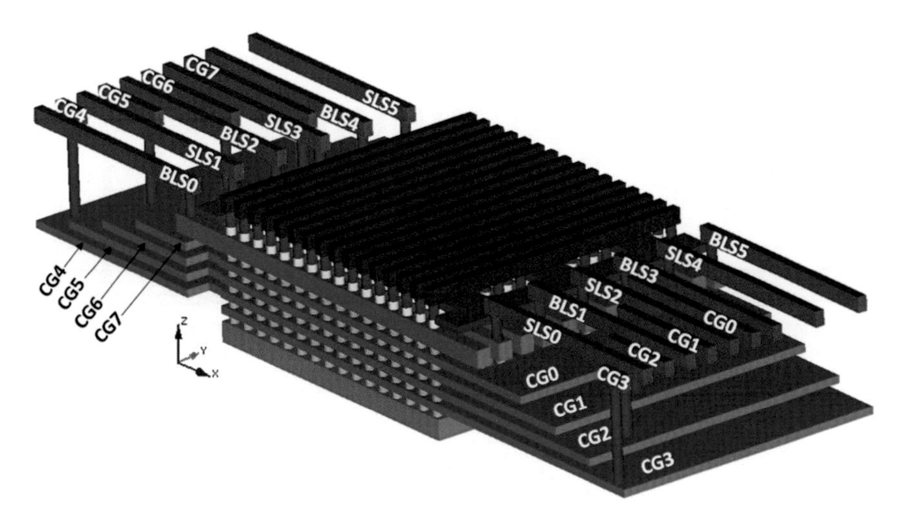

Fig. 4.25 Bird's-eye view of P-BiCS with array connections

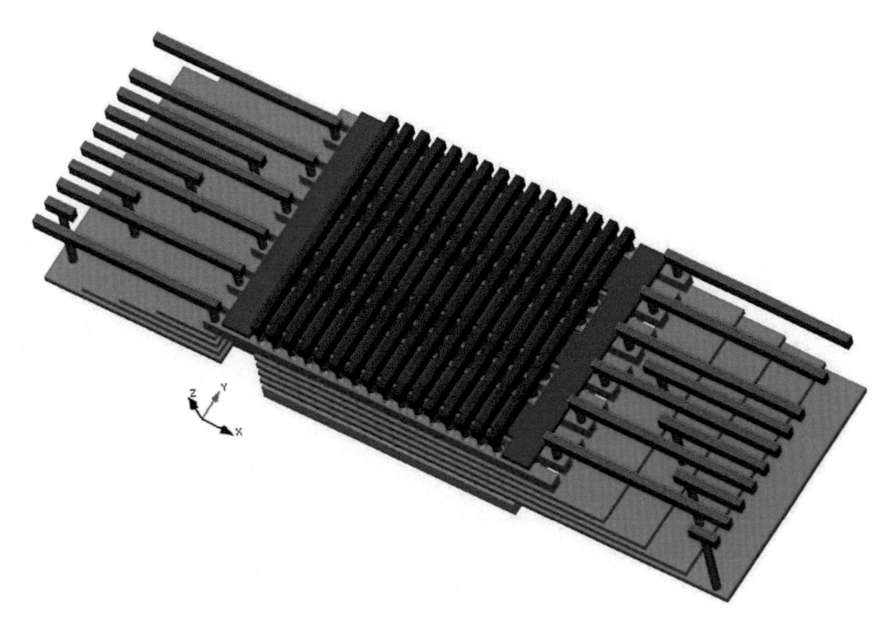

Fig. 4.26 Top view of Fig. 4.25

At the bottom of Fig. 4.30 it is possible to see some U-shapes (in pink) that have been placed to better identify NAND strings.

One other major difference of P-BiCS versus BiCS is the fact that the source line is placed at the top of the 3D structure [6]. In order to enhance noise immunity as much as possible, it is very important to have a source line with a very low resistance. For this reason, another level of source line connection is used on top of

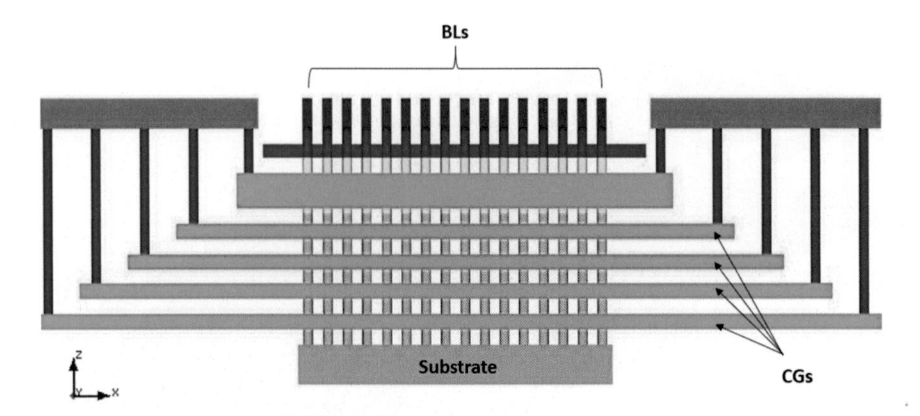

Fig. 4.27 X-Z cross section of Fig. 4.25

Fig. 4.28 Zoom of Fig. 4.25—top view

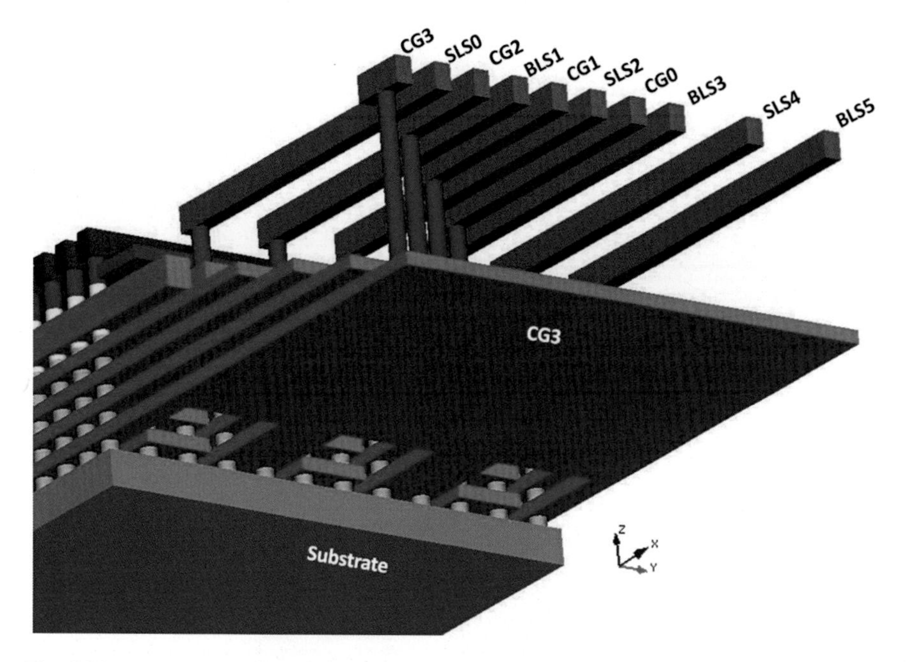

Fig. 4.29 Zoom of Fig. 4.25—bottom view

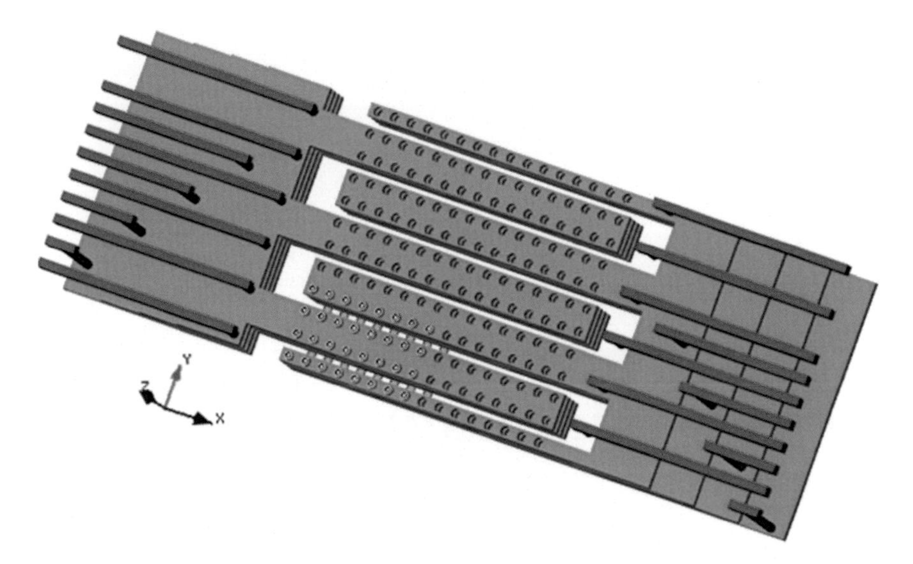

Fig. 4.30 P-BiCS: control gates are realized by fork-shaped plates

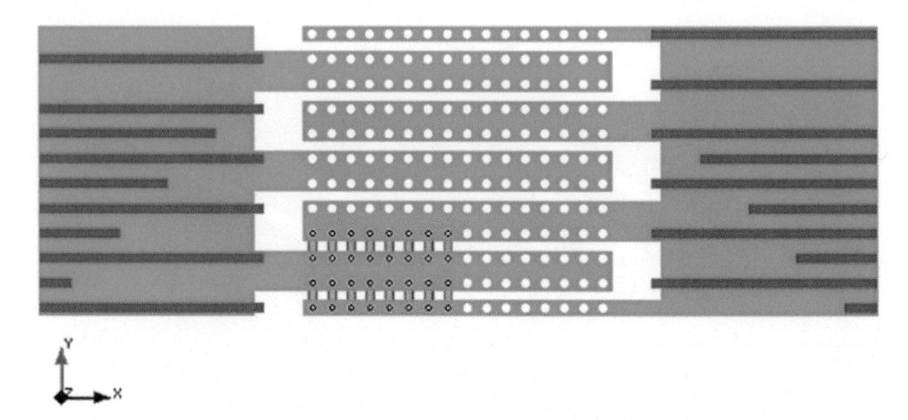

Fig. 4.31 Top view of Fig. 4.30

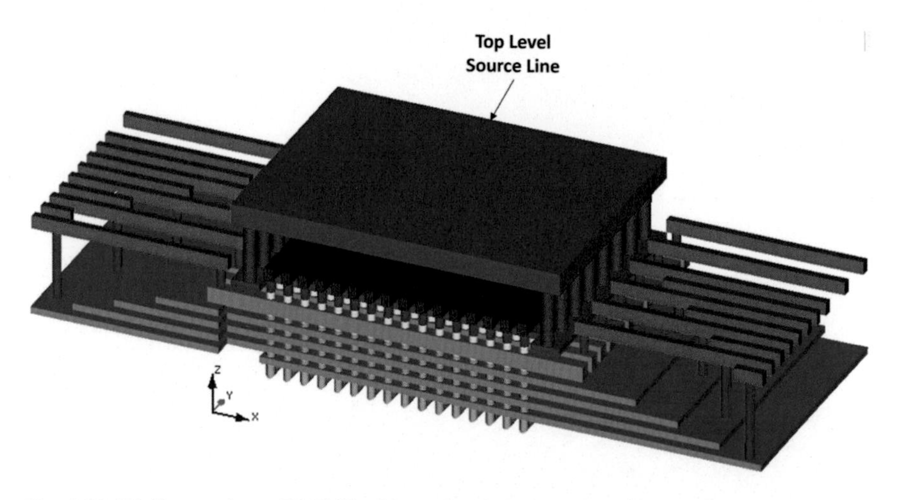

Fig. 4.32 Bird's-eye view of P-BiCS with top level source connection

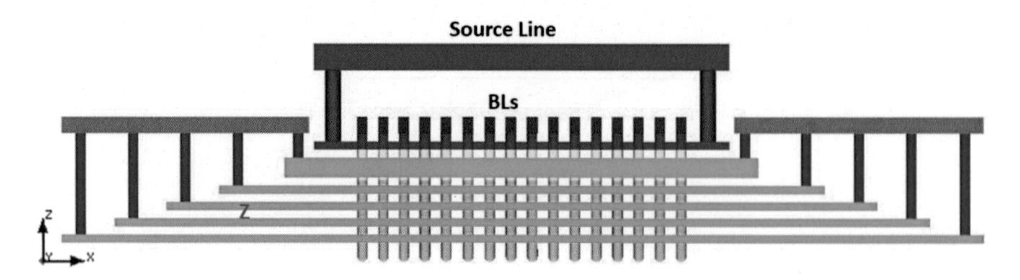

Fig. 4.33 X-Z cross section of Fig. 4.32

the one shown in Fig. 4.25. This additional layer is referred to as "Top Level Source Line" in Fig. 4.32. X-Z cross section and additional bird's-eye views are sketched in Figs. 4.33, 4.34 and 4.35.

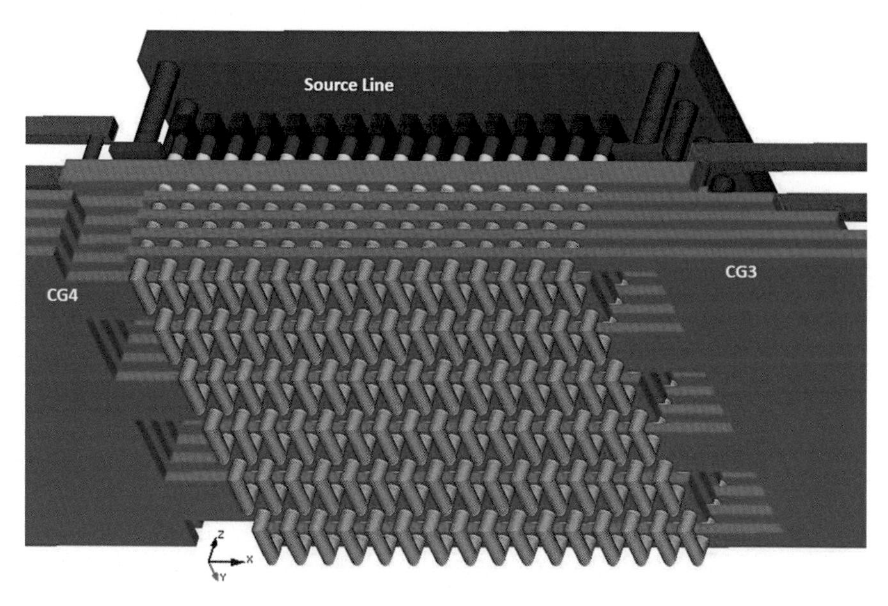

Fig. 4.34 Bottom view of Fig. 4.32

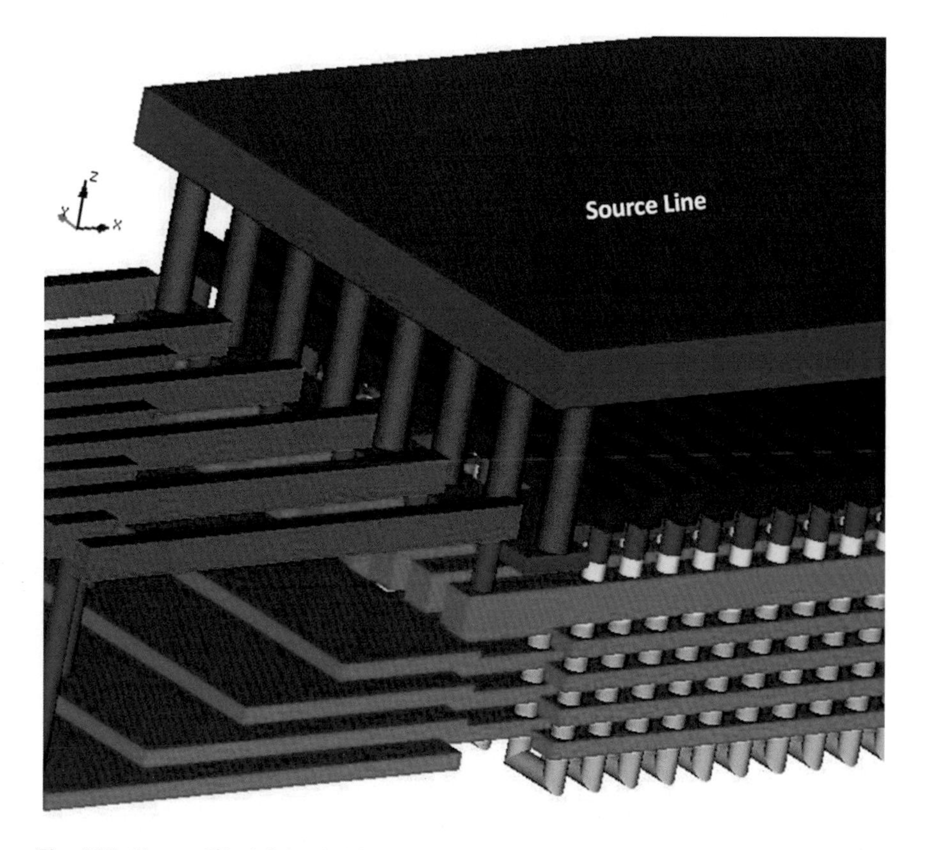

Fig. 4.35 Zoom of the left hand side of Fig. 4.32

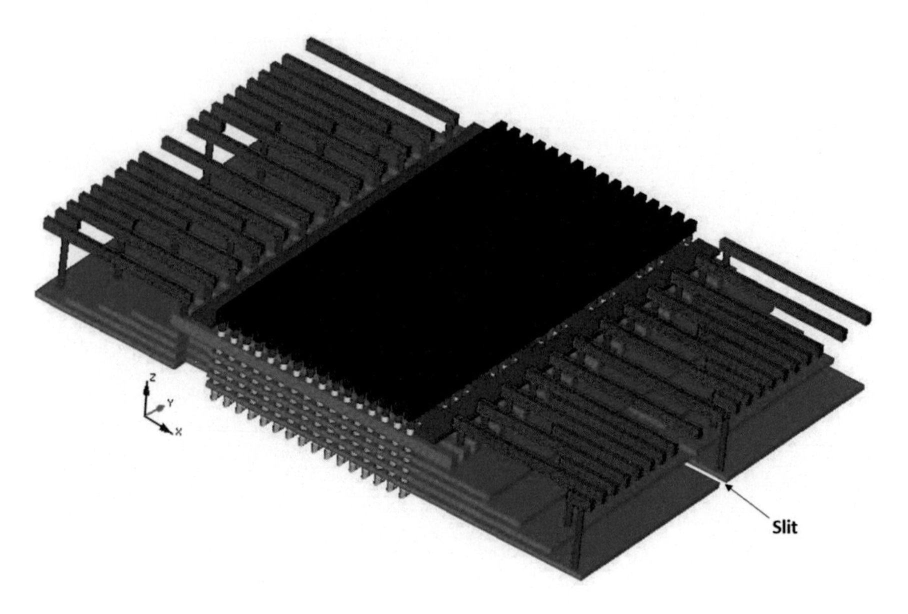

Fig. 4.36 P-BiCS array with slit

As we have seen with BiCS, it is important to regularly cut control gate plates by adding slits, in order to reduce program disturb and read disturb, as shown in Figs. 4.36 and 4.37.

Figure 4.38 is the top view of Fig. 4.36: it is clearly visible how the slit cuts the control gate plates in sub-plates. The corresponding Y-Z cross section is reported in Fig. 4.39.

4.4 VRAT and Z-VRAT

BiCS and P-BiCS were introduced by Toshiba. Samsung has followed a different path before reaching the architecture called V-NAND, which is the first 3D architecture with vertical channel brought to volume production. This section plus the following two describe the intermediate steps. Sect. 4.7 is devoted to V-NAND.

VRAT stands for *Vertical Recess Array Transistor* and was presented in 2008 [14]. Equivalent circuit schematic, vertical cross section and bird's-eye view of VRAT are shown in Fig. 4.40. Storage material is a nitride Charge Trap layer sandwiched, as usual, between tunnel and control oxides. 3D integration process for VRAT is called PIPE (*Planarized Integration on the same PlanE*) and the name has nothing to do with the pipe concept of P-BiCS. As this was one of the first attempts to build 3D structures, developing a simple unique integration scheme for vertical cells and interconnects was key; in other words, one of the most critical issues to

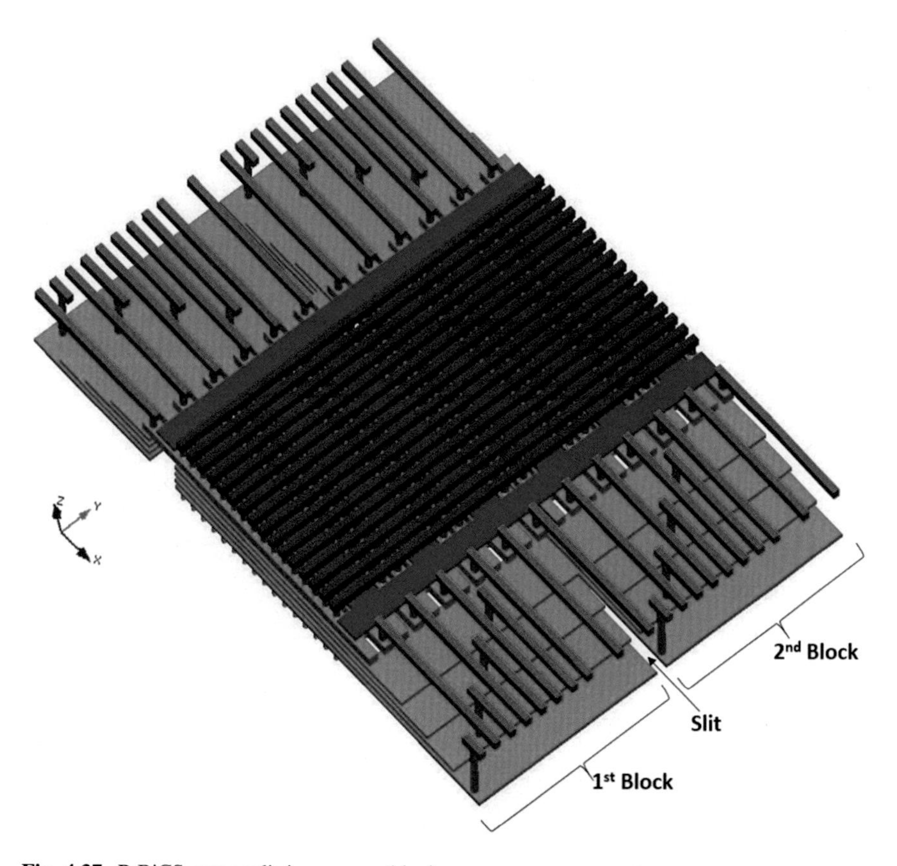

Fig. 4.37 P-BiCS array split in memory blocks

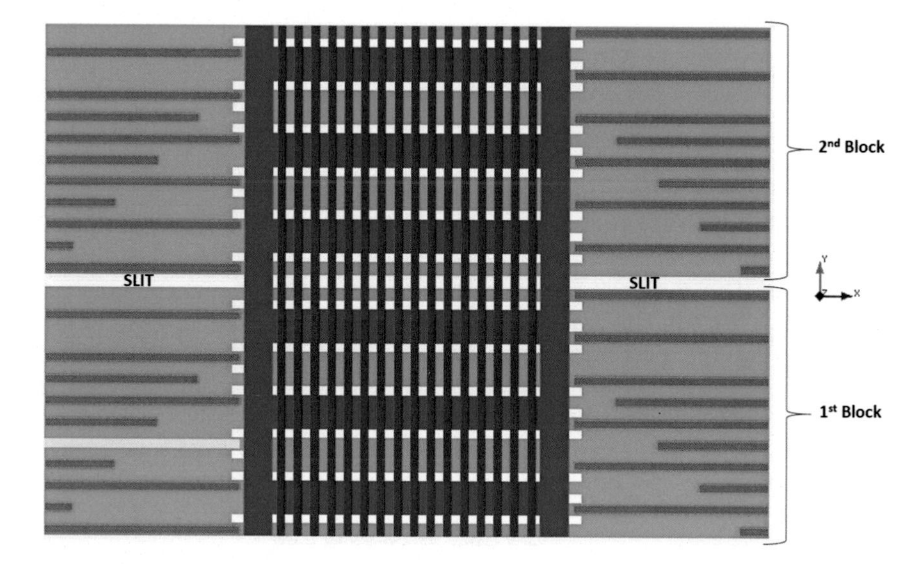

Fig. 4.38 Top view of Fig. 4.37

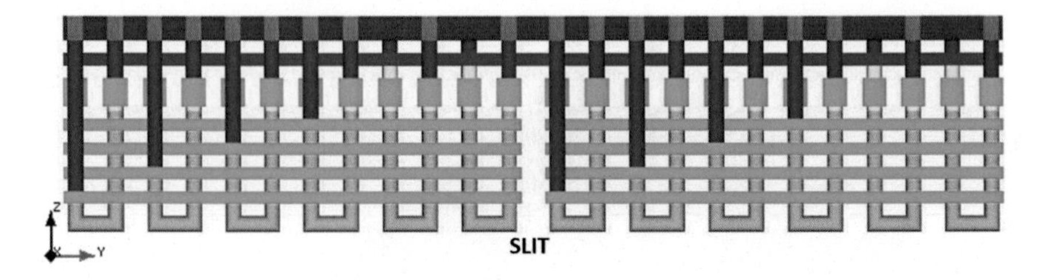

Fig. 4.39 Y-Z cross section of Fig. 4.37

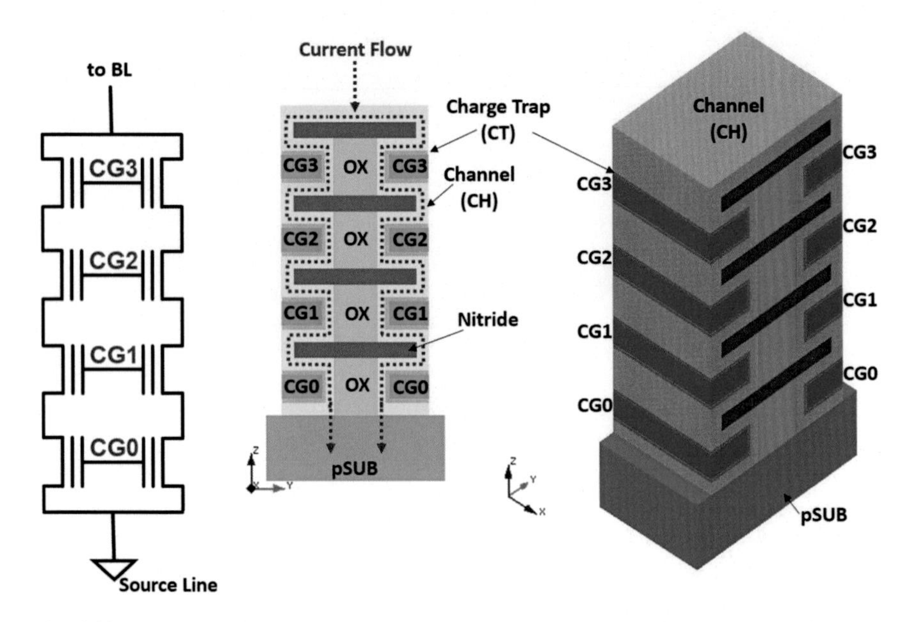

Fig. 4.40 VRAT NAND string

solve for 3D integration is how to connect memory cells from the matrix to the peripheral circuits.

Let's have a look at the main process steps for building VRAT structures. After deposition of multiple layers of nitride and oxide films, cell region is formed by wet etching process. Wordline electrode is deposited after cell's stack formation (oxides and CT material); at this point, an etch-back process removes the electrode from the sidewall, such that wordline layers are separated, as displayed in Fig. 4.40. At the end of the process, vertical strings are isolated (Fig. 4.41), and bitline and wordline contacts are added for connecting the array to the peripheral circuits (Fig. 4.42). Please note that wordline fan-out is not based on a stair-like structure, which can be seen as one of the merits of this solution; in fact, it doesn't require additional lithography steps (for staircase formation) and it is, therefore, cheaper (at least, in principle).

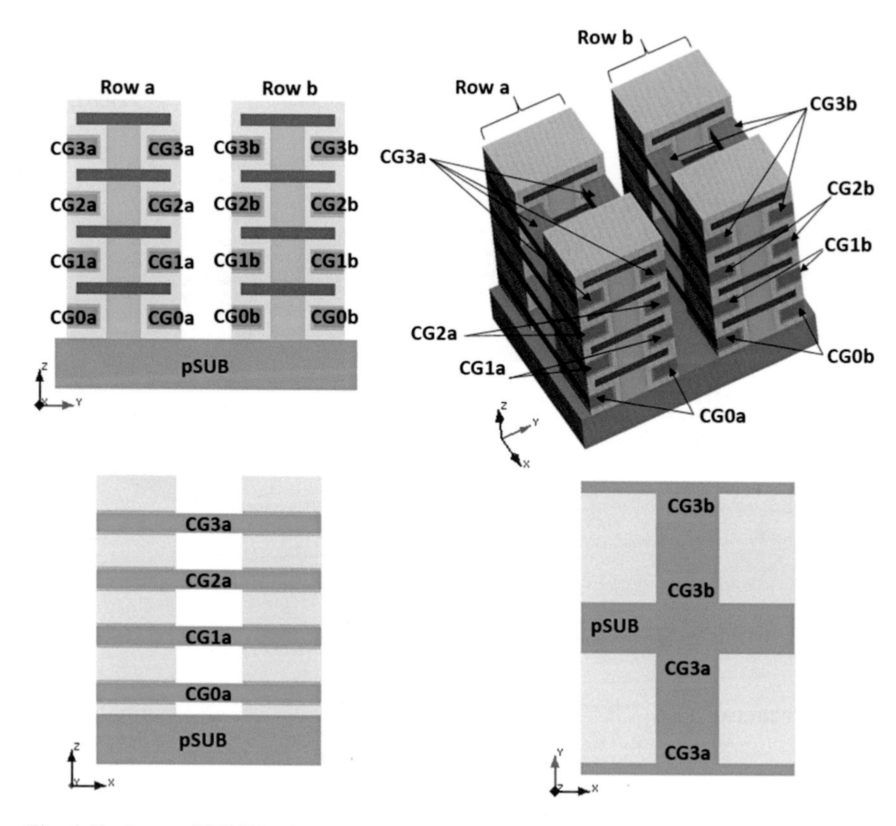

Fig. 4.41 Array of VRAT strings

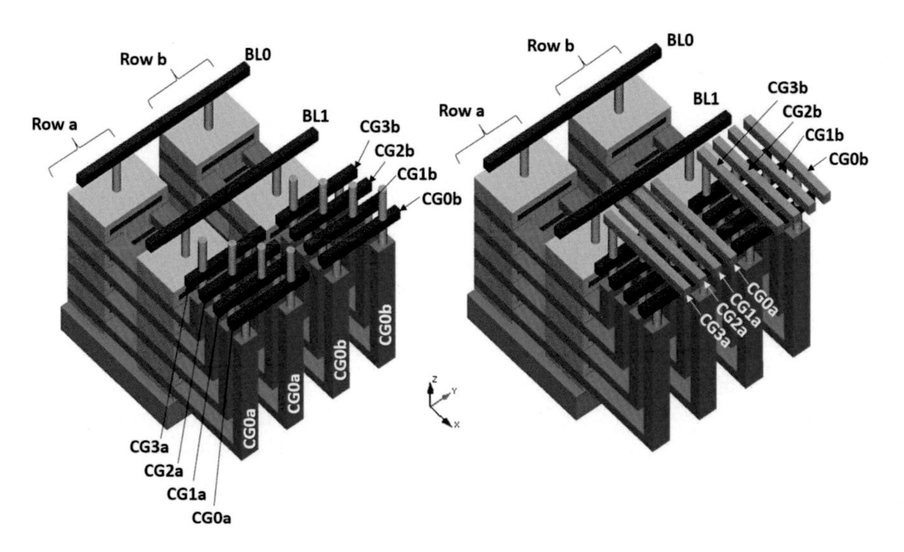

Fig. 4.42 VRAT array with BL and WL connections

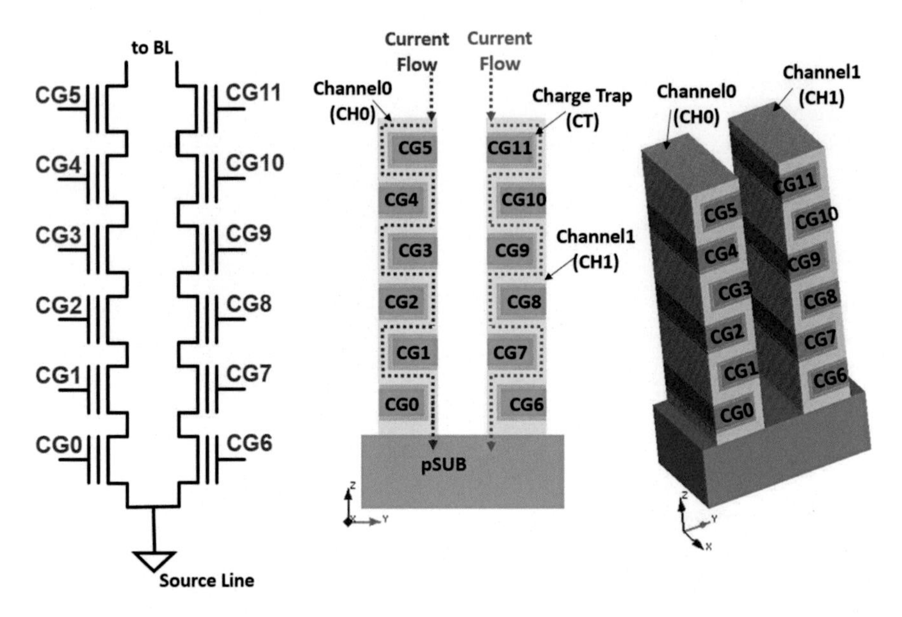

Fig. 4.43 Z-VRAT NAND strings

As a matter of fact, VRAT is a twin structure in the sense that current flows through 2 different paths because of the symmetry of the structure. Z-VRAT (*Zigzag* VRAT) is a further step towards a higher bit density [14]. In fact, each VRAT vertical structure is split into 2 narrower strings, as sketched in Figs. 4.43 and 4.44. The concept of the PIPE integration scheme stays intact; of course, the separation of the strings requires some additional process steps.

VRAT evolved in a different structure called VSAT, which is analyzed in the following section.

4.5 VSAT and A-VSAT

VRAT manufacturing is based on 2 basic concepts: gate-last and channel-first [15]. The challenge of this approach is in the undercut space for gate electrodes, which needs to be created and filled, thus requiring complex fabrication steps. VSAT (*Vertical Stacked Array Transistor*) revers the order: gate-first and channel-last. The basic building block of VSAT is displayed in Fig. 4.45. Let's look at the fabrication process. Multiple layers of polysilicon and nitride films are deposited one on top of each other; polysilicon acts as a gate, while nitride is the isolation material. The active region is defined after patterning the multiple layers, and a subsequent etching process. All gate electrodes are exposed on the same plane after a CMP process, thus allowing easy access to gate electrodes. Tunnel oxide, CT material,

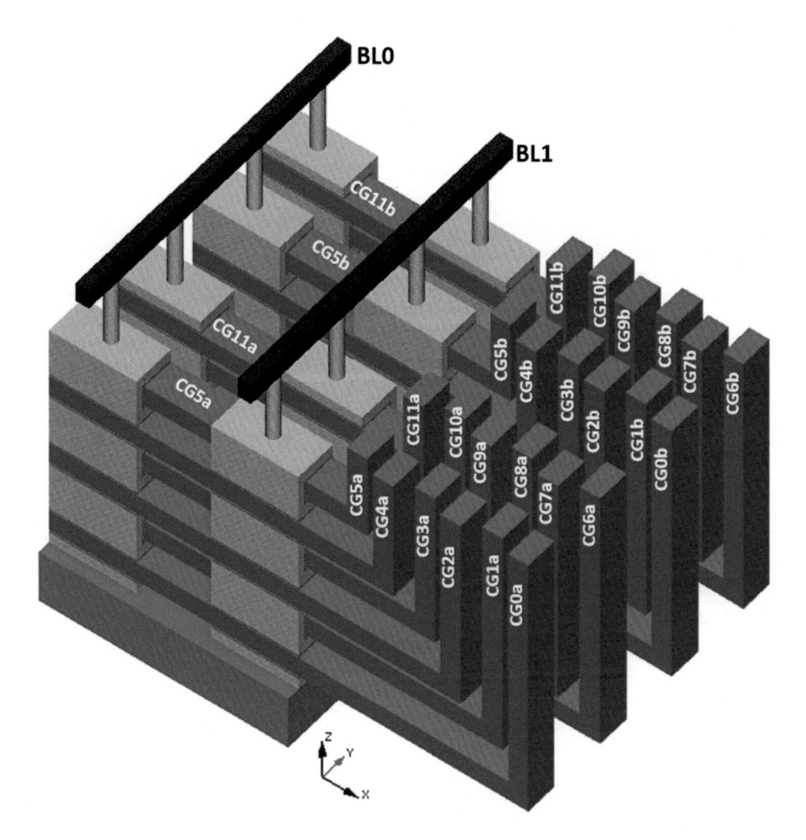

Fig. 4.44 Z-VRAT array with BL and WL connections

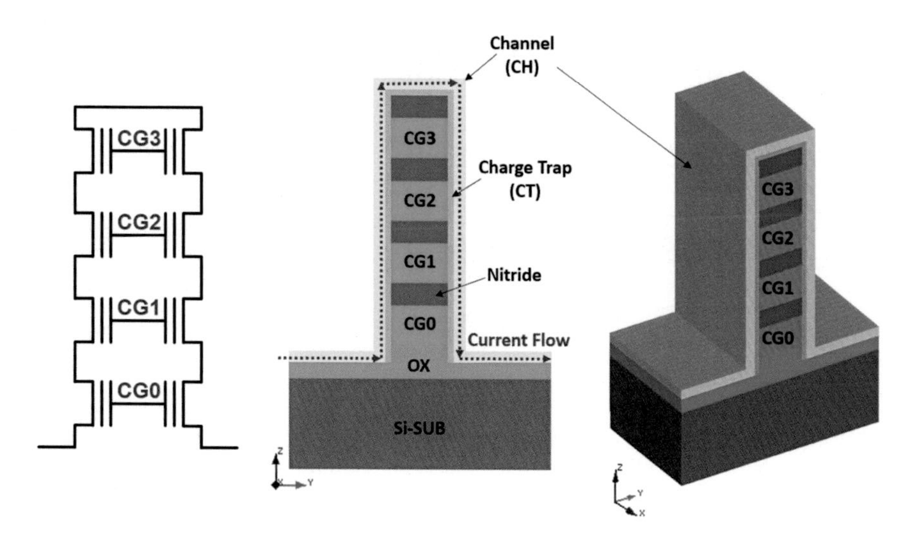

Fig. 4.45 VSAT basic building block

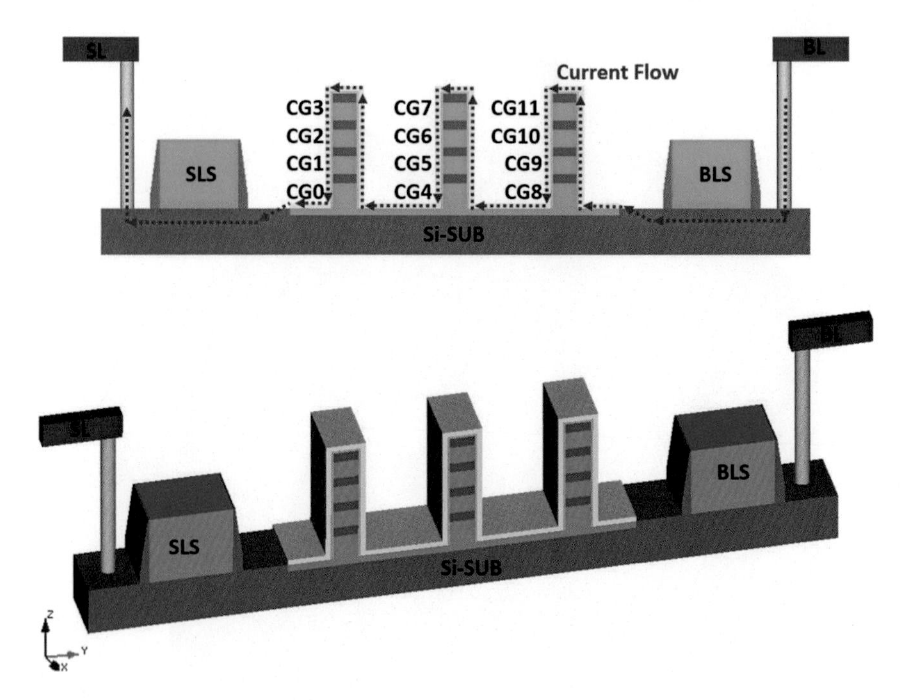

Fig. 4.46 VSAT NAND string

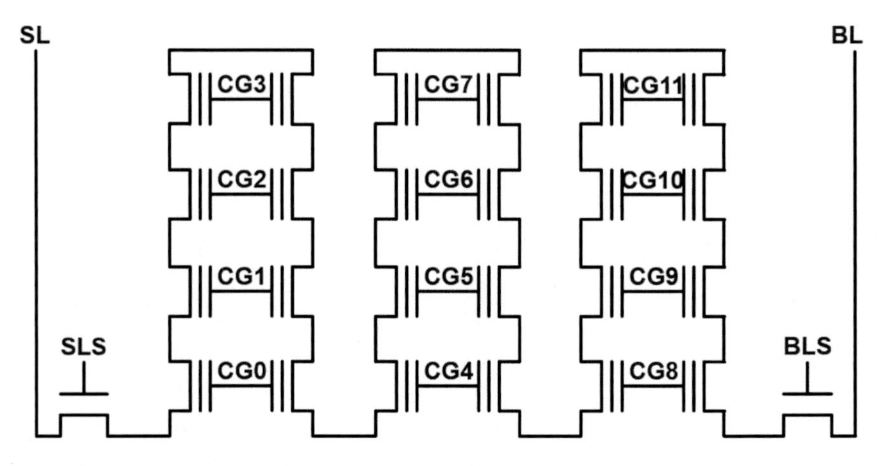

Fig. 4.47 Equivalent circuit of Fig. 4.46

and blocking oxide are then deposited on the active region. Channel is formed by a polysilicon deposition. Of course, final step is an etching process which is used to isolate vertical strings.

Multiple Fig. 4.45 can be combined together to form a NAND string, as shown in Figs. 4.46 and 4.47.

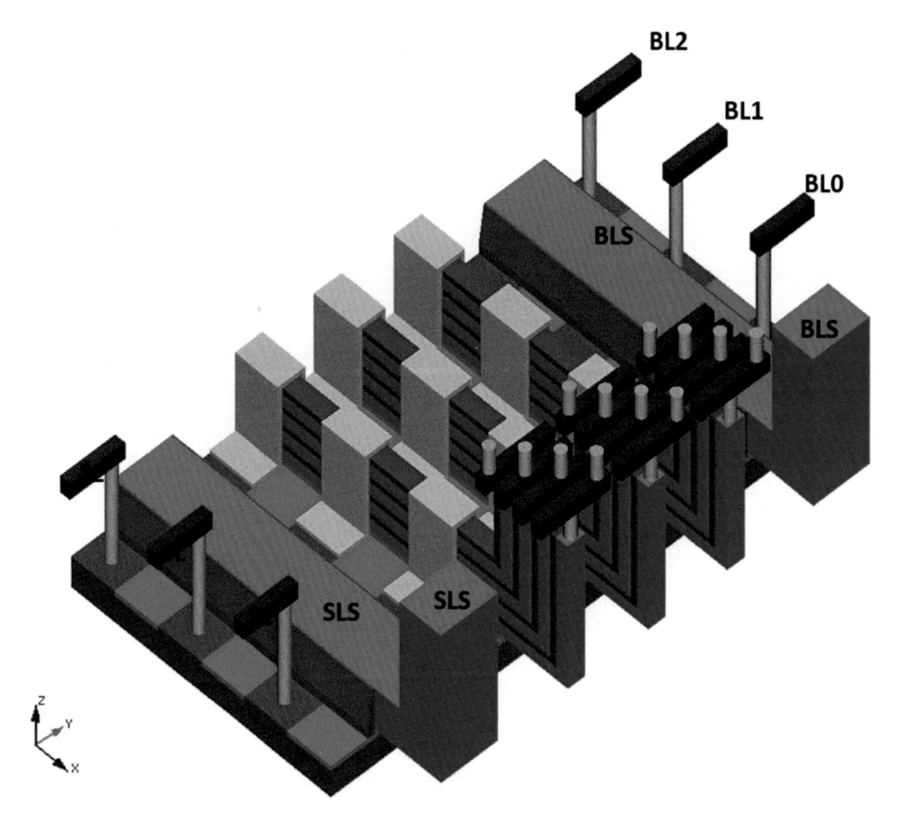

Fig. 4.48 VSAT fan-out

Vertical channel architectures usually have SL selectors at the bottom of the 3D stack; as we have seen in previous sections, this poses some challenges as SLS might have different gate oxide, gate length, and doping density than the memory cells along the string. Figure 4.46 shows that, in VSAT, SL selectors are not part of the vertical string as they are located in the peripheral region, like in the 2D case, thus simplifying the whole manufacturing process. All the merits of the PIPE process in terms of fan-out remain unchanged, as clearly visible in Fig. 4.48.

By comparing the schematic of VRAT and VSAT, we can notice that they are both "twin" cells in the sense that they are split in 2 parts. The difference is that in VRAT the 2 halves are in parallel, while they are in series in VSAT. In other words, this is like saying that in VSAT the current flows through the "same" cell twice, thus degrading the bit density. To solve this problem, an additional wordline cut process was proposed by Macronix in 2015 [16]. This improved vertical architecture is called A-VSAT, which stands for *Asymmetrical VSAT*: basic building block and A-VSAT NAND string are shown in Figs. 4.49 and 4.50, respectively.

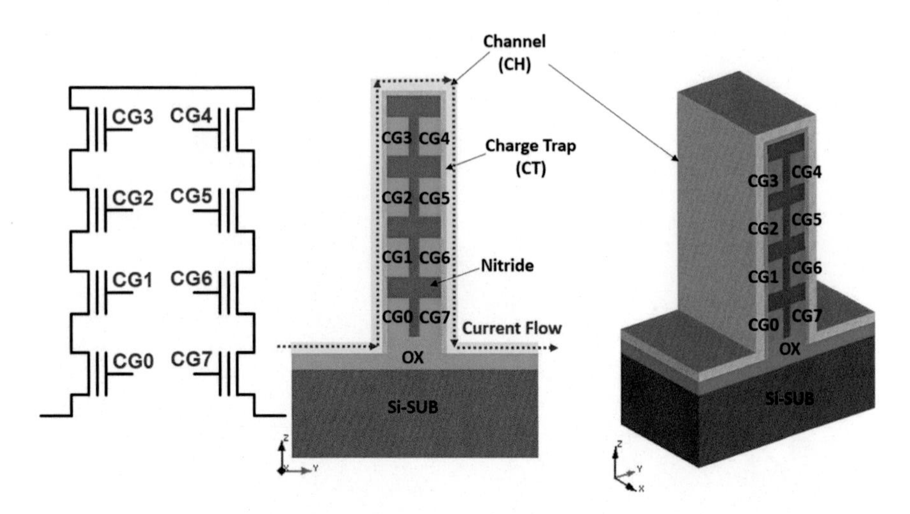

Fig. 4.49 A-VSAT basic building block

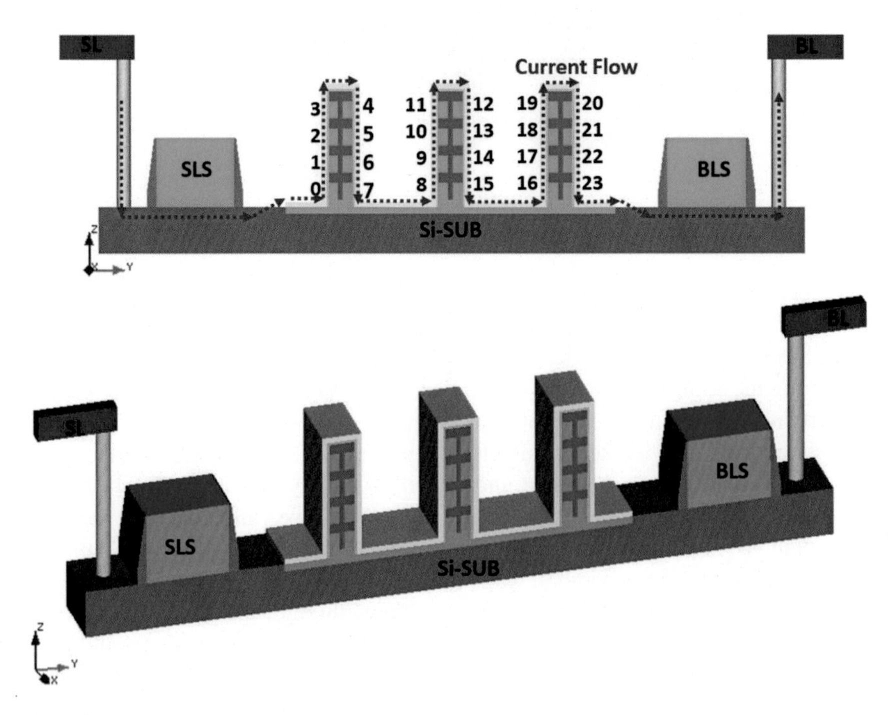

Fig. 4.50 A-VSAT NAND string

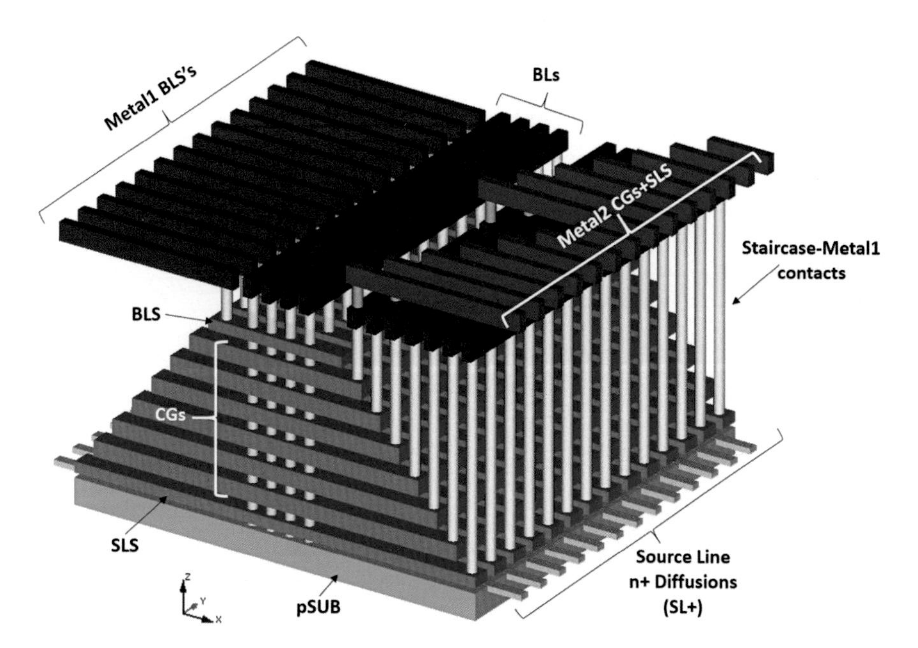

Fig. 4.51 TCAT NAND flash array

4.6 TCAT

TCAT (Terabit Cell Array Transistor) was proposed in 2009 [17]. A bird's-eye view of the TCAT array is presented in Fig. 4.51. The equivalent circuit is sketched in Figs. 4.52 and 4.53. Besides SL+ lines, the TCAT equivalent circuit is identical to the BiCS equivalent circuit shown in Fig. 4.4. All SL+ lines (which are n+ diffusions) are shorted together outside the array to form a common Source Line. Two level of metals are used to fan-out BLS's and CGs+SLS, respectively. Top and side views of TCAT can help understanding better the overall architecture (Figs. 4.54, 4.55, and 4.56).

There are 6 CG layers and 2 NAND blocks. Each block has 7 wordlines per layer. These 7 wordlines are shorted together at the Metal1 level, as displayed in Fig. 4.57. In order to clearly understand how many contacts are underneath the Metal1 layer, another layer removal is needed, as shown in Fig. 4.58. Metal2 is used for wordline decoding, while Metal1 is used to decode the NAND string. It is worth highlighting that, compared to BiCS, in the TCAT architecture the slit is a cut in the bitline layer, as it can be seen in Fig. 4.57; in fact, wordlines are already separated by construction (Fig. 4.59).

After the review of the architectural aspects of TCAT, we can now look at the differences with respect to BiCS. First of all, TCAT makes use of *gate-replacement* [17]; in other words, the gate layer is deposited only at the end of the stack formation, while BiCS is based on a gate-first process. Let's briefly analyze the

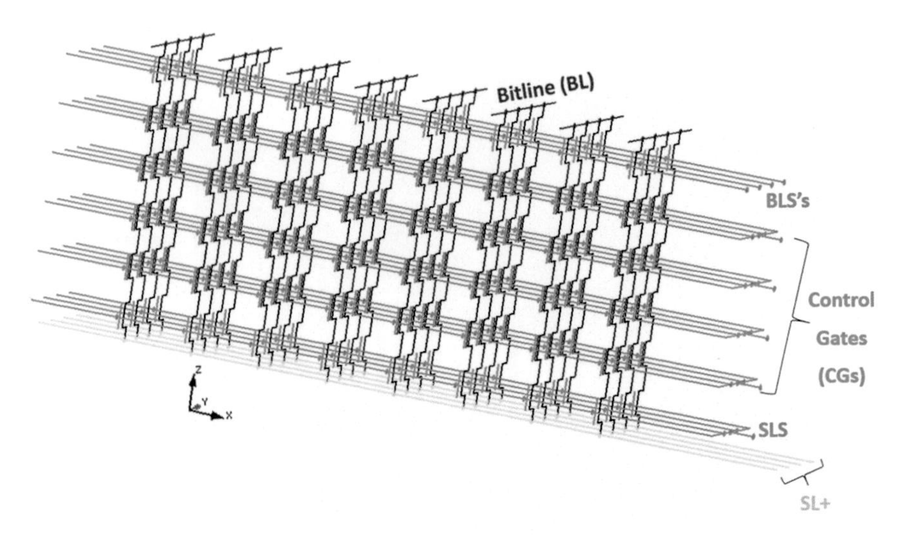

Fig. 4.52 TCAT circuit schematic

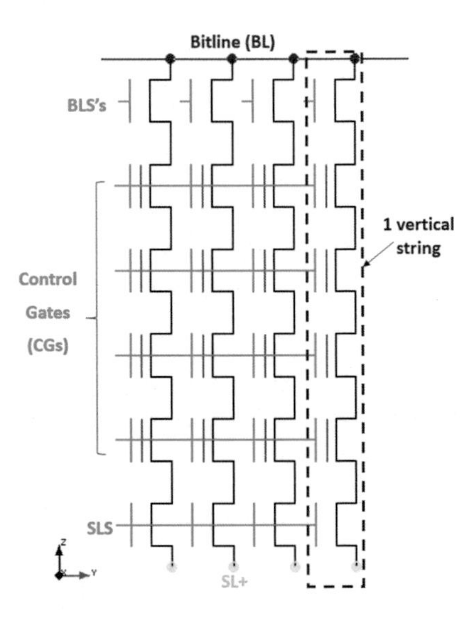

Fig. 4.53 Y-Z view of TCAT circuit schematic

gate-replacement technology. Multiple layers of oxide and sacrificial nitride layers are initially deposited. The whole stack is then etched between each row of pillars (holes) and nitride is removed. At this point, gate dielectric layers, and metal gates are deposited in the conventional order by filling the space between wordlines with tungsten. After that, an etching process is used to separate the gates. In this way, a metal gate SONOS memory cell is fabricated; this kind of cells allows faster erase speed, longer retention, and a wider V_{TH} window. Of course, the other big advantage of the metal gate is the reduced parasitic resistance of the wordline, which translates in faster operations.

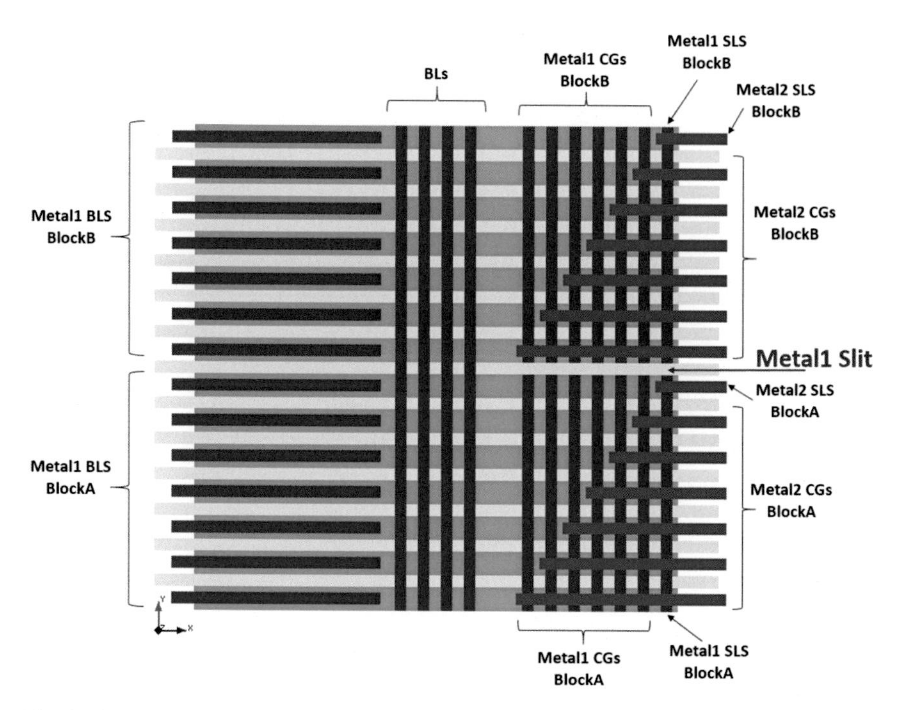

Fig. 4.54 Top view of Fig. 4.51

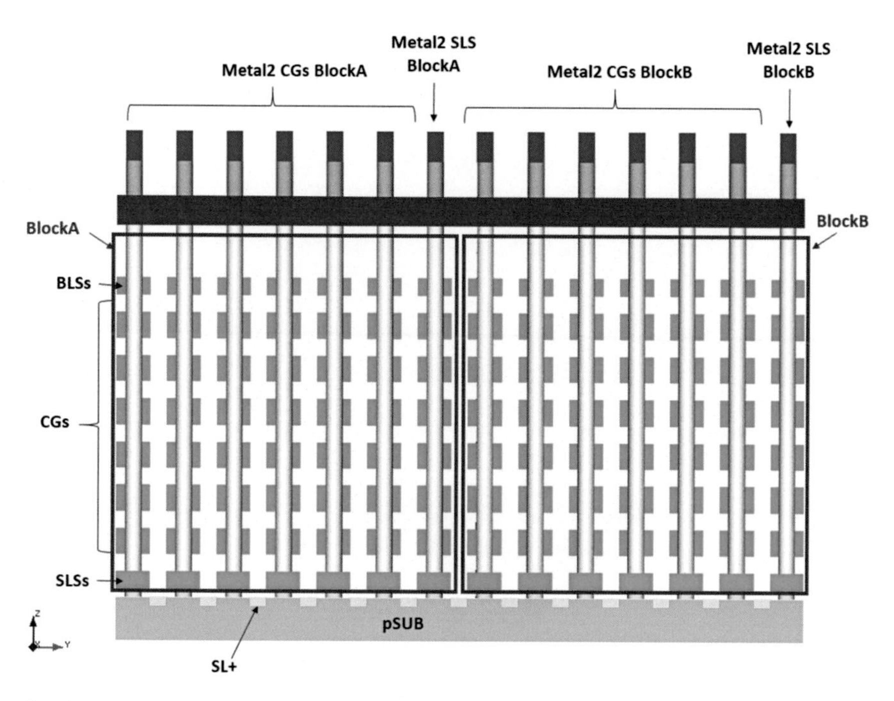

Fig. 4.55 Z-Y side view of Fig. 4.51

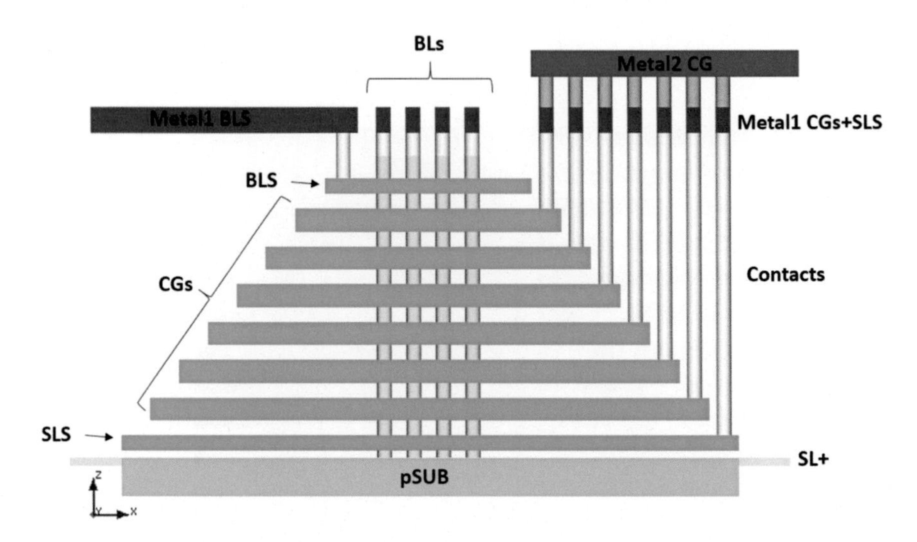

Fig. 4.56 Z-X side view of Fig. 4.51

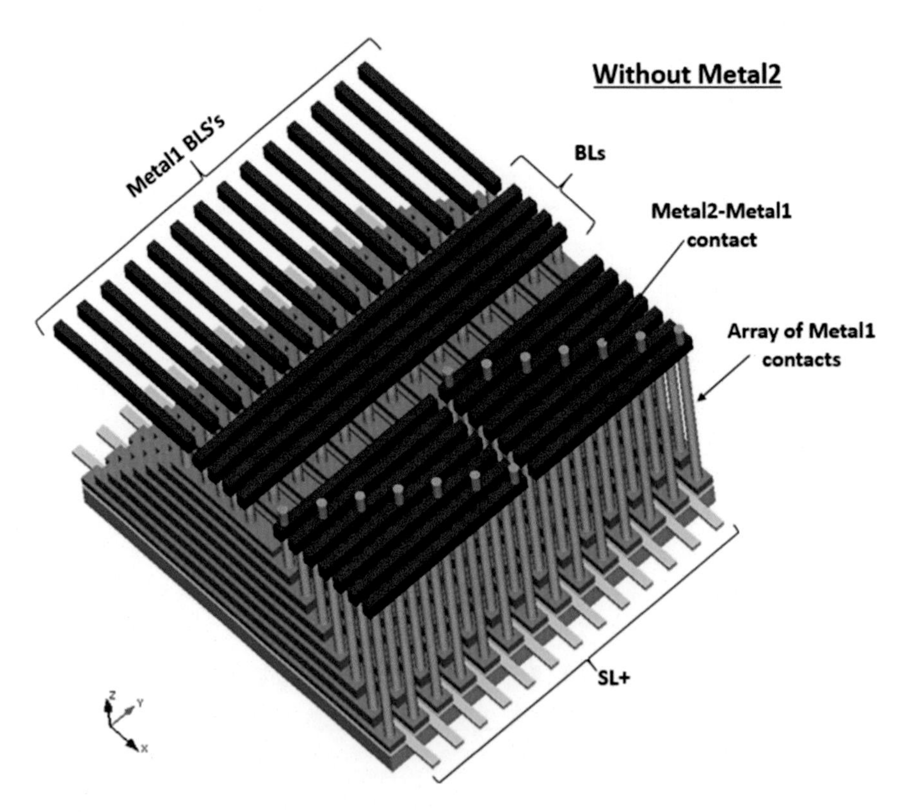

Fig. 4.57 TCAT NAND array without Metal2 layer

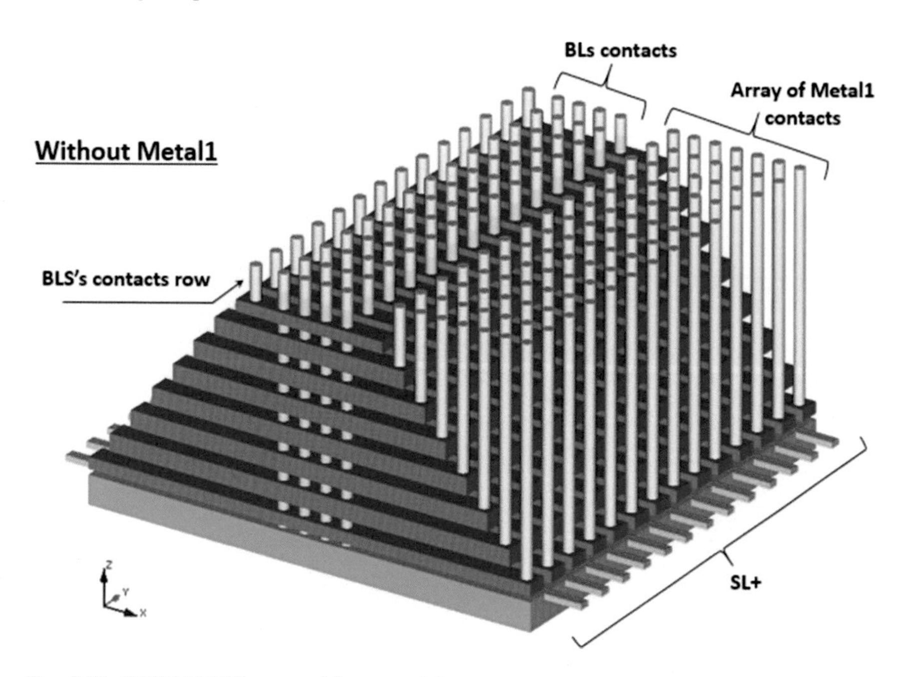

Fig. 4.58 TCAT NAND array without metal layers

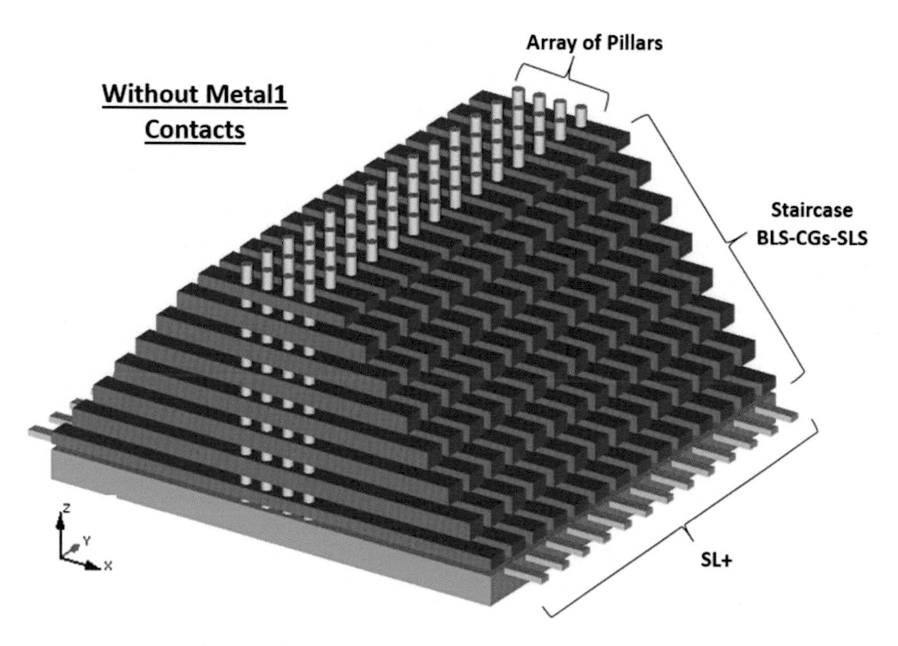

Fig. 4.59 TCAT NAND array without metal layers and bitline contacts

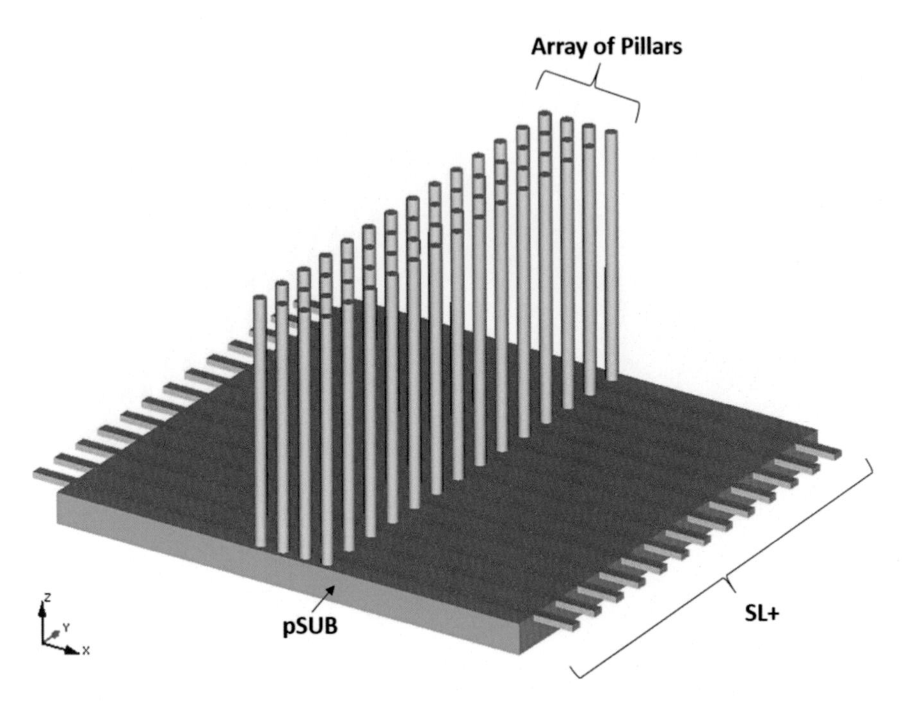

Fig. 4.60 TCAT pillars

Another difference is the bulk erase operation. As shown in Fig. 4.60, the vertical pillar is directly connected to the p-substrate and not to a n+ diffusion. Close to each NAND string there is a n+ region used to drain the cell's current. During erase, holes are directly provided by the substrate, without the need for GIDL (*Gate Induced Drain leakage*) generation at SLS, which is one of the main concerns of the BiCS architecture.

Last but not least, we can take a look at the NAND memory cell: a direct comparison between BiCS and TCAT is reported in Fig. 4.61. Because of the gate-last process adopted by TCAT, the charge trap layer is biconcave, which results in a reduced charge spreading effect. In fact, in a string of the conventional 3D BiCS, the charge trap nitride layer is continuously connected across all gates along the channel side and, as a matter of fact, it acts as a charge spreading path. As discussed in Chap. 2, this causes degradation of data retention characteristics. Thanks to the biconcave shape of the charge trap layer, in the TCAT case it is harder for electrical charges to move from one cell to another, as there is not straight connection of the charge trap layers between them [18].

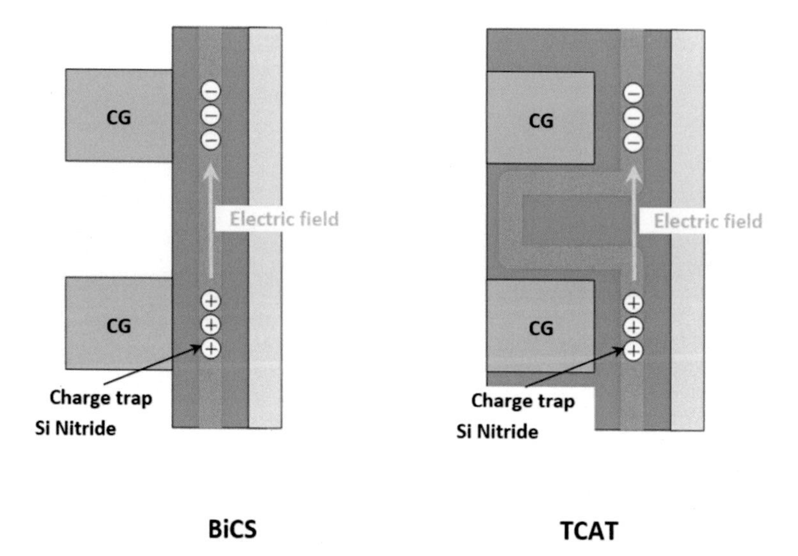

Fig. 4.61 BiCS and TCAT flash memory cells

4.7 V-NAND

TCAT turned out to be the foundation for the first 3D NAND architecture brought to volume production, which is called V-NAND. This technology was launched at Flash Memory Summit 2013 [19]. As summarized in Fig. 4.62, Samsung officially introduced the first generation of V-NAND in 2014 [20–22]. The first product was a 128 Gb MLC NAND, based on damascened metal-gate SONOS-type (CT) cell. The vertical stack of V-NAND Gen1 is made of 24 wordline layers, whose equivalent circuit is shown in Figs. 4.63 and 4.64. In terms of schematic diagram, the main difference with respect to Figs. 4.4 and 4.53 is the presence of dummy wordline layers (dummy CG).

In 3D architectures with vertical channel, memory cells have floating body. During programming, high channel boosting can generate hot carries at the edge of the string, because of the high lateral electric field. As such, the channel potential does not boost up as it should when WL0 is programmed (i.e. hot carriers discharge the channel). Of course, this behavior translates into Program Disturb. Therefore, dummy WLs are inserted between the selection transistors and WLs to prevent the above mentioned effect [23, 24].

To save silicon area, V-NAND Gen1 adopts a special layout for the array of pillars, known as *Staggered Pillars* (or Holes). In practice, even and odd rows of pillars are staggered, without changing the center-to-center distance between adjacent pillars. With reference to the left hand side of Fig. 4.65, please note that in Gen1 each bitline has to fit in 1 channel hole (pillar) pitch.

In V-NAND Gen2 layout of bitlines is different, as sketched on the right hand side of Fig. 4.65: we refer to this layout as *Staggered Bitline Contacts*. In this case,

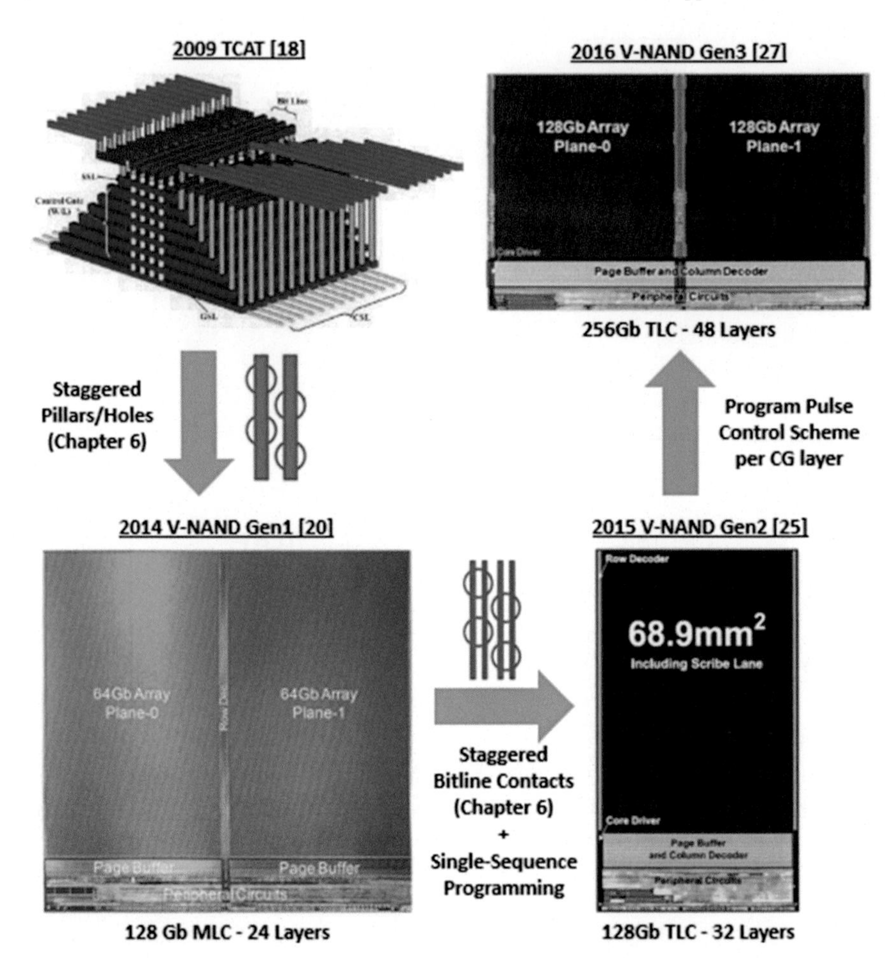

Fig. 4.62 From TCAT to V-NAND (pictures not in scale) [17, 20, 25, 27]

2 bitlines are arranged in 1 pillar pitch [25]. Of course, BL density doubles (NAND page size goes from 8 to 16 kB), but the number of contacts to the bottom Source Line plate is halved. The overall number of pillars is the same of the previous generation. Figure 4.66 displays a side by side comparison of the array cross sections of the 2 generations (not in scale).

The reader can find a detailed explanation of the above mentioned layout techniques in Chap. 6, which is devoted to the analysis of the most advanced 3D architectures for vertical channel Flash arrays.

V-NAND Gen2 was presented in 2015 [25, 26] in the form of 128 Gb TLC. Compared to the previous generation, there hasn't been macroscopic changes in the memory cell itself, but it is worth mentioning that the number of control gate layers is 32 instead of the previous 24.

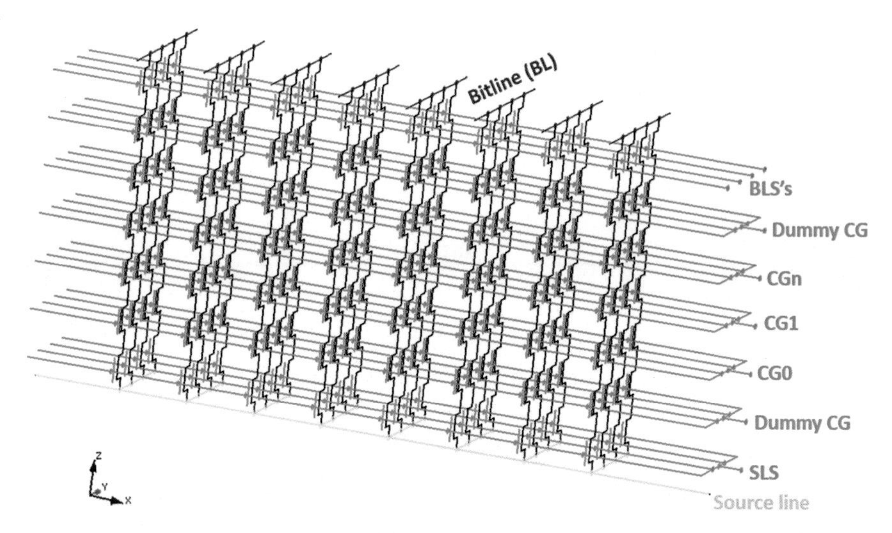

Fig. 4.63 Circuit schematic of a 3D V-NAND array

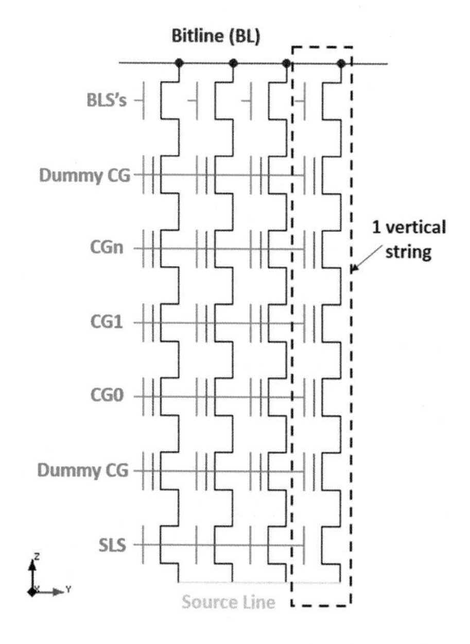

Fig. 4.64 Y-Z view of V-NAND circuit schematic

The other highlight of this device is the Single-Sequence Programming operation. TLC functionality (i.e. 3 bit/cell) requires the ability of squeezing 8 V_{TH} distributions instead of 4, within almost the same V_{TH} window. In planar NAND, conventional TLC programming algorithms are based on repeating the programming sequence 3 times per wordline: V_{TH} distributions become smaller and smaller after each sequence. In V-NAND Gen2, tight cell's V_{TH} distribution is achieved by

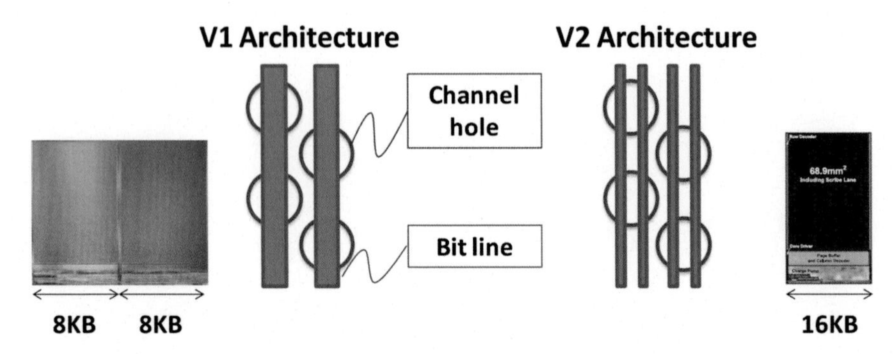

Fig. 4.65 First and second generations of V-NAND bitline architectures

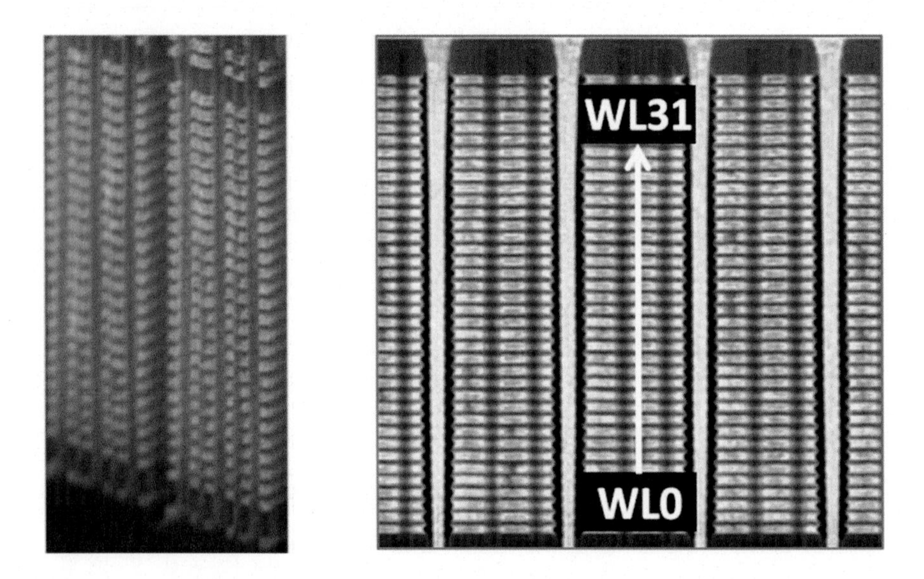

Fig. 4.66 Cross section of V-NAND Gen1 (*left*) and Gen2 (*right*), not in scale

exploiting the advantages of CT, i.e. small cell-to-cell interference and narrow natural V_{TH} distribution. This is coupled with an improved programming algorithm: V-NAND Gen2 asks for the 3 TLC pages of data at the start of programming and it writes the 3 pages at once. Of course, this approach translates into faster programming operations and lower power consumption.

V-NAND Gen3 becomes public at the *IEEE International Solid State Circuits Conference* (ISSCC) in 2016 [27]. It is again a TLC device, but this time it is a 256 Gb based on a vertical stack of 48 control gate layers. When increasing the number of layers, the etching technology can become a serious issue because of the aspect ratio of the pillar. Therefore, it is almost unavoidable to reduce the thickness of each single control gate layer. While the overall stack manufacturability benefits from thinner layers, the situation is totally different for wordlines: in fact, parasitic

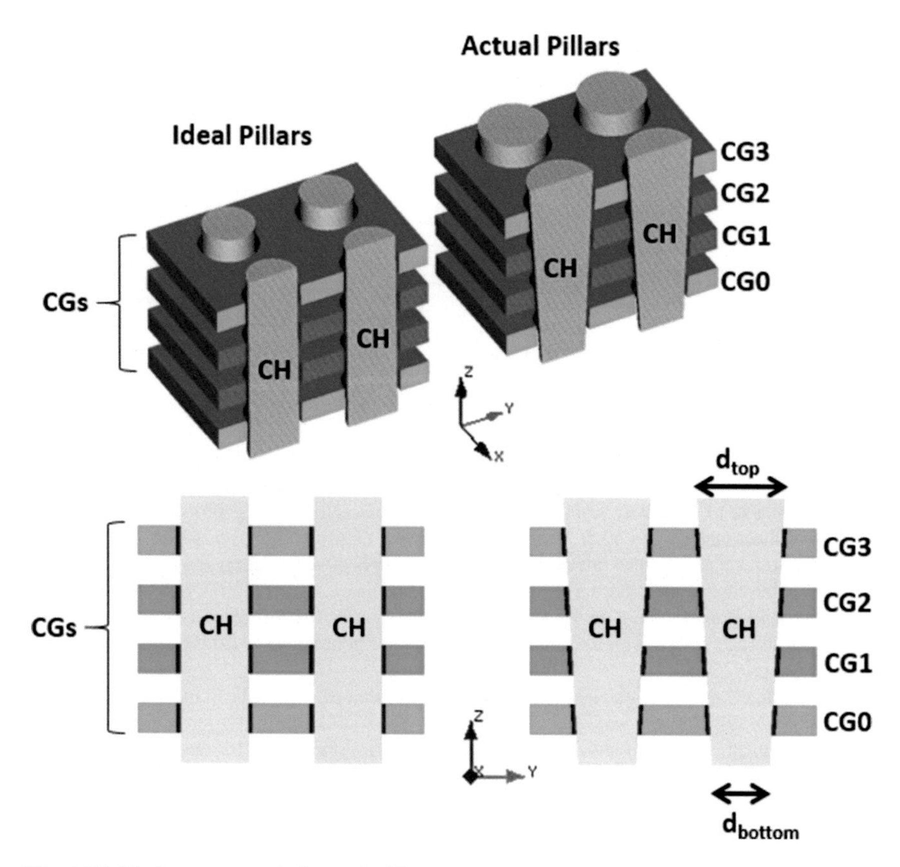

Fig. 4.67 Ideal versus actual shape of pillars

resistance and capacitance increase, thus slowing down both read and program operations. Not only that. Because of the increased resistance, channel hole size fluctuations become even more important. As a matter of fact, channel hole can be seen as a physical barrier to the charge flow along the wordline: a variation of the hole size translates into a variation of the parasitic resistance. This fact needs to be coupled with the actual shape of the pillar (Chap. 2), which is sketched in Fig. 4.67. As a result, the voltage transient is a function of the layer. By using the conventional approach of applying the same program step duration to each single wordline (Chap. 3), the NAND raw BER would become higher for the slower wordlines. As such, an adaptive program pulse scheme is adopted: basically, it varies the program pulse duration according to the target WL's characteristics.

As the number of layers in the 3D NAND stack grows up, we expect these non-uniformity effects to become more and more the focus for circuit designers and process technology engineers.

At this point the reader should have all the architectural elements to understand the challenges of stacking several memory layers, one on top of each other.

Even though BiCS and V-NAND with Charge Trap storage have been a major breakthrough in going to 3D, they are not the only possible approach, and Floating Gate is still alive. The 3D world is full of options! In the next chapter we'll see many different types of 3D Flash memory cells based on Floating Gate.

References

1. http://www.samsung.com/us/business/oem-solutions/pdfs/V-NAND_technology_WP.pdf. Samsung V-NAND technology, White Paper, Sept 2014
2. R. Micheloni, L. Crippa, Chapter 3, Multi-bit NAND flash memories for ultra high density storage devices, in *Advances in Non-volatile Memory and Storage Technology*, ed. by Y. Nishi (Woodhead Publishing, Sawston, 2014)
3. R. Micheloni et al., Chapter 7, High-capacity NAND flash memories: XLC storage and single-die 3D, in *Memory Mass Storage*, ed. by G. Campardo et al. (Springer, Berlin, 2011)
4. H. Tanaka et al., Bit cost scalable technology with punch and plug process for ultra high density flash memory, in *VLSI Symposium Technical Digest* (2007), pp. 14–15
5. Y. Fukuzumi et al., Optimal integration and characteristics of vertical array devices for ultra-high density, bit-cost scalable flash memory, in *IEDM Technical. Digest* (2007), pp. 449–452
6. M. Ishiduki et al., Optimal device structure for pipe-shaped BiCS flash memory for ultra high density storage device with excellent performance and reliability, in *IEDM Technical Digest* (2009), pp. 625–628
7. T. Maeda et al., Multi-stacked 1G cell/layer pipe-shaped BiCS flash memory, in *Digest Symposium on VLSI Circuits*, June 2009, pp. 22–23
8. R. Katsumata et al., Pipe-shaped BiCS flash memory with 16 stacked layers and multi-level-cell operation for ultra high density storage devices, in *2009 Symposium on VLSI Technology* (2009), pp. 136–137
9. H. Aochi, BiCS flash as a future 3-D non-volatile memory technology for ultra high density storage devices, in *Proceedings of International Memory Workshop* (2009), pp. 1–2
10. Y. Yanagihara et al., Control gate length, spacing and stacked layers number design for 3D-Stackable NAND flash memory, in *IEEE IMW* (2012), pp. 84–87
11. K. Takeuchi, Scaling challenges of NAND flash memory and hybrid memory system with storage class memory and NAND flash memory, in *IEEE Custom Integrated Circuits Conference (CICC)* (2013), pp. 1–6
12. A. Nitayama et al., Bit Cost Scalable (BiCS) flash technology for future ultra high density storage devices, in *2010 International Symposium on VLSI Technology Systems and Applications (VLSI TSA),* Apr. 2010, pp. 130–131
13. Y. Komori et al., Disturbless flash memory due to high boost efficiency on BiCS structure and optimal memory film stack for ultra high density storage device, in *IEDM Technical Digest* (2008), pp. 851–854
14. J. Kim et al., Novel 3-D structure for ultra high density flash memory with VRAT (vertical-recess-array-transistor) and PIPE (planarized integration on the same plane), in *2008 IEEE Symposium on VLSI Technology* (2008)
15. J. Kim et al., Novel vertical-stacked-array-transistor (VSAT) for ultra-high-density and cost-effective NAND flash memory devices and SSD (solid state drive), in *2009 IEEE Symposium on VLSI Technology* (2009)
16. Y.-H. Hsiao, Ultra-high bit density 3D NAND flash-featuring-assisted gate operation. IEEE Elect. Dev. Lett. **36**(10), 1015–1017 (2015)
17. J. Jang et al., Vertical cell array using TCAT (terabit cell array transistor) technology for ultra high density NAND flash memory, in *2009 IEEE Symposium on VLSI Technology* (2009)

18. W. Cho et al., Highly reliable vertical NAND technology with biconcave shaped storage layer and leakage controllable offset structure, in *2010 Symposium on VLSI Technology (VLSIT)* (2010), pp. 173–174
19. J. Elliott, E.S. Jung, Ushering in the 3D memory era with V-NAND, in *Proceedings of Flash Memory Summit*, www.flashmemorysummit.com, Santa Clara, CA, Aug 2013
20. K.-T. Park, Three-dimensional 128 Gb MLC vertical NAND flash memory with 24-WL stacked layers and 50 MB/s high-speed programming, in *IEEE ISSCC, Digest Technical Papers*, Feb 2014, pp. 334–335
21. K.-T. Park, Three-dimensional 128 Gb MLC vertical NAND flash memory with 24-WL stacked layers and 50 MB/s high-speed programming. IEEE J. Solid-State Circ. **50**(1), (2015)
22. K.T. Park, A world's first product of three-dimensional vertical NAND flash memory and beyond, in *NVMTS*, 27–29 Oct 2014
23. E. Choi et al., Device considerations for high density and highly reliable 3D NAND flash cell in near future, in *IEEE International Electron Devices Meeting* (2012), pp. 211–214
24. K. Shim et al., Inherent issues and challenges of program disturbance of 3D NAND flash cell, in *IEEE International Memory Workshop* (2012), pp. 95–98
25. J.-W. Im, 128 Gb 3b/cell V-NAND flash memory with 1 Gb/s I/O rate, in *IEEE International Solid-State Circuits Conference*, Feb 2015, pp. 130–131
26. J.-W. Im, 128 Gb 3b/cell V-NAND flash memory with 1 Gb/s I/O rate. J. Solid-State Circ. **51**(1) (2016)
27. D. Kang et al., 256 Gb 3b/Cell V-NAND flash memory with 48 stacked WL layers, in *IEEE International Solid-State Circuits Conference (ISSCC), Digest Technical Papers*, Feb 2016, pp. 130–131

Chapter 5
3D Floating Gate NAND Flash Memories

Rino Micheloni and Luca Crippa

5.1 Introduction

Planar NAND Flash memories (commercially available) are based on Floating Gate, which has been developed and engineered for many decades. Therefore, there have been many attempts to develop 3D Floating Gate cells in order to re-use all the know-how cumulated over time. Figure 5.1 is a summary of the Floating Gate variants based on vertical channel described in the present chapter. There is also one FG NAND string with horizontal channel, which is briefly described in Sect. 5.7.

5.2 Conventional Floating Gate (C-FG) Flash Cell

The first 3D vertical NAND structure based on Floating Gate was proposed back in 2001 [1]. This array is built around a 3D vertical FG cell, which is usually referred to as *3D Conventional FG* (C-FG) or S-SGT (*Stacked-Surrounding Gate Transistor*) [1–3]. The basic cell is shown in Fig. 5.2. Floating Gate (FG) and Control Gate (CG) completely surround the vertical *Channel* (CH) ; *Tunnel Oxide* (TOX) and *Inter Poly Dielectric* (IPD) complete the floating gate structure. Figure 5.3 offers top and side views of Fig. 5.2. X–Y and X–Z cross sections are sketched in Fig. 5.4.

Single memory cells can be vertically stacked to form a NAND Flash string, as shown in Fig. 5.5. For sake of simplicity, drawings are based on a stack of 6

R. Micheloni (✉) · L. Crippa
Performance Storage BU, Microsemi Corporation, Vimercate, Italy
e-mail: rino.micheloni@ieee.org

L. Crippa
e-mail: luca.crippa@ieee.org

R. Micheloni (ed.), *3D Flash Memories*, DOI 10.1007/978-94-017-7512-0_5

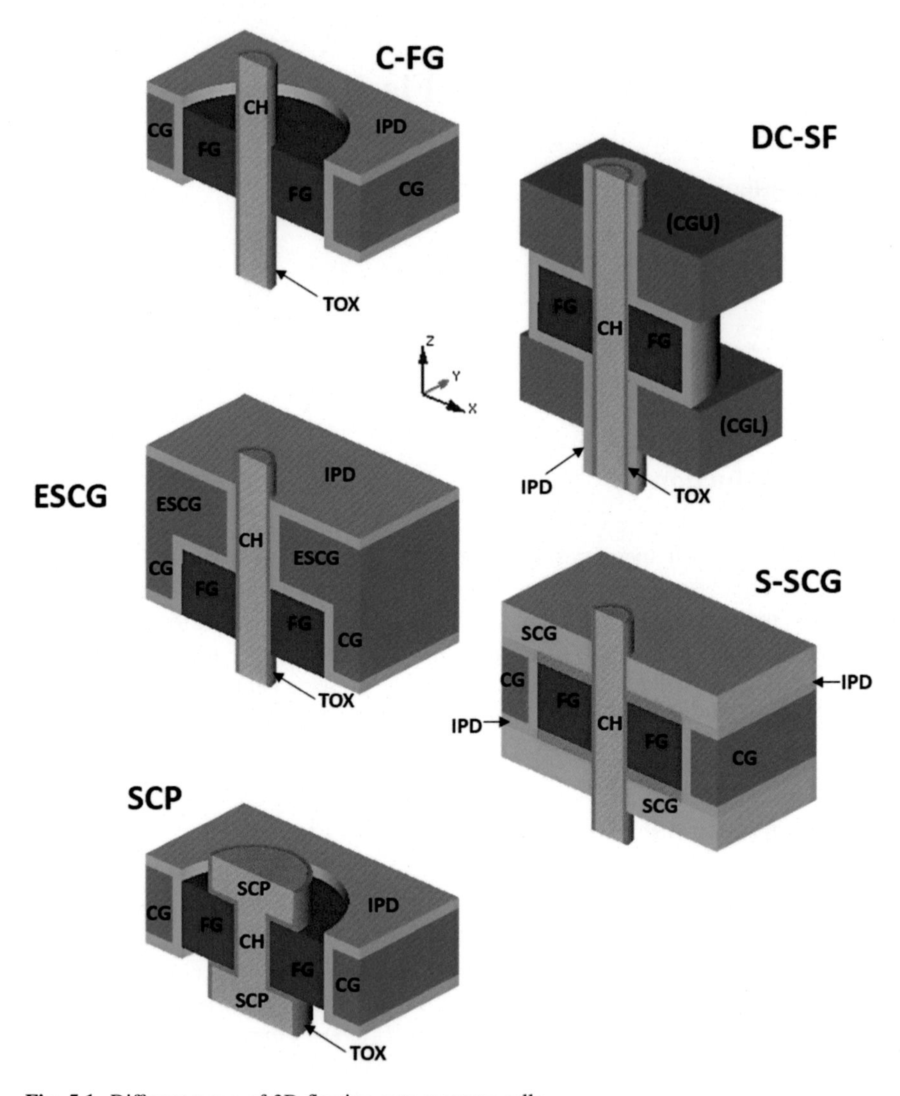

Fig. 5.1 Different types of 3D floating gate memory cells

memory cells; of course, more cells can be stacked together. Figure 5.6 displays various cross sections of Fig. 5.5. At the top of the vertical structure we can found the *Bitline Selector* (BLS) which is connected to the *Bitline* (BL) through the bitline contact. At the bottom, the *Source Line Selector* (SLS) connects the NAND string to the *Source Line* (SL). Because of their functionality, BLS and SLS are standard transistors, i.e. they are fabricated without a floating gate. In fact, cross sections reported in Fig. 5.7 reveal that both transistors use TOX as gate oxide. This is not the only option: IPD or any other type of oxide would do the job.

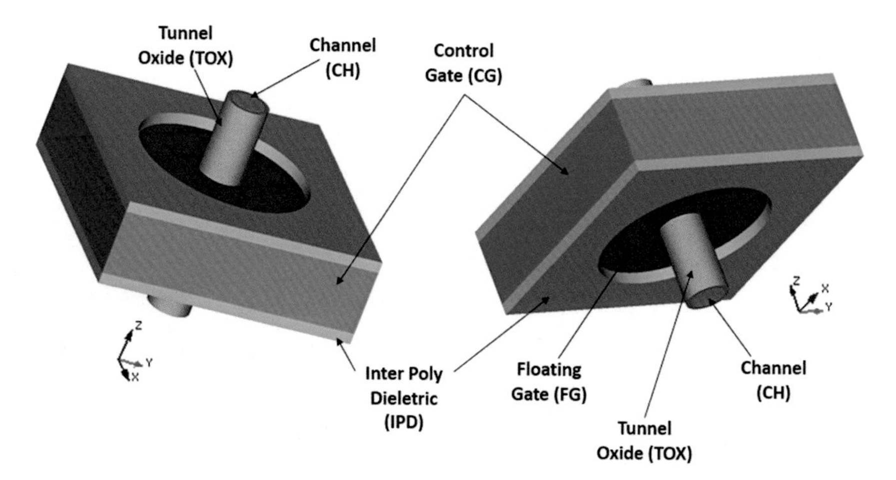

Fig. 5.2 Bird's-eye views of the 3D conventional floating gate cell (C-FG)

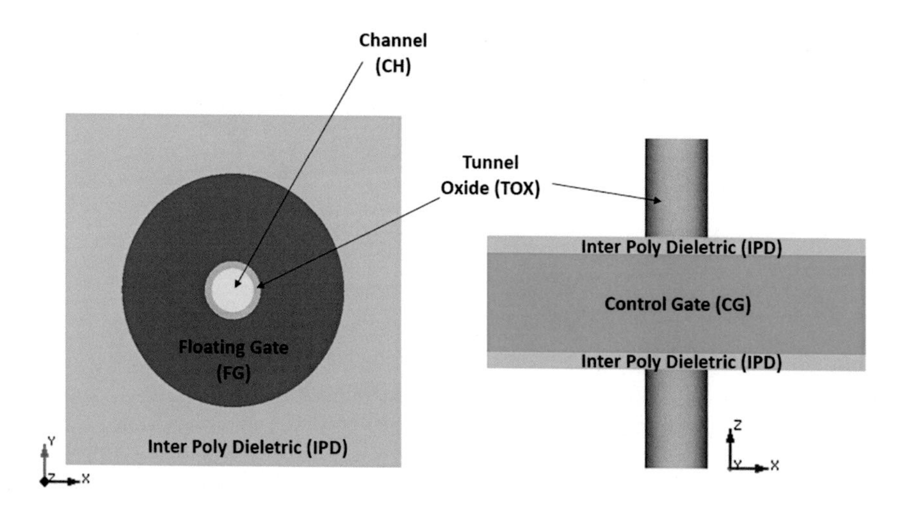

Fig. 5.3 Top and side views of the C-FG memory cell

Starting from a single NAND string (Fig. 5.5) we can build a complete memory array, as shown in Fig. 5.8.

All the Control Gates of the cells belonging to the same NAND page are shorted together to form a single wordline. This is true for BLS and SLS transistors too. Bitlines and wordlines are orthogonal to each other as in a standard planar array. In this specific example, 8 cells are addressed in parallel (8 bitlines), and there are 48 wordlines in total. In order to better understand all the details of such a complex structure, a layer-by-layer diagram is sketched in Fig. 5.9.

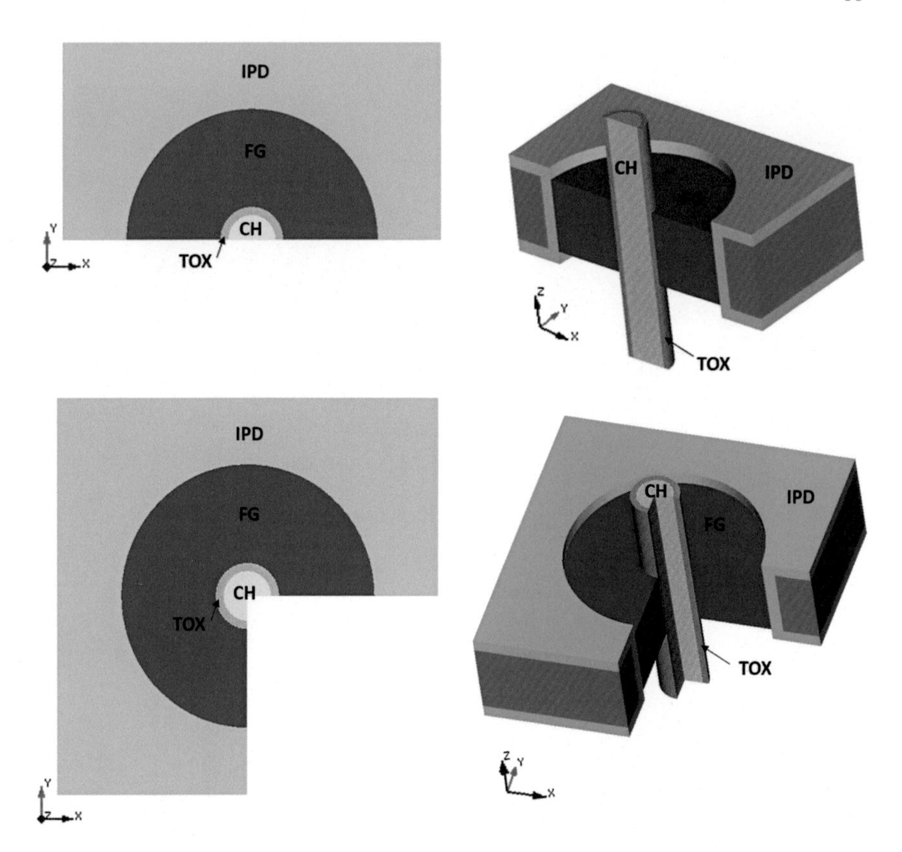

Fig. 5.4 X–Y and X–Z cross sections of the C-FG memory cell

Along the Y direction, there is a separation between wordlines: this is used during manufacturing to perform all the necessary process technology steps. Same applies to SLS lines. In reality, from an electric point of view, things are different. Figures 5.10, 5.11 and 5.12 add all the peripheral connections to the memory array. It is worth highlighting that wordlines at the same z-height are shorted together. In other words, the 48 wordlines form a single NAND memory block. Of course, BLS lines can't be connected together as they need to select a single NAND page out of the 8 belonging to the same "CG layer" (or simply "layer" in the following). Because SLS transistors need to be selective only when the NAND block is erased, then all the SLS lines of a specific block can be biased at the same voltage; therefore, SLS lines can be shorted.

The benefit of shorting lines together is a reduced decoding complexity which, as a matter of fact, translates into a saving of both power and silicon area.

With all the most recent process technology developments, CG and SLS plates can now be manufactured without gaps, even within the memory array, as sketched in Figs. 5.13 and 5.14.

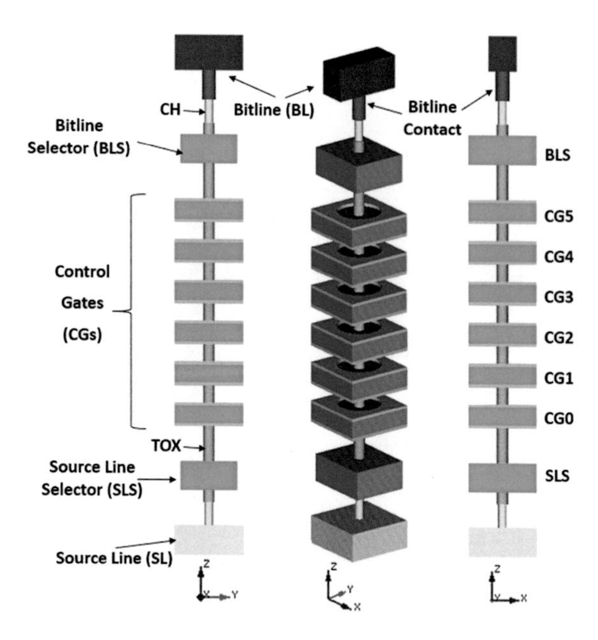

Fig. 5.5 Bird's-eye view and side views of the C-FG NAND Flash string

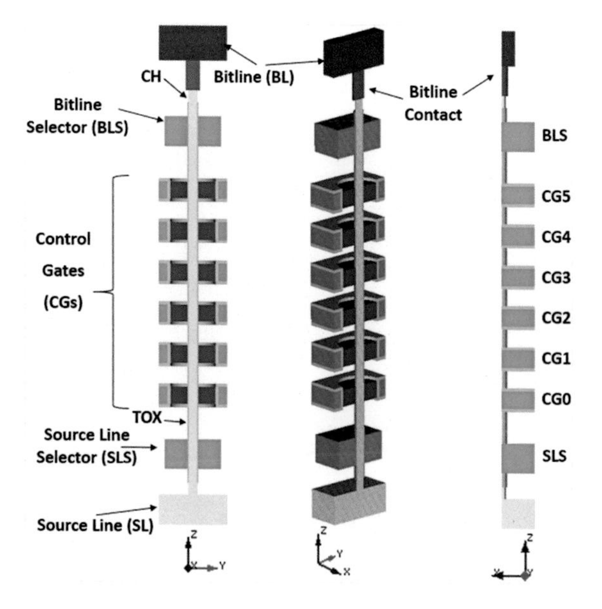

Fig. 5.6 Cross sections of Fig. 5.5

Given the complexity of the drawings, in order to focus on the details of the overall architecture, IPD and TOX will not be drawn in the remainder of this section. After this removal, the memory array looks like the one in Fig. 5.15.

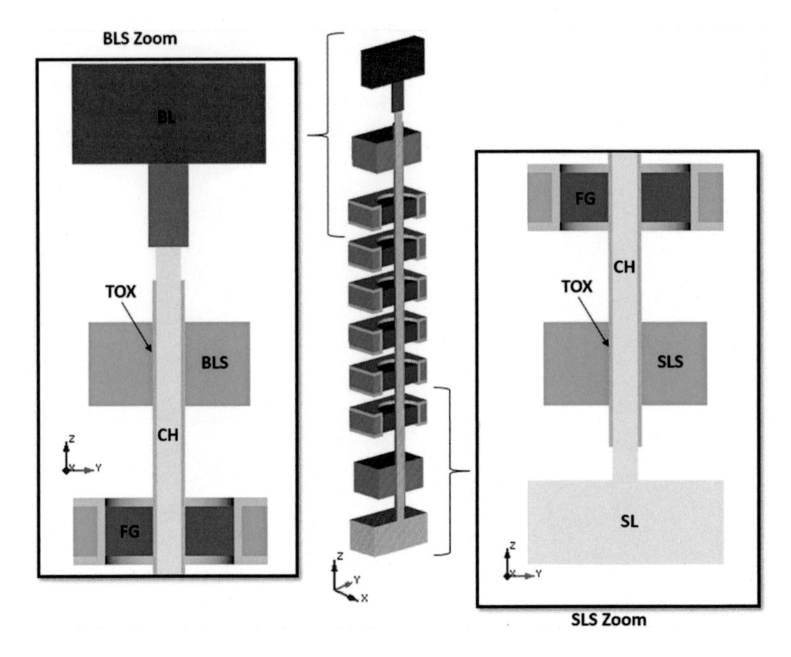

Fig. 5.7 Cross sections of transistors BLS and SLS

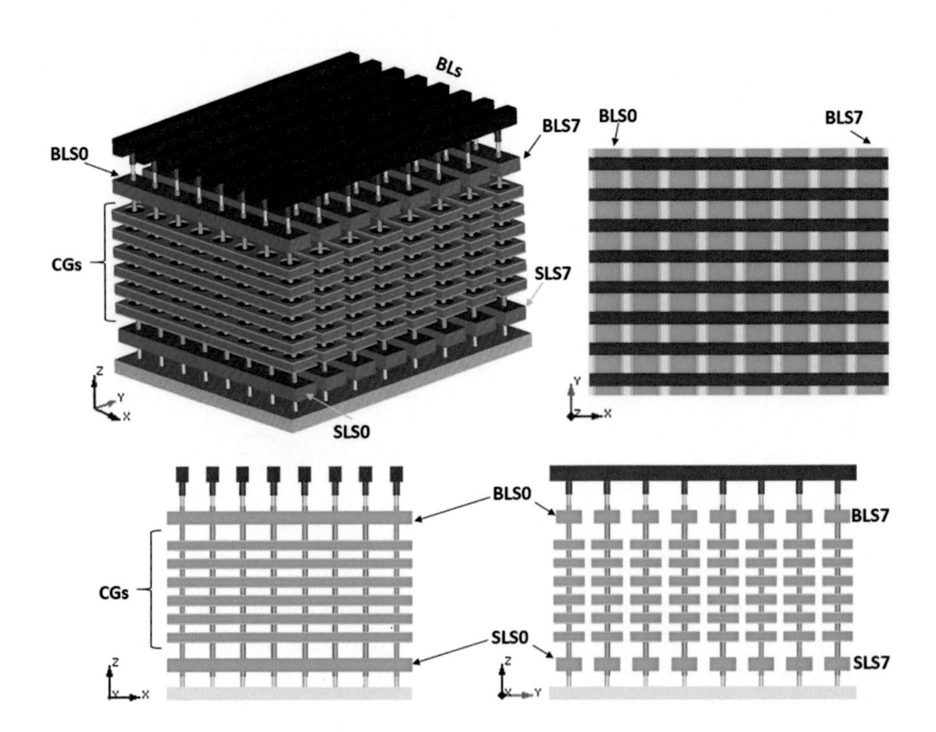

Fig. 5.8 Bird's-eye view and side views of the C-FG NAND Flash array

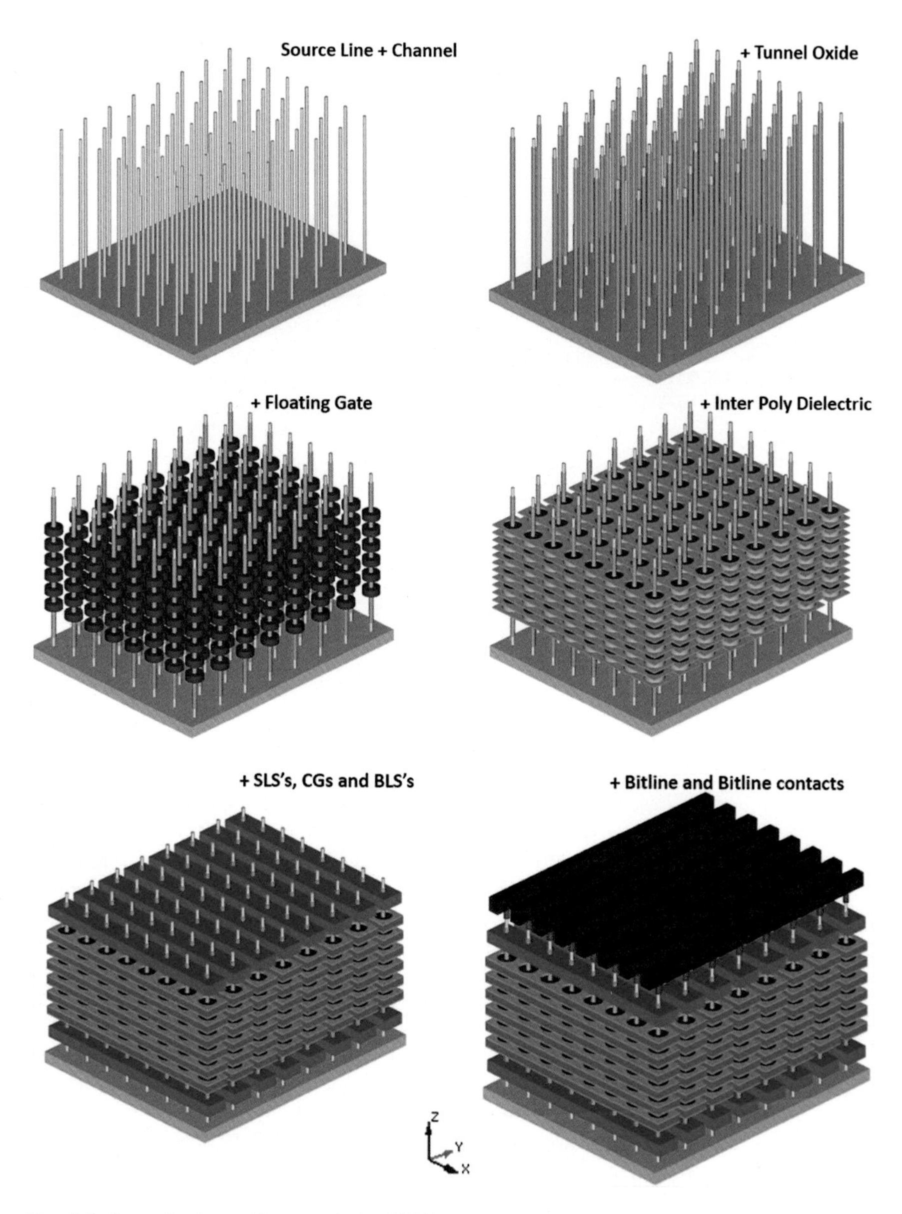

Fig. 5.9 Layer-by-layer diagram of the C-FG memory array

From a design perspective, it is critical to minimize the parasitic resistance of the Source Line as it plays the role of the "local ground" for each memory cell. Because of this, it is not possible to build a big Source Line plate at the bottom of the stack with just few contacts: when addressing tens of thousands of cells in parallel, the voltage fluctuations would inject too much noise during array operations. As such,

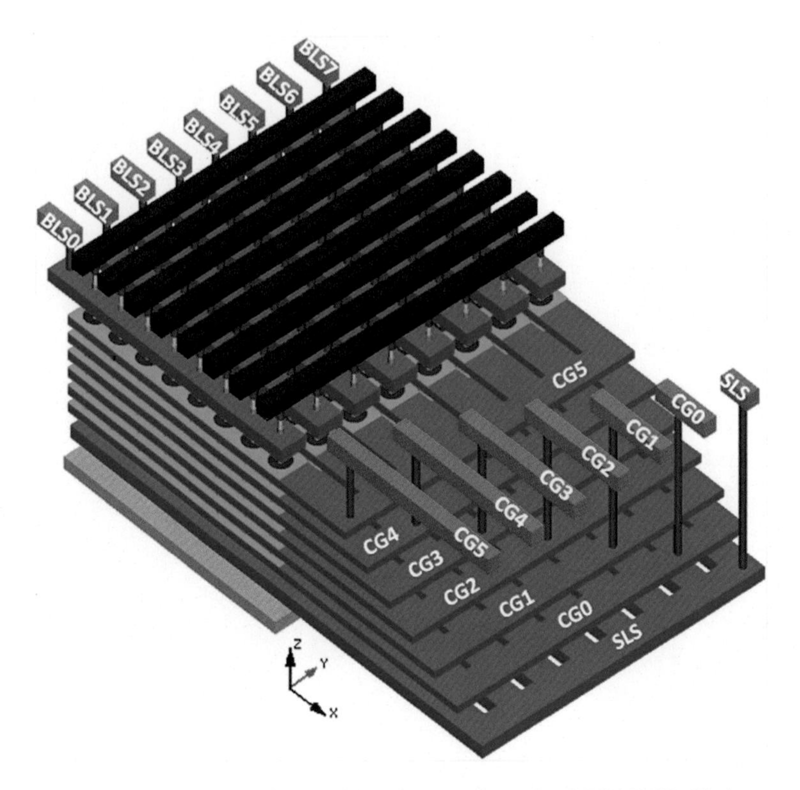

Fig. 5.10 Top bird's-eye view of the peripheral connections of a C-FG NAND Flash array

the memory array needs to be modified in order to accommodate additional contacts to the Source Line plate, as shown in Fig. 5.16. Please note that this modification has an impact on the bitline pattern too, as there is a need for an intra-bitline space to go down to the bottom of the structure.

At this point it is possible to build a metal mesh by using an additional metal layer and, therefore, decrease the parasitic resistance of the cell's source contact. This additional layer is referred to as "Top Source Line" in Figs. 5.17 and 5.18. Of course, this is just one example of metal mesh. Design might look different depending on the number of available metal layers and their parasitic resistance.

In order to minimize program and read disturbs (Chap. 2), and reduce parasitic loads, NAND blocks are usually separated by a slit, as shown in Figs. 5.19, 5.20 and 5.21. Of course, there is a need to cut CGs and SLS, but not bitlines and Top Source Lines, as they can be common to the entire memory plane. Fundamentally, this is the same approach we have seen with BiCS in Chap. 4.

It is worth highlighting that the size along the X axis is defined by the NAND logical page, while the size along the Y axis is defined by the number of blocks of the NAND device.

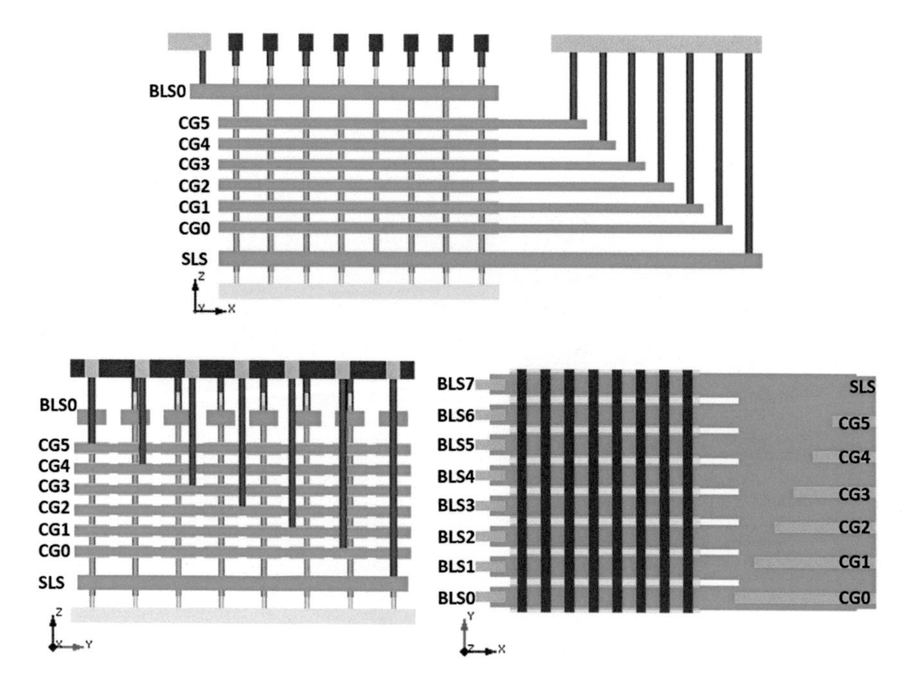

Fig. 5.11 Top and side views of Fig. 5.10

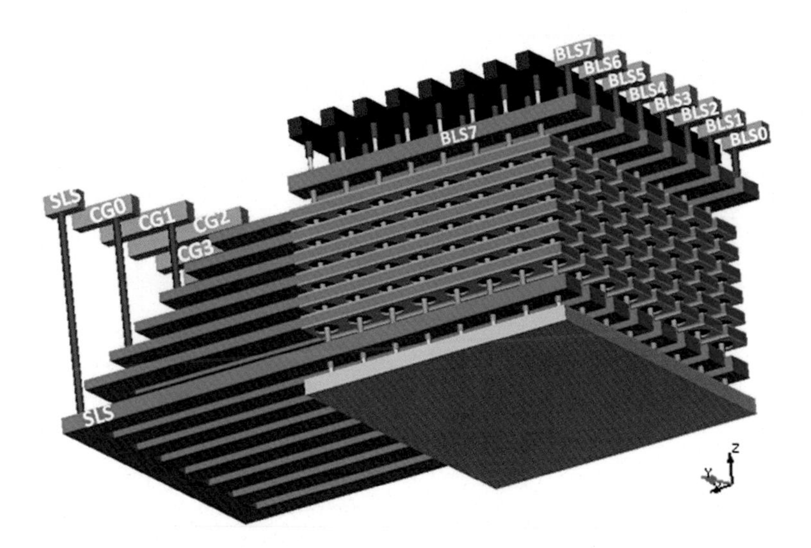

Fig. 5.12 Bottom bird's-eye view of the peripheral connections of a C-FG NAND Flash array

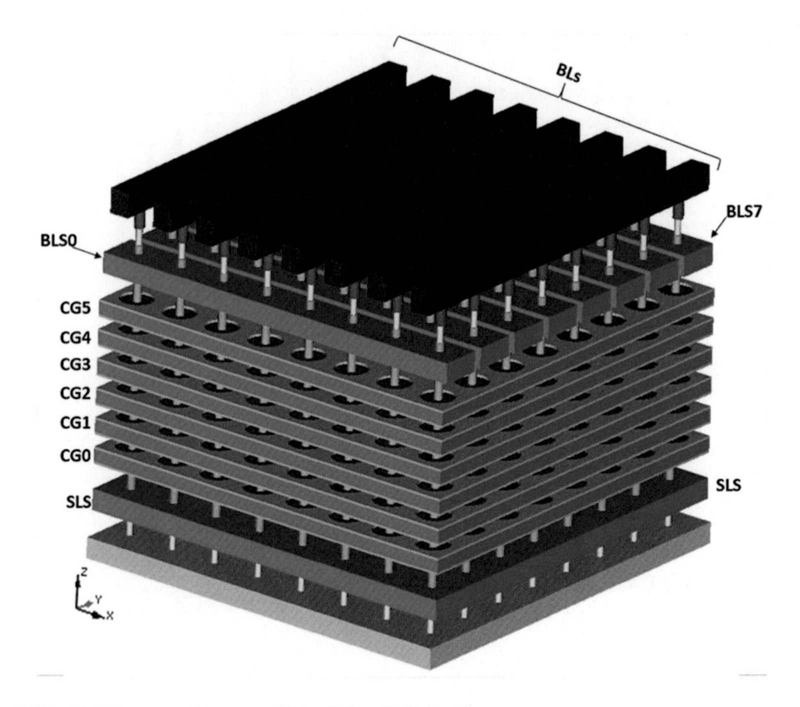

Fig. 5.13 C-FG array with monolithic CG and SLS plates

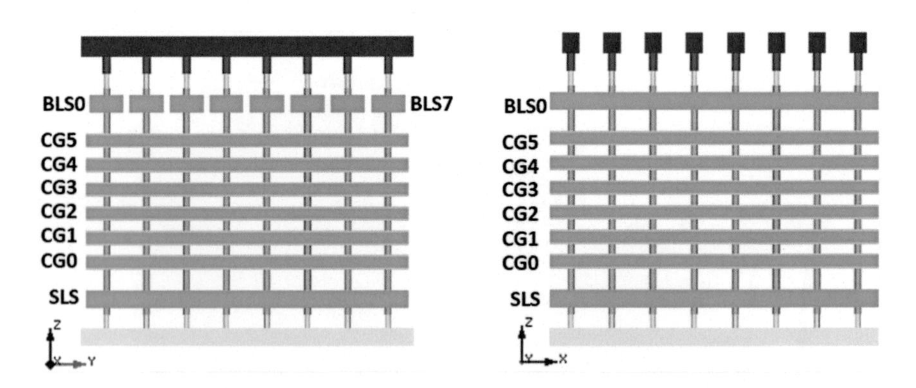

Fig. 5.14 Cross sections of Fig. 5.13

In all the drawing of this section we have used 6 CG layers, but this number can be scaled up depending on the capabilities of the process technology. Because we are talking about Floating Gate cells, the vertical scaling is limited by interference issues with neighboring cells; not only that, other effects must be carefully handled. This is why different types of 3D FG cells have been developed over time, and they are described in the following sections.

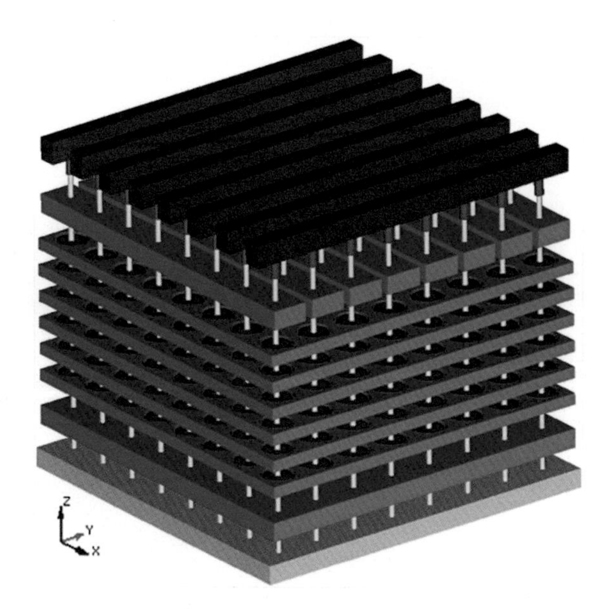

Fig. 5.15 Memory array without IPD and TOX

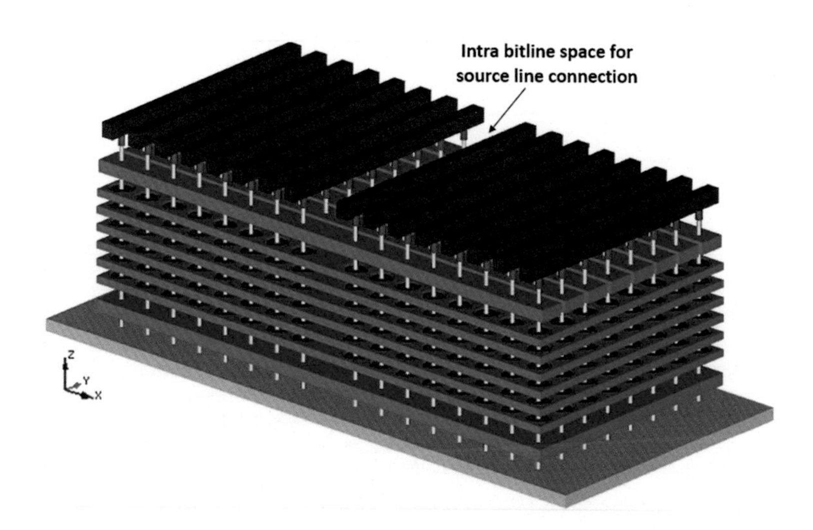

Fig. 5.16 Modified array to enable more contacts to the source line plate

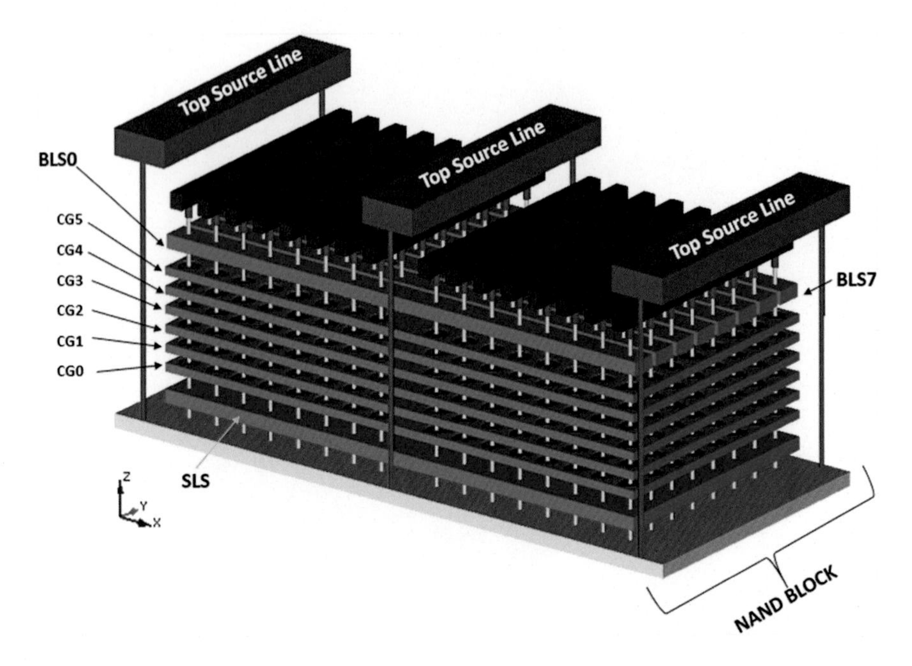

Fig. 5.17 C-FG array with top source line

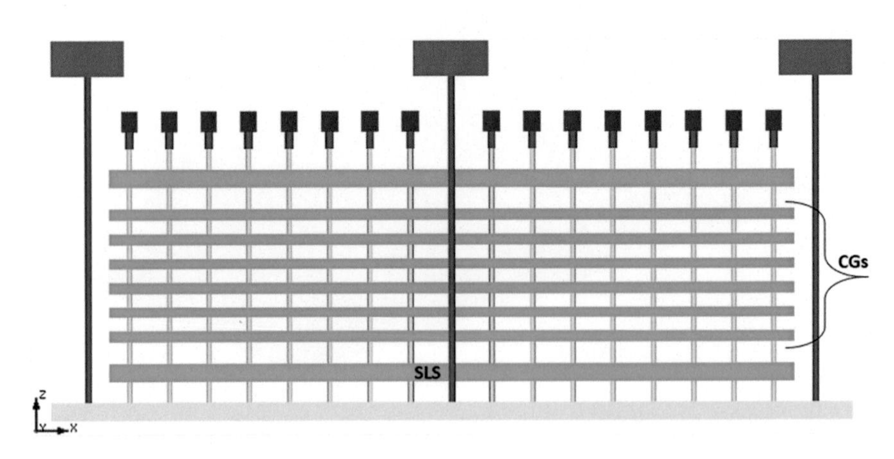

Fig. 5.18 X–Z cross section of Fig. 5.17

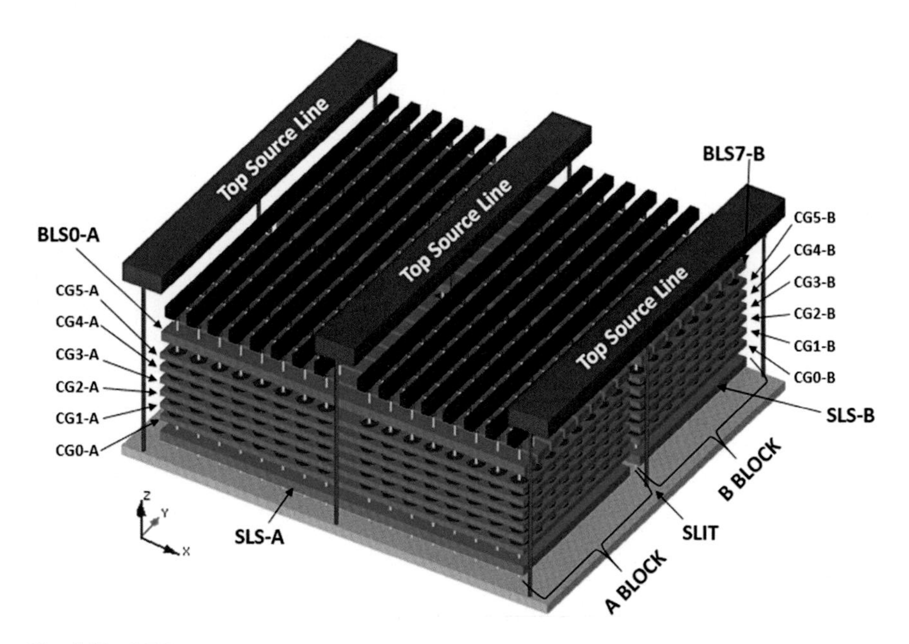

Fig. 5.19 C-FG array with 2 NAND blocks

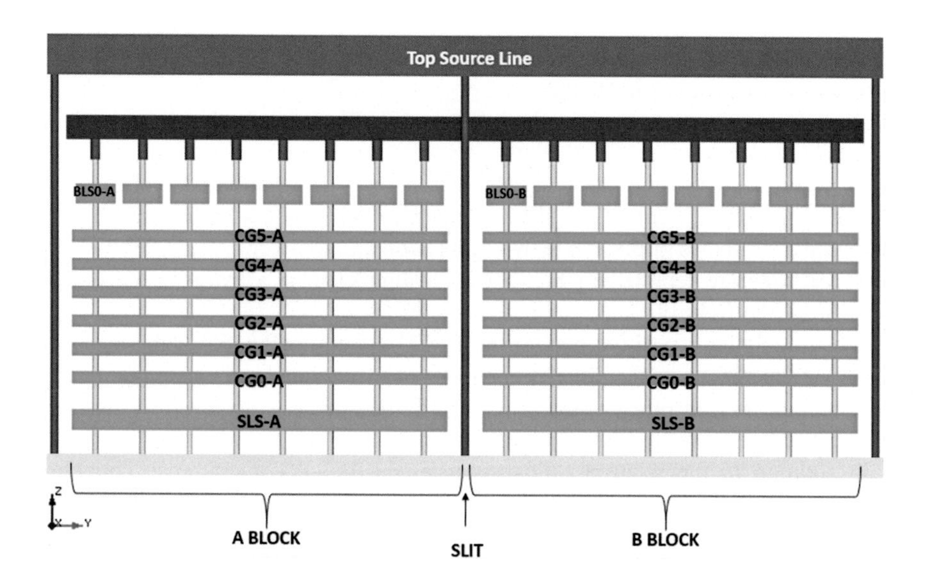

Fig. 5.20 Y–Z cross section of Fig. 5.19

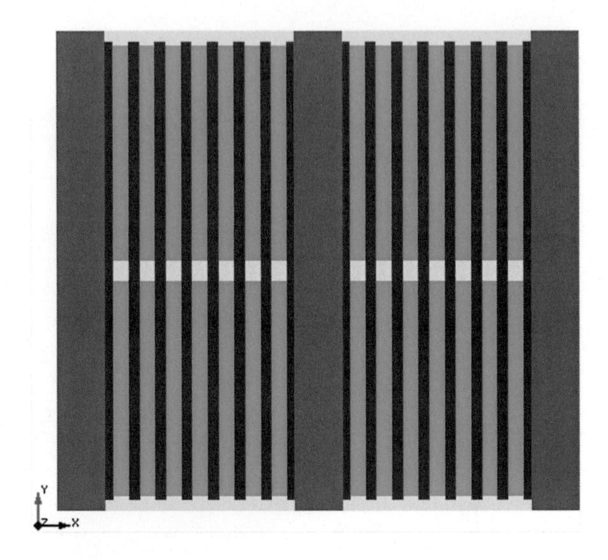

Fig. 5.21 Top view of Fig. 5.19

5.3 Extended Sidewall Control Gate (ESCG) Flash Cell

When looking at 3D cells, another concern is the high resistance of the source/drain (S/D) regions for enhancement-mode operations; as a result, these regions require high-doping, but this is very difficult to achieve with a polycrystalline-silicon channel. Moreover, this diffused S/D areas lead to short channel effects and disturbance of the conventional bulk erase operation. Therefore, a higher voltage during read is required to electrically invert the S/D layer: this is almost impossible with C-FG because of the FG thickness.

In order to overcome these issues and reduce the interference effect, the *Extended Sidewall Control Gate* (ESCG) memory cell was introduced [4].

Figure 5.22 shows bird's-eye views, and lateral and top views of the ESCG cell. Cross sections are reported in Fig. 5.23. Floating Gate is a cylinder, fully surrounded by the Control Gate.

When applying a positive bias to the ESCG structure (Fig. 5.24) the electron density at the surface of the pillar is one order of magnitude higher than that of the Conventional FG. In other words, a lower S/D resistance is achieved by using a highly inverted electrical S/D scheme.

In addition to the reduced S/D resistance, the FG–FG interference coupling capacitance is also reduced, thanks to the ESCG shielding structure. Please note that the ESCG region is not floating. Moreover, the CG coupling capacitance (C_{CG}) is remarkably increased due to the increased area between Control Gate and Floating Gate. As a result, a better CG coupling ratio can be achieved, which is very important for high-speed NAND Flash operations [5].

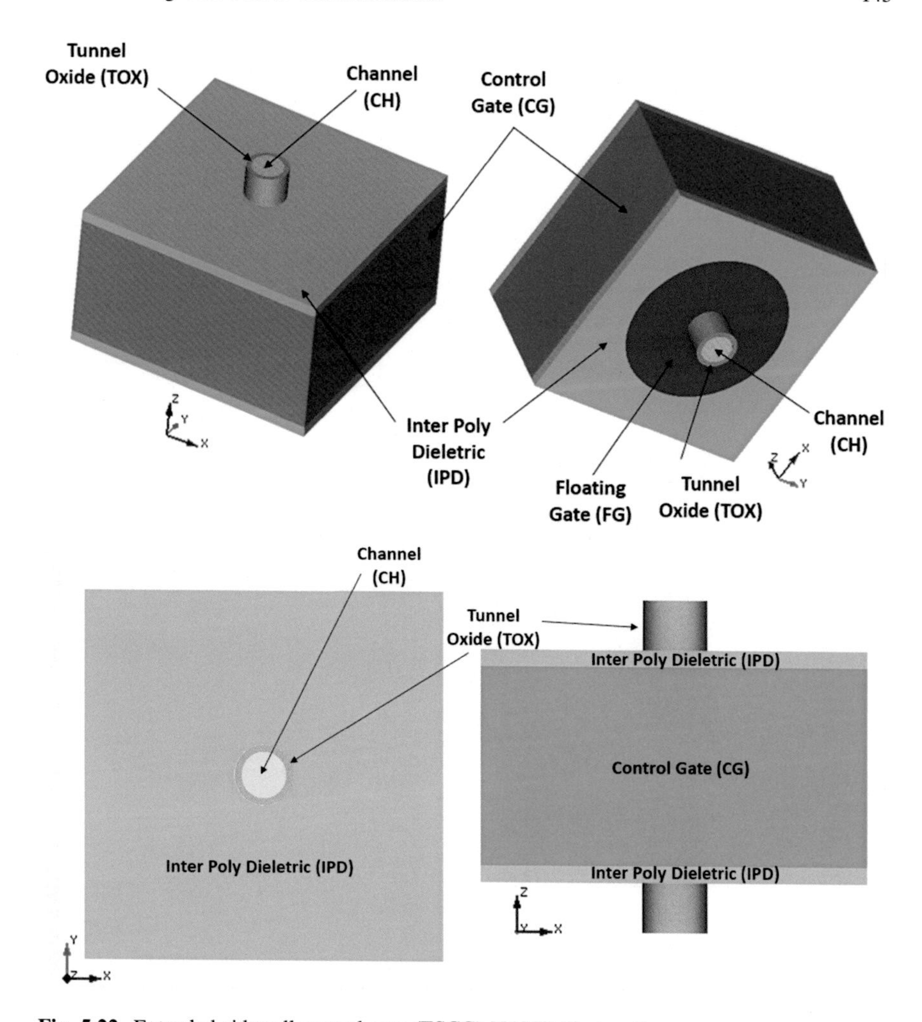

Fig. 5.22 Extended sidewall control gate (ESCG) NAND Flash cell

As usual, multiple memory cells can be connected to form a NAND string, as sketched in Figs. 5.25 and 5.26. For sake of clarity, a string of 6 memory cells is used.

For SLS and BLS selectors, the same considerations we made for C-FG (Fig. 5.7) apply here: SLS e BLS don't have a floating gate. Starting from Fig. 5.25 a core of the memory array can be built, as sketched in Fig. 5.27, which is the equivalent of Fig. 5.8 for C-FG. At this point, all the following steps, in terms of building multiple blocks, and adding peripheral connections and source line mesh are basically the same that were described in the previous section.

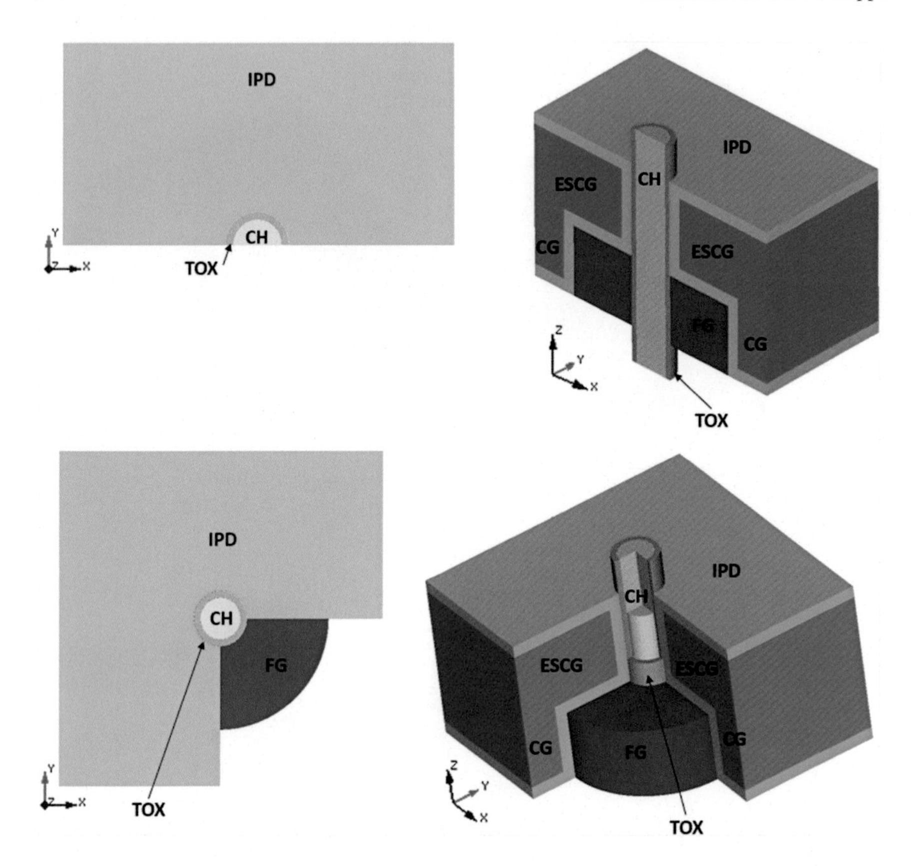

Fig. 5.23 Cross sections of the ESCG NAND Flash cell

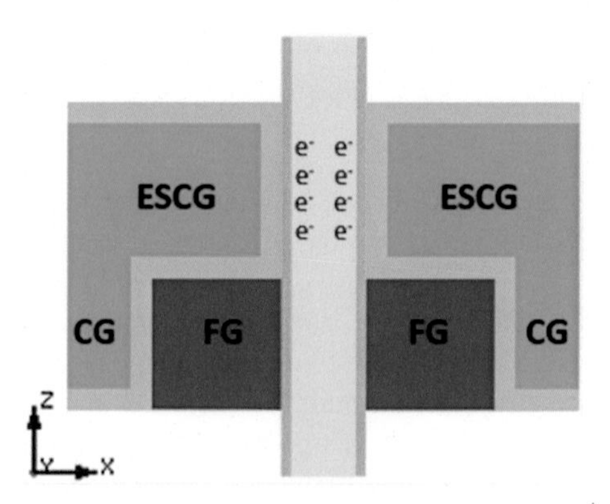

Fig. 5.24 Electron density at the surface of the pillar

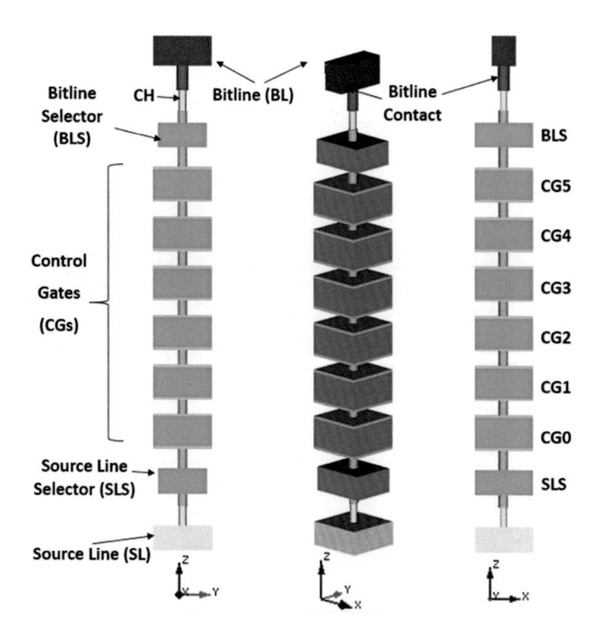

Fig. 5.25 Bird's-eye view and side views of the ESCG NAND Flash string

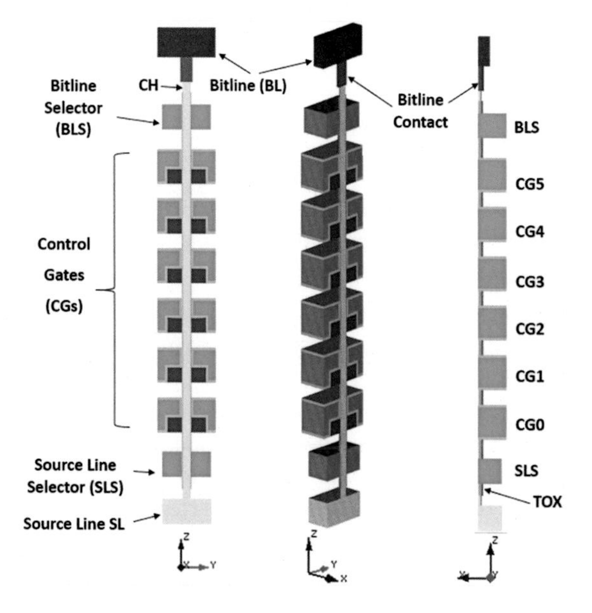

Fig. 5.26 Cross sections of Fig. 5.25

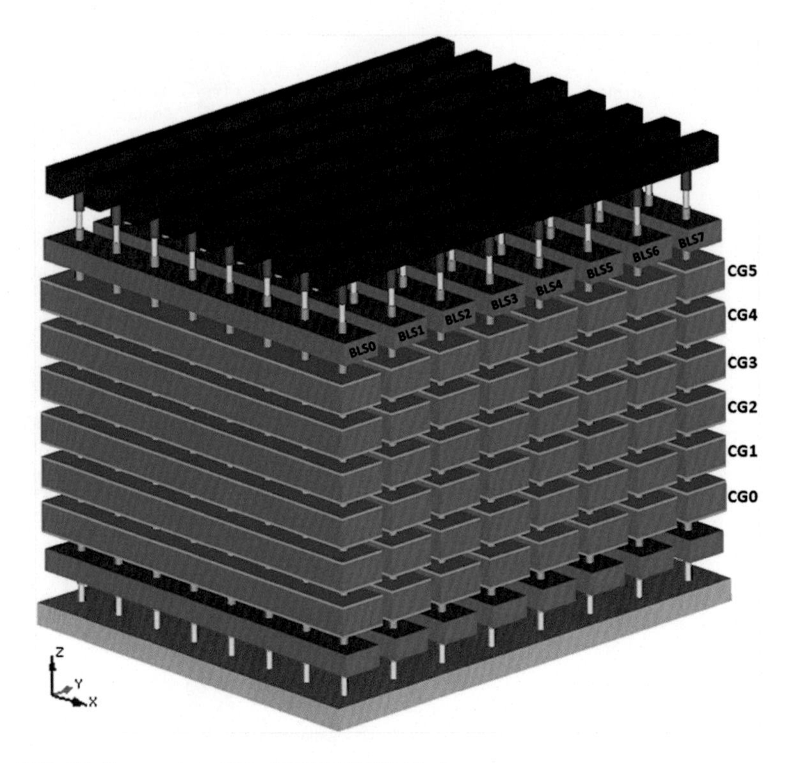

Fig. 5.27 Bird's-eye view of the ESCG NAND Flash array

5.4 Dual Control-Gate with Surrounding Floating Gate (DC-SF) Flash Cell

Another possible NAND cell based on Floating Gate is named DC-SF (Fig. 5.28), which stands for *Dual Control-Gate with Surrounding Floating Gate* [6]. This time the Floating Gate is controlled by means of two control gates (CGs). The immediate benefit of this structure is an increased coupling ratio FG/CG due to the enlargement of surface area between FG and CGs. Of course, this combination enables lower voltage biasing for program and erase operations. Another key advantage of DC-SF is the absence of FG-FG interference as the Control Gate sitting between 2 Floating Gates plays the role of electrostatic shield. As a result, this memory cells allows wide Program/Erase (P/E) threshold voltage window, which is key for multilevel storage [7].

Cross sections in Fig. 5.29 show that the surrounding FG is fully isolated by the inter poly dielectric (IPD), and it is capacitive coupled to both Upper (CGU) and Lower (CGL) control gates along the vertical direction Z.

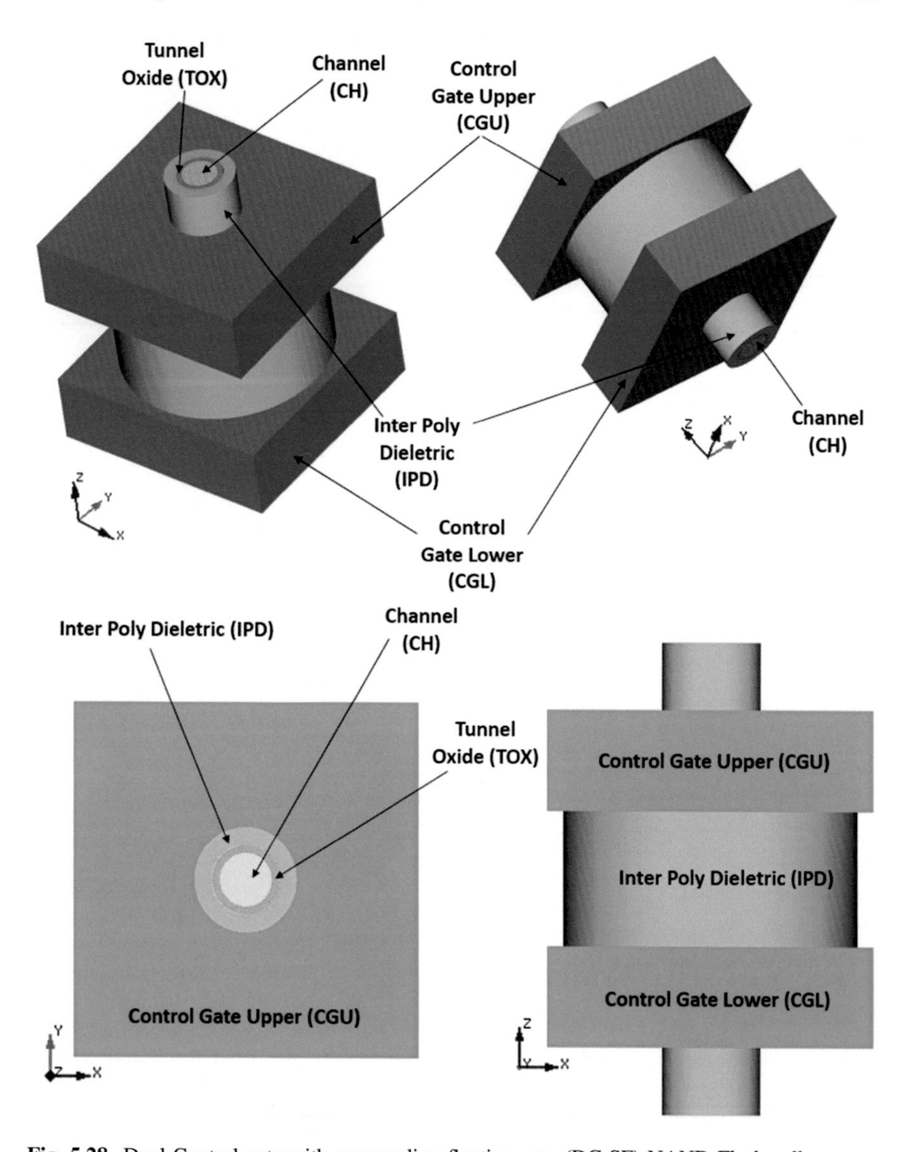

Fig. 5.28 Dual Control-gate with surrounding floating-gate (DC-SF) NAND Flash cell

The tunnel oxide is formed only in the region between CH polysilicon and FG, while IPD is added on the sidewall of the CG, thus forming a thicker dielectric layer on the CG. As a consequence, charges can only move through the tunnel oxide without any tunneling to the control gates.

Figures 5.30 and 5.31 show a NAND Flash string based on DC-SF and few cross sections, respectively. It is worth highlighting that, inside the NAND string, control gates are shared between 2 floating gates, such that the overall numbers of

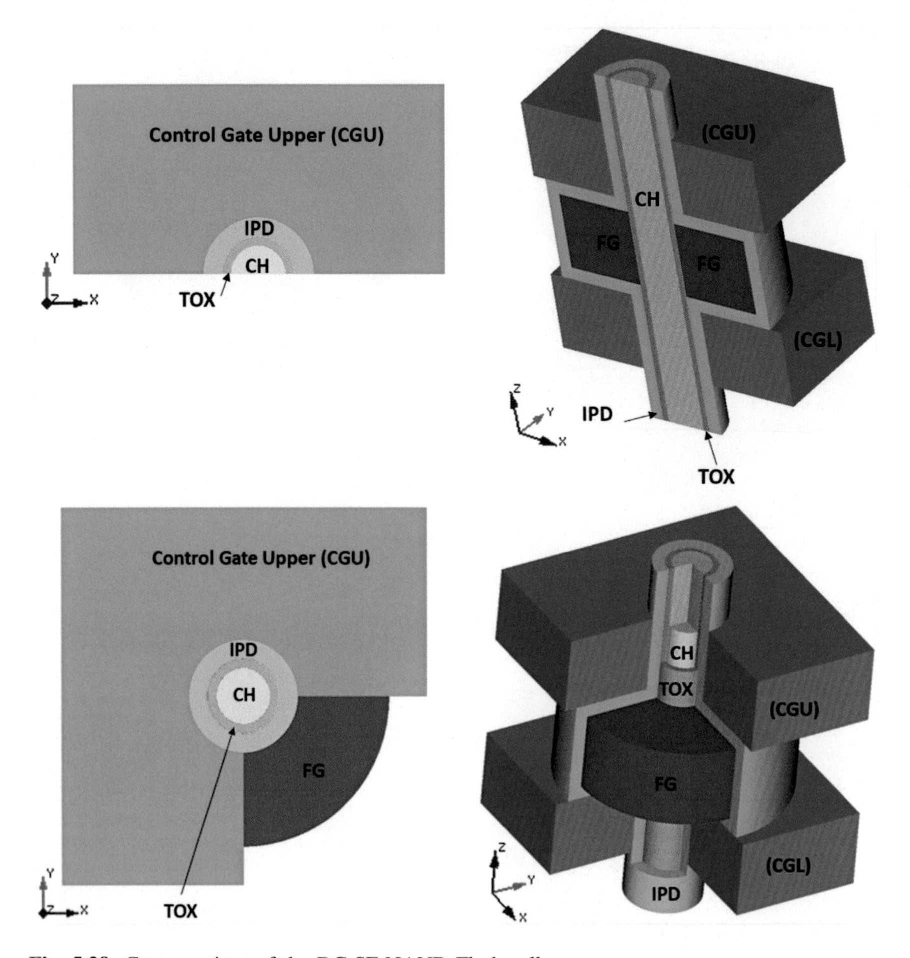

Fig. 5.29 Cross sections of the DC-SF NAND Flash cell

layers can be significantly reduced. SLS e BLS selectors don't use the floating gate, as we have already seen for C-FG and ESCG: they are just normal nMOS transistors.

Starting from their physical structures, it is possible to compare BiCs and DC-SF in order to understand how prone they are to retention problems. Strings of the 2 types are shown in Fig. 5.32. In a string of the conventional 3D BiCS, the charge trap nitride layer is continuously connected across all gates along the channel side and, as a matter of fact, it acts as a charge spreading path. As discussed in Chap. 2, this causes degradation of data retention characteristics. On the contrary, the surrounding FG is completely isolated by IPD and tunnel oxide, implying that DC-SF has a much easier job in retaining electrons [8, 9].

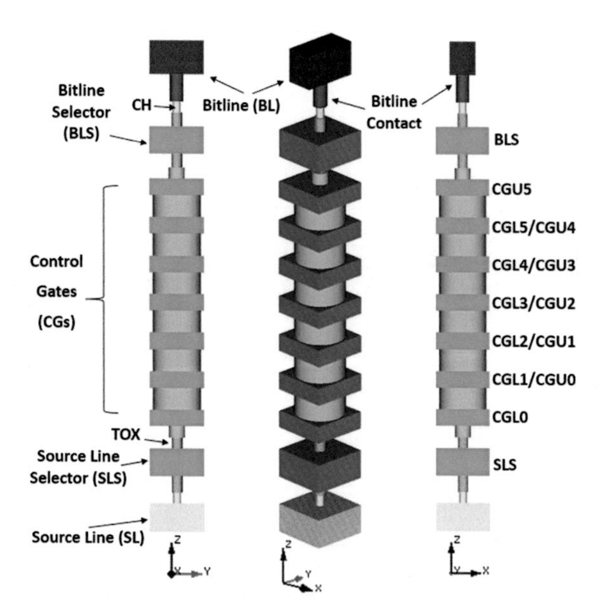

Fig. 5.30 Bird's-eye view and side views of the DC-SF NAND Flash string

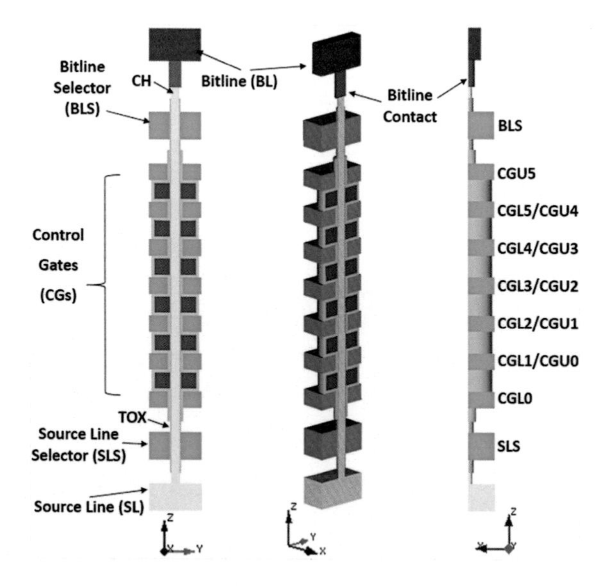

Fig. 5.31 Cross sections of Fig. 5.30

The downside of a memory cell with 2 control gates is the increased complexity of the biasing schemes [10, 11], and the additional 2 gate layers at the top and bottom of the NAND string.

As usual, starting from a single NAND string, an entire memory array can be built, as shown in Fig. 5.33. Because Fig. 5.33 is basically identical to Fig. 5.8 (of course, except for the memory cell itself), all the considerations made for C-FG in terms of peripheral connections and source line resistance hold true here.

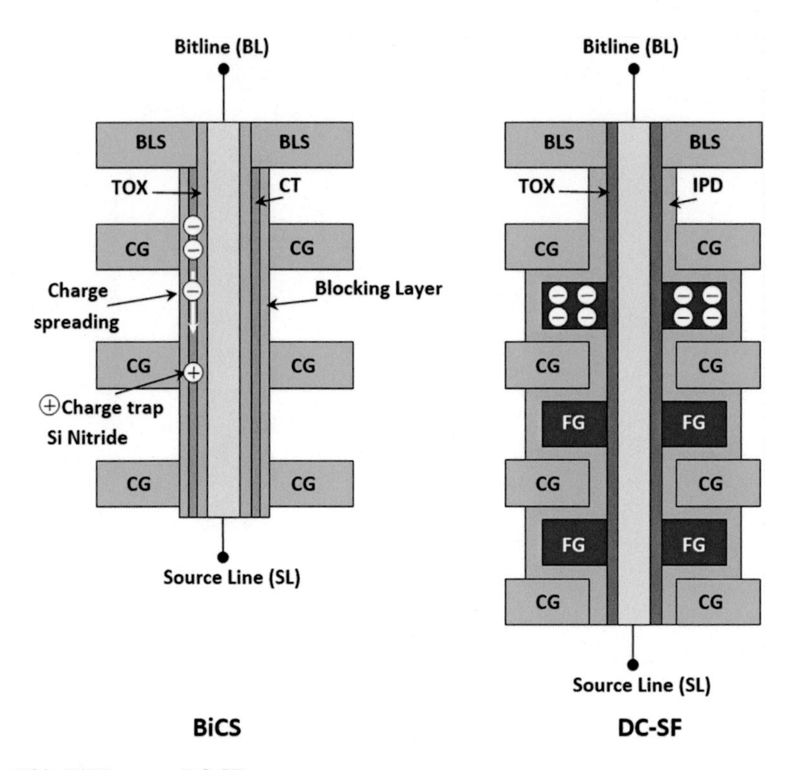

Fig. 5.32 BiCS versus DC-SF

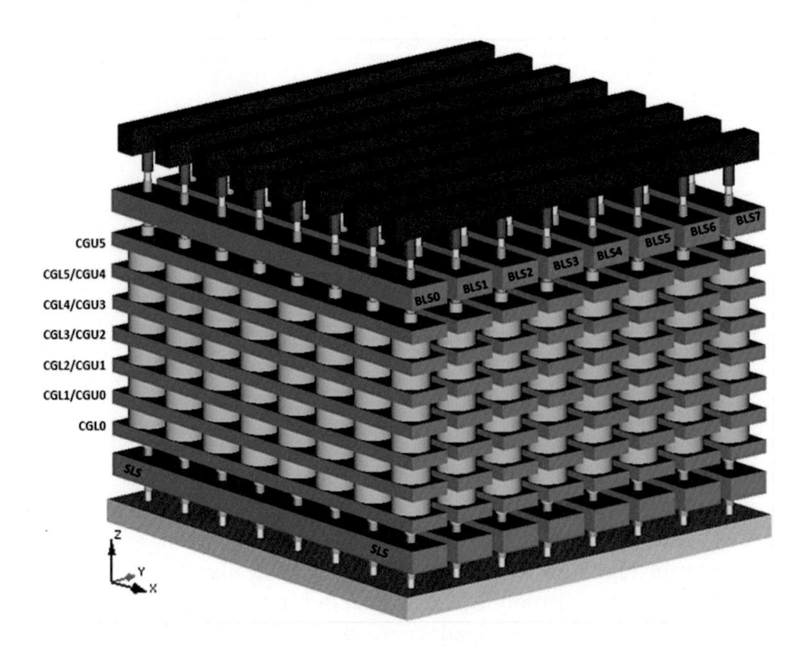

Fig. 5.33 Bird's-eye view of the DC-SF NAND Flash array

5.5 Separated Sidewall Control Gate (S-SCG) Flash Cell

Another 3D floating gate option that leverages the sidewall concept is the *Separated Sidewall Control Gate* (S-SCG) Flash cell [12], which is shown in Figs. 5.34 and 5.35. As we have seen with DC-SF and ESCG, the adoption of a *Sidewall Control Gate* (SCG) structure has multiple benefits, including a significant reduction of the interference coupling effect, and a high CG-FG coupling capacitance.

S-SCG cells can be combined in NAND string by sharing sidewall gates, as sketched in Figs. 5.36 and 5.37; sharing is done to reduce complexity and minimize the number of layers, which is critical for 3D integration.

One of the most critical issues of this architecture is the "direct" disturb to the neighboring passing cells, as a result of the high coupling capacitance between SCG and FG. In this case we are talking about "direct" disturb because the sidewall

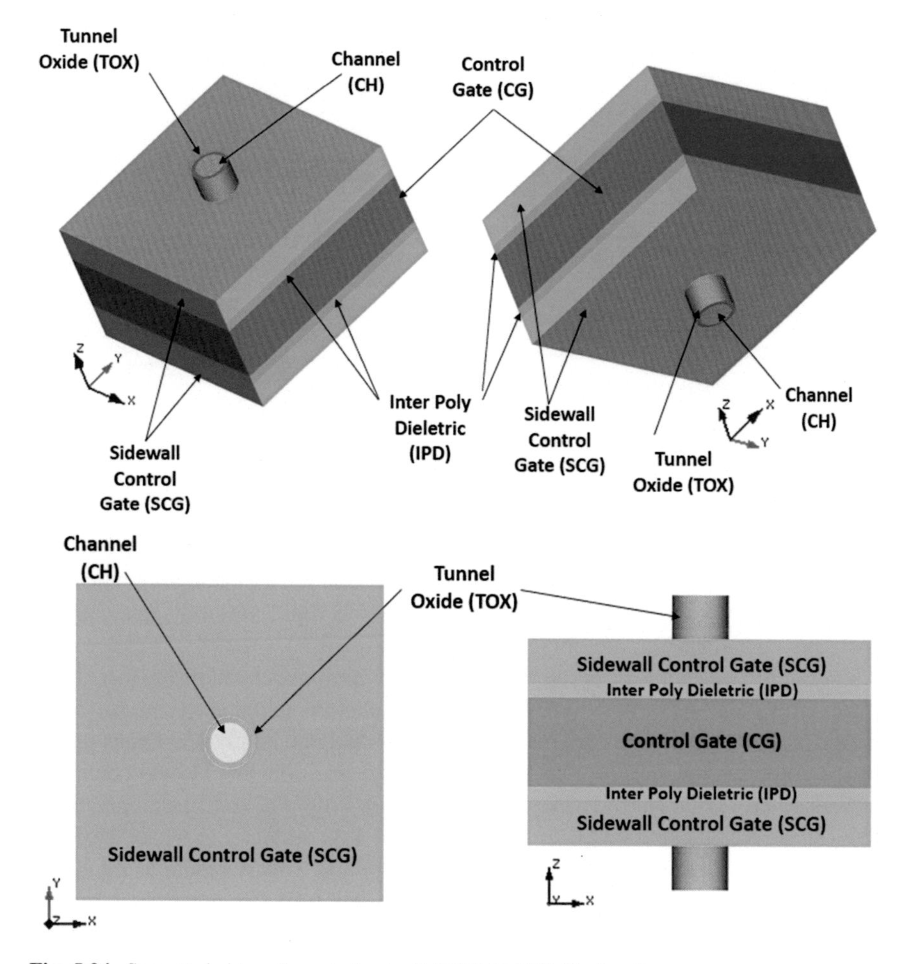

Fig. 5.34 Separated sidewall control gate (S-SCG) NAND Flash cell

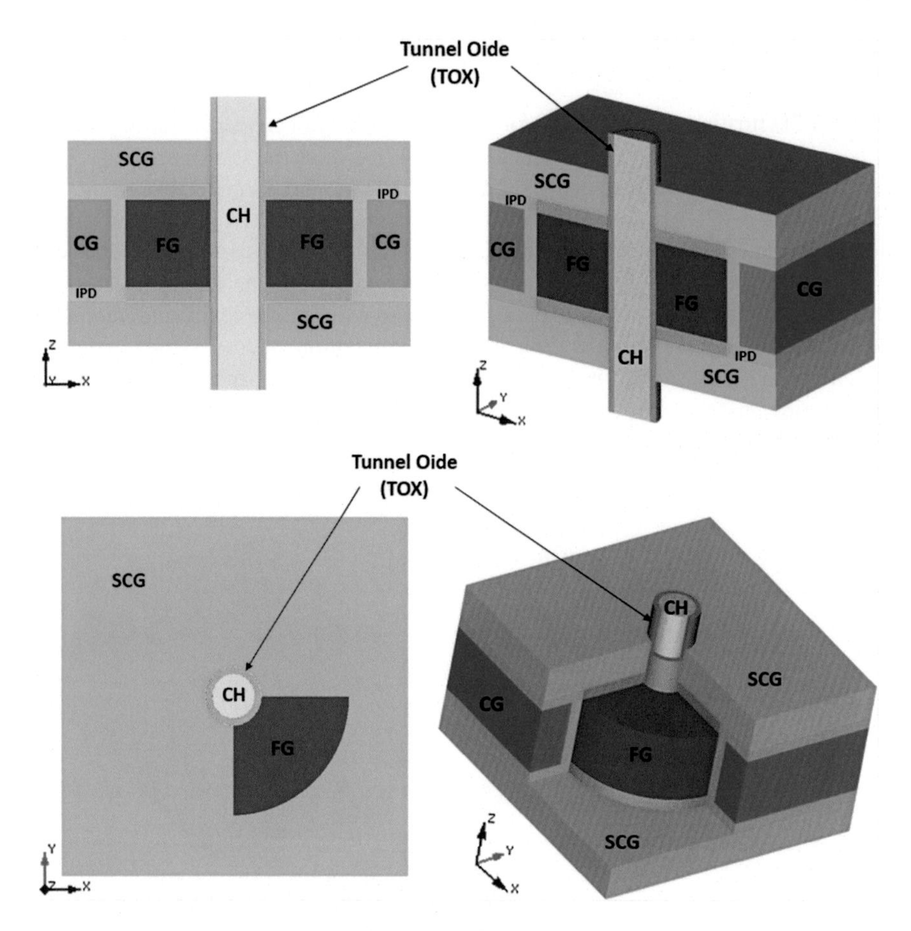

Fig. 5.35 Cross sections of the S-SCG NAND Flash cell

control gate is shared between adjacent cells. In fact, a voltage applied to SCG is actually applied to both floating gates (again, because of the high coupling ratio). Of course, this is worse in the DC-SF case because of the 2 sidewall gates, while ESCG has one sidewall gate only.

In order to reduce the decoding complexity and minimize the number of contacts in the array, all SCGs of a NAND block are shorted together (common SCG scheme) and, therefore, they are biased at the same voltage in any operations of the array. In other words, besides being an electrostatic shield for the FG-FG coupling, sidewall gates can be properly biased to help Read, Program, and Erase operations [13]. During Read, common SCG is biased around 1 V, similarly to what described for ESCG, thus electrically inverting the channel (pillar). This is similar to what was described in the previous section for ESCG, but this time the inversion happens simultaneously on both source and drain, thanks to the double sidewall gate architecture (Fig. 5.38).

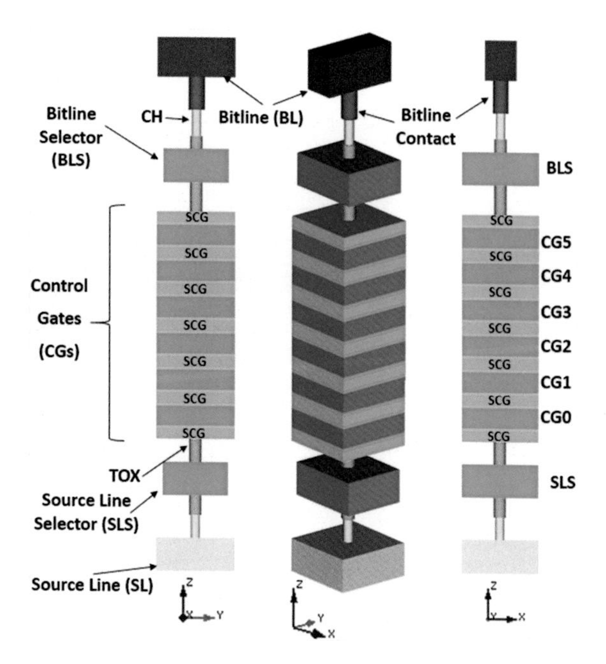

Fig. 5.36 Bird's-eye view and side views of the S-SCG NAND Flash string

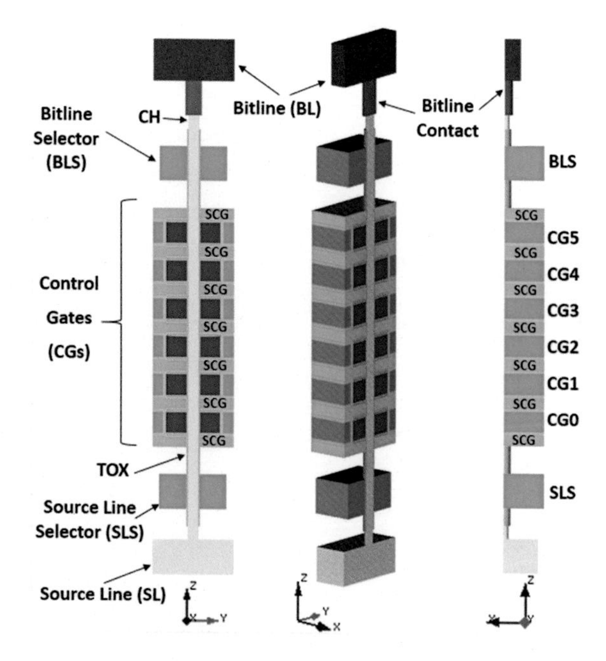

Fig. 5.37 Cross sections of Fig. 5.36

In a similar way, the common SCG is biased at a medium voltage (around 11 V) during programming to enhance the channel boosting efficiency (Chap. 3). Figure 5.39 displays a bird's-eye views of the S-SCG NAND Flash array.

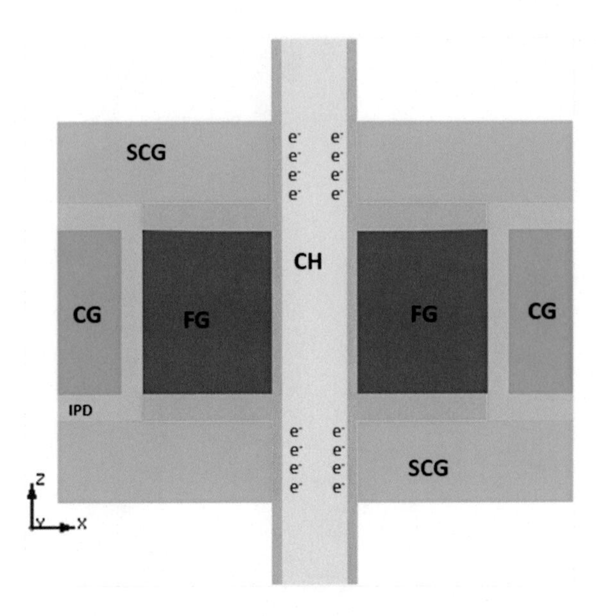

Fig. 5.38 Source and drain inversion thanks to the common SCG approach

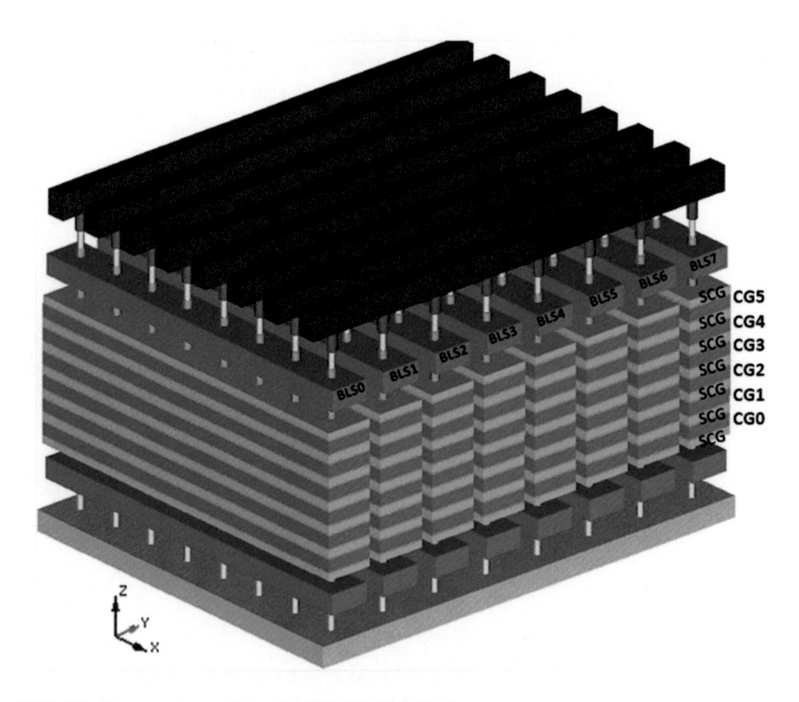

Fig. 5.39 Bird's-eye view of the S-SCG NAND Flash array

5.6 Sidewall Control Pillar (SCP) Flash Cell

As discussed, the drawback of the Sidewall Control Gate approach is the direct disturb induced by SCG on adjacent memory cells. Moreover, there might be reliability issues arising from the IPD close to SCG, because of the high voltages applied during program and erase operations. In addition, there is an intrinsic limitation to vertical scaling (let's say below 30 nm) because, again, the thickness of both SCG and IPD can't be scaled too much due the voltage budget they need to handle.

To overcome the above mentioned issues, in 2012 a memory cell called *Sidewall Control Pillar* (SCP) was proposed [14]. Please refer to Figs. 5.40 and 5.41 bird's-eye views, and top and side views of the SCP Flash cell.

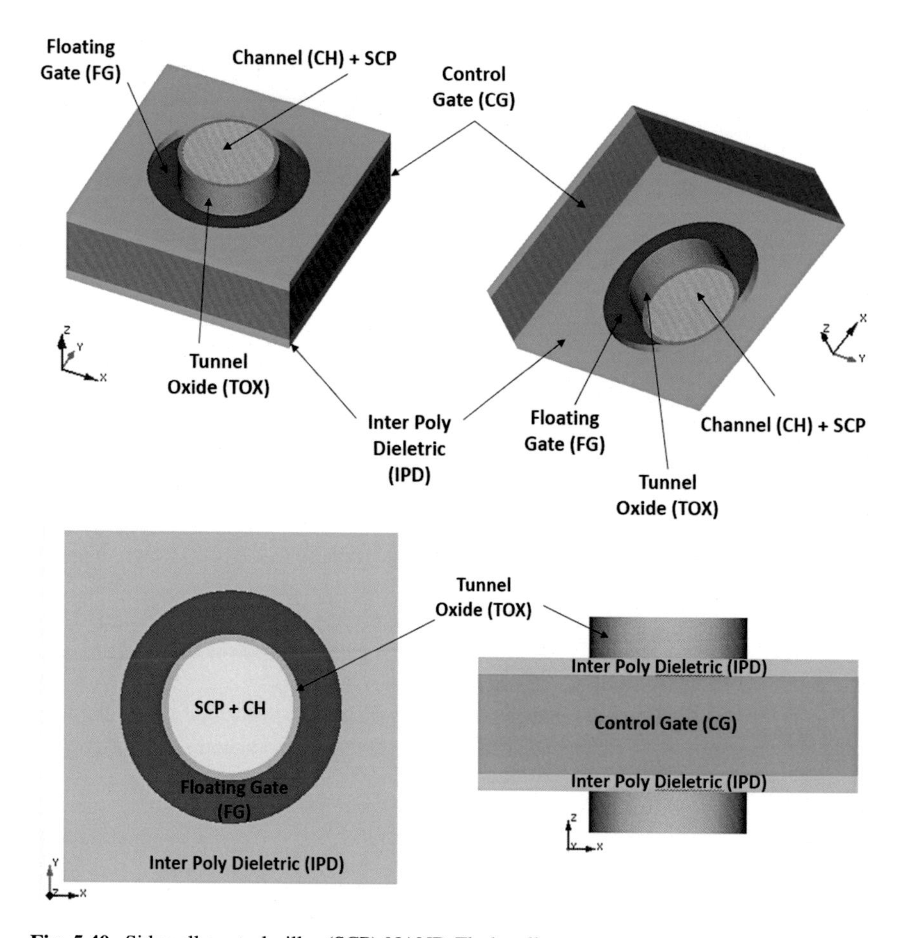

Fig. 5.40 Sidewall control pillar (SCP) NAND Flash cell

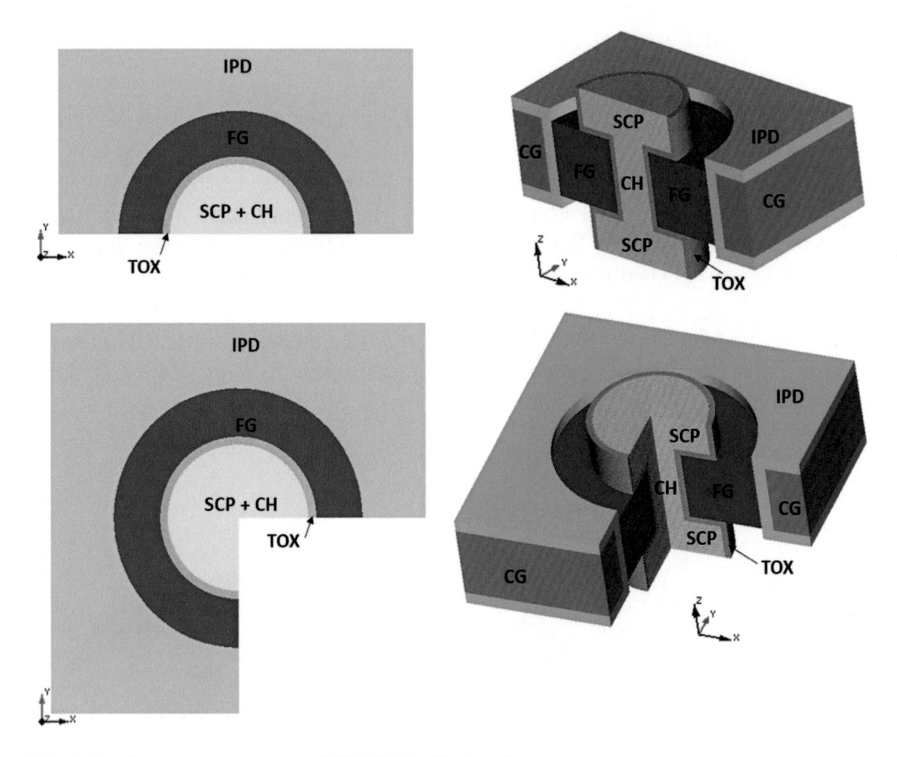

Fig. 5.41 Cross sections of the SCP NAND Flash cell

This time the isolation effect of the Sidewall approach is realized by using the polysilicon pillar itself. In fact, top and bottom sides of each FG are partially covered by the pillar. Furthermore, the thickness of SCP can be reduced in the 20 nm range, with a significant improvement from a scaling perspective. It is worth highlighting that, being designed without a Sidewall Gate, SCP can be operated in the same way as C-FG, thus leveraging all the benefits of well-known biasing schemes.

SCP NAND string and its cross sections are shown in Figs. 5.42 and 5.43, respectively. Please note that SCP is shared between two adjacent cells, in order to reduce both complexity and stack's height.

This is the last floating gate option with vertical channel described in the chapter. In the next section, the reader can find a Flash cell with a rotated, i.e. horizontal, channel.

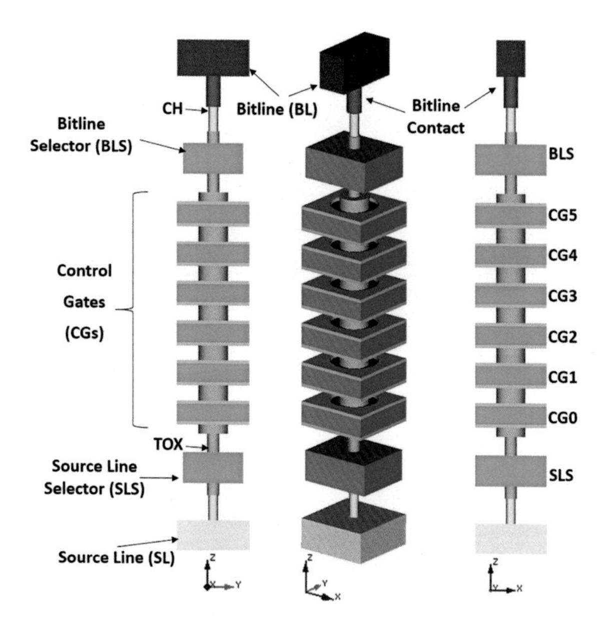

Fig. 5.42 Bird's-eye view and side views of the SCP NAND Flash string

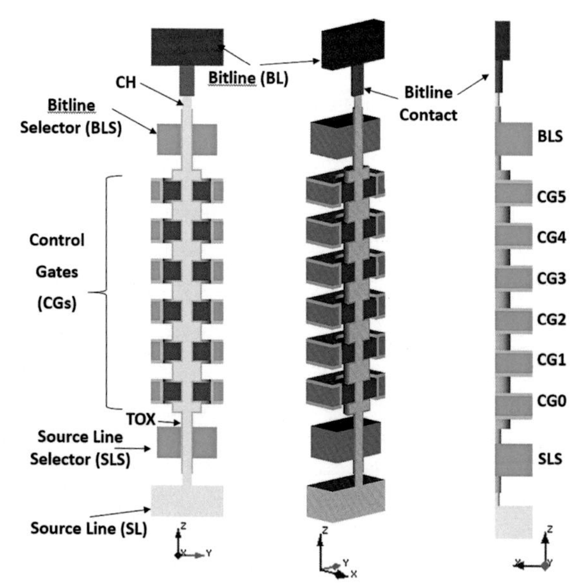

Fig. 5.43 Cross sections of Fig. 5.42

5.7 Horizontal Channel (HC-FG) Flash Cell

In all the floating gate cells analyzed in the previous sections, cell's size is relatively large because of the gate-all-around approach. Moreover, the additional sidewall gates complicate all the operations on the cell itself. Please note that all the above mentioned cells have a vertical channel.

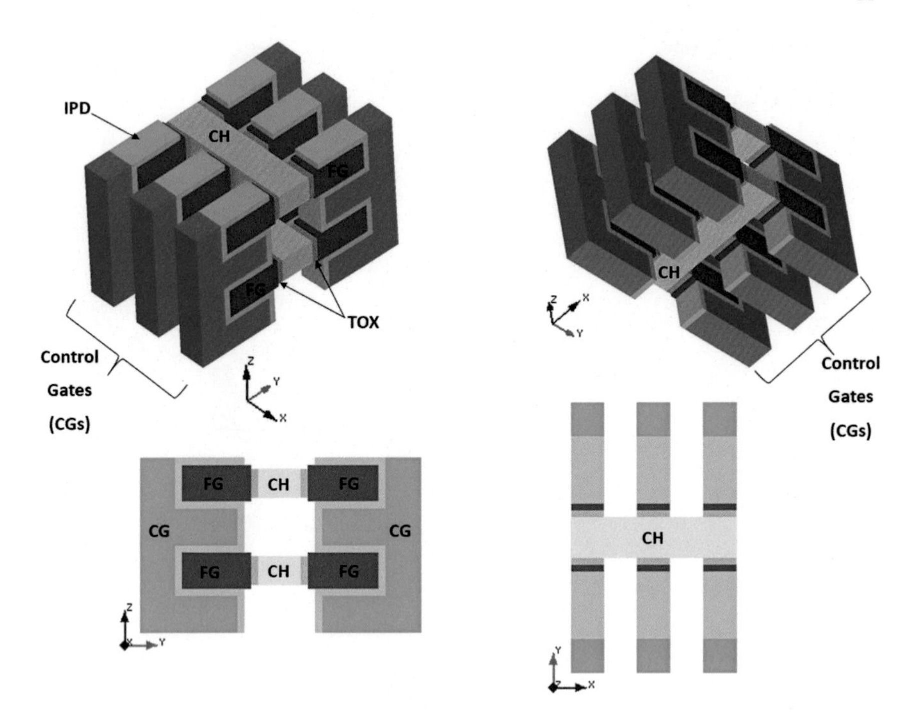

Fig. 5.44 Bird's-eye views and side views of the horizontal channel floating gate (HC-FG) cell

Floating gate can also be vertically integrated by using a horizontal channel [15]. Bird's-eye views and side views of the *Horizontal Channel Floating Gate* (HC-FG) Flash cell are shown in Fig. 5.44.

The starting point is the 3D Stacked NAND architecture presented in Chap. 3, where conventional FG cells are stacked perpendicular to the substrate. In Stacked NAND devices, memory layers can be independently decoded because control gates are split by layers.

With HC-FG, NAND strings are again stacked perpendicular to the substrate, but this time with a double-gate structure, which removes the risk of short circuiting floating gates along the vertical direction; of course, one of the two gates is shared between adjacent cells. Additionally, control gates are shorted across all memory layers.

From a process technology point of view, the floating gate cell is formed by a channel-first technique, very similar to a 2D process. Another advantage is that program and erase operations can be carried out at the same voltage conditions as for 2D FG cells, thus re-using all the existing know-how in terms of cell's biasing schemes and tunnel oxide development.

For 3D NAND devices with horizontal channels, NAND channel decoding is more difficult than what we have seen in Chap. 4 for vertical channel architectures. Main reason being the fact that channels need to be decoded along the NAND string

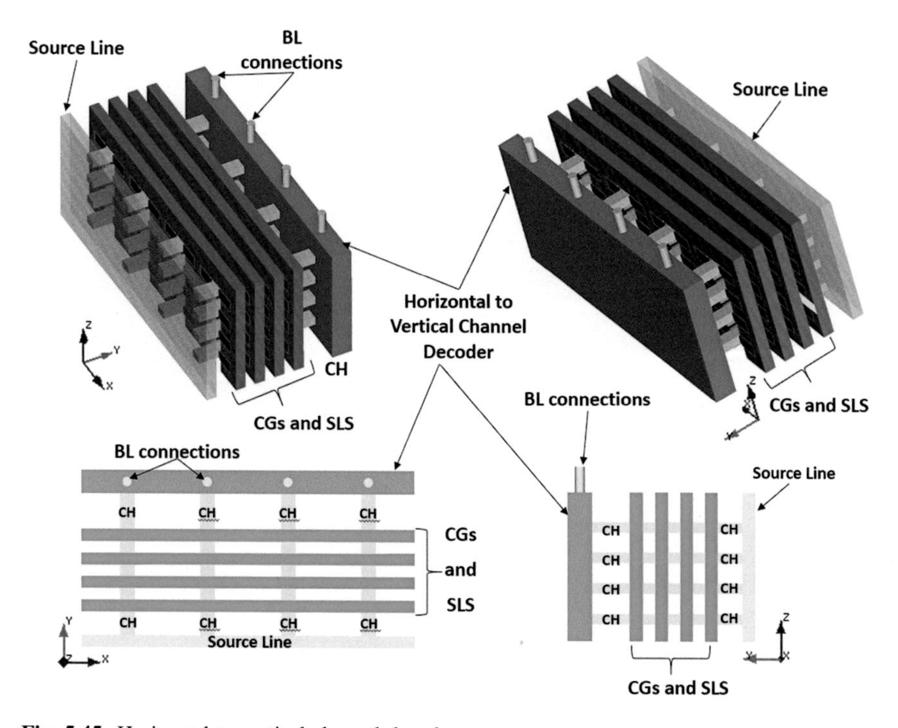

Fig. 5.45 Horizontal-to-vertical channel decoder

direction and then vertically connected to bitlines, which run on top of the array. There are several different ways of implementing this channel decoding, but the basic concept is summarized in the block called *Horizontal to Vertical Channel Decoder* (HVCD) in Fig. 5.45. Of course, each NAND string (channel) goes through all the CGs layers, having the Source line on one side, and HCVD on the opposite side. In Fig. 5.45 there are 16 strings but only 4 BL contacts, which means that a 16:4 decoder has to be implemented, at the minimum cost.

Readers can refer to Chap. 7 for some detailed implementations, like PN diode and normally-ON SSL, which are typically used in 3D VG-type NAND.

This last architecture completes the overview of the most studied alternatives for 3D integration of Floating Gate cells.

5.8 3D FG NAND in the Industry

When looking at FG based 3D NAND memory arrays of hundreds of Gb, Micron has definitely been the most active company. As summarized in Fig. 5.46, the first 3D FG device was presented at the *International Electron Devices Meeting Conference* (IEDM) in 2015 [16]. It is a NAND chip which can be configured as

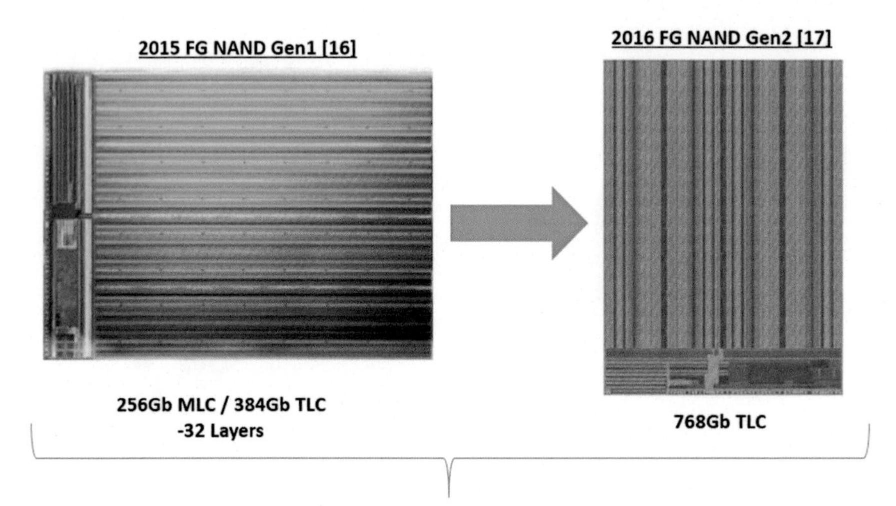

Fig. 5.46 3D FG NAND devices presented by Micron (pictures not in scale) [16, 17]

either 256 Gb MLC or 384 Gb TLC. The vertical string is based on a conventional 3D C-FG cell. In terms of the number of control gate layers, there are 32 memory layers plus additional layers for dummy wordlines and selectors (similar to what discussed in Chap. 4). It is worth pointing out that source and drain selectors are single oxide transistors.

Because the entire stack is built above the silicon, the area under the array is used for circuits, thus saving silicon area and improving bit density. There are 4 metal layers: 2 at the bottom for connecting CMOS circuits underneath the array, and two at the top, one for bitline and one for power and global interconnects, respectively.

A 3D FG NAND with the record density of 768 Gb becomes public at the *IEEE International Solid State Circuits Conference* (ISSCC) in 2016 [17]. This is again based on C-FG vertical strings but in [17] there are more details about how peripheral circuits are laid out under the memory array.

Let's start from the conventional approach. Usually, circuits are placed at the same level of the top of the stack, and they sit along the sides of the matrix, as shown in Figs. 5.47 and 5.48.

As mentioned, in [17] a bunch of circuits are placed under the matrix itself, including Page Buffers, wordline drivers, data path, and both block and column redundancies. For a better understanding, let's consider the simple case where only Page Buffers are moved: the array would look like Figs. 5.49 and 5.50. A bottom view might also be helpful: Fig. 5.51 shows the pool of Page Buffers, and it also highlights that not necessarily Page Buffers need to occupy all the space. The remaining area can be utilized for other CMOS circuits, as displayed in Fig. 5.52.

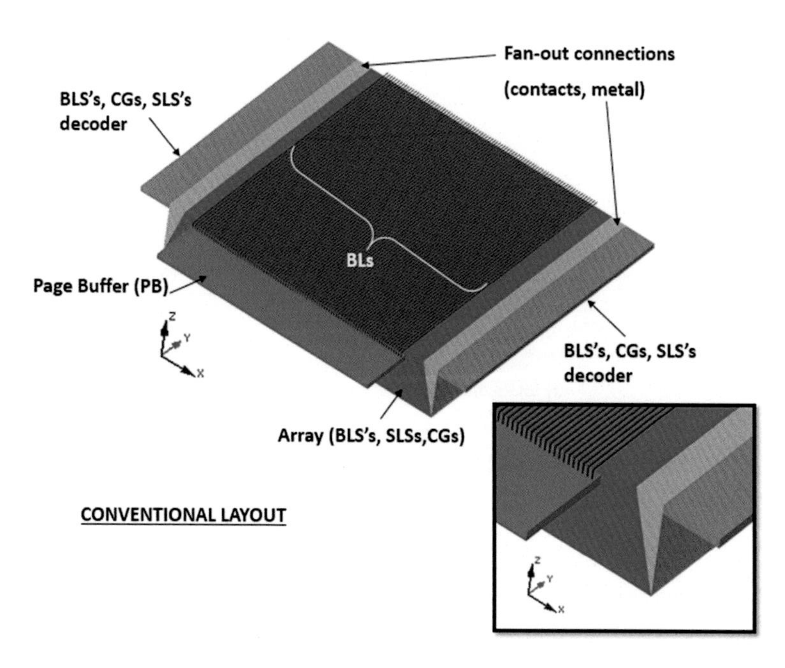

Fig. 5.47 Conventional layout of peripheral circuits at the top of the 3D stack

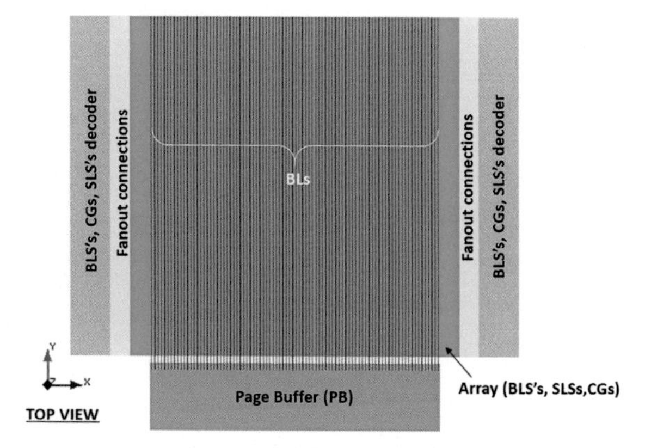

Fig. 5.48 Top view of Fig. 5.47

The benefit of this approach is not only the area reduction. Being close (or "local") to the specific portion of the array they need to serve, Page Buffers and wordline drivers under the array allow cutting bitlines and wordlines, respectively. The resulting line segmentation helps reducing timings, especially for wordlines, which are made of polysilicon with high intrinsic resistivity.

By adopting the same philosophy, the Source Line plate is also "local" to a specific area of the array; in this way, voltage bounces on cell's source are minimized, thus making the all-bitline sensing more reliable.

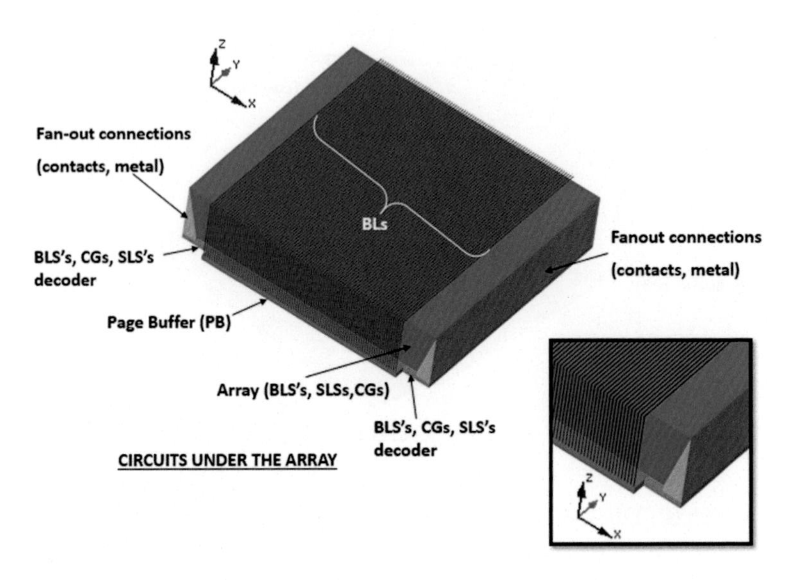

Fig. 5.49 Page buffers under the array [17]

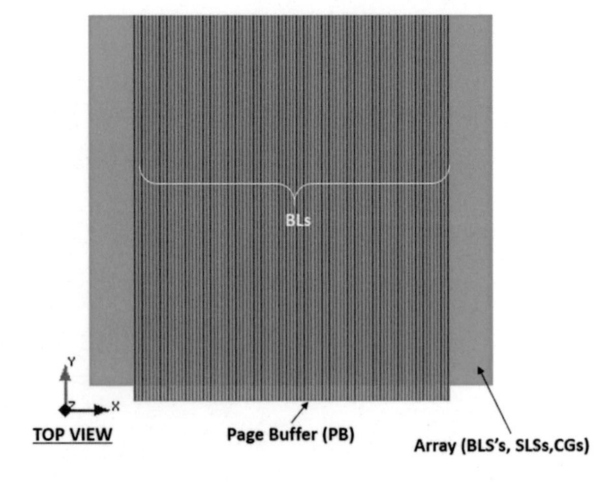

Fig. 5.50 Top view of Fig. 5.49

This innovation on the circuit front is another proof that 3D integration opens the door to completely new opportunities, and the competition between Charge Trap and Floating Gate is definitely not over.

As a lesson learned from Chaps. 4 and 5, there are two aspects to keep in mind when looking at 3D integration. The first one is the functionality of the memory cell itself, and its interaction with the other memory cells in the NAND string and in the whole array. Of course, Charge Trap and Floating Gate give a totally different starting point.

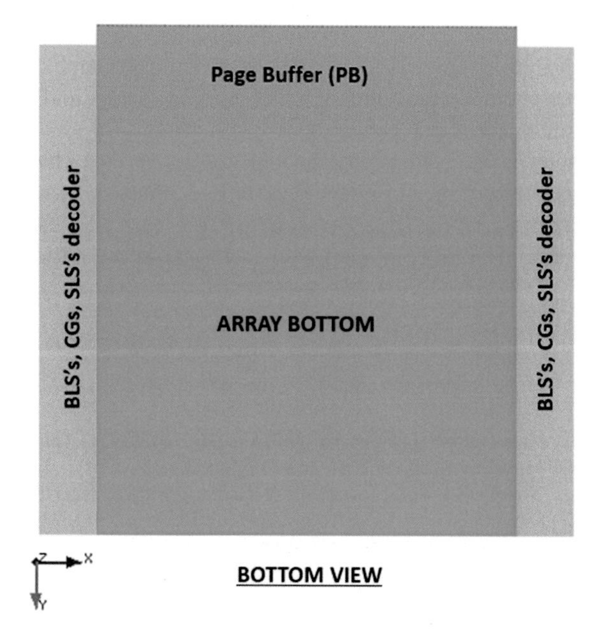

Fig. 5.51 Bottom view of Fig. 5.49

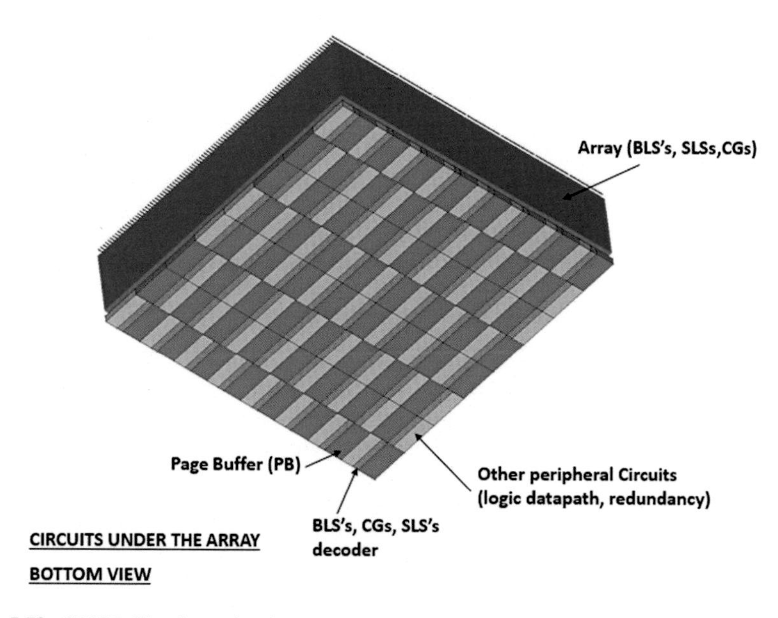

Fig. 5.52 CMOS Circuits under the array [17]

The second one is how to connect wordlines, bitlines, NAND string selectors, and source line to the peripheral circuits, such as decoders and sense amplifiers (Chap. 3). The peripheral connections described in this chapter and in the previous one are just examples.

The next chapter is devoted to the most advanced vertical channel 3D architectures, such as staggering of pillars and bitline contacts, which have been developed to maximize bit density. The good thing is that all these architectures, one way or the other, can be combined with each type of memory cells.

References

1. T. Endoh et al., Novel ultra high density flash memory with a stacked-surrounding gate transistor (S-SGT) structured cell. IEDM Tech. Dig. pp. 33–36 (2001)
2. T. Endoh et al., Novel ultra high density flash memory with a stacked-surrounding gate transistor (S-SGT) structured cell. IEEE Trans. Electron Devices **50**(4), 945–951 (2003)
3. T. Endoh et al., Floating channel type SGT flash memory. in *The 1999 Joint International Meeting*, Hawaii, vol. 99–2, Abstract No. 1323, 17–22 Oct 1999
4. M.S. Seo et al., The 3-dimensional vertical FG nand flash memory cell arrays with the novel electrical S/D technique using the extended sidewall control gate (ESCG). in *Proceedings of IEEE International Memory Workshop* (2010), pp. 1–4
5. M.S. Seo et al., 3-D Vertical FG NAND flash memory with a novel electrical S/D technique using the extended sidewall control gate. IEEE Trans. Electron Devices **58**(9) (2011)
6. S. Whang et al., Novel 3-dimensional dual control gate with surrounding floating-gate (DC-SF) NAND flash cell for 1 Tb file storage application. in *Proceedings of International Electron Devices Meeting (IEDM)* (2010), pp. 668–671
7. Y. Noh et al., A new metal control gate last process (MCGL process) for high performance DC-SF (dual control gate with surrounding floating gate) 3D NAND flash memory. in *Symposium on VLSI Technology* (2012), pp. 19–20
8. R. Micheloni, L. Crippa, Multi-bit NAND flash memories for ultra high density storage devices (chapter 3). in Y. Nishi (ed.) *Advances in Non-volatile Memory and Storage Technology* (Woodhead Publishing, 2014)
9. R. Micheloni et al., High-capacity NAND flash memories: XLC storage and single-die 3D (chapter 7). in G. Campardo et al. (ed.) *Memory Mass Storage* (Springer, 2011)
10. H. Yoo et al., New read scheme of variable Vpass-read for dual control gate with surrounding floating gate (DC-SF) NAND flash cell. in *Proceedings of 3rd IEEE International Memory Workshop* (2011), pp. 1–4
11. S. Aritome et al., Advanced DC-SF cell technology for 3-D NAND flash. IEEE Trans. Electron Devices **60**(4), 1327–1333 (2013)
12. M.S. Seo et al., A novel 3-D vertical FG nand flash memory cell arrays using the separated sidewall control gate (S-SCG) for highly reliable MLC operation. in *Proceedings of 3rd IEEE International Memory Workshop (IMW)* (2011), pp. 1–4
13. M.S. Seo et al., Novel concept of the three-dimensional vertical FG nand flash memory using the separated-sidewall control gate. IEEE Trans. Electron Devices **59**(8), 2078–2084 (2012)
14. M.S. Seo, Highly scalable 3-D vertical FG NAND cell arrays using the sidewall control pillar (SCP). in *Proceedings of 4th IEEE International Memory Workshop (IMW)* (2012), pp. 1–4
15. K. Sakuma et al., Highly scalable horizontal channel 3-D NAND memory excellent in compatibility with conventional fabrication technology. IEEE Electron Device Lett. **34**(9) (2013)

16. K. Parat, C. Dennison, A floating gate based 3D NAND technology with CMOS under array. *in Conference on International Electron Devices Meeting (IEDM)*, San Francisco (USA), Dec 2015
17. T. Tanaka et al., A 768 Gb 3 b/cell 3D-floating-gate NAND flash memory. in *2016 IEEE International Solid-State Circuits Conference (ISSCC), Digest of Technical Papers*, (San Francisco, USA, 2016), pp. 142–143

Chapter 6
Advanced Architectures for 3D NAND Flash Memories with Vertical Channel

Luca Crippa and Rino Micheloni

6.1 Introduction

One of the key metrics to benchmark different 3D architectures is the storage density, which is here indicated with *Bit_Density* [1]. Given a specific Flash memory die, this density is defined as the ratio between the storage capacity of the die, *Die_Capacity* [bit], and its silicon area, *Die_Size* [mm^2].

$$Bit\_Density = \frac{Die\_Capacity}{Die\_Size} \tag{6.1}$$

Die_Size can be calculated as the sum of the area of the memory matrix, A_{MAT}, with the area of the peripheral circuits (e.g. row and column decoders, charge pumps, sense amplifiers, etc.), A_{PERI}:

$$Die\_Size = A_{MAT} + A_{PERI} \tag{6.2}$$

Generally speaking, A_{MAT} is bigger than A_{PERI}. For a planar memory device, A_{MAT} [mm^2] is:

$$A_{MAT} = \frac{Die\_Capacity \cdot A_{CELL}}{n_{bitpercell}} \tag{6.3}$$

where A_{CELL} [µm^2] is the area of a single memory cell, and $n_{bitpercell}$ is the number of logic bits stored in it (multi-level storage, Chap. 3).

L. Crippa (✉) · R. Micheloni
Performance Storage BU, Microsemi Corporation, Vimercate, Italy
e-mail: luca.crippa@ieee.org

R. Micheloni
e-mail: rino.micheloni@ieee.org

R. Micheloni (ed.), *3D Flash Memories*, DOI 10.1007/978-94-017-7512-0_6

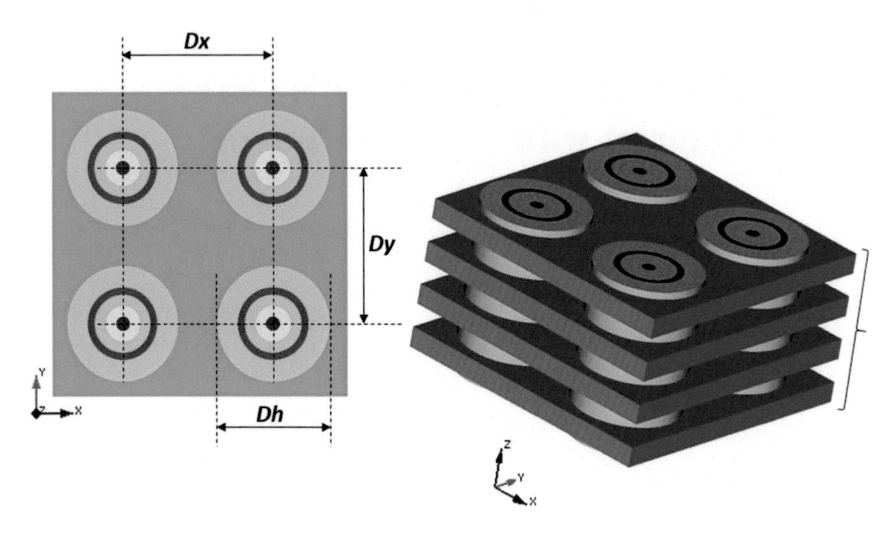

Fig. 6.1 3D memory array

For a generic 3D memory array, as shown in Fig. 6.1, Eq. (6.3) becomes

$$A_{MAT} = \frac{Die\_Capacity \cdot A_{CELL}}{Ln \cdot n_{bitpercell}} \tag{6.4}$$

where Ln is the number of memory layers, i.e. the number of Control Gate planes, and A_{CELL} can be computed as

$$A_{cell} = Dx \cdot Dy \tag{6.5}$$

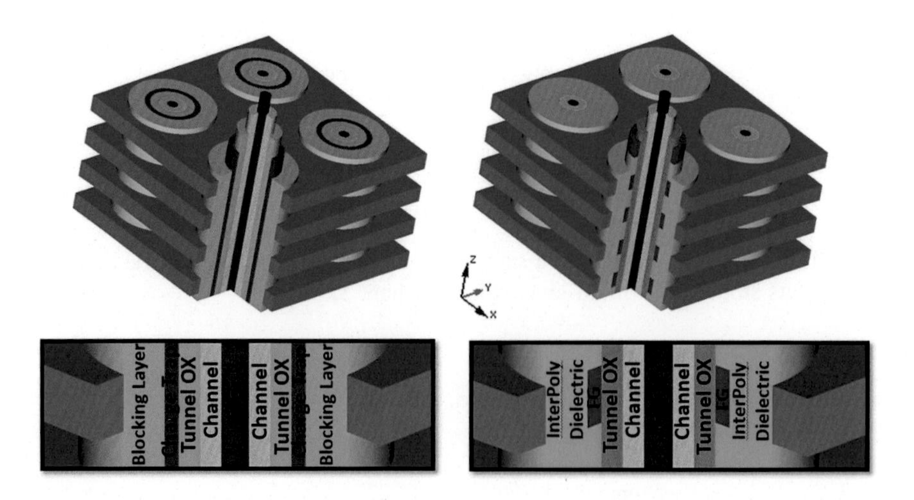

Fig. 6.2 3D NAND flash cells: charge trap (CT) and floating gate (FG)

Therefore, the overall area can be reduced by either increasing the number of layers, or by shrinking the linear dimensions *Dx* and *Dy*. It is worth highlighting that Eq. (6.5) is not taking into account any array overhead.

All the architectural solutions described in this chapter can be used in conjunction with both floating gate and charge trap Flash cells, as displayed in Fig. 6.2.

6.2 Arrays of Conventional Pillars (Holes)

When building an entire 3D NAND Flash memory matrix, Eq. (6.4) remains valid only if we assume that a single cell is replicated along both X and Y axes without any interruption, as shown in Figs. 6.3 and 6.4. The addition of Source Line, SLS and BLS transistors, and 16 CG layers doesn't change the equation because the cross section X-Y is not impacted.

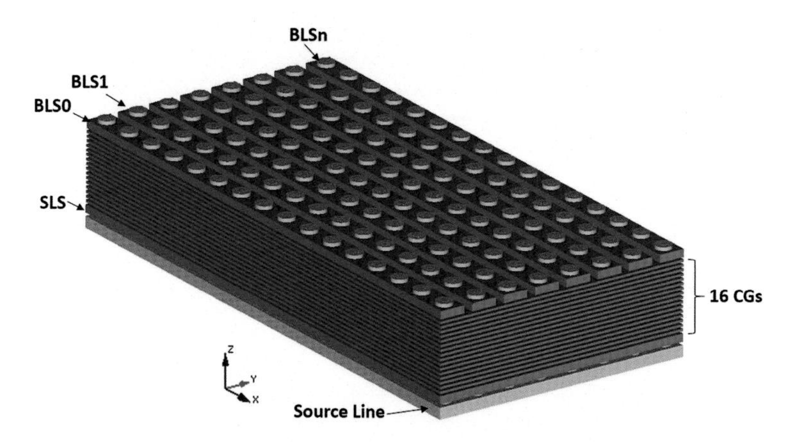

Fig. 6.3 Array of pillars in a 3D NAND flash matrix

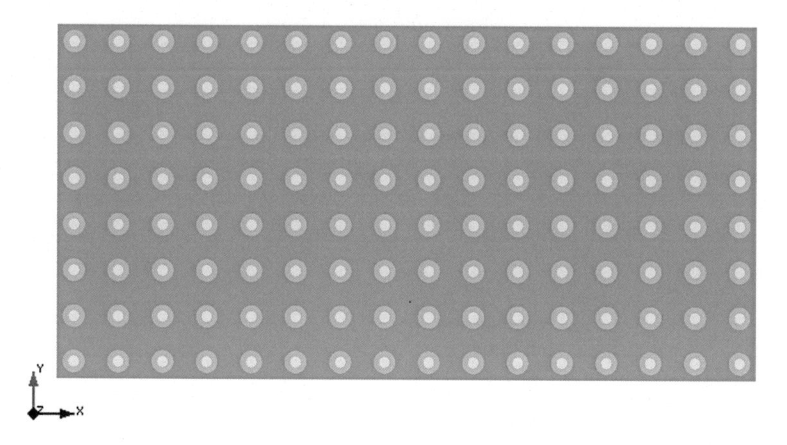

Fig. 6.4 Top view of Fig. 6.3

As we have discussed in the previous chapters, such a structure has two main issues:

- Program and Read disturbs (Chap. 2);
- Source Line layer can't be a single big layer because its parasitic resistance would be too high and, therefore, there is a need for a metal mesh.

Let's now refer to Fig. 6.5, which is Fig. 6.3 with bitlines and bitline contacts.

Compared to planar arrays, 3D NAND is more prone to Program Disturb because more pages sit on the same CG layer [2, 3]. An example might help to understand better the problem. With reference to Fig. 6.6, the memory cell to be

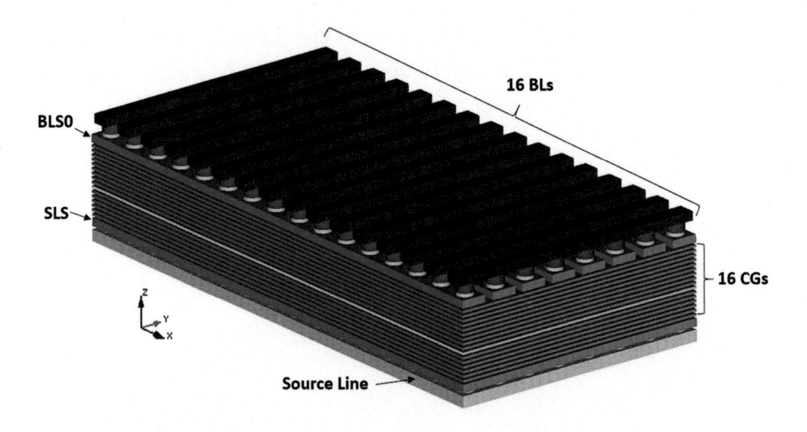

Fig. 6.5 3D NAND flash matrix with vertical channels (pillars)

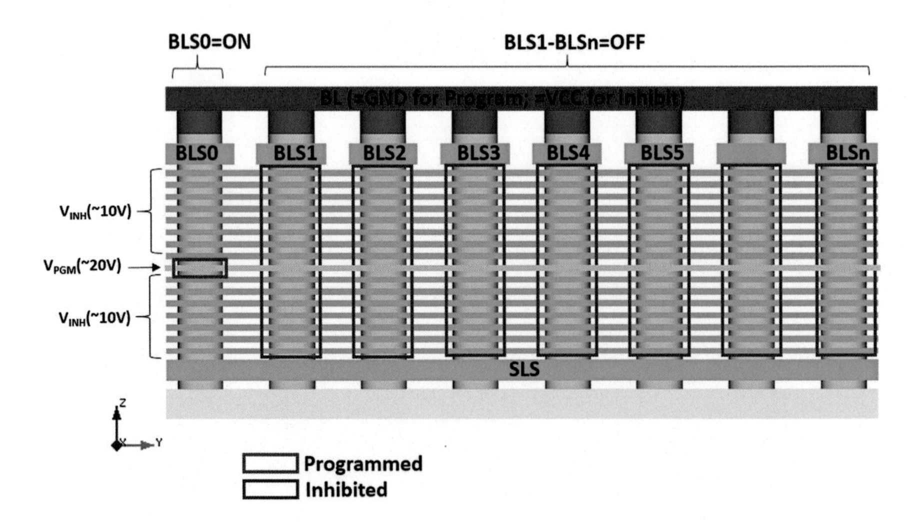

Fig. 6.6 Program disturb

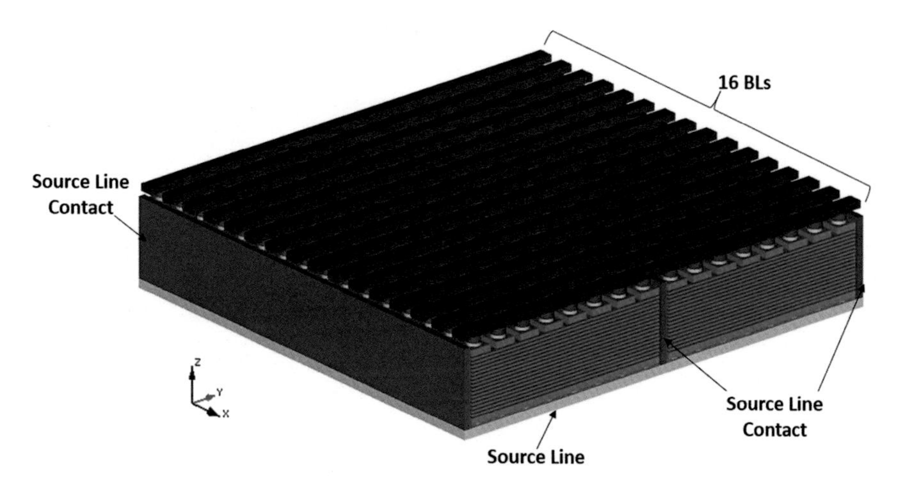

Fig. 6.7 Array slits

programmed is the one inside the red box: its gate is biased at V_{PGM}, the corresponding bitline is at GND, BLS0 is ON, and all the remaining BLSs are OFF such that pillars are inhibited (thanks to the self-boost effect, Chap. 2). In other words, while in a planar array the Program Disturb is limited to the X direction, in 3D it propagates along the Y direction too. As a result, the overall Disturb is proportional to the number of pillar per CG layer.

Of course, the same behavior applies to Read Disturb.

In order to limit the propagation of disturbs, it is necessary to reduce the size of CG layers. This is usually accomplished by introducing slits inside the array, as displayed in Fig. 6.7 and Fig. 6.8. Once the CG layer is cut, the space of the slit itself can be used to help reducing the Source Line resistance, as already discussed in Chap. 4. It is worth highlighting that a smaller CG layer is also equivalent to a smaller logic Block size, which is extremely important when looking at data management inside a Solid State Drive [4].

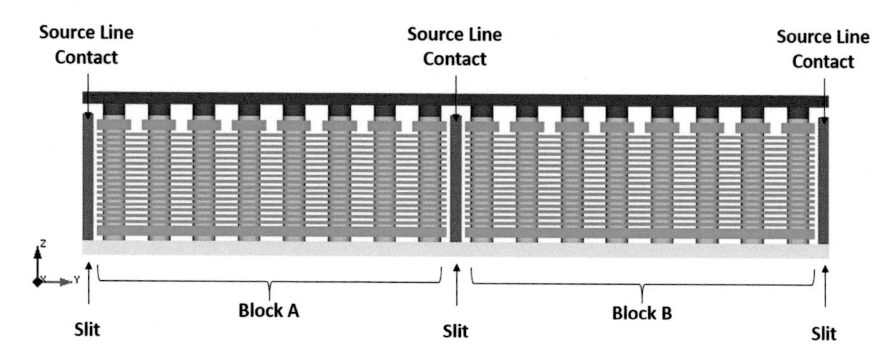

Fig. 6.8 Side view of Fig. 6.7

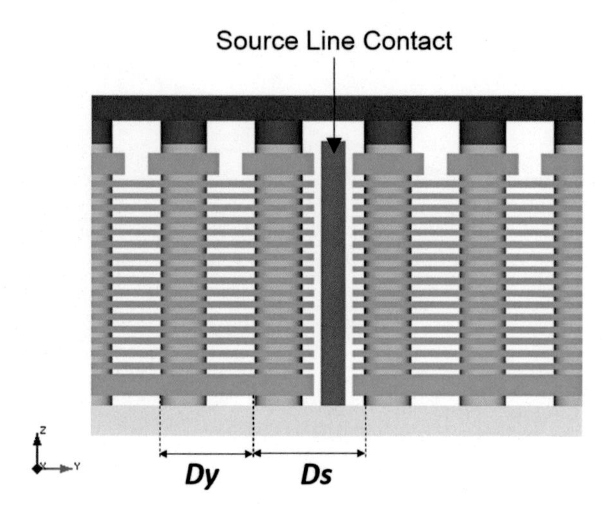

Fig. 6.9 Array overhead introduced by slits

The downside of array slits is their cost in terms of silicon area, because the matrix becomes wider along the Y direction. With reference to Fig. 6.9, the slit overhead per NAND block *Doh* is

$$Doh = Ds - Dy \tag{6.6}$$

where *Ds* is the pitch of pillars with a slit in between.

This overhead is shared among the *p* pillars of a NAND block along the Y axis, such that the effective overhead per memory cell *Doheff* is

$$Doheff = \frac{Ds - Dy}{p} \tag{6.7}$$

As a result, the effective cell size A_{cell_eff} becomes

$$A_{cell\_eff} = Dx \cdot (Dy + Doheff) \tag{6.8}$$

A_{cell_eff} can replace A_{cell} in Eq. (6.4). As a matter of fact, a higher number of pillars per CG layer improves the bit density but it worsens Program and Read Disturbs.

6.3 Arrays of Staggered Pillars (Holes)

In Fig. 6.1 NAND string pillars are laid out as a simple matrix; in other words, pillars of 2 adjacent rows (columns) are aligned along the Y (X) axis. In this section we present a different layout, called *staggered pillars*, such that the array size can be reduced along one direction [5–7].

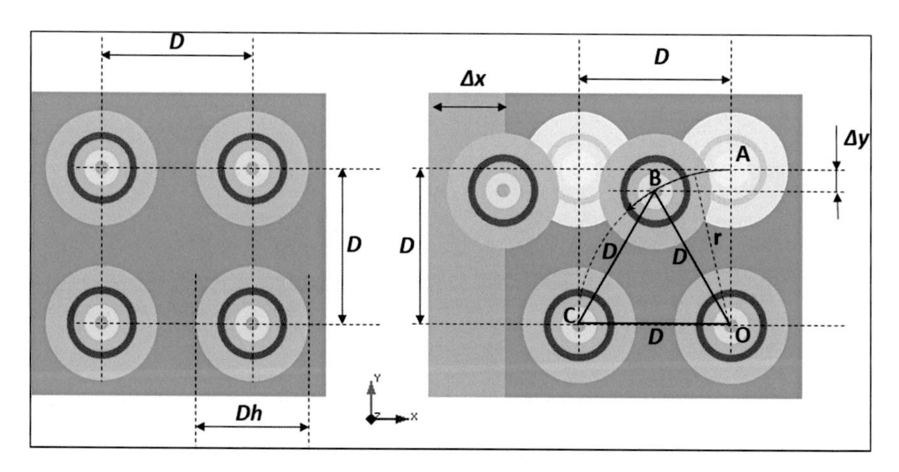

Fig. 6.10 Conventional (*left*) and staggered (*right*) pillar architectures

Let's start from the left hand side of Fig. 6.10, which is Fig. 6.1 with $Dx = Dy = D$, just for sake of simplicity. On the right hand side we draw a circle centered in "O" with radius r = D, and we rotate the center "A" of the pillar of the second row from the bottom to point "B", where $\overline{BC} = D$.

In this way, center-to-center distance among all pillars is still D, but pitch along Y axis, i.e. the distance between the first row and the second row (the staggered one), has been reduced by Δy.

At the same time, the matrix gets bigger in the X direction by Δx. Because X is the wordline direction (Fig. 6.7), Δx is paid per wordline. Nowadays, each wordline is made of 16k bytes and, therefore, Δx becomes negligible.

At this point Eq. (6.5) can be re-written as

$$A_{cell} = D \cdot (D - \Delta y) \tag{6.9}$$

Δy can be easily calculated by looking at Fig. 6.11

$$\Delta y = D - D\cos\alpha = D\left(1 - \cos\frac{\pi}{6}\right) \tag{6.10}$$

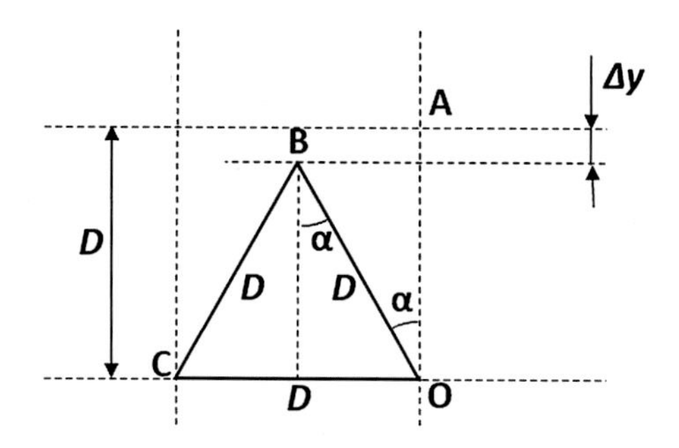

Fig. 6.11 Calculation of *Dry* with $Dx = Dy = D$

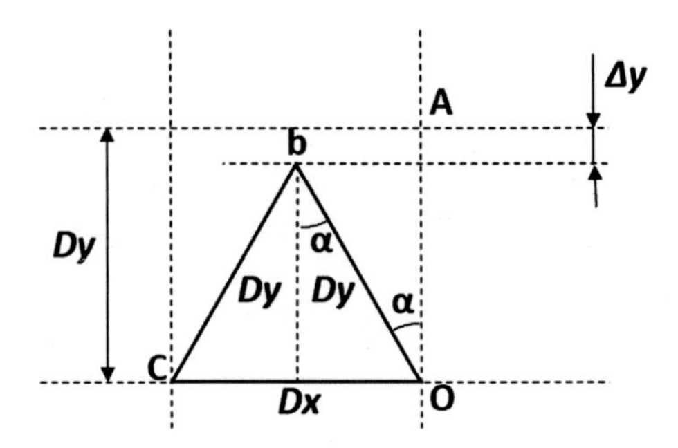

Fig. 6.12 Calculation of *Dry* with $Dx \neq Dy$

All in all, this layout change translates into a significant 13.5 % reduction of the matrix size along the Y direction.

Of course, all the calculations above can be repeated for an array where *Dx* and *Dy* have different values, as sketched in Fig. 6.12. In this case, Δy is equal to

$$\Delta y = Dy \cdot \left(1 - \sqrt{1 - \left(\frac{Dx}{2Dy}\right)^2}\right) \tag{6.11}$$

Therefore, Eq. (6.8) becomes

$$\Delta_{cell\_eff} = Dx \cdot \left(Dy\sqrt{1 - \left(\frac{Dx}{2Dy}\right)^2} + Doheff\right) \tag{6.12}$$

While saving silicon area, the above described staggering technique has a major impact on the bitline density, as sketched in Fig. 6.13 (pillars only) and Fig. 6.14 (bitlines): this is due to the misalignment of even and odd rows of pillars along the X axis. As usual, X corresponds to the wordline direction, while bitlines run orthogonal (Y axis).

Because the number of bitlines is doubled compared to the conventional approach, pillar staggering simplifies BLS decoding. In fact, odd pillars (Fig. 6.13) are connected to odd bitlines, while even pillars are connected to even bitlines. Hence, each couple of even and odd adjacent bitline selectors can be driven in parallel, i.e. transistors can be shorted together, as shown in Fig. 6.15.

If we keep the number of pillars per NAND block fixed (8 × 16 in Fig. 6.13), then page size doubles, while block size stays the same. Program disturb is halved because of the number of BLS is halved. Of course, the number of pillars along the X direction can be changed to fit NAND product specification.

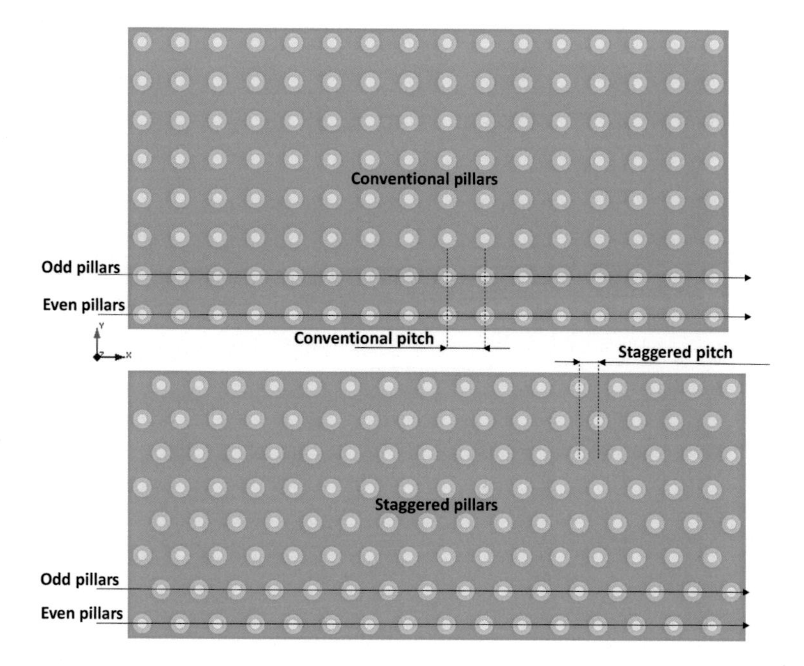

Fig. 6.13 Conventional versus staggered pillars

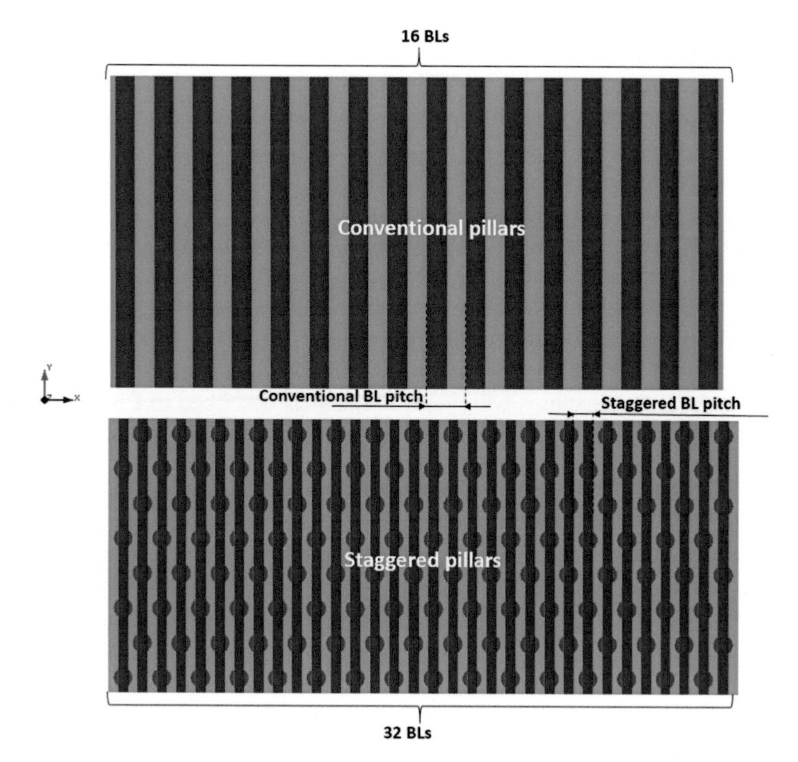

Fig. 6.14 Impact of staggered pillars on bitlines density

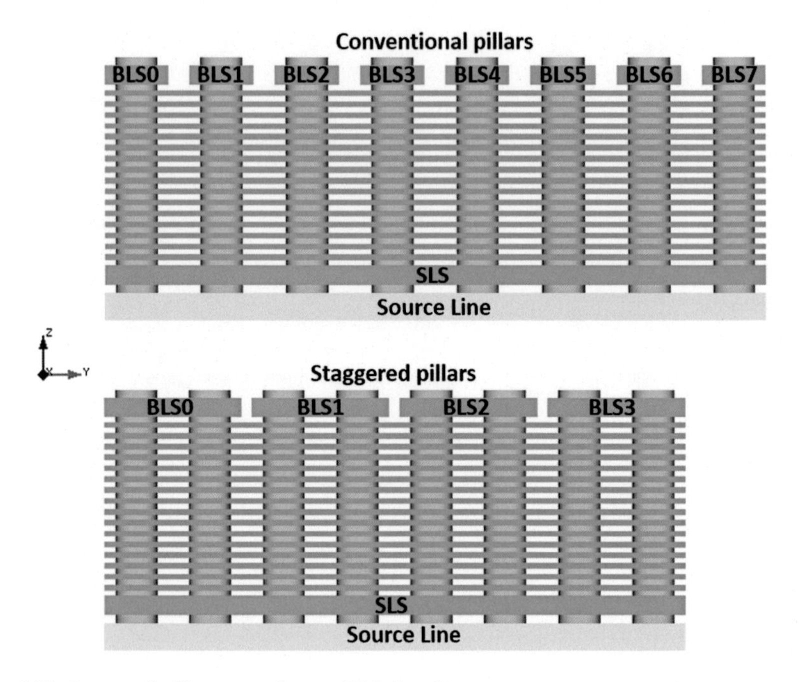

Fig. 6.15 Impact of pillar staggering on BLS decoding

At this point we are ready to re-draw the NAND array of Fig. 6.5 after staggering the pillars along the wordline direction, as sketched in Fig. 6.16. The reader can notice that there are 4 BLS selectors instead of 8, and 32 bitlines instead of 16. To better appreciate the details, Fig. 6.16 includes 2 zoom boxes: the one on the left hand side shows a top view, where it is clearly visible how 2 adjacent BLS selectors are shorted together; the other zoom presents a view from the bottom, after removing all the layers except for bitlines and contacts to bitlines.

Of course, this staggering technique can be combined with the slit architecture described in the previous section, resulting in the memory array drawn in Figs. 6.17 and 6.18, where 2 NAND Blocks have been used as an example.

All 3D NAND architectures with vertical channels described in Chaps. 4 and 5 can modified by staggering pillars, as we'll see in the following sections.

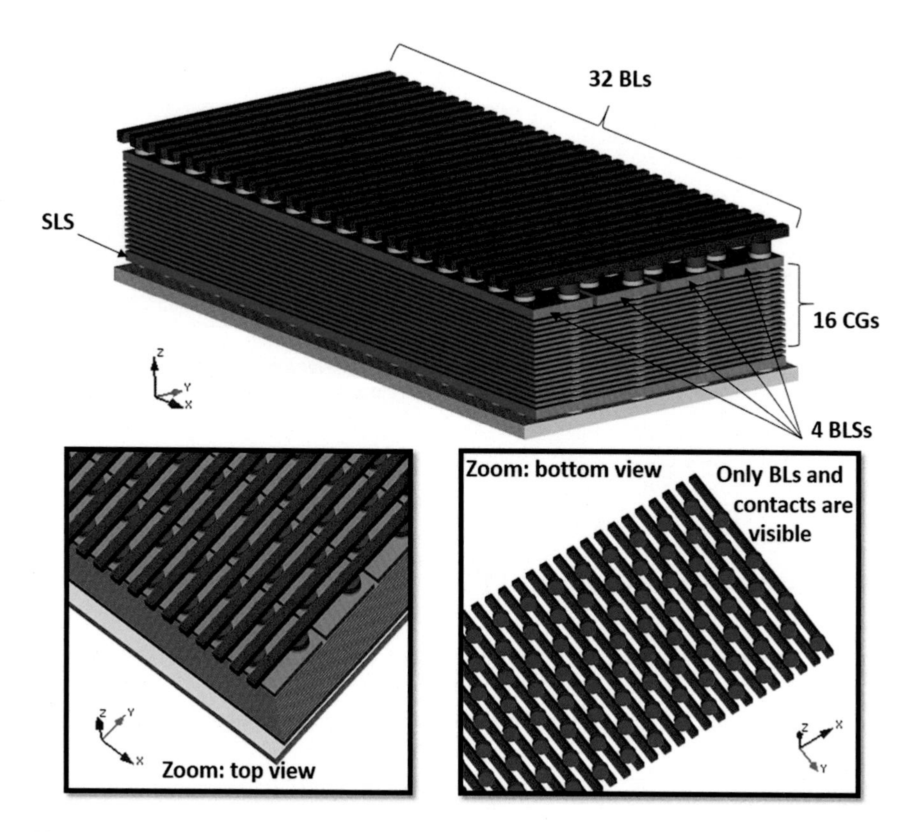

Fig. 6.16 3D NAND flash matrix with staggered pillars

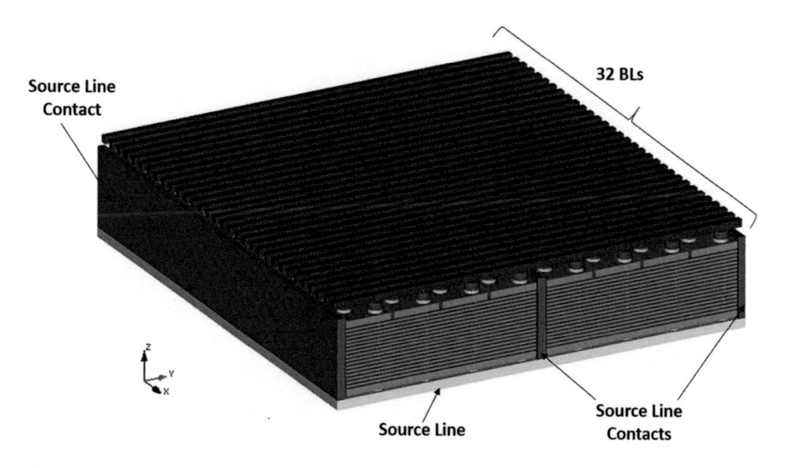

Fig. 6.17 3D NAND array with staggered pillars and slit (2 blocks)

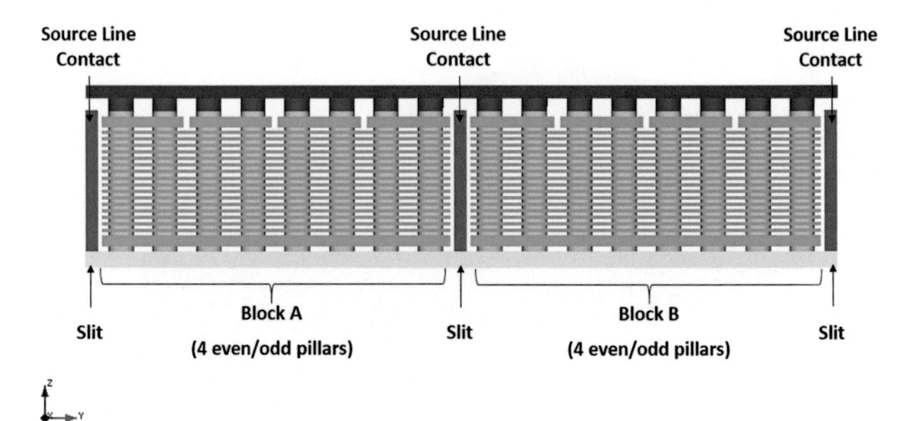

Fig. 6.18 Y-Z cross section of Fig. 6.17

6.4 P-BiCS with Staggered Pillars

Figure 6.19 shows a P-BICS array with staggered pillars. For a detailed description of this 3D vertical channel architecture, please refer to Chap. 4.

After removing bitlines and Source Line, the array looks like Fig. 6.20: pillar staggering is visible on both SLS and BLS groups of layers. The Pipe structure is hardly visible from the top; therefore, it is necessary to take a look at the array from the bottom, as displayed in Fig. 6.21. As expected, staggering of NAND strings is also visible from this side.

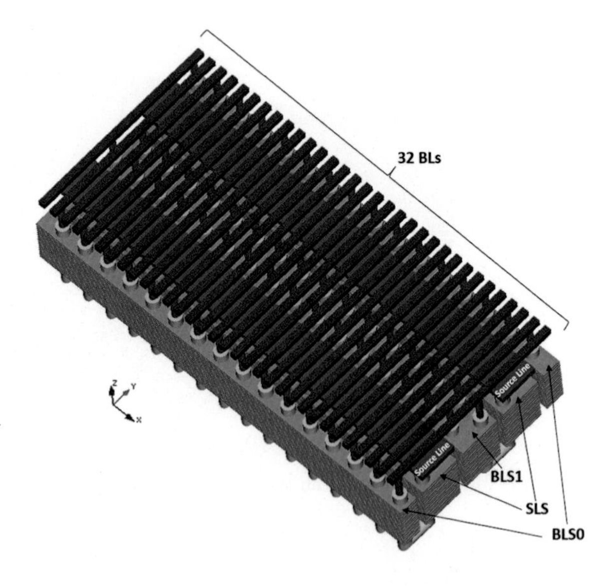

Fig. 6.19 P-BiCS with staggered pillars

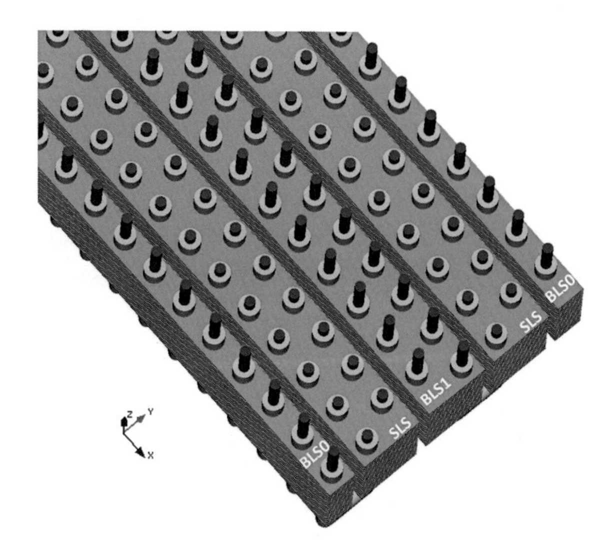

Fig. 6.20 P-BiCS with staggered pillars after removal of bitlines and source line

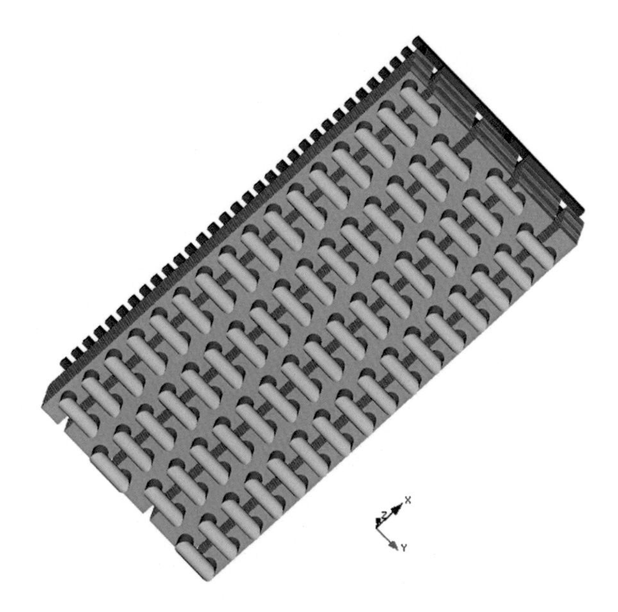

Fig. 6.21 Bottom bird's-eye view of P-BiCS with staggered pillars

Figures 6.22 and 6.23 are frontal and lateral views of Fig. 6.19, respectively. Each NAND string is made of 32 memory cells, 16 on each side of the Pipe. In fact, there are 32 CG layers, from CG0 (adjacent to BLS) to CG31 (adjacent to Source Line). As described in Chap. 4, control gates with the same number (i.e. CG31) might be shorted together in the fan-out region just outside the array.

It is worth underlining that, in this diagram, SLS and BLS selectors have the same thickness of the control gates. Of course, this is a big advantage in terms of 3D integration, but they need to be carefully designed as they are cells.

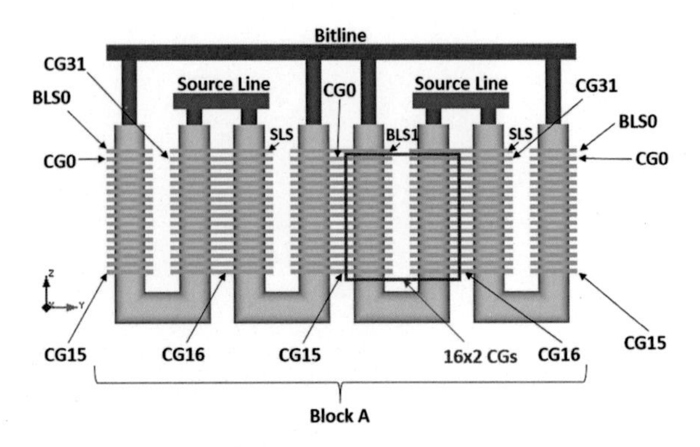

Fig. 6.22 Y-Z view of Fig. 6.19

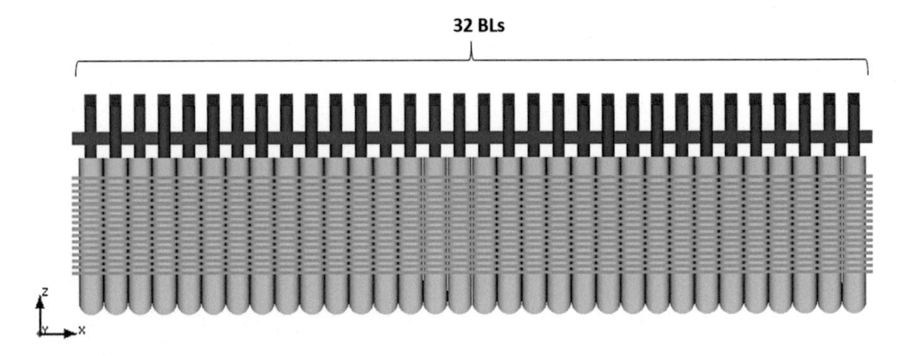

Fig. 6.23 Y-Z view of Fig. 6.19

Selectors may also be built by using 2 or 3 cells in series (*Cell-Type String Select Transistor* [8]); in this case, cells used as selectors need to be programmed like a normal cell. Dummy layers at the top of the stack might be inserted to prevent the n + region of bitline contacts from diffusing into the channel underneath the selectors: this is mandatory to hold the boosted voltage when the channel is inhibited during programming. This approach is not restricted to P-BiCS: it can be used in conjunction with other types of charge trap cells.

In Fig. 6.22 there are 2 signals to drive bitline selectors, i.e. BLS0 and BLS1, and one signal, i.e. SLS, for source line selectors: please keep in mind that layers biased by signals with the same name are shorted together in the fan-out region outside the memory array.

It is worth taking a closer look at the BLS0 selector which, as a matter of fact, can be viewed as a split selector. BLS0 on the left hand side of Fig. 6.22 is used to select a row of even pillars, while BLS0 on the right hand side drives a row of odd pillars. Figure 6.24 might help getting a better understanding. When a page is addressed, all bitlines need to carry a bit of information. As such, BLS0 at both top and bottom sides of Fig. 6.24 need to be ON. Same applies to BLS1; perhaps, this

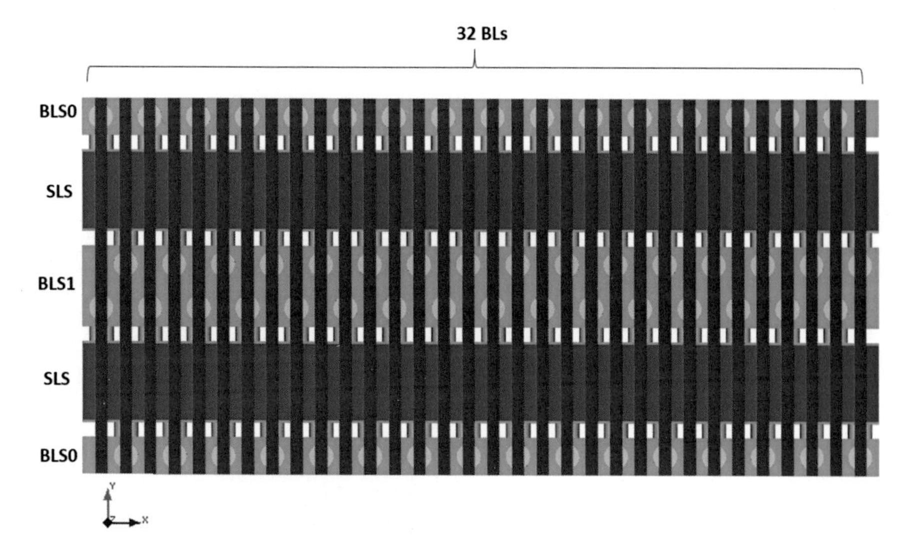

Fig. 6.24 Top view of Fig. 6.19

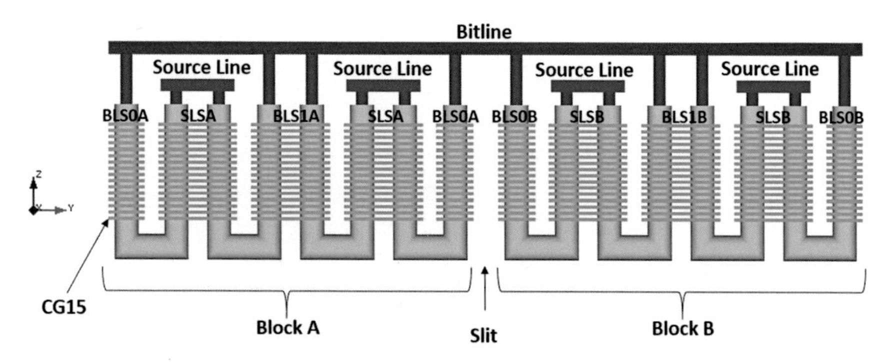

Fig. 6.25 P-BiCS array with 2 NAND blocks and staggered pillars

is more evident given the topology. Obviously, SLS can be common to the entire NAND Block, as BLS0 and BLS1 select which strings are selected.

By replicating Fig. 6.22 along the Y axis, we can now build a memory array with 2 NAND Blocks, as shown in Fig. 6.25. Please note that bitline is shared, while BLSs are driven by different signals (BLS0A, BLS1A, BLS0B, and BLS1B), as it is for SLSs (SLSA and SLSB). Control gates are split as well, to reduce disturbs and parasitic load.

Figure 6.26 is the bird's-eye view of Fig. 6.25. As usual, a slit is used to isolate NAND Blocks. Bottom view sketched in Fig. 6.27 highlights the logic meaning of the control gate layer at the bottom of the 3D NAND stack: it can be either CG15 or CG16 depending on the position inside the string.

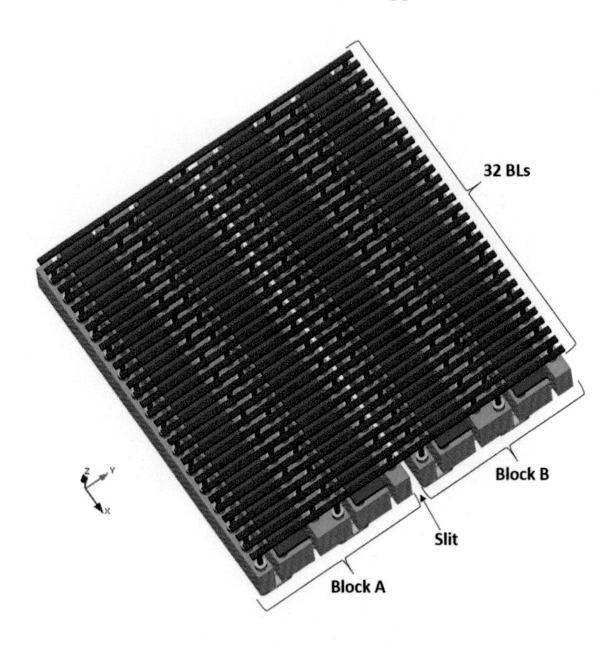

Fig. 6.26 Bird's-eye view of Fig. 6.25

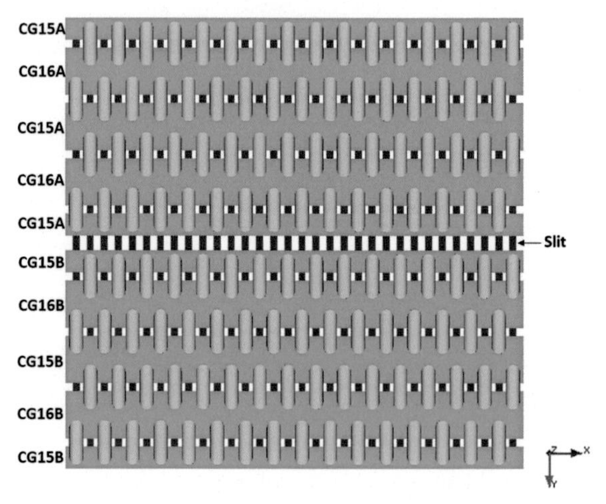

Fig. 6.27 Bottom view of Fig. 6.26

6.5 Monolithic Even-Odd Rows of Pillars

The best possible configuration for minimizing Program and Read disturbs is shown in Fig. 6.28. Basically, each control gate layer is used for one pair of even and odd rows of pillars. In other words, there is a slit along the Y axis every 2 rows of pillars. We'll refer to this architecture as MEOP, *Monolithic Even-Odd rows of Pillars*.

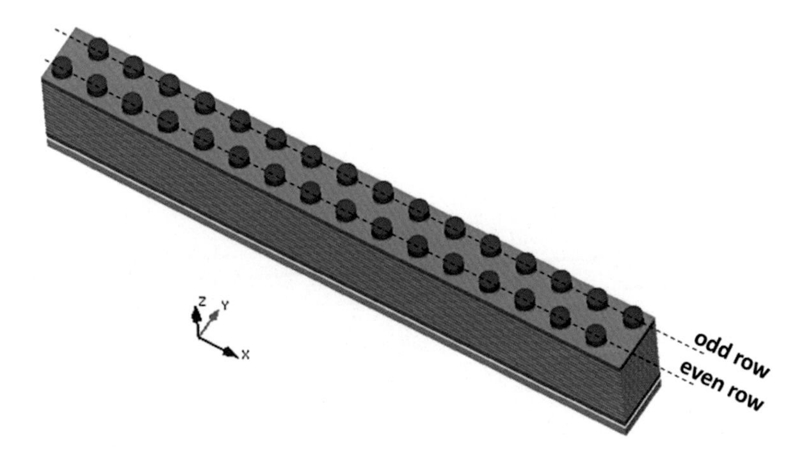

Fig. 6.28 Row of paired even-odd pillars with bitline contacts

This configuration allows the introduction of a very dense Source Line mesh, as each row of pillars can have one lateral contact to the Source Line, as sketched in Fig. 6.29. Of course, all these additional slits have a negative impact on the bit density.

Multiple monolithic even-odd pillars can be placed along the Y direction to build a 3D NAND array, as displayed in Fig. 6.30. As anticipated, the Source Line mesh (formed by vertical Source Line contacts) is pretty dense.

In this example there are 8 MEOPs, from MEOP-A to MEOP-H (Fig. 6.31). A NAND Flash Block can be made of multiple MEOPs, by shorting their control gates together. We can build 4 NAND Blocks (A′, B′, C′, D′) by connecting 2 adjacent MEOPs: program disturb doubles but there is no impact on the Source

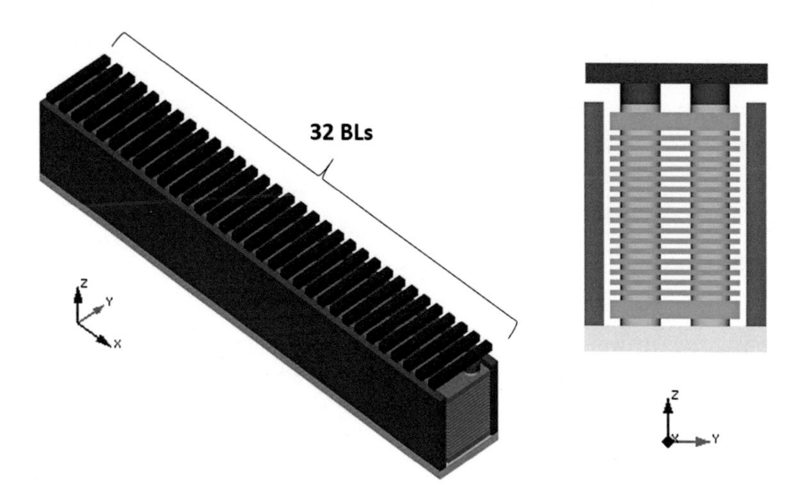

Fig. 6.29 Row of paired even-odd pillars with source line contacts

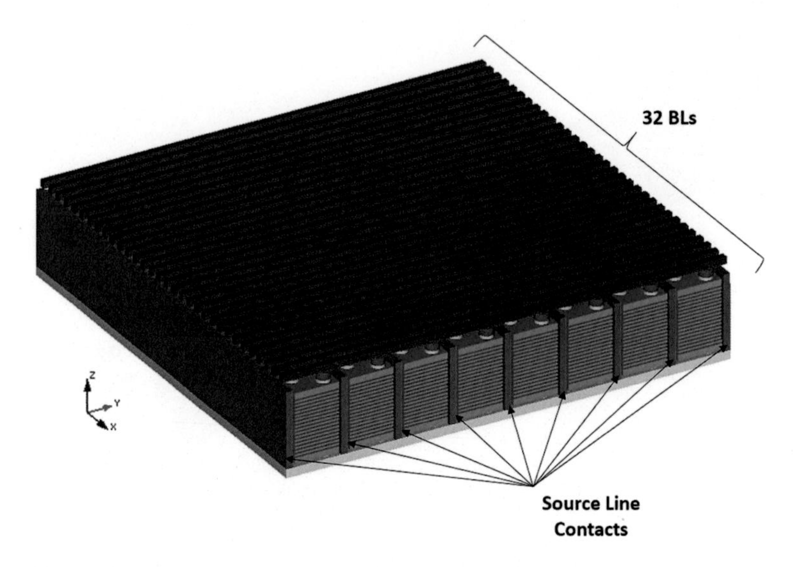

Fig. 6.30 3D NAND array based on monolithic even-odd pillars

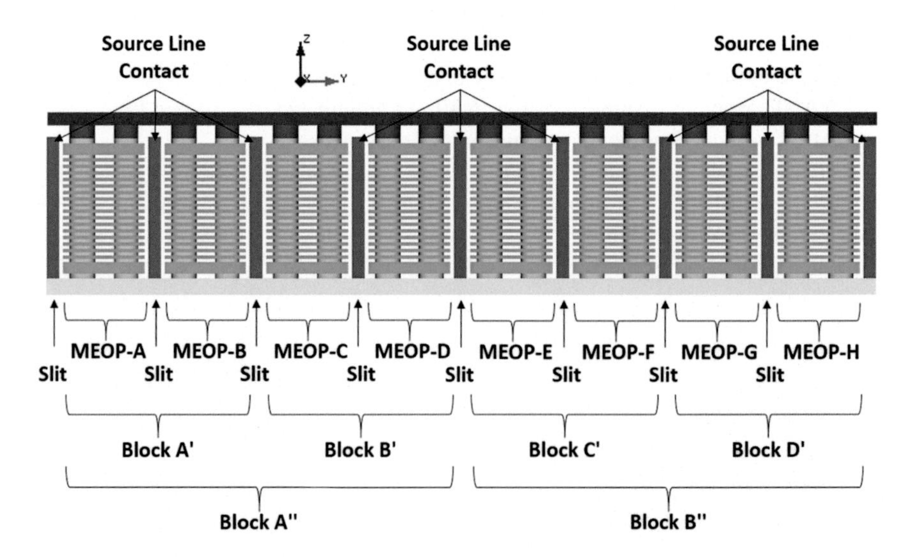

Fig. 6.31 MEOPs can be merged to form NAND flash blocks

Line resistance. If we connect 4 MEOPs, then Fig. 6.30 becomes similar to Fig. 6.17 (i.e. they have the same Program and Read disturbs), because there are just 2 logical NAND Blocks, A″ and B″, but the number of contacts to the Source Line plate is definitely higher.

Control gates of different MEOPs are shorted together in the fan-out region. A typical staircase solution is sketched in Figs. 6.32 and 6.33.

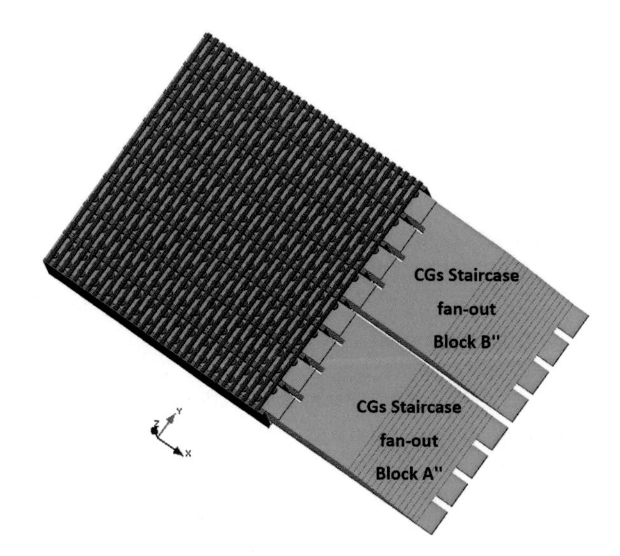

Fig. 6.32 Top view of the fan-out region of Fig. 6.30

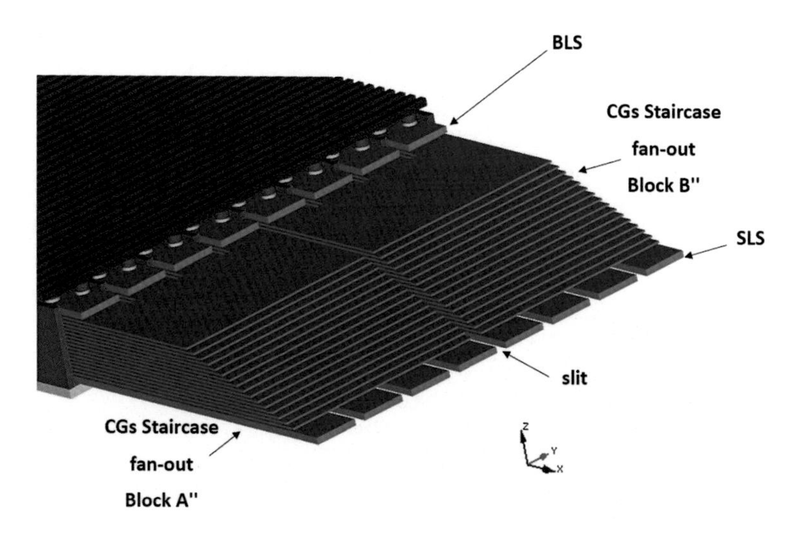

Fig. 6.33 Bird's-eye view of Fig. 6.32

Why shorting MEOPs together if Program and Read disturbs get worse? The reason can be found in the fan-out connections of control gates [3]. In fact, each of the 16 layers of the 3D stack shown in Fig. 6.33 need to be contacted and routed to the wordline drivers: connecting 16 layers in the pitch of 2 rows of pillars is totally different from connecting 16 layers in the pitch of 8 rows of pillars! The latter is definitely more challenging in terms of line (metal) density. Of course, this gets even worse as the number of CG layers gets bigger.

6.6 Staggered Bitline Contacts

As discussed in Sect. 6.3, staggered pillars imply using 2 rows of pillars to build a NAND page (Fig. 6.13): there is an area saving at the expense of a doubled bitline density.

This architecture can be pushed even further by staggering bitline contacts. Figure 6.34 shows a NAND page built by using 2 pairs of even-odd rows of pillars. In all the 3D memory arrays that we have studied so far, we always had 1 bitline per column of pillars, but this time we need 2 bitlines per column of pillars: in fact, one bitline is connected to the even (odd) row of pair0, while the other one is for the even (odd) row of pair1 (Fig. 6.34).

Figure 6.35 illustrates the concept of staggered bitline contacts. Starting from the top of the figure, the first section is the top view of the 3D stack of Fig. 6.34. The section in the middle shows the bitline comb; the reader can clearly see that there are 2 bitlines (red lines) that need to fit within a single column of pillars. Bitlines hide all the contacts to the pillars underneath. In the third section of Fig. 6.35 bitlines are made transparent: in this way, bitline contact (orange) and pillar contact (blue) become visible.

Figure 6.36 is a zoom of Fig. 6.35 to better appreciate the staggering of bitline contacts.

Figure 6.37 shows 3D views of Fig. 6.34: the bottom view helps visualizing the staggering of bitline contacts over pillar contacts. Please note that there are 64 bitlines instead of the 32 bitlines of Fig. 6.30. By adding Source Line contacts the matrix becomes the one in Fig. 6.38, which is the equivalent of Fig. 6.29.

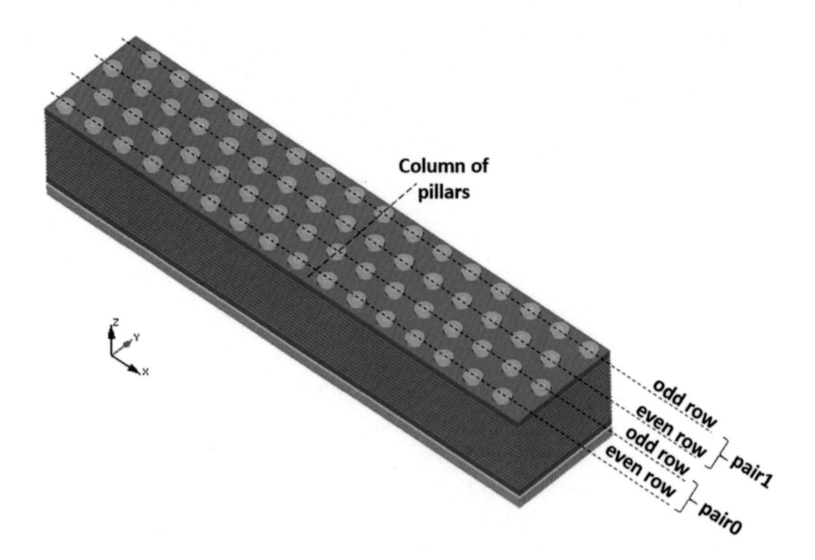

Fig. 6.34 NAND page made of 4 rows of pillars

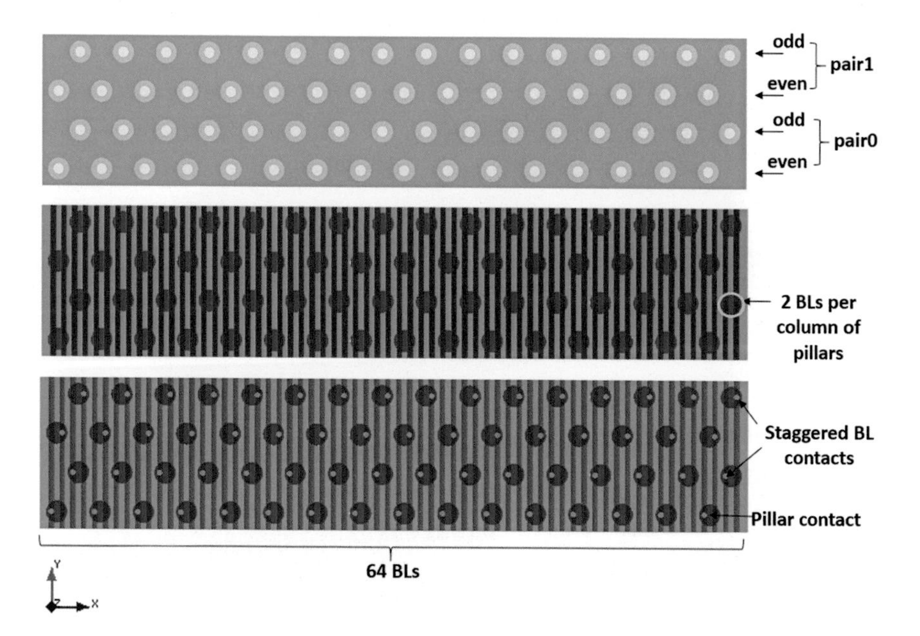

Fig. 6.35 Staggered bitline contacts

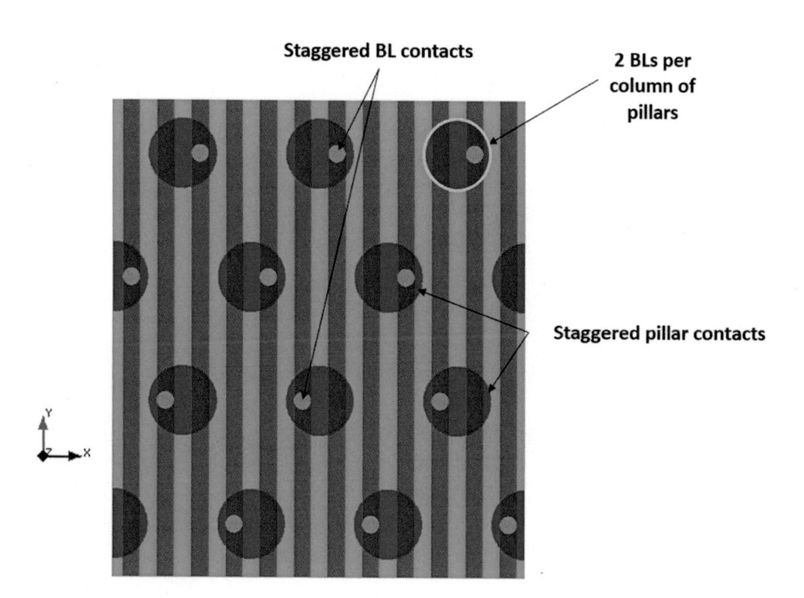

Fig. 6.36 Staggered BL contacts over staggered pillars

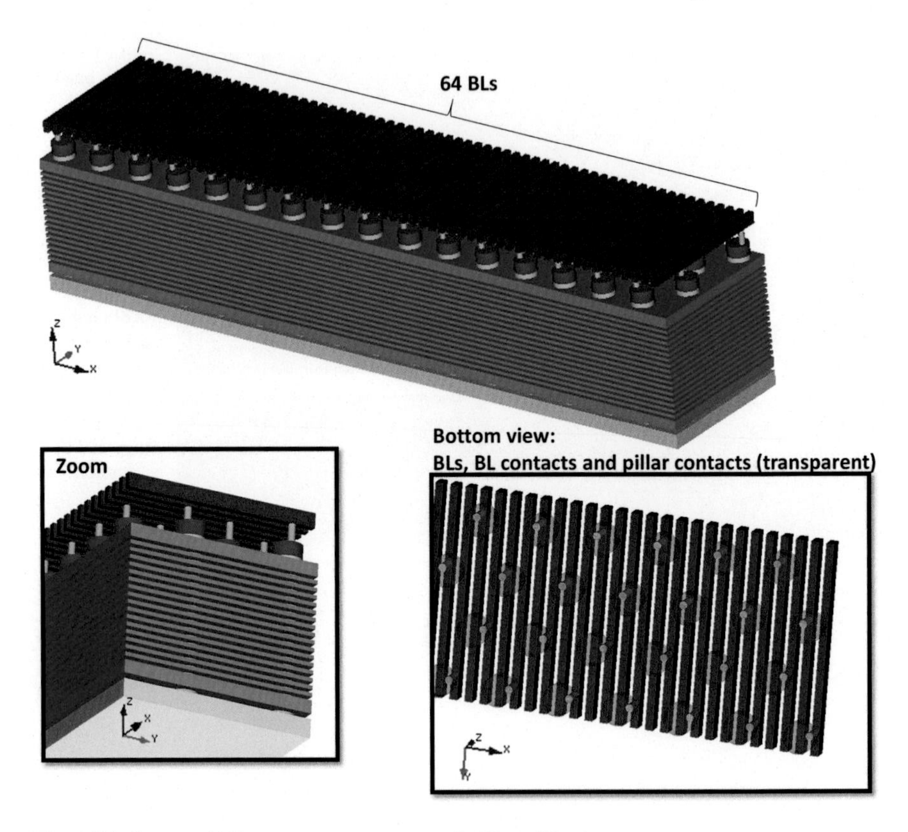

Fig. 6.37 Staggered BL contacts over staggered pillars: 3D views

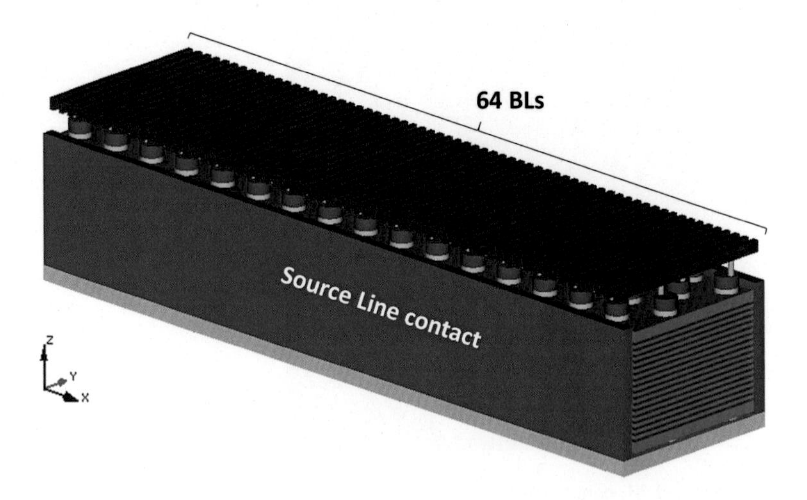

Fig. 6.38 NAND array of Fig. 6.37 with the addition of vertical source line contacts

To make a comparison, let's rebuild the matrix of Fig. 6.30 with staggered BL contacts. Of course, we assume a fixed number of layers (16), 8 pairs of rows of pillars and 32 columns of pillars, such that the overall number of cells is the same. As a result of this operation, matrix of Fig. 6.39 is the matrix of Fig. 6.30 re-drawn with staggered BL contacts. We assume 2 NAND Blocks in both matrixes, by adopting the proper fan-out connections.

We can notice:

- The NAND page of Fig. 6.39 is doubled: there are 64 bitlines instead of the 32;
- The matrix with staggered BL contacts has 4 × 16 (CGs) pages, while the other one has 8 × 16 pages, resulting in a difference in terms of Program and Read disturbs;
- In both cases there is one vertical Source Line contact per page, but the overall number of contacts is halved when staggering of BL contacts is adopted, with a clear advantage in terms of silicon area.
- Figure 6.39 requires a tighter bitline pitch.

This last point is not trivial, even with "big" memory cells. Let's make a numerical example. Let's assume *Dx* and *Dy* in the range of 160 nm. To fit 4 bitlines, width and spacing of metal lines is already down to 20 nm!

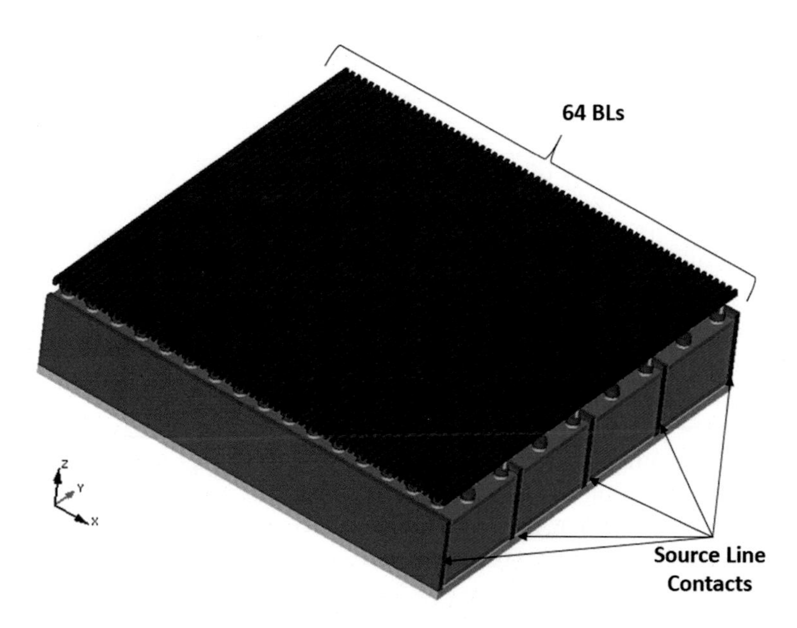

Fig. 6.39 NAND array of Fig. 6.30 re-drawn with staggered BL contacts

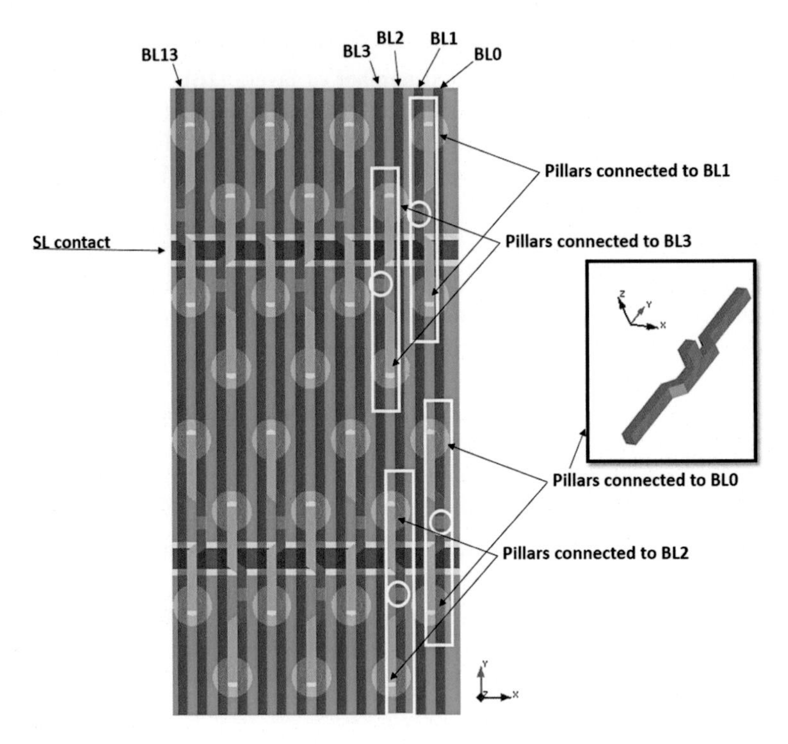

Fig. 6.40 Bracket-shape staggering of BL contacts

There are multiple ways of staggering bitline contacts; in the following we see a different implementation, which has a bracket-shape as shown in Fig. 6.40 [9, 10]. In the drawing bitlines are transparent to show what happens underneath. There are 3 MEOPs, which are separated by SL contacts.

Pairs of pillars belonging to adjacent MEOPs are connected to the same bitline by using the bracket-shape structure drawn in the inset of the figure. The pairs we are referring to are highlighted with white rectangles. Most part of the bracket-shape structure is for routing: the actual contact to the bitline is inside the white circle. In this example, 4 pairs of pillars are connected to 4 bitlines, from BL0 to BL3.

The memory array of Fig. 6.39 can be re-drawn by using bracket-shape BL contacts, as displayed in Figs. 6.41 and 6.42.

To better appreciate how the bracket-shape BL contacts are laid out, Figs. 6.43 and 6.44 show the memory array of Fig. 6.41 without bitlines.

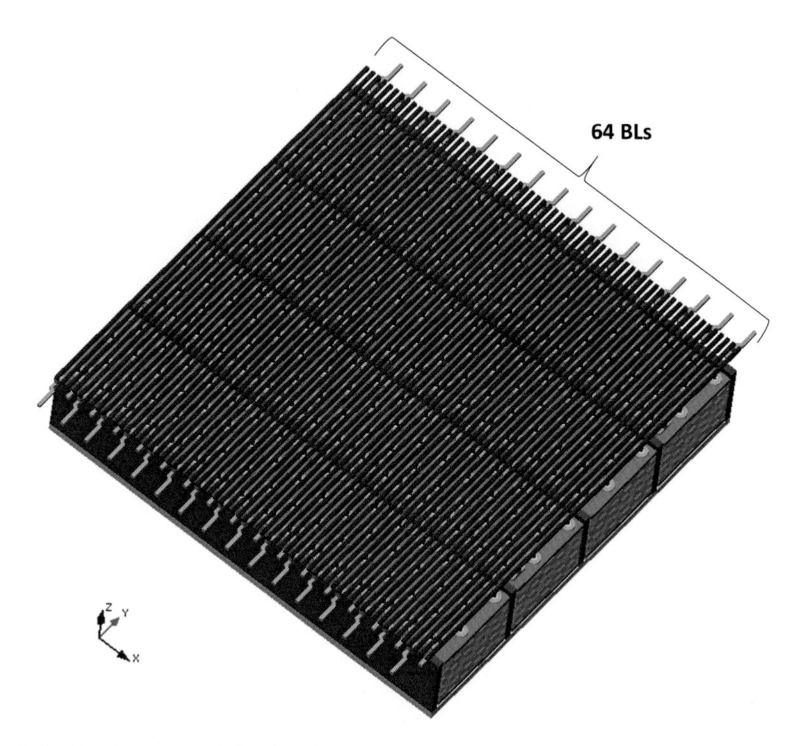

Fig. 6.41 NAND array with bracket-shape BL contacts

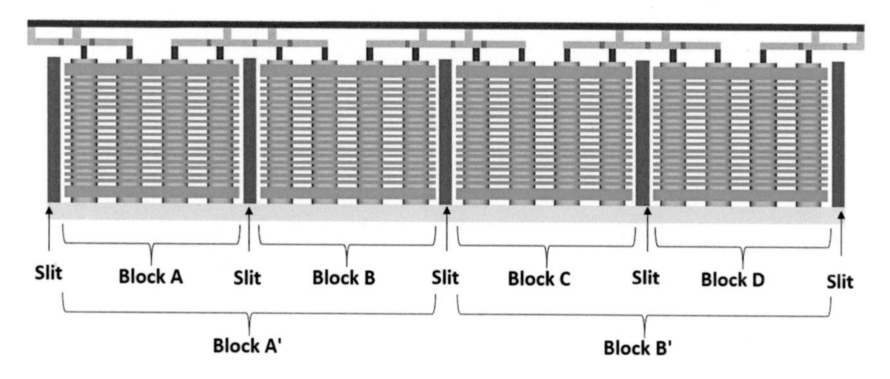

Fig. 6.42 Lateral view of Fig. 6.41

It is worth highlighting that with the bracket-shape approach, the contact between pillar and bitline can be centered with respect to the pillar itself, which might represent an advantage from 3D integration point of view.

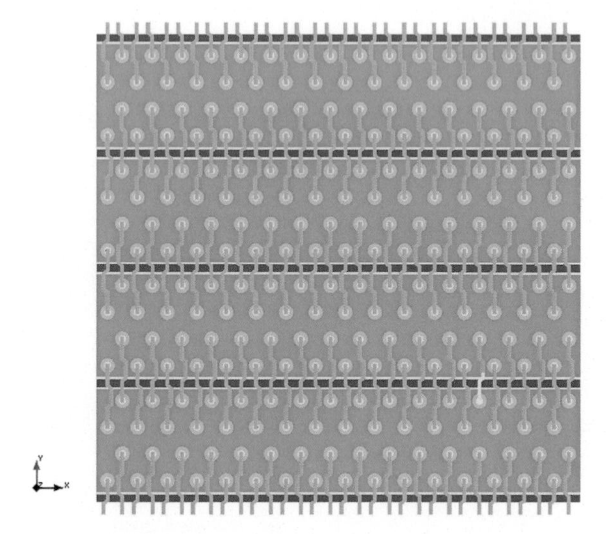

Fig. 6.43 Top view of Fig. 6.41 without bitlines

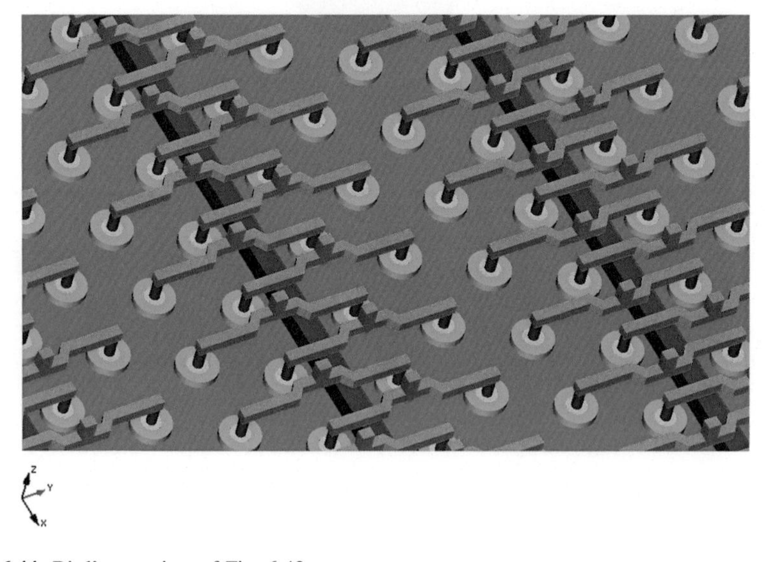

Fig. 6.44 Bird's-eye view of Fig. 6.43

6.7 Summary

In this chapter we have reviewed some of the most advanced architectures for 3D integration of NAND Flash memories with vertical channel: Fig. 6.45 collects their lateral views.

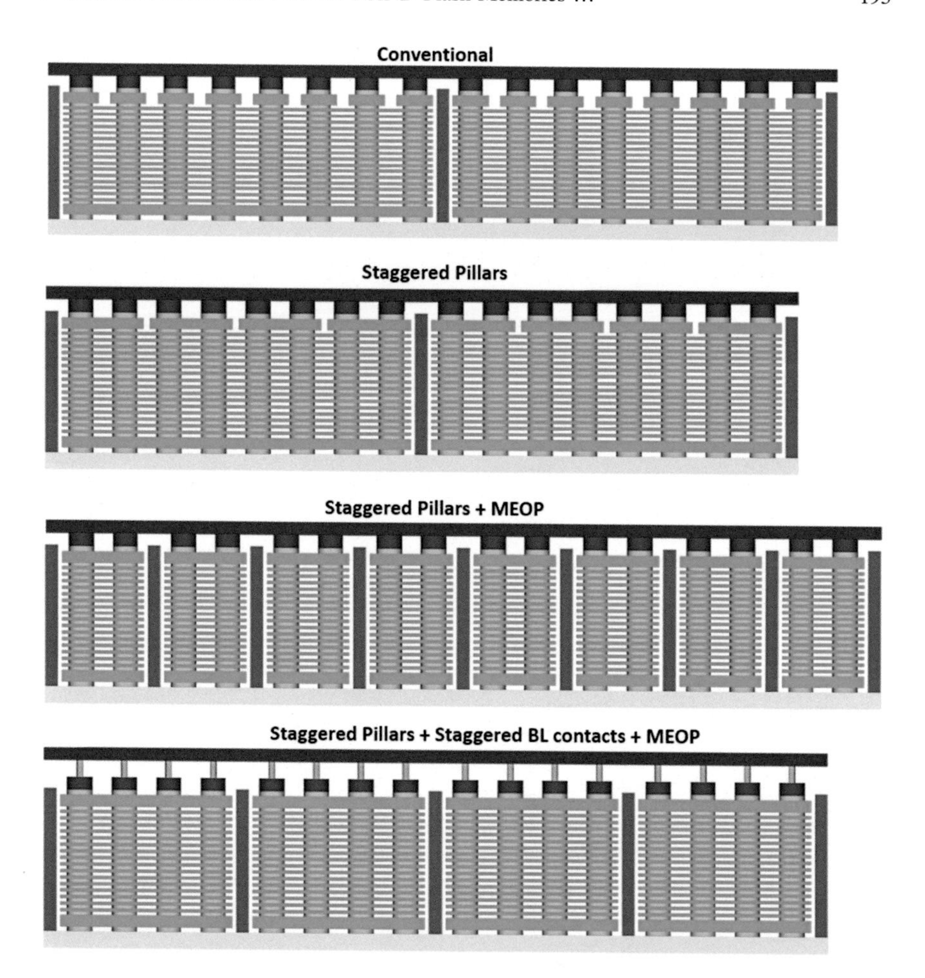

Fig. 6.45 Lateral views of the 3D architectural solutions presented in this chapter

Compared to the conventional architecture, staggered pillars save around 12 % of silicon area. MEOP implies an area overhead because the increased density of vertical contacts to the Source Line plate; staggering of bitline contacts helps mitigating the issue. In terms of Program and Read disturbs, MEOP is the best approach but the implications on the fan-out of CGs should be analyzed.

Table 6.1 is a summary of the 3D architectural solutions presented in this chapter. SLOH stands for *Source Line Overhead.*

We are just at the start of the 3D journey and we do expect more and more advancements in the coming years. Most of the Flash vendors are already talking about reaching up to 100 vertical layers and, for sure, this will require a lot of innovation in process technology, materials, circuit design, Flash management algorithms, ECC and, last but not least, 3D architectures.

Table 6.1 Comparison of the 3D advanced architectures presented in this chapter

	Conventional	Staggered pillars	Staggered pillars + MEOP	Staggered pillars + staggered BL contacts + MEOP
Cell area A = ideal	A + SLOH	A * 0.865 + SLOH	A * 0.865 + 4 * SLOH	A * 0.865 + 2 * SLOH
BL density [a.u]	1	2	2	4
Pgm disturb [a.u]	8	4	1	1
Page size [a.u]	P	2P	2P	4P
Block size [a.u]	B	B	B/4	B/2
Source plate (bottom of the 3D stack) resistance	R	R * 0.865	R * 0.865/4	R * 0.865/2

References

1. R. Micheloni, L. Crippa, A. Marelli, *Inside NAND Flash Memories* (Springer, Berlin, 2010)
2. K. Parat, C. Dennison, A floating gate based 3D NAND technology with CMOS under array, in *IEDM*, 7 Dec 2015
3. Y. Komori et al., Disturbless flash memory due to high boost efficiency on BiCS structure and optimal memory film stack for ultra high density storage device, in *IEDM Technical Digest* (2008), pp. 851–854
4. R. Micheloni, A. Marelli, K. Eshghi, *Inside Solid State Drives (SSDs)* (Springer, Berlin, 2013)
5. K.-T. Park, Three-dimensional 128 Gb MLC vertical NAND flash memory with 24-WL stacked layers and 50 MB/s high-speed programming, in *IEEE ISSCC Digest Technical Papers*, pp. 334–335, 2014
6. K.-T. Park, Three-dimensional 128 Gb MLC vertical NAND flash memory with 24-WL stacked layers and 50 MB/s high-speed programming. IEEE J. Solid-State Circ. **50**(1) (2015)
7. K.T. Park, *A World's First Product of Three-Dimensional Vertical NAND Flash Memory and Beyond*, NVMTS 27–29 Oct 2014
8. D.-H. Lee, A new cell-type string select transistor in NAND flash memories for under 20 nm node, in *4th IEEE International Memory Workshop (IMW)*, Milan, May 2012, pp. 1–3
9. J.-W. Im, 128 Gb 3b/cell V-NAND flash memory with 1 Gb/s I/O rate, in *2015 IEEE International Solid-State Circuits Conference* (2015) pp. 130–131
10. W. Jeong, 128 Gb 3b/cell V-NAND flash memory with 1 Gb/s I/O rate. J. Solid-State Circ. **51** (1) (2016)

References

Chapter 7
3D VG-Type NAND Flash Memories

Andrea Silvagni

7.1 Introduction

Solid State Disk applications have been the main drivers for the cost reduction of NAND Flash memory. Technology scaling limits have been almost reached and new directions were investigated for a decade. 3D NAND appeared as the most promising and short term viable solution to the scaling problem. 3D NAND offers the possibility to keep scaling memory density beyond 2D NAND 1z nm technology and opens new landscapes if the technology hurdles can be definitely overcome. Additionally Multi bit per cell technology is applicable to 3D NAND as well; therefore, the storage capacity growth trend is not going to meet the inflection point soon. Big Data storage demand remains certainly the driver for the rapidity of 3D NAND to become industry mainstream.

With respect to the flat 2D NAND Flash, three-dimensional architectures demand a more challenging approach to deal with process, design architectures, parasitic effects, testing. At the time of writing this chapter, 3D technology is moving into production inside solid state storage solutions; however, those NAND chips aren't commercially available as raw components. 3D NAND chips are used by vertically integrated manufacturers and few partners to build the storage systems under a strict control on the usage. Technology solutions are still diversified among the few players in the 3D arena and no unique mainstream solution is emerged yet.

In this chapter the principal architectural pieces of 3D NAND will be reviewed putting the focus on one of them: VG-Type, Vertical Gate 3D NAND.

A. Silvagni (✉)
Macronix, Milan, Italy
e-mail: andreasilvagni@macronix.com

R. Micheloni (ed.), *3D Flash Memories*, DOI 10.1007/978-94-017-7512-0_7

7.2 3D NAND Architectures

Several options for a true 3D NAND have been proposed for a decade in the technical symposiums before the first use in production. The main contributors have been Toshiba [1–3], Samsung [4–7], Macronix [8–10], Hynix [11–13], Micron [14] amongst the Memory companies. The main scope of 3D development was lowering the cost per bit and increase the chip density without shrinking the 2D lithography node. In the following Fig. 7.1, the main concepts presented in technology symposiums are summarized. The key structures have been presented in the range of only nine years, from 2007 to 2015. Most of the structures are based on charge-trap devices, only few are based on floating-gate, and the choice of one specific technology over another remains closely linked to the experience and confidence of each player. The production of products based on 3D NAND Flash cell started in 2013 and by the end of 2015 have been announced 3D-NAND chips with capacity of 256 Gb with 48 layers and 3bit per cell technology. Breakthrough of 1 Tb in one single chip is expected in the coming years. Vertical channel architecture type is currently the main choice, while charge trapping and floating gate technologies seem to be both pursued for production but with a preference on the first for most players.

The common feature among the different 3D NAND solutions is constituted by very deep vertical (z direction) etching steps that define the Flash cells geometries simultaneously. Transistor geometries are formed by the deep trench through a

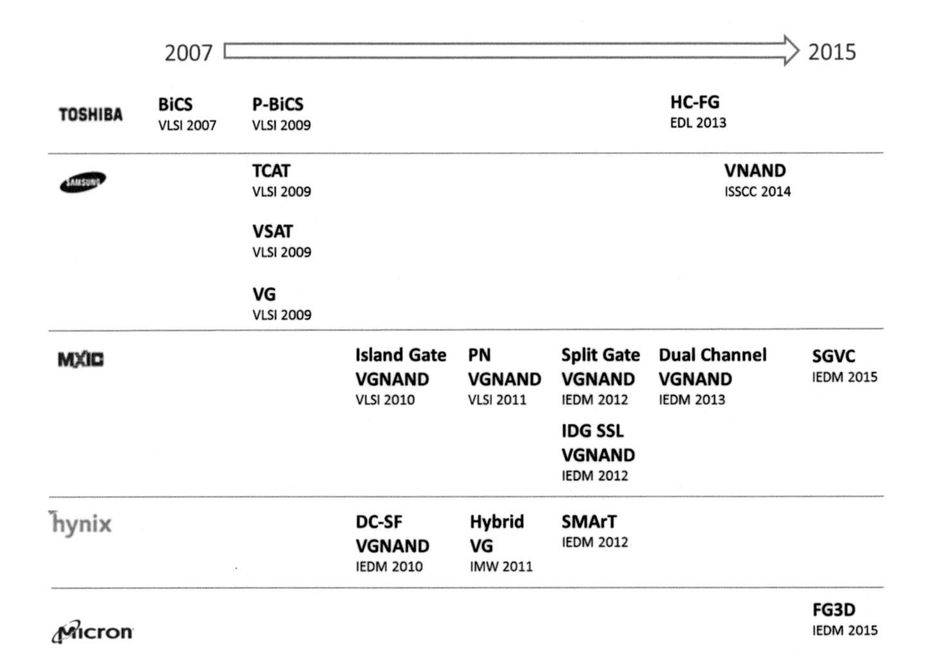

Fig. 7.1 3D Architectures proposed in technical symposiums from 2007 to 2015

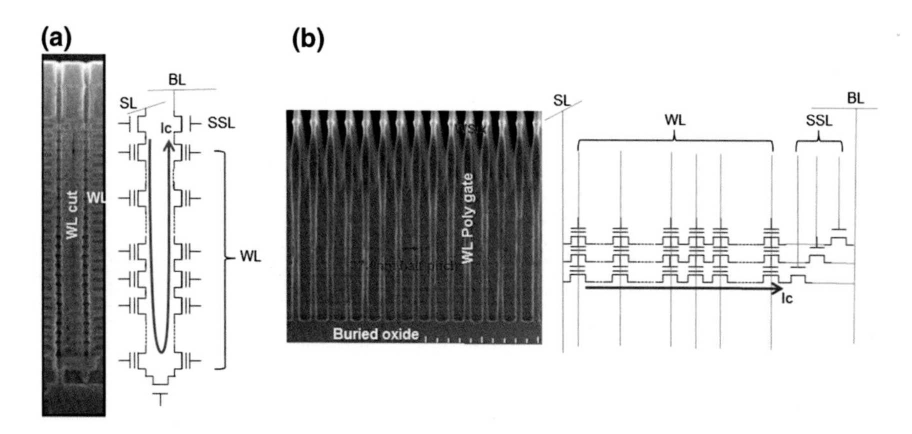

Fig. 7.2 **a** VC- type P-BICS 3D NAND, **b** VG-Type 3D NAND [2, 8]

multiple polysilicon/oxide stack. The cells stack shows an impressive high aspect ratio such as the one shown in Fig. 7.2, where one example of *Vertical-Channel* (VC) and one of *Vertical-Gate* (VG) type are shown. In VC gate-all-around type, the channel is realized by etching a hole through the layers stack in one single step, and then forming the transistor structure with deposition of its ONO charge trapping layers, tunnel oxide and the polysilicon channel fill in the middle. The cell gates are constituted by the polysilicon horizontal layer surrounding the vertical channel forming a *Gate-All-Around* (GAA) structure. The string current flows in the cells in the vertical direction. An additional wordline cut is necessary to separate the wordlines in the X direction. The channel is connected by a bridge at the bottom to another adjacent stack so that the resulting string is constituted by two times the stack and thus both the bitlines and the sourceline can be connected through the top metal backend and dedicated selector transistors. In the VG-type the vertical etching is necessary to separate the strings in one direction and to separate the wordlines in the other direction. The wordlines and the horizontal strings can be identified in the figure. The current flows in horizontal direction and each layer must be connected to the top metal bitlines and source lines by a proper connecting structure (not shown here).

3D NAND architectures need to be proven cost effective in terms of process costs and product yield because the process complexity is not trivial if compared to 2D NAND. The cost must be evaluated considering other factors rather than simply the number of layers: the overhead of decoding structures and circuits plus the yield determine the cost advantage. The non-uniformity of the critical dimensions through the layer stack plays a significant a role in the determination of the effective cost per bit. It is recognized that in order to reach cost effectiveness in 3D NAND the number of layers must be roughly above 32 which is already an impressive number. Currently 48 layers have reached production stage. About 64 layers might be already a limit, considering the etching process difficulties of a stack with an aspect ratio of 40 [15].

Most of the proposed 3D NAND devices are based on *Charge Trapping* (CT) SONOS (Silicon-Oxide-Nitride-Oxide-Silicon) devices but also traditional floating gate structures have been proposed. CT SONOS devices did not become the mainstream in 2D NAND Flash where Floating gate cells are still dominant; however several drawbacks of the CT technology can be minimized by engineering the charge trap stack, Read/Write algorithms and advanced error correction schemes managed by the controller processor in Solid State Disk products.

Figure 7.3 shows the top view of 3D NAND cells for vertical channel and vertical gate types. The drilled circular hole imposes some limitations on both X and Y directions for VC-type 3D NAND. The constraints are the minimum diameter size of the channel determining the cell current and the minimum thickness of the ONO trapping layers for sustainable reliability. The resulting minimum hole size must be greater than roughly 50 nm, although much bigger hole size such as 120 nm would be necessary to control some negative effects induced by the circular structure and critical dimensions along the stack [16]. This hard limitation in the 2D shrink-ability of GAA VC-type needs to be compensated by the increase of the number of layers. A vertical etch (wordline slit) is also needed to separate the different strings control gates. In order to optimize the flash cell array density, the hole trenches can also be arranged in a staggered fashion [16], controlling two strings with the same wordlines and select gate, thus requiring one less wordline cut.

For VG-type NAND, the cell geometrical limitations are restricted mainly to one direction due to ONO stack and to the polysilicon gate fill-in. Hence, the overall cell size of VG-type can shrink better than VC-type in 2D dimensions. The double-Gate VG-type NAND can achieve a $4F^2$ cell size in 2x nm and better scalability than VC-type; therefore, generally speaking, less layers are necessary to obtain the same cost with benefits on the manufacturability.

Figure 7.3c, d show a simplified 3D view of a Vertical-Channel hole-type 3D NAND and a Vertical-Gate 3D NAND. The VC-type cell gate all-around structure with the channel built on a poly structure is shown. Gate all-around structures leverage the very effective electric field enhancement effect resulting from the circular structure, thus obtaining optimized programming conditions. However, field enhancement also enhances disturb effects such as program disturbs and Vpass disturb. Due to the field enhancement, the circular cell structure has a strong sensitivity to curvature variations present along the stack that may limit the practical minimum cell size, as already mentioned. *Macaroni* body structures [17], constituted by a thin polysilicon channel with hollow center filled by dielectric, have been utilized for better gate controllability and to reduce the trap density at the boundary.

In the 3D-VG-type NAND shown in Fig. 7.3d the channel is constituted by the polysilicon layer which is defined by first etching of the poly/oxide stack. Afterwards, the tunnel oxide and the ONO charge trapping stack is deposited and finally the wordlines are formed. The limits of scalability in VG-type 3D NAND are only constrained by the ONO structure and the poly channel and not by the field enhancement constraints. Because there is no field enhancement effect due to the planar structure in VG-type 3D NAND, the resulting planar FinFET cell transistor

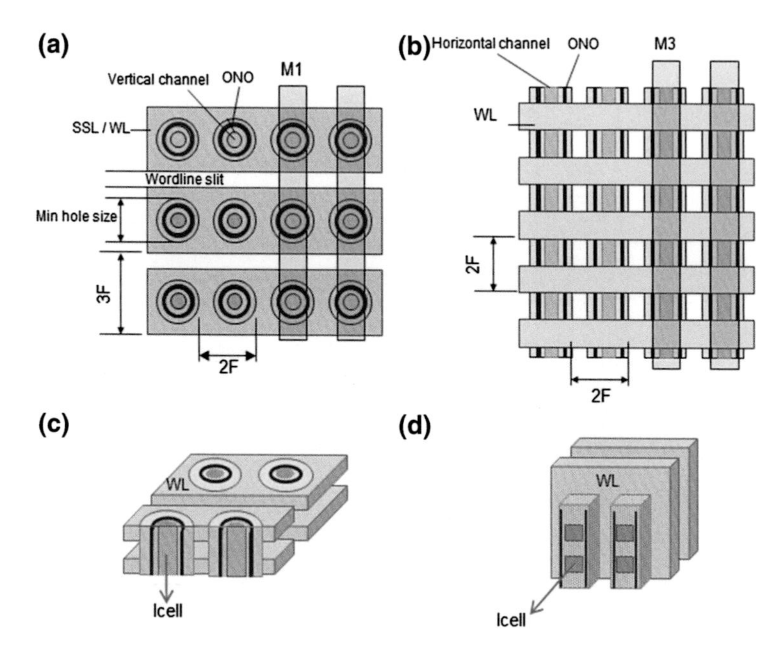

Fig. 7.3 3D-SONOS VC and VG architectures: top view and 3D view

needs to be engineered; however, the excellent planarity of the structure makes the engineering of the oxide and nitride stack easier, by utilizing the extensive knowledge available for barrier engineered BE-SONOS 2D structures. The channel current flows horizontally in VG-type 3D NAND and, therefore, there is no degradation when the number of layers is increased; on the contrary, each layer may exhibit a different behavior from string current performance point of view. Even if the channel lateral dimension of the FinFET cell structure is as small as 8 nm, the channel current is not degraded because the effective channel width is determined by the poly thickness in the Z direction, which does not necessarily need to scale. In contrast, VC-type NAND channel effective width is determined by the planar dimension of the poly channel.

A *Single Gate Vertical Channel* (SGVC) architecture not based on drilled-hole channel with GAA structure has been presented [9]. The SGVC structure shown in Fig. 7.4 can achieve a better memory density than other drilled-hole VC-type structures, with the same advantages of the cell's planar structure of the VG-type 3D NAND.

SGVC implements a single-gate flat CT-TFT (*Charge Trapping Thin Film Transistor*) cell with an ultra-thin body. The vertical etching doesn't need to be precisely controlled because the flat cell is tolerant to critical dimension variations; this is different from GAA structures which are susceptible to field-enhancement variations due to curvature variations. A U-turn string is naturally formed by polysilicon deposition on the trench sidewall. The U-turn structure allows a

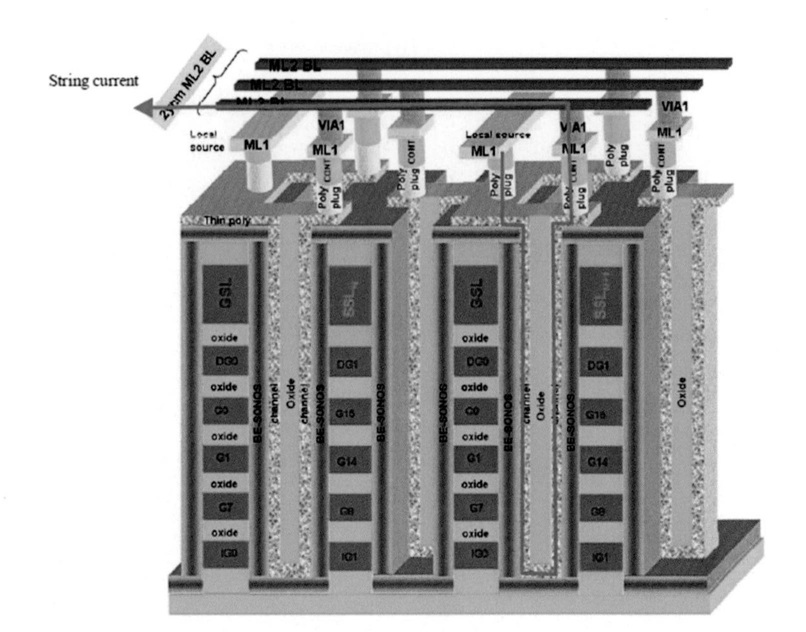

Fig. 7.4 SGVC: a vertical channel approach with planar cell structure [9]

straightforward connection to the metal backend. Important characteristic of SGVC compared to GAA structure is that each *Wordline* (WL) controls two independent cells instead of a single one, thus doubling the cell density in a natural way. In the X direction a bitline cut is necessary to separate each string, but it is not critical for the ultra-thin TFT structure.

Although the SGVC structure requires a wordline and sourceline metal strapping in order to reduce the resistance, the 2D layout is very compact and shows 90 % array efficiency. A reasoned calculation of the memory density of SGVC compared to GAA structure shows that SVGC can achieve from 2 to 4 times the bit density, given the same number of layers. In other words, there is more margin to play with the critical dimensions and Z-dimension complications.

7.3 VG-Type 3D NAND Architecture

The 3D Vertical Gate (3D VG-type) NAND architecture (Horizontal Channel structure) has been investigated and extensively studied, being the following references only a small example of related publications [8, 10, 18–25]. The scalability of the architecture has been studied down to the 2× nm node.

Figure 7.5 shows a vertical cross section of 3D-VG CT-NAND with 8 layers [10]; the polysilicon layer stack is shown, each layer is labeled PLx; the inset figure shows the double-gate FinFET cell and the charge trapping layers between the

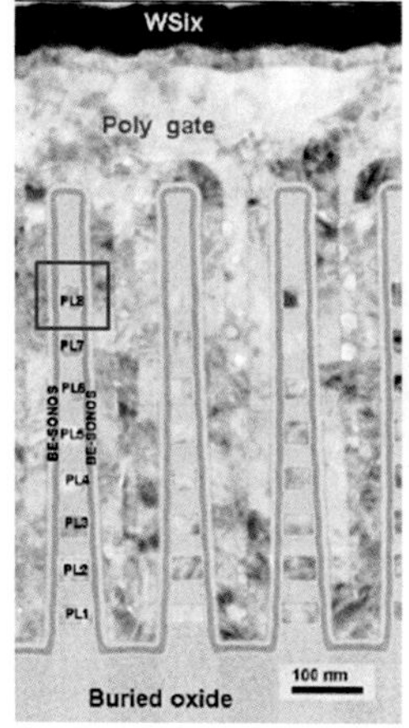

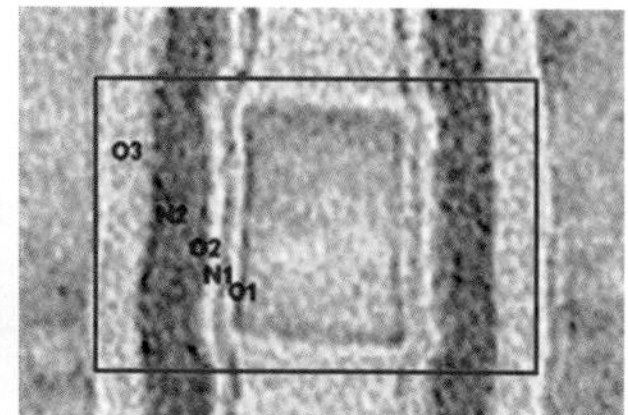

Fig. 7.5 Vertical section of 8 layers VG-type 3D NAND [10]

polysilicon gate and the polysilicon channel. The charge trapping medium is a very planar BE-SONOS stack (ONONO).

The NAND cell string runs horizontally and is made of several wordlines plus the dummy wordlines and the select transistors. A silicide layer can be easily deposited on the top of the polysilicon gate to achieve a lower wordline resistance and a very fast access time despite the long wordlines.

A wordline pitch of 2 times 38 nm and a bitlines pitch of 2 times 75 nm have been studied [10]. With such a relaxed pitch and 8 layers, it is already possible to produce an equivalent cell efficiency of a 2D NAND in 1× technology node. A scaled down version already demonstrated the equivalent 1y nm technology node feasibility.

Figure 7.6 shows a 3-dimensional view of the building block of VG-type split-gate 3D NAND structure. Each string is made of 64 wordlines. At the end of the string the channel is selected by an independent SSL (*String Select Line*) transistor *(island-gate* formed transistor) and then connected to a common poly plate (BL pad) indicated as PLn. Each BL pad is used to connect all the strings of the corresponding layer, and then each BLpad is connected to the top Metal 3 bitlines in order to achieve the connection of the BL pad polysilicon plate. Staircase contacts (BL poly plugs) are realized by drilling the BL pads progressively. Poly plugs are finally realized in order to connect each pad to the corresponding top bitline.

On the opposite side of the SSL select transistor, NAND strings are vertically connected to a poly plug. In this way, the source line of the stack is directly connected to the metal backend with a low resistance.

Figure 7.6 is the basic building block of the memory array: it is worth noting that SSL select transistors and source plugs are alternated on both sides of the block in an even/odd fashion. With this arrangement, it is possible to keep a relaxed double pitch between SSL transistors and the corresponding backend connections, at a cost of a slightly lower array density.

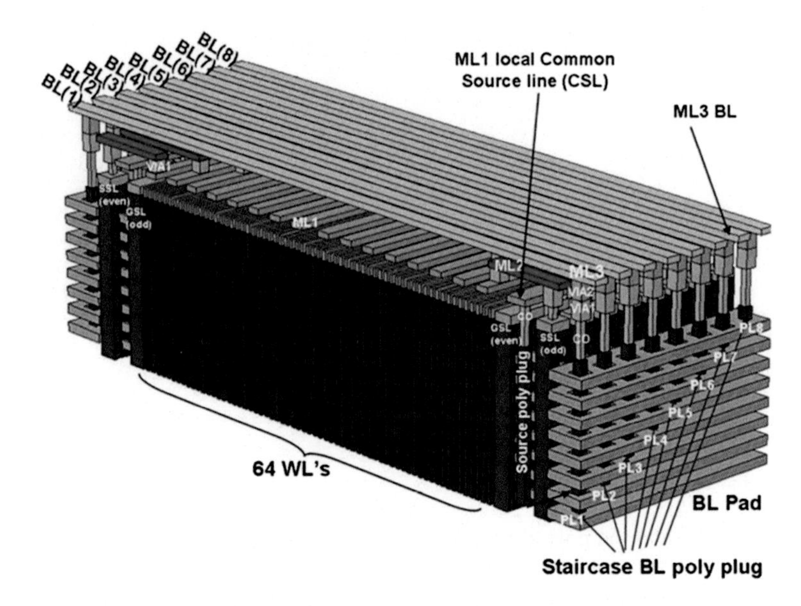

Fig. 7.6 Horizontal channel/vertical gate structure 8-layers split page [10]

Figure 7.7 shows the top view of the VG-type NAND unit of the 8 layer split-bitline structure. Each unit is composed of 16 strings for each layer. Since each string has its own SSL select transistor, the main metal bitlines are connected to a pair of poly BL pads from each side. Each string also has two GSL (*Ground Select Line*) due to the split-page architecture. In total there are 16 pages in a unit. The full page is built by repeating the unit block several times (e.g. 8 k times to obtain an 8 kB page size). When the first page (SSL0) is selected, all the strings corresponding to the same select transistors are connected to the 8 Metal bitlines: cells from all the layers are present in a page.

The layer selector used to connect each layer to the top bitlines is critical for VG-type 3D NAND in terms of area overhead. Other approaches for the layer selector scheme and SSL arrangement have also been proposed: PN diode selection method [19] and staggered SSL arrangement [26]. In particular, the proposed staggered SSL arrangement allows achieving a block efficiency of 80 %, which is similar to that of 2D NAND.

The characteristics of 3D-VG type can be summarized as follows:

1. Planar Scalability: studied down to 25 nm;
2. Equivalent 2D technology feasibility well below 10 nm;
3. Horizontal planar BE-SONOS flash cell that allows mitigate the CD variability along the layers;
4. Low wordline and Source Line CSL resistance;
5. Constant array efficiency independently from the number of layers number;
6. Multilevel functionality.

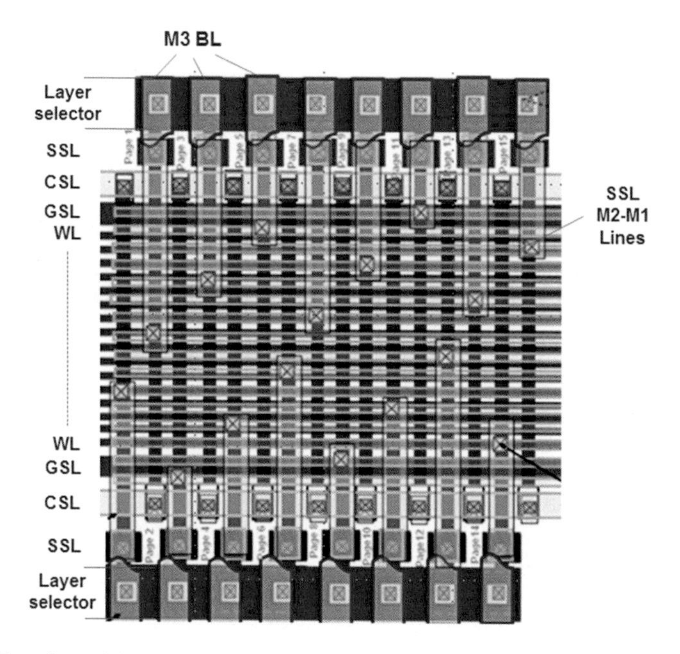

Fig. 7.7 Top view of the VG-3D NAND split-page unit block [10]

7.4 Key Architectural Considerations for VG-Type 3D NAND

The main challenge of 3D NAND is related to vertically stacking more devices in order to increase the bit density per 2D unit area. Obviously, the more layers can be realized, the lower should be the overall cost; however, several aspects contribute to the reduction of the theoretical gain. One of the issues is related to the yield loss derived from complexity of vertical trench or hole etching with many layers. Besides yield issue impacting on cost, the area efficiency of the 3D architecture is dependent on the overhead of side structures such as layer connecting structures or decoders. Architectural solutions are required to optimize the overall array efficiency.

From a performance perspective, there are also few aspects that call for architectural innovations. The use of charge trapping cells requires complex programming and erasing algorithms, and the engineering of the trapping layers. The wordline and source line resistance must be kept low without impacting the array size. RC delays in wordline may not only have an impact on AC performances but also on the efficiency of the programming sequences used to reduce disturbs. The long source lines need to be designed to minimize the parasitic voltage drops, which may reduce the margin for multilevel cell operations. Those problems are specific to each type of 3D architecture and key innovations are required.

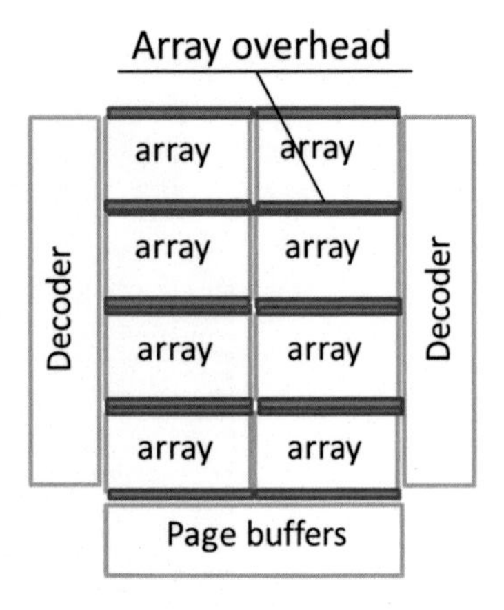

Fig. 7.8 Array overhead in 3D NAND

The first technology challenge in 3D structures is related to how accessing NAND strings or wordline layers. Specifically, in VG-type 3D NAND wordlines run vertically and are common to all layers; however, flash cell strings run horizontally on each layer, and each string of cells must be connected to metal bitlines running at the top. On the contrary, in VC-type NAND the horizontal wordlines need to be connected to the decoders. In both cases the connection to each layer is the key step that has an impact on the array efficiency.

Figure 7.8 shows an example of the array architecture in a 3D NAND. Each array block needs a special structure for connecting the horizontal layers (the strings in the VG-type NAND) to the top metal lines. From the pure area point of view it is necessary to keep this overhead as small as possible and in general to avoid a linear increase of its area with the number of layers. Another overhead is represented by the WL Decoders.

A method proposed for connecting the horizontal strings to the main bitline of a VG-type NAND is shown in Fig. 7.9 [18]. All the strings from each layer are simply connected together by vertical poly plugs. In order to select the proper string (layer), a series of SSL is realized in each string. Some of the SSL transistors need to be normally-ON SSL transistors. Each string can be selected by applying proper biasing to the SSL gates, as shown in the decoding table. The series of SSL constitutes the overhead of the structure. If the number of layers is increased, more SSL series transistors are required and this will increase the overhead area. Moreover, the depletion type SSLs require special implantation steps.

Another straightforward method to access the layers is to realize a connecting "staircase" structure. Each polysilicon step of the staircase is connected to one of the layers and it can be connected to the top metals by landing a contact down to the step.

Figure 7.10 shows a progressive "etch and trim" method for realizing the staircase structure [1]. The method consists in multiple etching steps with

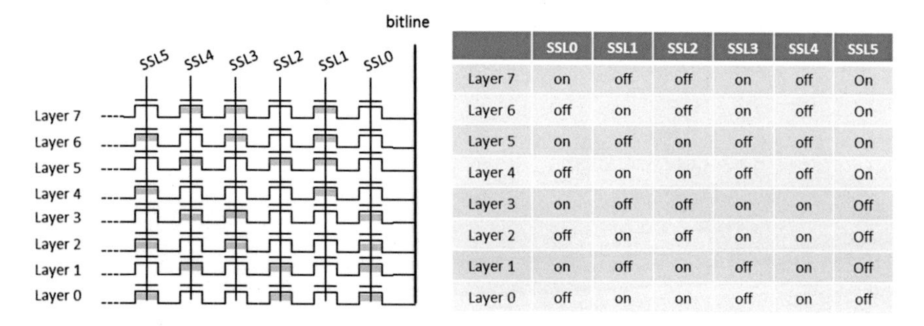

	SSL0	SSL1	SSL2	SSL3	SSL4	SSL5
Layer 7	on	off	off	on	off	On
Layer 6	off	on	off	on	off	On
Layer 5	on	off	on	off	off	On
Layer 4	off	on	on	off	off	On
Layer 3	on	off	off	on	on	Off
Layer 2	off	on	off	on	on	Off
Layer 1	on	off	on	off	on	Off
Layer 0	off	on	on	off	on	off

Fig. 7.9 Layer selection in VG-NAND with multiple SSL [18]

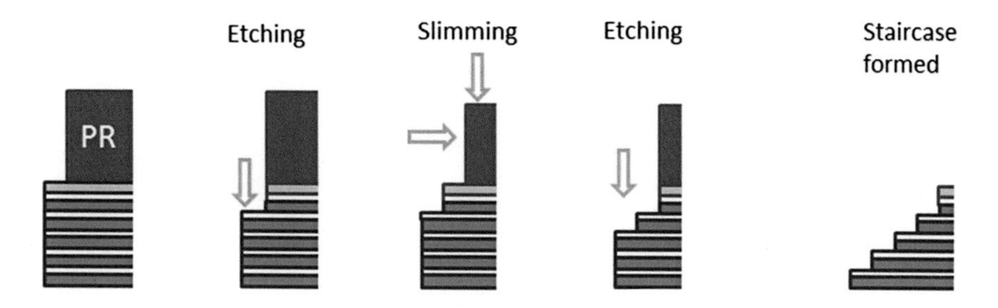

Fig. 7.10 Fabrication method for staircase formation in 3D NAND [1]

photoresist trimming until each layer is exposed and a contact can land on it later on. This method could theoretically be realized with one single lithographic mask; however, the control of this process might become very difficult as the number of layers grows up.

Trim and etch method increases the connecting structure area linearly with the number of layers and, therefore, it is not very efficient.

Figure 7.11 shows another simple method for connecting wordlines in VSAT architecture [5]: layers for multiple gates are deposited over the substrate *mesas* and simultaneously form the "PIPE" structure at the edge, without any additional complex fabrication step. Also in this case the space required to connect the layers is proportional to the number of layers itself.

It has been shown by different sources that the linear increase of die area to build staircase contacts is not cost effective. More in general, the process costs cannot increase linearly with the number of layers [27]. The staircase contact structure is critical because a precise landing of the contacts, and also because it requires multiple process steps including lithographic steps. The tradeoff between complexity of the process flow and the die size must be found.

A proposal to use a *Minimum Incremental Layer Cost* (MiLC) for 3D VG NAND [10] is shown in Fig. 7.12. The proposal is a lithographic method for realizing the staircase structure with minimum lithography steps. When increasing

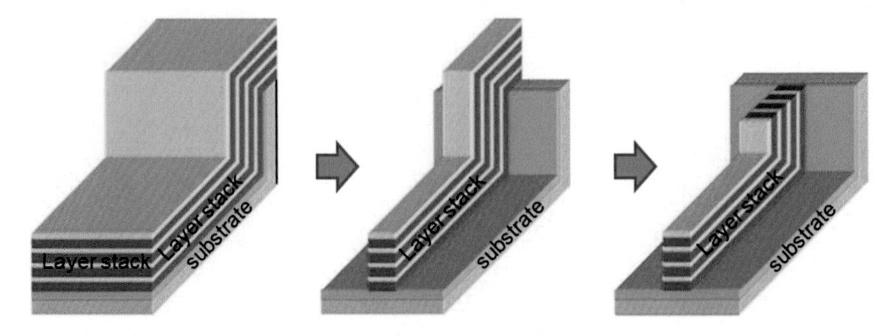

Fig. 7.11 PIPE WL connection in VSAT [5]

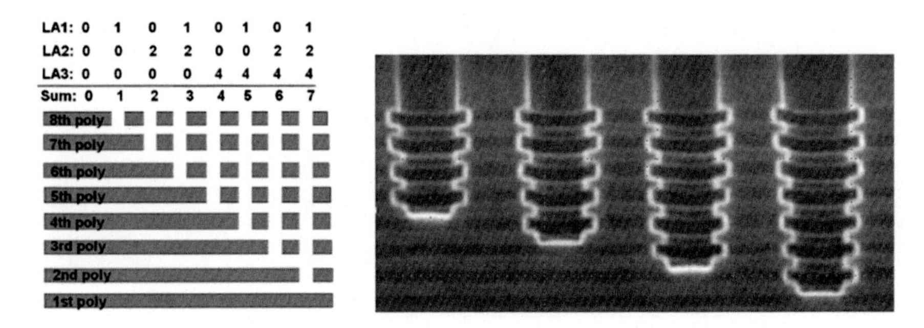

Fig. 7.12 Formation of the MiLC staircase contacts in 3D-VG NAND [10]

memory layers, the array efficiency is kept constant due to the special structure. The cost due to the additional masks and process steps is a logarithmic function of the number of layers, instead of a linear function. The log increase of the cost is sustainable in the overall cost structure when the number of layer is increased. For 8 layers, a superimposition of three masks LA1, LA2, LA3 and related etching steps allow drilling down to each polysilicon layer. The final staircase result is shown in the figure, where the precise landing on the poly pads is shown. An isolation spacer is deposited for isolating the contact from the unwanted layers along the hole. With MILC it is possible to double the number of layers and contacts by adding only one mask.

Even if the polysilicon bitline pad in the VG-TYPE 3D NAND structure can be connected efficiently to the top metal through the MiLC approach, it is still necessary to select one single string out of the many connected to the same poly pad.

A proposal for string selection has been done by using PN diodes as select device, as displayed in Fig. 7.13 [19]. PN diodes can be self-aligned to the source side. Each layer has horizontal source line that can be biased independently (e.g. by using a staircase structure), while the drain side of the strings is vertically connected by poly plugs. String selection is done by proper biasing of horizontal *Source Lines* (SL) and bitlines.

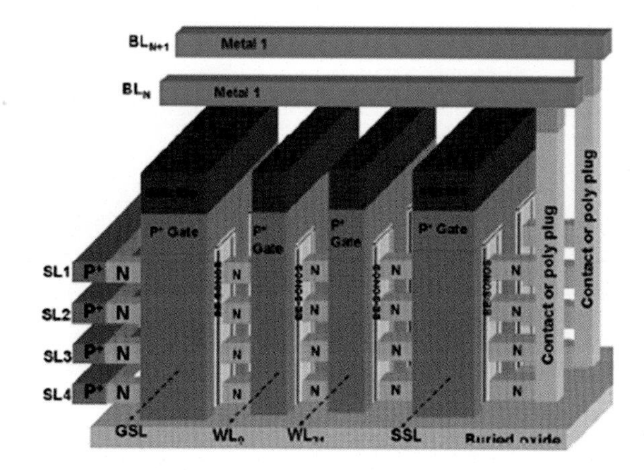

Fig. 7.13 PN-diode decoding structure in 3D-VG NAND [19]

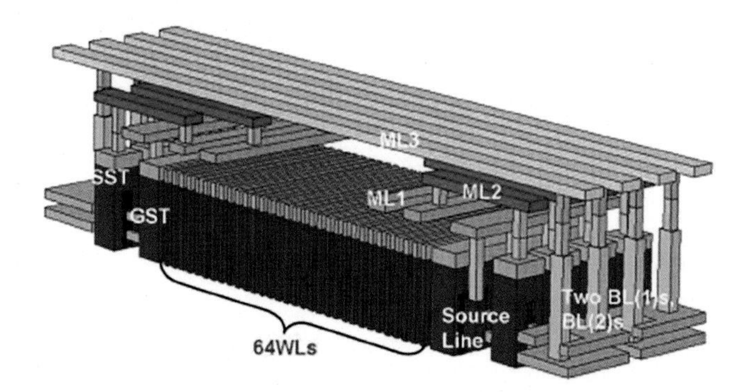

Fig. 7.14 Simplified 2 layer split page VG NAND [28]

A single island-gate SSL selection for VG-type 3D NAND has been realized to individually identify the strings connected to the same poly pad. In order to relax the pitch of SSLs to twice the pitch of bitlines, a split-bitline layout has been proposed [10] and it is shown in Fig. 7.14. The split BL layout architecture consists in having SSL of even and odd strings on the opposite sides of the block; thus, the string current flows in opposite directions for even and odd strings. The description of this architecture is done in the next section. The split-bitline approach improves the process window as SSL are realized with double pitch compared to bitlines. BL contacts are also at double pitch of the channel BL.

In 3D NAND, the string select transistor and the ground select transistor gate dielectrics may make use of the cell's trapping dielectric for simplicity of the process architecture. In this case, the select transistors may suffer unwanted charge injection that causes V_{TH} shift during operation. For example, holes injection during erasing may lower the V_{TH} of the select transistor. If V_{TH} is lowered, then it

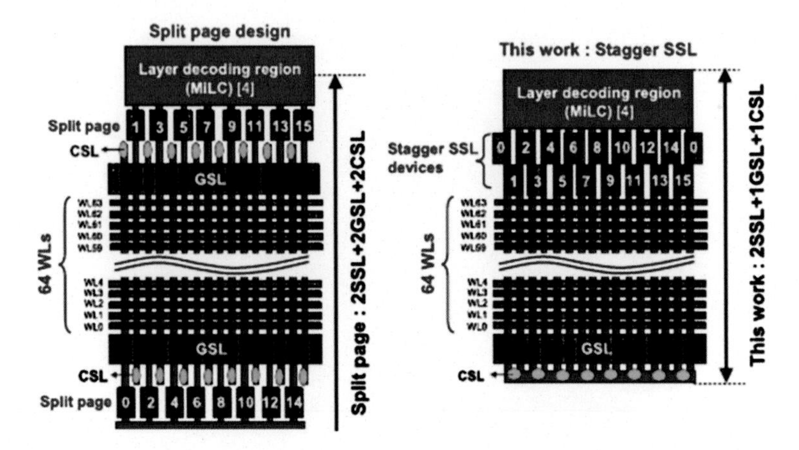

Fig. 7.15 Optimization of split-page string decoding in 3D-VG-Type NAND [26]

is not possible to properly isolate the strings during self-boosting (programming), thus causing failures. To avoid the malfunction, it is necessary to properly bias the select transistor for deselection, by using slightly negative voltages. Moreover, it might be possible to implement V_{TH} adjustments of the select transistors by special algorithms. Because the use of negative voltages complicates the decoding of the control lines, the use of charge-trapping free gate dielectric has been proposed [29]. Charge trapping free dielectric can be realized by removing the trapping oxide (O3) in the BE-SONOS Stack (O1/N/O2/N/O3) and replacing it with an additional oxide (O4) to reinforce the gate dielectric.

In order to reduce the overhead of the additional CSL and GSL lines of the split-page design and optimize the space occupied by the island-gate SSLs, a new decoding scheme with staggered SSL was proposed and it is shown in Fig. 7.15. In this arrangement, each SSL transistor selects 2 strings and a combination of 2 SSL in series is used to decode one string out of 16. Having SSL only on one side has also the benefit of eliminating one GSL and one source line connecting area, thus saving additional space; the resulting array efficiency is close 80 %, similar to that of 2D NAND. By using staggered SSL arrangement, VG-type 3D NAND becomes the most efficient among the 3D NAND solutions.

7.5 VG-Type 3D NAND Array Operations

In this section we will analyze the principal working modes of a VG-type 3D-NAND structure. 3D VG NAND with split page architecture [10] will be used for this purpose. In order to simplify the understanding of the 3D structure, an example with only two layers will be used. In Fig. 7.16 a 3D drawing of the two-layer structure is shown. The figure shows a structure having two basic units

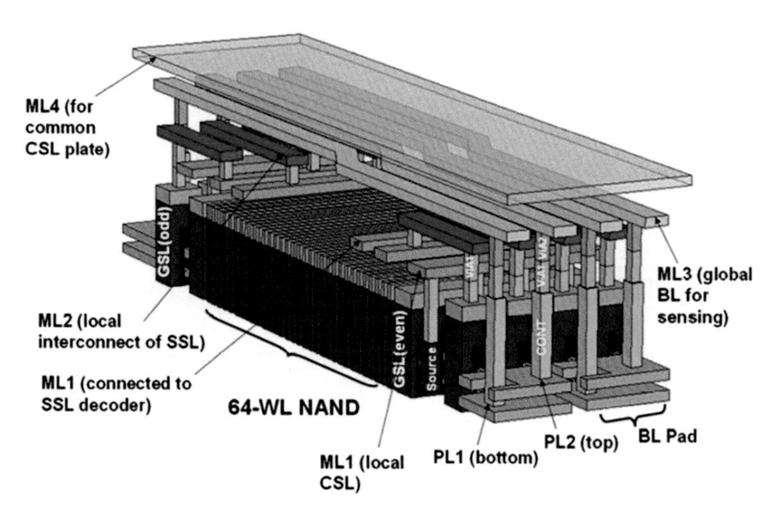

Fig. 7.16 2-layers structure example (2 units) [28]

side by side. Each unit is composed of four strings for each layer. Due to the split page architecture, SSLs of even and odd strings are placed on the opposite sides of the unit, thus allowing larger process window for SSL formation. Each single unit contains 4 pages that can be accessed from the two M3 bitlines. Each page contains cells that are on both layers. In the general case, each page contains N cells belonging to all N layers. Staircase bitline contacts are fabricated for the array decoding by using the MiLC approach [10]. ML1 and ML2 are used to decode the eight SSL devices. The source contact is built at the line end of each channel, and it is directly connected to a local ML1 common source line (local CSL). A top Metal layer can be used to connect M1 local CSL to reduce the overall resistance. Each memory cell is accessed by selecting the corresponding WL, ML3 BL (corresponding to the memory layer), and SSL (corresponding to one channel BL).

Page programming and reading are performed by simultaneously operating SSLs in parallel, in different units for a larger page. A simplified schematic of this basic unit will be used in the next sections to describe the 3D NAND flash operations.

7.5.1 Read Operation

In Fig. 7.17 it is shown a simplified circuital schematic of 3D-VG-type Flash array. There are two layers and four strings for each layer, in a split page organization. The polysilicon BL pad structure is visualized in the schematic for a better understanding of the architecture. Reading a Flash cell is achieved by using read schemes as in a standard 2D NAND. Forward read sensing can be used also in 3D NAND; the Flash string must be properly biased in order to read out the current

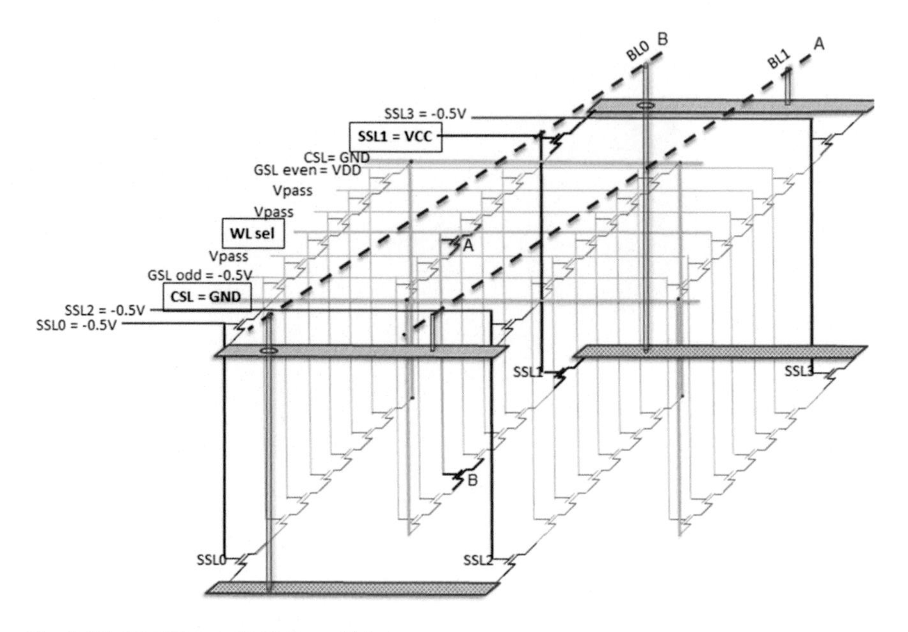

Fig. 7.17 3D VG-type NAND read biasing

value of the selected cell. The forward-read sensing method is not described here but is deeply studied in literature [30].

Reading a cell requires the corresponding SSL to be turned ON with VCC voltage, while other SSLs need to be turned OFF. In the example of Fig. 7.17, the selected cells are A and B of Page1. The corresponding SSL1 is at VCC. Turning OFF of other SSLs is achieved by applying a negative voltage, e.g. −0.5 V, so that SSLs are turned OFF even if their V_{TH} has drifted. The VGSL even transistors are turned ON by VCC and the selected wordline stays at read voltage of 0 V or even slightly negative. All other wordlines are turned ON at the Vpass voltage of about 6 V. As shown by labels, cell 'B' is read out through bitline BL0, while cell 'A' is read out through bitline BL1.

Figure 7.18 shows a detailed timing diagram of the read phase with forward voltage sensing. To avoid "local self-boosting" at the cells between SSL and selected WL, that may lead to hot-carrier induced read disturb, a proper discharge of the channel is done by pulsing SSL during wordlines turn on [28]. The selected wordline is biased slightly negative during the read phase.

Other read methods such as "reverse read" can be used as well in VG-type 3D NAND. Reverse read can be useful to cope with specific characteristics of the VG-type 3D NAND described in this section. In VG-type 3D NAND, cells belonging to each of the layers are in the same page and, hence, are read out at the same time. Every layer is connected to a separate M3 bitline, but every layer also share the same SSL select line in the vertical direction. Due to the etching profile of the stack there are geometrical differences among layers. As a consequence, the

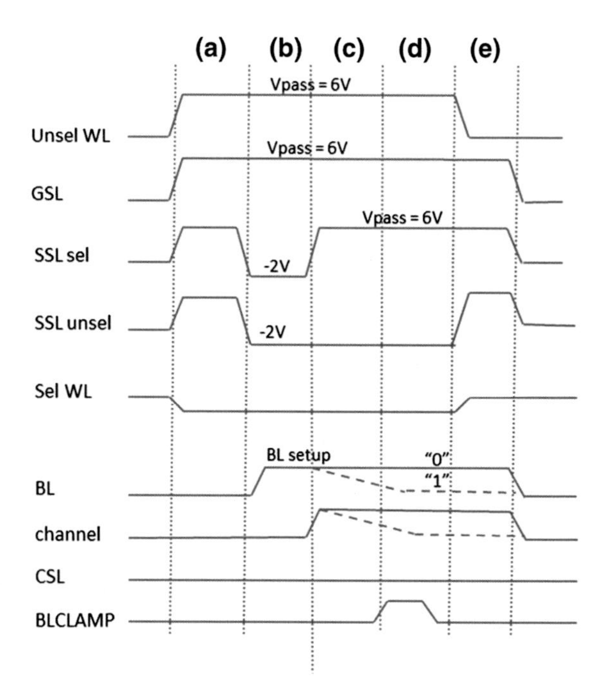

Fig. 7.18 Forward read with bitline discharge to avoid read disturb [28]

V_{TH} distribution of a specific layer is intrinsically different from the one of other layers, and each layer also shows a different BL bias dependence during read. The proposed layer-aware read method uses reverse read with different judging level for each bitline, to compensate for the layer differences.

Figure 7.19 shows a schematic example of the reverse-read sensing with multi V_{TH} reading [28]. BL compensation consists in biasing the BLCLAMP transistors of the page buffer at a proper voltage, depending on the layer being read. Reverse read consists in the following sequence: firstly, bitlines BL are discharged to Ground (GND) through the BLBIAS transistor; afterwards, CSL is biased at a positive voltage (reverse read) and BLC is turned OFF to isolate the sensing node SEN from the bitline; WL is raised to the Read or Program Verify voltage (in the example the read voltage), then SSL is turned ON, and BL capacitance charges through the cell and CSL until it reaches voltage V_{WL}-V_{THC}, where V_{THC} is the selected cell's threshold voltage. During the sensing phase, BLC is set to the discriminating voltage V_{BLCL} (which varies according to bitline/layer L). If the cell's V_{TH}, V_{THC}, is lower than the target V_{TH}, then the BL voltage is higher than V_{BLC}-V_{THBLC}, thus the BLC transistor is OFF and the voltage of SEN node remains high. Otherwise, if the V_{THC} is equal or greater than target V_{TH}, then the BLC transistor is ON, SEN discharges, and the sense amplifier output switches to '0' value.

The proposed method for read and verify can optimize the reading conditions based on the characteristics of the cells in each layer. Therefore, it is possible to

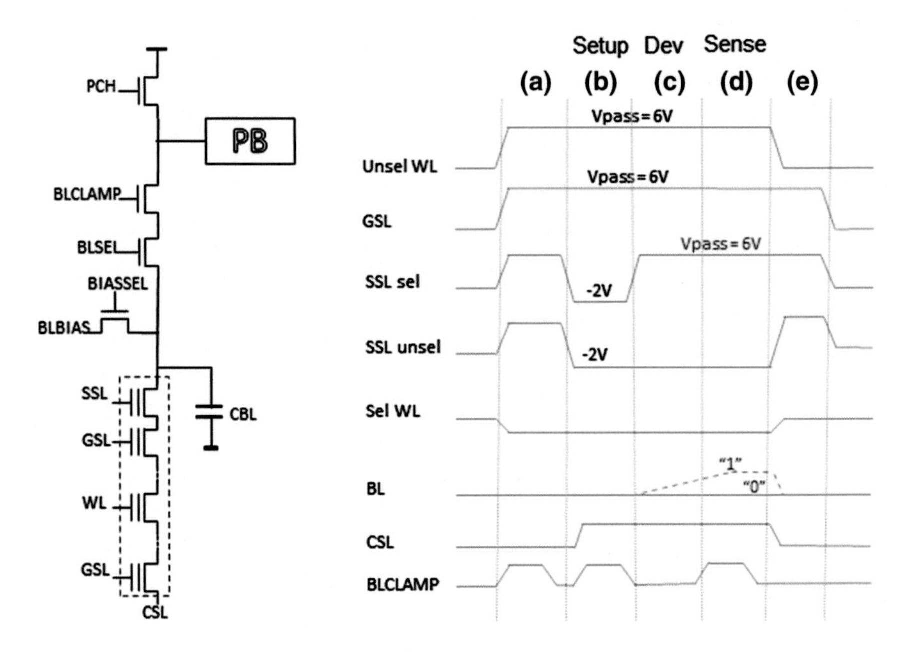

Fig. 7.19 Reverse read operation with BL clamp compensation [28]

optimize the target Program Verify PV for each layer instead of using a common PV level; the programming time can significantly be reduced, and also the program disturbance is reduced by avoiding unnecessary over programming of the lower layers. A drawback of the layer-aware read/verify method is the difference in the bitline voltage swing between layers. The top layer will have higher V_{TH} than other layers and, therefore, the greatest BL swing. This happens because, during reverse read, the BL is charged from GND to the target voltage, which is different for each layer/bitline. The layers with the higher operating point will result in higher background pattern dependency. In order to level this kind of dependency, a method was proposed where each bitline is pre-charged to specific layer-dependent voltage instead of discharging all the bitlines to GND during the setup phase [31].

7.5.2 Program Operation

Programming 3D-VG NAND is obtained by optimized ISPP (*Incremental Step Programming Pulse*) algorithms. In Fig. 7.20 it is shown the equivalent circuit to illustrate the programming of page 0 (SSL1) and the corresponding program-inhibit method used on the other cells not being programmed. When programming page 1, the select line SSL1 is at VCC, while a slightly negative voltage is applied to other SSLs to guarantee turn-off. GSL (even) is turned-on, while GSL (odd) is turned-off.

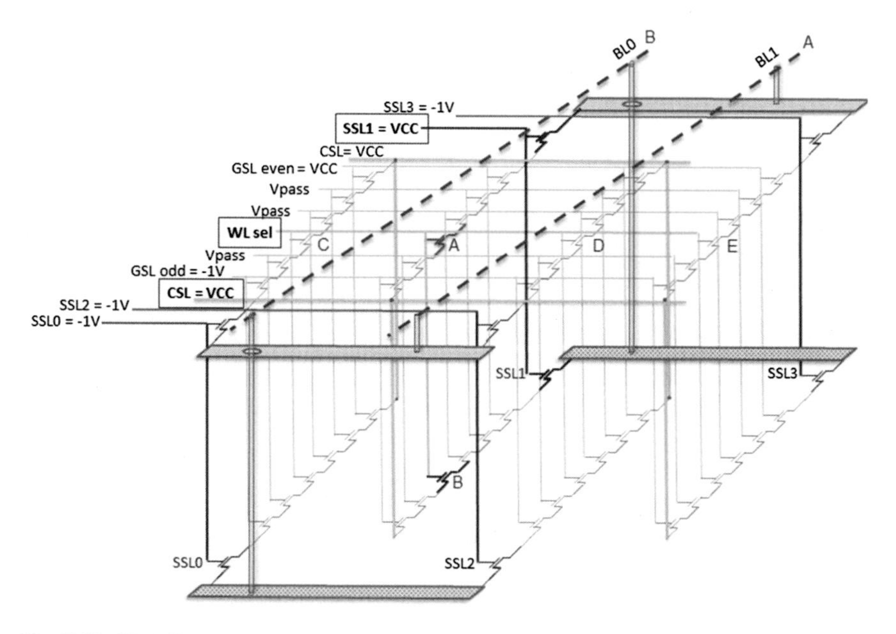

Fig. 7.20 3D VG-type program biasing

CSL is set to VCC in order to perform the inhibit operation. The selected wordline is biased with the programming voltage according to the ISPP scheme, while other wordlines are biased with Vpass voltage. Program Inhibit of unselected pages is achieved by using different methods. The cells belonging to Page 3 (SSL3) bitlines (e.g. Cell 'E') are inhibited by entirely floating the NAND string, because SSL3 and GSL (even) are both deselected; the cells belonging to even pages (SSL0 and SSL2), i.e. Cells 'C' and 'D', are inhibited by charging the channel via CSL and GSL, which are both at VCC [10].

7.5.3 Erase Operation

A *Barrier Engineered SONOS* (BE-SONOS) device is adopted in 3D VG-type NAND. Erasing of BE-SONOS devices is achieved by holes injection from the channel. BE-SONOS can overcome Electron Gate injection due to the high electric field during erase and reduce erase saturation with an excellent retention. Vertical Channel devices, which usually have a gate-all-around structure, can benefit from the *Electric Field Enhancement* (FE) due to the curvature effect which reduces the electric field through the gate, thus mitigating the electrons injection from the poly gate. However, FE method has several drawbacks and imposes limitations on the cell size [32, 23]. Another way to reduce gate injection would be through the use of Hi-K material which can minimize the electric field in the oxide. BE-SONOS

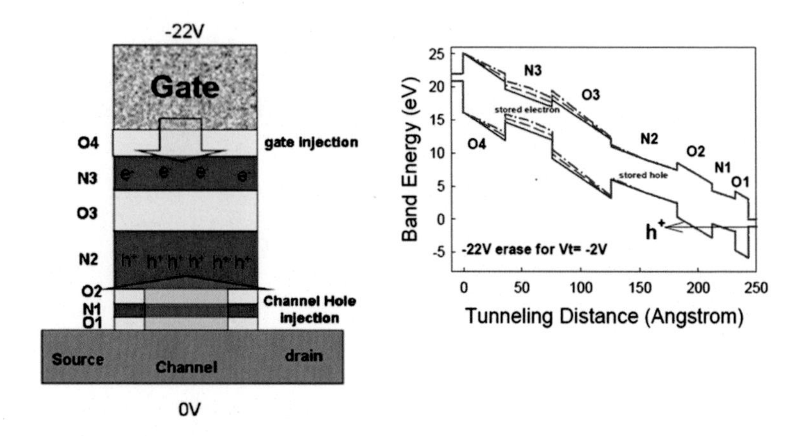

Fig. 7.21 Erase Mechanism in a BE-SONOS CT device [23]

devices allow avoiding the complication and drawbacks of FE or the use of Hi-K materials. In 3D-VG NAND, further engineering of the BE-SONOS structure is possible, thanks to the planar and very uniform structure [23]. In Fig. 7.21 erase through holes injection from the substrate in a BE-SONOS device and the parasitic electron injection from the polysilicon gate are shown. The displayed structure shown is BE-SONOS with additional nitride and oxide layers specifically added to stop electron injection at the additional nitride layer.

In 3D-NAND a junction-less cell with floating body is generally employed; therefore, there is no holes source for erasing from the substrate. Holes generation for erase can be obtained by GIDL (*Gate-Induced Drain Leakage*) at the select transistors. Figure 7.22 shows that erase is achieved by applying a large voltage to the CSL (*Common Source Line*) and MBL (*Main Bitline*), while an intermediate voltage of about 6 V is applied to the SSL (*String Select Line*) and GSL (*Ground Select Line*). The select transistors should be biased to pass the erase voltage (13 V) from the source lines to the channel; at the same time, the pass transistors must be guarded against parasitic charge injection that may damage them or make their threshold voltage drift. With appropriate biasing of SSL and GSL, the GIDL induced current that is built up at the drain junctions of the select transistors can transfer the erasing voltage to the string channel and minimize the unwanted hole injection that would result in erasing of the SSL and GSL themselves. To Erase a Block in 3D-VG Type NAND, a high voltage is applied to the source lines while the wordlines are grounded. To inhibit erasing of the unselected blocks, the wordlines are left floating so that they are self-boosted to a inhibit voltage. It is important to reduce any leakage on the wordlines that could reduce the efficiency of the boosted potential and, hence, the erase inhibit.

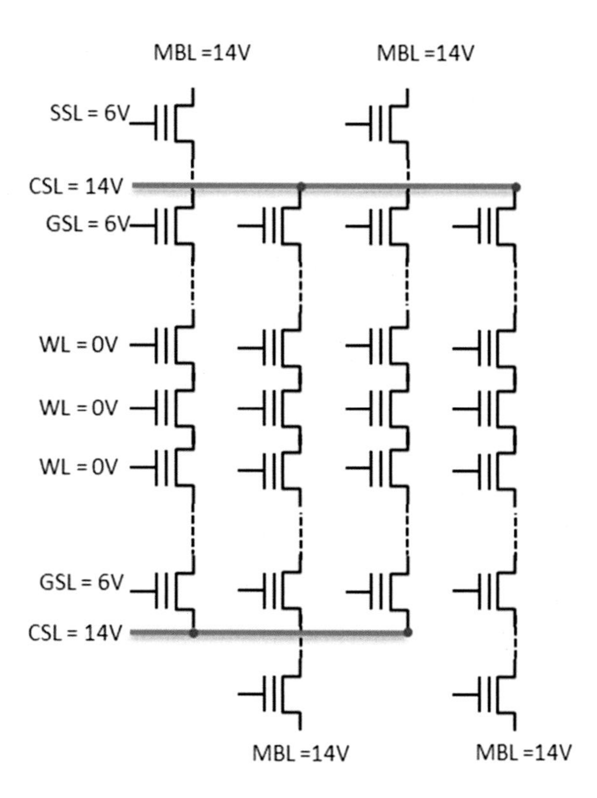

Fig. 7.22 VG-type block erase diagram (selected block)

7.6 Disturbs of VG-Type 3D NAND

The most peculiar disturbs and interference effects in 3D NAND are along the vertical (Z) direction, which does obviously not exist in 2D architectures. In VG-type 3D NAND, bitlines are completely shielded by wordline polysilicon plugs; therefore, there is no electrostatic interference in the X-direction (along the wordline direction) or even in diagonal direction between different layers, as shown in Fig. 7.23. Instead, Z interference can be observed between adjacent vertical layers: it is a back-gate bias effect caused by electrostatic interference between adjacent cells. The electrostatic interference is caused by electrons stored in the trapping layers during programming: the V_{TH} shift of the adjacent victim layer can be in the range of 150 mV. The interference is obviously dependent on the inter-layer distance and it can be optimized [24, 33].

Wordline interference has also been observed in VG-type 3D NAND when programming adjacent wordlines in the same layer due to the tight pitch. The WL interference is associated to channel potential variations in the junction free channel of the selected cell, which are induced by the charge accumulated in adjacent pass cells. WL interference results in 400 mV V_{TH} shift; this shift can be contained by proper algorithmic countermeasures. Programming algorithm with Pre-PV

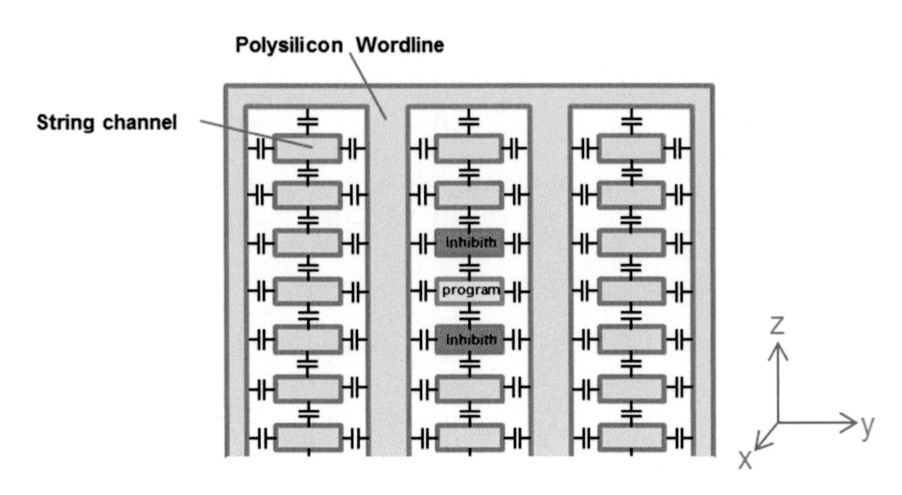

Fig. 7.23 Z disturbs in VG-type 3D NAND

(*Pre-Program Verify*) level can be used to alleviate WL-interference. The algorithm consists of programming WL(n) to a pre-PV level lower than the target PV level, then programming WL(N + 1) to the pre-PV level, going back to program WL(n) to the final PV value, and so on.

Since all the layers are programmed simultaneously, as part of the same page in 3D-VG NAND, the interference and other effects caused by the programming pattern must be considered.

A program disturb effect has been studied. As shown in Fig. 7.23, when a certain layer is programmed, the related channel is kept at 0 V, while adjacent channels, if not programmed, are boosted to the inhibit voltage. Due to the capacitive coupling between layers, the resulting boosted value is lowered with respect to the case where all the channels are in the inhibit condition. The inhibit function is therefore degraded and a program disturb can be observed. The threshold shift due to Z-disturb can result in a shift of 0.6 V of the V_{TH} distribution, because of the reduced inhibit efficiency [24]. Moreover, when different layers are programmed at different times, an impact of the first programmed cells on the following programming operations can be observed, due to the electrostatic potential change. This effect is substantially changing the V_{TH} distribution.

Another Z-effect, called enhanced programming Z-disturb, can be observed in VG-type 3D NAND. When adjacent layers are programmed (biased to 0 V), the programming speed is reduced with respect to the case where adjacent layers are inhibited (boosted to inhibit high voltage). This effect is explained by the inversion electron density function at the polysilicon channel that is increased in the second case, while in the first case it limits the F-N tunneling current.

It is necessary to consider the impact of the Z-effects when the programming pattern changes during the ISSP steps. The described Z-disturbs will cause the

Layer	1st STEP	2nd STEP	3rd STEP
8	Inhibit	Program	Inhibit
7	Program	Inhibit	Inibith
6	Inhibit	Inhibit	Program
5	Inhibit	Program	Inibith
4	Program	Inhibit	Inibith
3	Inhibit	Inhibit	[illegible]
2	Inhibit	Program	Inibith
1	Program	Inhibit	Inibith

Fig. 7.24 An algorithmic programming method to level out Z-disturbs [24]

enlargement of the programmed distribution and reduce the margin for multilevel operation.

Z-interference (electrostatic) effects are obviously dependent on the thickness of the buried oxide between layers; therefore, they are also controllable, to some extent, by augmenting the inter distance between layers [8]. Algorithmic methods can also be adopted to control the Z-disturbs. In the example shown in Fig. 7.24 the programming sequence is split into a sequence involving only certain layers at a time. In this example eight layers are divided into three groups and programming is accomplished in three programming steps. During each of the programming steps, each inhibited layer has only one adjacent layer (in Z-direction) which is programmed, thus limiting the interference to one neighbor only. At the same time, each programmed layer has two inhibited adjacent layers on both sides so that the enhanced Z-programming effect is constant [25].

Vpass disturb and Read disturb are also critical in 3D Flash. In VG-type 3D Flash there are 2N pages per each wordline, where N is the number of layers. Thus programming the entire block results in a very large Vpass disturb time. Read disturb is also worsened due to the large number of pages in a WL.

Compared to Vertical Channel nanowires type of 3D NAND, substantially, the effect of disturbs is similar. In VC-type 3D NAND the Field Enhancement Effect allows achieving shorter programming time; however, enhanced field increases the Vpass disturbance. On the contrary, the planar structure of VG-type 3D NAND can minimize the Vpass disturb.

By implementing proper algorithmic sequences, as described in this section, it is possible to obtain a V_{TH} window large enough for MLC operations, as shown in

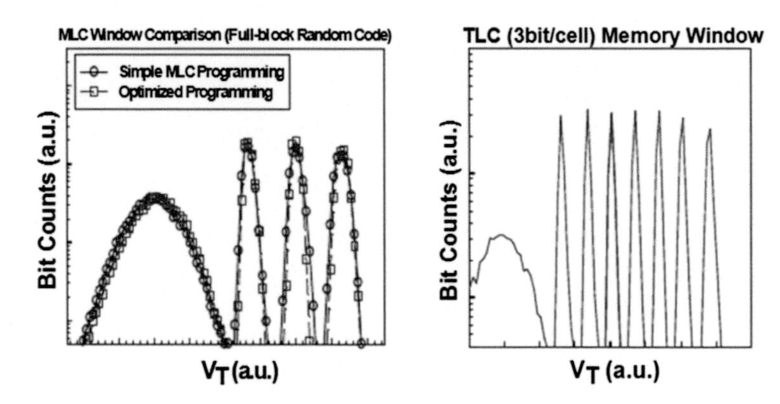

Fig. 7.25 MLC and TLC capability of VG-type 3D NAND [24]

Fig. 7.25. TLC window has been also demonstrated. With proper optimization of the structure and Error Correcting techniques, VG-type 3D NAND is a viable candidate for realizing high density products.

7.7 Conclusions

Vertical Gate (VG-type) 3D NAND technology capability has been demonstrated with several studies and test chips. The technology can achieve cost effectiveness through multiple memory layers and multilevel operations. The planarity of the CT SONOS Flash cell is the key element because it avoids all the issues of the non-ideal vertical etching required by Vertical Channel (VC-type) 3D NAND architectures based on gate-all-around structures.

Acknowledgements The author would like to thank Macronix International for the use of publications which have been cited in this chapter.

References

1. H. Tanaka et al., Bit cost scalable technology with punch and plug process for ultra high density flash memory, in *VLSI Symposium Technical Digest* (2007)
2. R. Katsumata, M. Kito et al., Pipe-shaped BiCS flash memory with 16 stacked layers and multi-level-cell operation for ultra high density storage device, in *Symposium on VLSI Technology* (2009)
3. K. Sakuna et al., Highly scalable horizontal channel 3-D NAND memory excellent in compatibility with conventional fabrication technology, in *EDL*, vol. 34 (2013)
4. J. Jang, H.S. Kim et al., Vertical cell array using TCAT(terabit cell array transistor) technology for ultra high density NAND flash memory, in *IEDM* (2009)

5. J. Kim, A.J. Hong et al., Novel vertical-stacked-array-transistor (VSAT) for ultra-high-density and cost-effective NAND Flash memory devices and SSD (solid state drive), in *Symposium on VLSI Technology* (2009)
6. K,T. Park, D.S. Byeon, A world's first product of three-dimensional vertical NAND Flash memory and beyond, in *NVMTS* (2014)
7. K.T Park et al., Three-dimensional 128 Gb MLC vertical NAND flash-memory with 24-WL stacked layers and 50 MB/s high-speed programming, in *ISCC* (2014)
8. H.T. Lue, T.H. Hsu et al., A highly scalable 8-layer 3D vertical-gate (VG) TFT NAND flash using junction-free buried channel BE-SONOS device, in *VLSI Symposia on Technology* (2010)
9. H.T. Lue, T.H. Hsu et al., A novel double-density single gate vertical channel (SGVC) 3D NAND that is tolerant to deep vertical etching CD variation and possesses robust read-disturb immunity, in *IEDM* (2015)
10. S.H. Chen, H.T. Lue et al., A highly scalable 8-layer vertical gate 3D NAND with split-page bit line layout and efficient binary-sum MiLC (minimal incremental layer cost) staircase contacts, in *International Electron Device Meeting (IEDM)*, session 2-3 (2012)
11. S.J. Whang, K.J. Lee et al., A novel three-dimensional dual control-gate with surrounding floating-gate (DC-SF) NAND flash cell, in *IEDM* (2010)
12. E.S. Choi et al., A novel 3D cell array architecture for terra-bit NAND flash memory, in *IMW* (2011)
13. E.S. Choi, S.K. Park, Device considerations for high-density and highly reliable 3D NAND Flash cell in near future, in *International Electron Device Meeting (IEDM)*, session 9-4 (2012)
14. K. Parat, C. Dennison, A floating gate based 3D NAND technology with CMOS under array, in *IEDM 2015*
15. J. Choi, K.S. Seol, 3D approaches for non-volatile memory, in *VLSI* (2011)
16. K.T. Park, J.M. Han et al., tree-dimensional 128 Gb MLC vertical NAND flash memory with 24WL stacked layers and 50 MB/s high speed programming, in *ISSCC* (2014)
17. Y. Fukuzumi, R. Katsumata, Optimal integration and characteristics of vertical array devices for ultra-high density, bit-cost scalable flash memory, in *IEDM* (2007)
18. W. Kim, S. Choi et al., Multi-layered vertical gate NAND flash overcoming stacking limit for terabit density storage, in *Symposium on VLSI Technology* (2009)
19. C.H. Hung, H.T. Lue et al., A highly scalable vertical gate (VG) 3D NAND flash with robust program disturb immunity using a novel PN diode decoding structure, in *VLSI Symposia on Technology*, session 4B-1 (2011)
20. C.P. Chen, H.T. Lue et al., A highly pitch scalable 3D vertical gate (VG) NAND flash decoded by a novel self-aligned independently controlled double gate (IDG) string select transistor (SSL), in *VLSI Symposia on Technology* (2012)
21. K.P. Chang, H.T. Lue et al., An efficient memory architecture for 3D vertical gate (3DVG) NAND flash using plural island-gate SSL decoding and study of its program inhibit characteristics, in *International Memory Workshop (IMW)* (2012)
22. T.H. Yeh, P.Y. Du, Increasing VG-type 3D NAND flash cell density by using ultrathin polysilicon channels, in *IMW* (2013)
23. H.-T. Lue, R. Lo, A novel double-trapping BE-SONOS charge-trapping NAND flash device to overcome the erase saturation without using curvature-induced field enhancement effect or high-K (HK)/metal gate (MG) materials, in *IEEE IEDM* (2014)
24. C.C. Hsieh, H.T. Lue et al., Study of the interference and disturb mechanisms of split-page 3D vertical gate (VG) NAND flash and optimized programming algorithms for multi-level cell (MLC) storage, in *VLSI Symposia on Technology*, session 11-3 (2013)
25. W.C. Chen, H.T. Lue, Study of the programming sequence induced back-pattern effect in split-page 3D vertical-gate (VG) NAND flash, in *VLSI-TSA* (2014)
26. T.H. Yeh, C.J. Wu et al., A new string decoding scheme for enhancing array block efficiency of vertical gate type (VG-Type) 3-D NAND. EDL **36**(4), 2 (2015)
27. H.T. Lue, Tutorial 3D NAND, in *IMW* (2014)

28. C.H. Hung, H.T. Lue et al., Design innovations to optimize the 3D stackable vertical gate (VG) NAND flash, in *International Electron Device Meeting (IEDM)*, session 10-1 (2012)
29. Y.-R. Chen, C.-J. Wu, Trapping-free string select transistors and ground select transistors for VG-type 3D NAND flash memory, in *IMW* (2014)
30. R. Micheloni et al., *Inside NAND Flash Memories* (Springer, Netherlands, 2010)
31. C.H. Hung, M.F. Chang et al., Layer-aware program and read schemes for 3D stackable Vertical-Gate BE-SONOS NAND flash against cross-layer process variations. IEEE JSSC **50**(6) (2015)
32. A. Maconi, C.M. Compagnoni et al., A new erase saturation issue in cylindrical junction-less charge-trap memory arrays, in *IEDM* (2012)
33. Y.H. Hsiao, H.T. Lue, A critical examination of 3D stackable NAND flash memory architectures by simulation study of the scaling capability, in *IMW* (2010)
34. E.S. Choi, H.-S. Yoo, A novel 3D cell array architecture for terra-bit NAND flash memory, in *IMW* (2013)

Chapter 8
RRAM Cross-Point Arrays

Huaqiang Wu, Yan Liao, Bin Gao, Debanjan Jana and He Qian

8.1 Introduction of RRAM

In the age of Big Data, it has been always a dream for researchers to find the next generation nonvolatile memory with higher density, lower latency and lower cost. As a matter of fact, the last mainstream memory, NAND Flash memory, was created decades ago. Nowadays, NAND Flash memory, based on metal-oxide-semiconductor field-effect-transistor with an additional floating gate, is still one of the most popular nonvolatile memories. However, its speed and density are now approaching the physical limits of its basic structure. It takes longer than 1 μsec to store electrons into the floating gate [1], and it seems impossible to scale planar technologies below 10 nm. This is not only due to the cost of lithography, but also to the crosstalk and parasitic effects caused by the thick gate. In fact, to make sure that electrons don't leak away from the floating gate, gate thickness can't be too thin. Moreover, since the number of electrons trapped inside the floating gate is less than 100 at 20 nm, losing few electrons can cause severe reliability issues [2].

H. Wu (✉) · Y. Liao · B. Gao · H. Qian
Institute of Microelectronics, Tsinghua University, Beijing, China
e-mail: wuhq@tsinghua.edu.cn

Y. Liao
e-mail: lyth031@gmail.com

B. Gao
e-mail: gaob1@tsinghua.edu.cn

H. Qian
e-mail: qianh@mail.tsinghua.edu.cn

D. Jana
Micron Technology Inc., Boise, USA
e-mail: janadebanjan@micron.com

R. Micheloni (ed.), *3D Flash Memories*, DOI 10.1007/978-94-017-7512-0_8

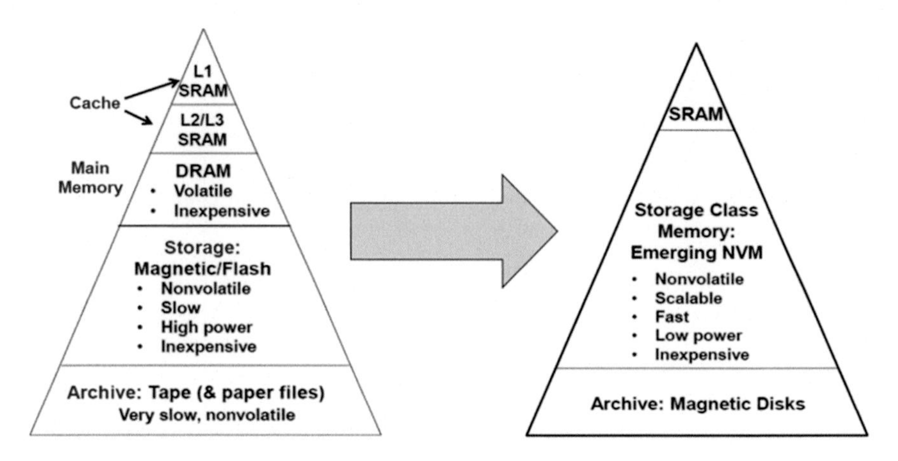

Fig. 8.1 Possible revolution in the memory hierarchy

In such a situation, 3D vertical NAND is the emerging technology, and it has recently achieved ultra-high density by stacking 48 memory layers. 3D NAND replaces planar technology and extends the life of NAND Flash in the semiconductor industry. But how many layers can we stack? 128, 256 or more? There are many challenges to face with 3D vertical NAND technology and the speed limitation still exists.

Recently, resistive random access memory (RRAM) has attracted a lot of attention because of its excellent single-cell performance, like scalability (<10 nm), latency (<10 ns), power consumption (voltage <3 V), improved reliability, endurance ($1E^6$–$1E^{12}$) and retention (>10 years) [3–5]. The 3D integration of RRAM, especially the 3D Cross-point RRAM array with an effective memory cell area of $4F^2/n$ (F is the minimum feature size, n is the number of 3-D-stacked memory layers), can achieve ultra-high density, comparable with 3D vertical NAND technology [6]. RRAM is one of the most promising candidates for the next generation nonvolatile memory [6]. More than this, it may even overthrow the memory hierarchy, as shown in Fig. 8.1, and it may play an important role in nonvolatile logic computing with its excellent combination of speed and density [7]. In both cases we are talking about a revolution in semiconductor industry.

8.1.1 History and Development

It has been more than 50 years since the first observations of the resistive switching phenomena. In the 60s, some researchers, like Hickmott, Gibbons and Beadle, described the switching properties of thin oxide films [8, 9]. However, these early reported resistive switching phenomena were not adequately robust for memory application. It was only in the late 90 s when there was a renewed interest for resistive switching as an alternative to conventional silicon-based memories. Many types of material, such as complex metal oxides [8], binary metal oxides [9] and organics [10] show resistive switching properties. In 2004 Samsung demonstrated a NiO-based RRAM integrated with 0.18 μm CMOS technology at the International Electron Devices Meetings (IEDM); in this paper, some critical device performance metrics for memory application, such as endurance (more than 10^6 times of set/reset cycles and 10^{12} reading cycles) and programming characteristics (operating below 3 V and 2 mA), were discussed [11]. Nowadays, the study of RRAMs is a hot area with hundreds of publications per year.

Recently, great progress has been made in the performance of RRAM, including speed, scalability and endurance. Lee et al. [12] demonstrated a HfO_x-based bipolar RRAM with a switching time of 300 ps, while excellent scaling properties of TiN/HfO_x/Hf/TiN bipolar RRAM were reported by Govoreanu et al. [13] with a cell's area smaller than 10×10 nm^2. Moreover, TaO_x-based RRAM showed high switching endurance (>10^{12} cycles), as proposed by Lee et al. [14]. Significant breakthrough was also made with 3D architectures and high-density integration, as it will be discussed in detail later in this chapter.

8.1.2 Structure and Mechanism of RRAM

In general, the basic structure of RRAM is a switching medium sandwiched between two electrodes, like in Fig. 8.2a. Nonvolatile storage is based on resistive switching between a low-resistance state (LRS or ON state) and a high-resistance state (HRS or OFF state), under voltage or current stimulation. However, there are multiple variations of this simple structure. There is a variety of materials that can serve as a switching medium, like HfO_x, AlO_x and TaO_x, and there are many choices for electrodes, such as Pt, TiN and Ti. In fact, resistive switching does not only depend on the switching medium, but also on the electrodes and their interfacial properties [6].

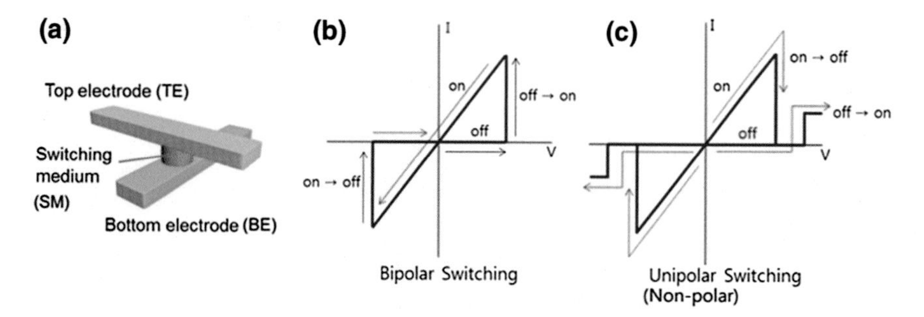

Fig. 8.2 **a** Schematic of top-electrode/switching medium/bottom-electrode sandwich structure. Schematic I–V curves of **b** bipolar and **c** unipolar switching [2]

Different switching materials combined with different electrodes generate two kinds of switching modes and various switching mechanisms. Conventionally, the switching event from LRS to HRS is called *reset process* and the corresponding voltage is called reset voltage (V_{reset}). Inversely, the switching event from HRS to LRS is called *set process* and the corresponding voltage is called set voltage (V_{set}). The forming process is similar to the set process, but it happens when a fresh sample is in its initial resistance state, with the goal of enabling resistive switching cycles. Usually, the forming voltage is higher than V_{set}. According to the electrical polarity required by set and reset processes, there are two kinds of switching modes: bipolar and unipolar. Bipolar switching means that set and reset processes are dependent on opposite electrical polarity voltage, while unipolar switching means that set and reset voltages are with the same electrical polarity, but different amplitude, as shown by Fig. 8.2b, c.

As mentioned above, resistive switching is not only dependent on switching medium, but it is also affected by electrodes and their interfacial properties [6]. A comprehensive overview of resistive switching mechanisms will not be discussed here; we briefly introduce the most common mechanisms with the highest scaling potential, like valence change and electrochemical metallization [2]. In these mechanisms, the formation and rupture of conductive filament in the bulk switching medium are the intrinsic physical phenomena causing resistive switching.

In valence change-based RRAM, the switching medium is typically a transition metal oxide, such as TiO_x, HfO_x and TaO_x [2]. The forming process creates a conductive filament of oxygen vacancies in the switching medium, and it enables resistive switching cycles thanks to the migration of oxygen ions under high electric field. After that, local rupture and re-formation of the conductive filament occur during the reset and set processes, respectively, which leads to the switching between LRS and HRS [15]. For a memory cell in LRS, the current flows through the conductive filament, while in HRS it flows through the filament gap region through tunneling (Fig. 8.3).

For electrochemical metallization-based RRAM, also known as *Conductive Bridge Random Access Memory* (CBRAM), there are various switching materials, such as chalcogenide, amorphous silicon and so on. In general, one of the electrodes is an active metal, like Ag and Cu, while the other one is usually an inert metal [2].

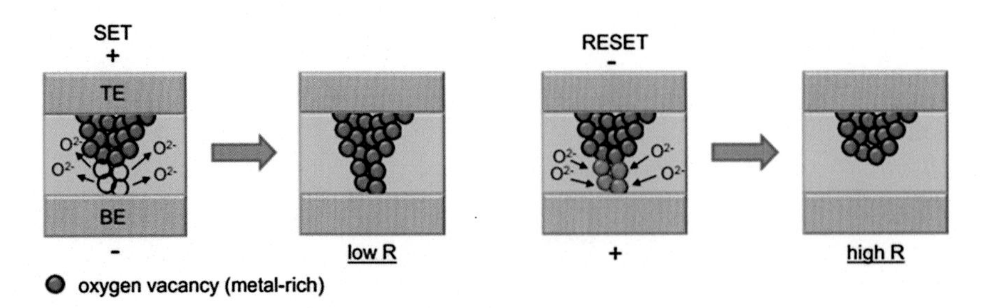

Fig. 8.3 Schematic illustration of the switching process in valence change-based RRAM [2]

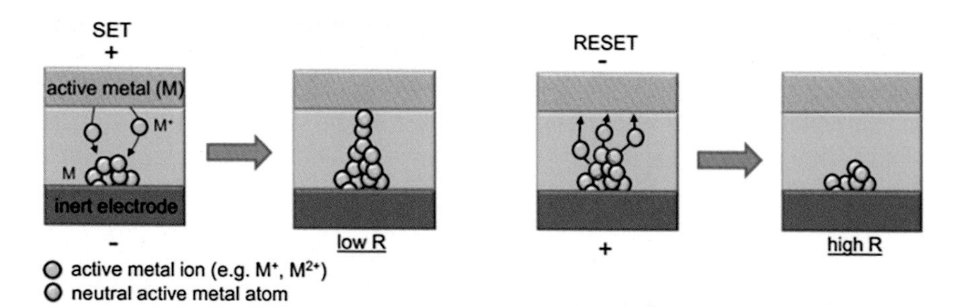

Fig. 8.4 Schematic illustration of the switching process in electrochemical metallization-based RRAM [2]

The key difference between valence change-based RRAM and electrochemical metallization-based RRAM is the filament type. As shown in Fig. 8.4, the metal ions dissociated from the active metal electrode constitute the conductive filament in electrochemical metallization-based RRAM. Conversely, its restoration under reverse voltage causes the reset process. In other words, the electrochemical metallization-based RRAM has to be bipolar. Electrochemical metallization-based RRAM is not well suited for 3D vertical integration because the active metal can easily diffuse across insulation layers. Till now, all the 3D vertical RRAMs reported in literature are based on valence change mechanism.

Another promising nonvolatile memory technology is *Phase Change Memory* (PCM); again, it utilizes the difference between a low-resistance state (crystalline phase, set state) and a high-resistance state (amorphous phase, reset state) of the phase change material. Figure 8.5a shows the typical structure of a PCM cell. The electrical current crowding at the "heater" results in a programmed region illustrated by the mushroom boundary [16]. To reset a PCM cell into the amorphous phase, a large electrical current pulse is applied for a short time, making the programming region first melted and then rapidly quenched. This amorphous region is in series

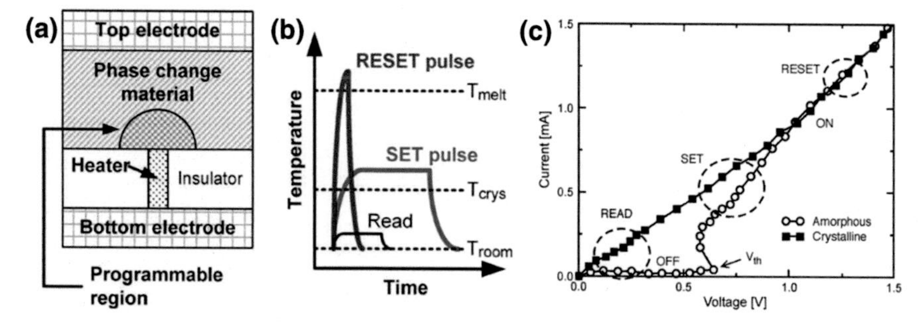

Fig. 8.5 **a** Cross-section of the conventional PCM cell. **b** PCM cells are programmed and read by applying electrical pulses which imply temperature change. **c** I–V characteristics of set and reset states. The reset state shows switching behavior at the threshold switching voltage (V_{th}) [16]

with the crystalline region of the PCM and dominates the resistance. To set a PCM cell into crystalline phase, a medium electrical current pulse is applied for a long time, annealing the programming region at a temperature between the crystallization temperature and the melting temperature. To read the state of the PCM cell, its electrical current is measured (to determine the resistance); of course, the absolute value of the current has to be low enough to avoid disturbing the current state itself.

The resistance difference between set and reset states is remarkable below the threshold switching voltage (V_{th}), while the reset state shows electronic threshold switching behavior when the voltage increases to V_{th}. If a voltage larger than V_{th} is applied for longer than the crystallization time, the PCM cell switches to the low-resistance state, and this is the critical point of the set process. Without the electronic threshold switching phenomenon, the resistance of the reset state would be too high to conduct enough current to provide Joule heating to crystallize the PCM cell [16].

8.2 3D RRAM

8.2.1 3D Architectures

Generally speaking, a 2D cross-point array includes bit lines (BL), word lines (WL), and memory cells located at the intersection of each BL and WL. This cross-point array can achieve the highest $4F^2$ (F is the feature size) integration density in 2D manner. To build a 3D RRAM cell, there are two possible 3D integration approaches: one is the conventional planar RRAM-based cross-point array [17, 18], stacked layer by layer (3D Cross-point RRAM); the other one is the novel 3D vertical RRAM [19–25] sandwiched between the vertical electrodes and multilayer horizontal electrodes (3D VRRAM). Figure 8.6 compares the structures of 3D Cross-point RRAM and 3D VRRAM [26].

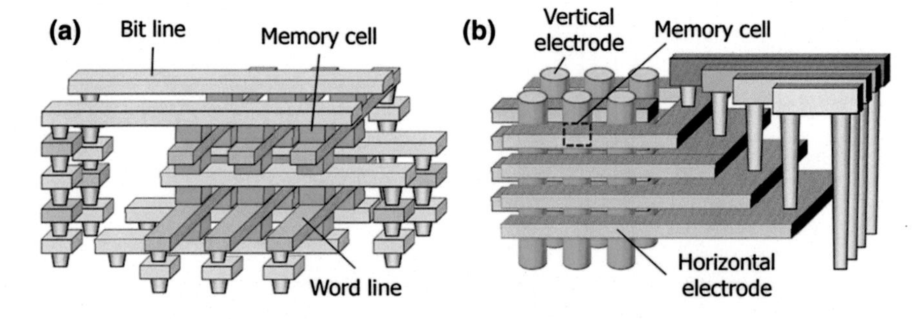

Fig. 8.6 Schematic drawing of **a** 3D Cross-point RRAM, and **b** Vertical RRAM [26]

8.2.1.1 3D Cross-Point RRAM

The 3D Cross-point RRAM, whose density can be as high as $4F^2/N$ (N is the stack number), has an advantage in lateral scaling compared to 3D VRRAM, since the thickness of RRAM cell and selector may take more area in the 3D VRRAM case. At the same time, the large resistance of vertical electrode affects the array performance of 3D VRRAM. However, 3D Cross-point RRAM, because it simply stacks planar RRAM layers, does not save lithography steps or masks, and, therefore, the bit-cost remains high. On the contrary, 3D VRRAM requires only one critical lithography step or mask for array fabrication (Fig. 8.7); as a result, it is a more promising approach for reducing bit-cost [27]. Of course, costs can be significantly reduced by stacking more layers.

For 3D Cross-point RRAM, Baek et al. [17] were the first to demonstrate a 2-layer 4 × 5 array. Lee et al. [18] have reported 2-stack 8 × 8 array 1D-1R (one diode-one resistor) structure with 0.5 μm × 0.5 μm cells as part of the high density stacked RRAM investigation. The schematic view of the fabricated 2 stacks cross-point structure is shown in Fig. 8.8a. One BL serves two adjacent cross-point layers. Therefore, the layer thickness and the number of critical lithographic steps can be reduced. In this cross-point memory stack, Ti-doped NiO was used for the storage node, and p-CuO_x/n-$InZnO_x$ heterojunction thin film was used as an oxide diode, as shown in Fig. 8.8b.

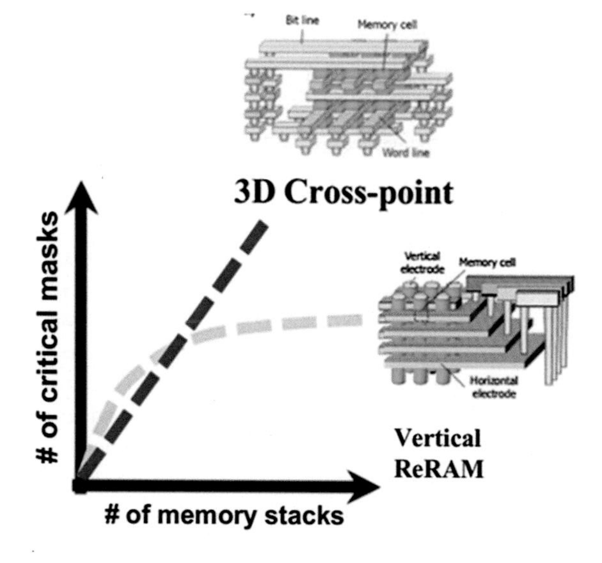

Fig. 8.7 The required number of masks as a function of the number of memory stacks for 3D Cross-point and Vertical RRAM [27]

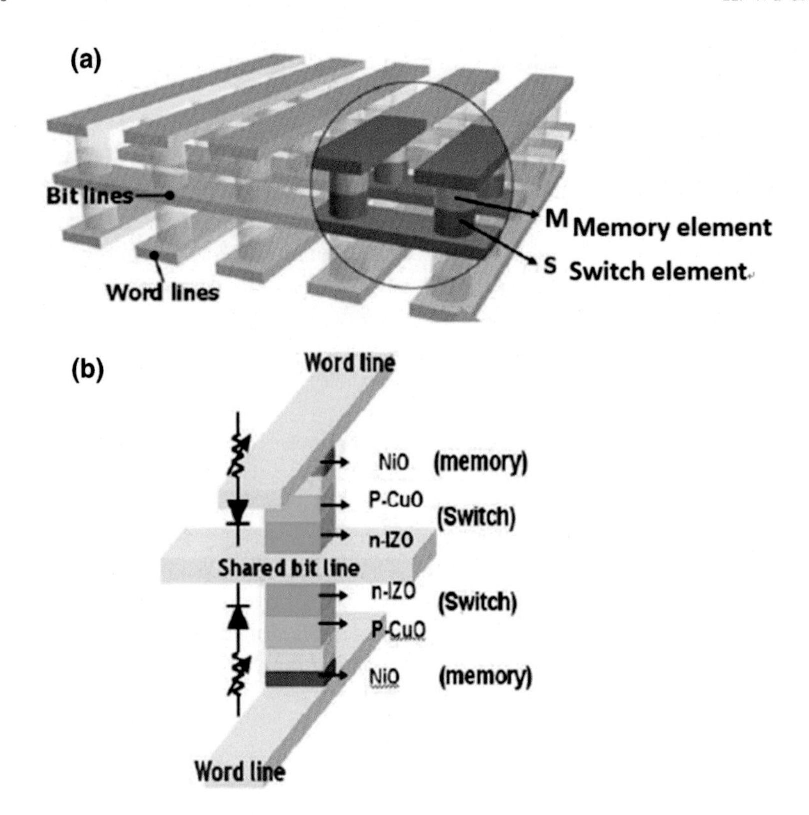

Fig. 8.8 **a** Generalized 2 layers cross-point memory structure. A single cell of the array consists of a memory element and a switch element between conductive lines on top (*word line*) and bottom (*bit line*). **b** Schematic diagram of a 2-stack 1D-1R memory cell with upper layers reversed to share the bit line [18]

8.2.1.2 Vertical RRAM

VRRAM was initially proposed by Yoon et al. in 2009 [28]. After that, several different architectures of 3D VRRAM have been proposed and demonstrated. Figure 8.9 shows two typical 3D VRRAM array architectures [29]. One of them uses metal lines (1D) as the horizontal electrode, and the other one uses metal plane (2D). Both of them use a metal pillar as the vertical electrode. Metal plane based VRRAM requires one critical lithographic step only, and it has much lower line resistance, which significantly reduces IR drop and RC delay effects. Metal line based VRRAM requires at least two critical lithographic steps, and it suffers from more serious IR drop and RC delay effects compared to metal plane based VRRAM and 3D Cross-point RRAM. However, the density of metal line based VRRAM can be two times higher than that of metal plane based VRRAM, since both left and right sides of the metal pillar can form an independent RRAM cell. Therefore, the theoretical limit of integration density of a metal line based VRRAM can reach $2F^2/N$.

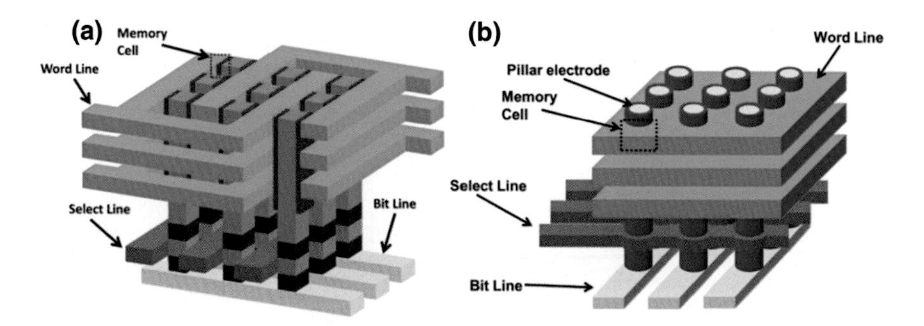

Fig. 8.9 Two typical 3D VRRAM array architectures: in **a** the memory cell sits between the horizontal word line and the vertical pillar, while in **b** the memory cell sits between the plane electrode and the vertical pillar [29]

Chen et al. [30] demonstrated a cost-effective fabrication process for metal plane based VRRAM. As shown in Fig. 8.10, multi-layered silicon oxide and metal are deposited initially, followed by a hole etching process. Then switching layer and vertical metal pillar are deposited. Finally, the plane electrodes of different layers and pillar electrodes are patterned. Actually, only the first pattern (forming a deep hole) is a critical lithographic step.

Baek et al. [26] were the first to demonstrate the metal line based 3D VRRAM. Since the aspect ratio of metal etching is much worse than oxide or nitride etching, they developed a fabrication process without etching metal, as illustrated in Fig. 8.11. Initially, multi-layered silicon oxide and silicon nitride are deposited, followed by deep hole etching and switching layer/vertical metal electrode deposition. Then, the vertical metal electrode and the stacked layers are etched to form horizontal line and vertical metal pillar. After that, the silicon nitride layers are wet etched and the remaining space is filled with metal. Finally, the new deposited metal is patterned to form the horizontal line electrode.

In addition, unlike NAND flash, the requirement for random access of each individual RRAM cell imposes a unique challenge to the design of the 3D VRRAM array architecture. To apply voltage on each vertical electrode, a transistor array is introduced below the VRRAM array [30], as illustrated in Fig. 8.12. A vertical transistor is proposed to achieve $4F^2/N$ integration density. In this architecture, a BL is connected to the source of each vertical transistor, and the select line (SL) is connected to the gate of each vertical transistor. By selecting BL and SL, one vertical electrode can be identified. At the same time, a plane electrode, which works as WL, is selected. In this case, each RRAM cell can be random accessed by using this 3D selection scheme. It should be noticed that it is very challenging to design an architecture that can randomly access each single cell in a metal line based VRRAM array without significant area penalty.

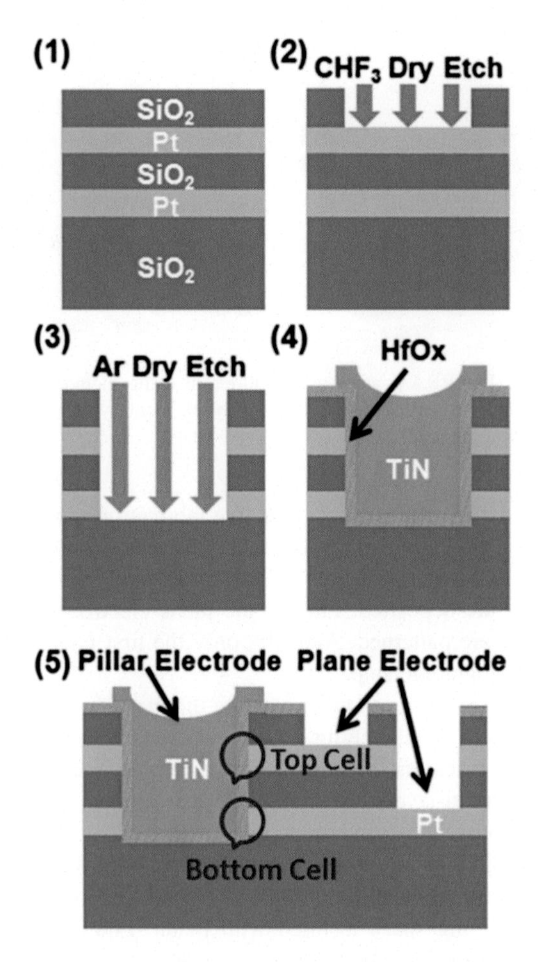

Fig. 8.10 The cost-effective process flow of 3D vertical RRAM: (*1*) Multiple Pt (20 nm)/SiO_2 (30 nm) are deposited by evaporation/LPCVD; (*2*) A trench (1–100 μm in size) is dry etched down to the bottom SiO_2 layer; (*3*) 5 nm HfO_x is deposited by ALD conformally covering the sidewall of the trench; (*4*) 150 nm TiN is deposited by sputtering to fill the trench as the pillar electrode; (*5*) the plane electrode Pt is exposed by dry etching [30]

8.2.1.3 RRAM Cells in 3D Array

The RRAM device in 3D cross-point array has structure and fabrication process similar to 2D cross-point arrays, and thus it doesn't suffer from any performance degradation. The typical device structure includes top electrode, bottom electrode, and a resistive switching dielectric layer between them. The switching layer can be deposited by using either PVD or ALD method. Low operation voltage (1–3 V), fast switching speed (0.3–50 ns), low peak current (25–100 μA), large HRS/LRS

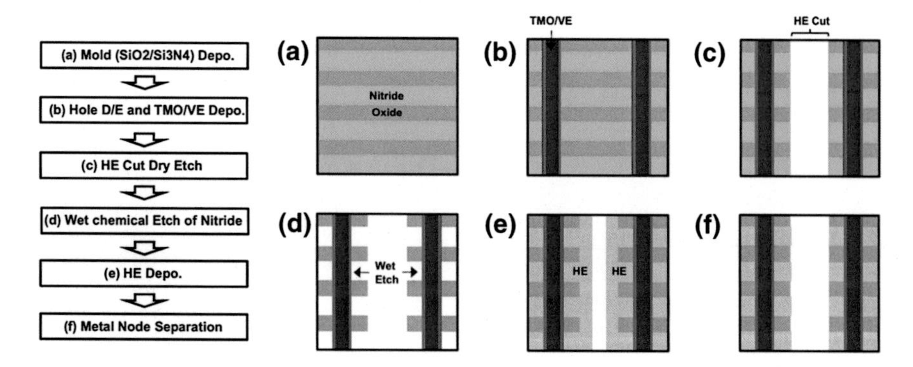

Fig. 8.11 Key process flow of VRRAM cell arrays [26]

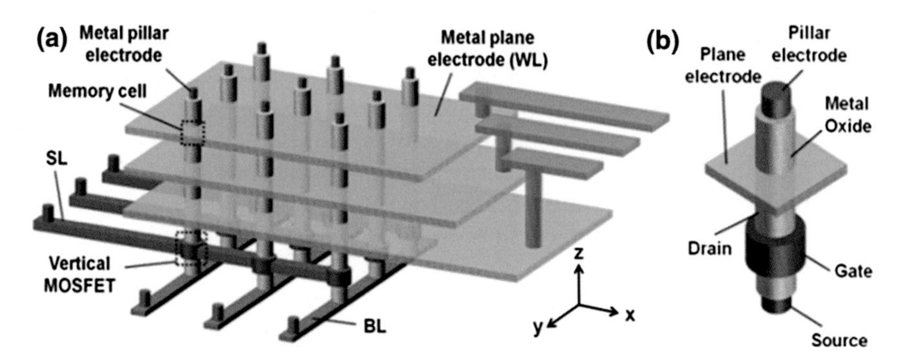

Fig. 8.12 Schematic view of the 3D cross-point architecture, **a** using the vertical RRAM cell [30] and a vertical MOSFET transistor **b** as the bit-line selector to enable the random access of each individual cell in the array

ratio (10–1000x), robust endurance (10^6 to 10^{12} SET/RESET cycles) and retention have been demonstrated with planar RRAM cells.

However, the performance may degrade for vertical RRAM due to the change of device structure and imperfect interface between electrode and switching layer. Park et al. [27] proposed using a barrier layer to improve RRAM's performance and achieve self-rectifying characteristics. Chen et al. [30] demonstrated Pt/HfO_x/TiON/TiN vertical RRAM cell with <50 μA peak current, >10 × HRS/LRS ratio, ~0 ns speed, >10^8 endurance, and >28 h @ 125 °C retention. Vertical RRAM can compete with planar RRAM on all memory metrics. However, the application of Pt electrode is not compatible with traditional CMOS technology. Cha et al. [31] developed a TiN/Ta_2O_5/Ta VRRAM. The SET/RESET voltage was lower than 1.1 V and RESET current was around 100 uA. 10^6 switching cycles were also demonstrated. Hsu et al. [32] presented a Ti/TiO2/TaO_x/Ta VRRAM with ultralow RESET current (~10^{-7} A). 10^{10} switching cycles were achieved under 6 V/1 μs

SET pulse and −6.5 V/1 μs RESET pulse. Looking forward, more efforts should be spent on structure and fabrication process of vertical RRAM cells.

8.2.2 Sneak Path Issues in Cross-Point RRAM

8.2.2.1 Misreading

There is no doubt that 3D RRAM is very promising in terms of ultra-high integration density; however, the sneak paths of RRAM arrays degrade the array performance, increase the total power consumption and limit the array size. As shown in Fig. 8.13, when a device is addressed, which means that BL and WL connected to this device are biased, while other BLs and WLs are floating, the current flows through the selected path (the purple one), in parallel with several paths through unselected RRAM cells (like the red one and the blue one) which are called sneak paths. The red path marked as "1" in Fig. 8.13 is a typical sneak path in a single layer cross-point array, which is composed of three unselected devices. There are $(m-1) \times (n-1)$ possible sneak paths in a single layer cross-point array with m rows and n columns; this means that the bigger the array size is the higher the undesired current on the BL and WL is. The issue is even more complex in a 3D arrays due to the additional introduced sneak paths, such as the blue one marked as "2" in Fig. 8.13.

As a result, the equivalent resistance of the selected device in HRS might be significant smaller than expected, making it difficult to distinguish HRS and LRS during read [33]. More intuitively, let's analyze the worst situation in a single layer cross-point array to see how the sneak paths limit the array size, as shown in Fig. 8.14. Let's assume that there are m rows (BL) and n columns (WL) in the cross-point array and the R_{off}/R_{on} ratio is β. The worst case for HRS read happens when the parasitic resistance caused by sneak paths is the smallest, which means all

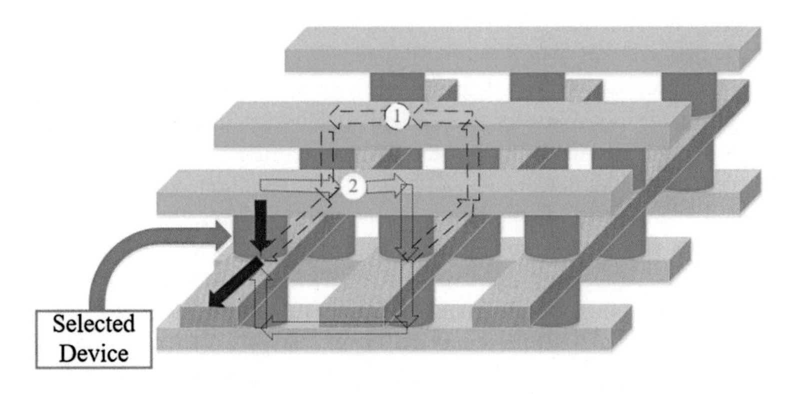

Fig. 8.13 Schematic of sneak paths in a cross-point array with multiple layers

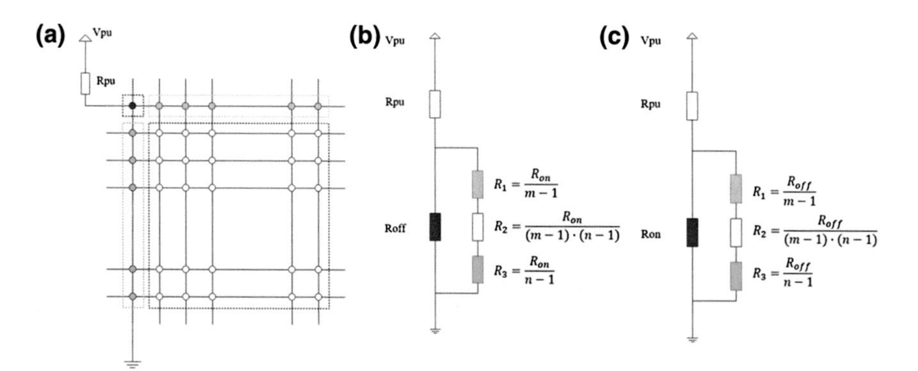

Fig. 8.14 A $m \times n$ single layer cross-point array (**a**) with its equivalent circuit in the worst situation (**b**) and (**c**). R_1, R_2 and R_3 are the equivalent resistors of bits on selected BL, bits on unselected WL and BL, bits on selected WL, respectively

the unselected devices in the array are set to LRS. Conversely, when all the unselected devices are set to HRS, the parasitic resistance is the biggest, leading to the worst case for LRS read. To avoid the misreading, the readout resistance of HRS under the worst case must be larger than the readout resistance of LRS, which means:

$$R_{off} // \left(\frac{R_{on}}{m-1} + \frac{R_{on}}{(m-1)\cdot(n-1)} + \frac{R_{on}}{n-1}\right) > R_{on} // \left(\frac{R_{off}}{m-1} + \frac{R_{off}}{(m-1)\cdot(n-1)} + \frac{R_{off}}{n-1}\right)$$

Then,

$$(1 - \frac{(m-1)\cdot(n-1)}{m+n-1}) \cdot (\beta - 1) > 0$$

Obviously, the R_{off}/R_{on}-ratio β is larger than 1, which results in:

$$\mathrm{n} < \frac{2\cdot(m-1)}{m-2} \Leftrightarrow \mathrm{m} < \frac{2\cdot(n-1)}{n-2}$$

That is to say, if either n or m is larger than 2, then the maximum array size can't be larger than 3×3, independently from the R_{off}/R_{on} ratio β [34]!

Moreover, sneak paths increase the current in word lines and bit lines too. When considering the line resistance, the misreading issues become more severe [35], and the power consumption goes up.

8.2.2.2 Write Crosstalk

Write crosstalk caused by sneak paths is a problem too. Let's take a single 2×2 cross-point array with its selected word line biased at the operating voltage (V), selected bit line grounded and others floated: its equivalent circuit is shown in Fig. 8.15a. When a set voltage is applied on the selected device (the red one in the figure), an almost equal voltage is applied on several unselected devices at the same time (e.g. the yellow one in the figure), which cause an undesired set or reset operation (write crosstalk). Luckily, the write crosstalk issues can be suppressed by using a proper write scheme, such as V/2 or V/3 write schemes, as shown in Fig. 8.15b, c. A sufficient voltage margin between the selected bit (V) and the unselected bits (V/2 or V/3) effectively suppresses the probability of unintentionally write of unselected cells [35]. However, this partial biasing scheme brings new reliability issues on unselected RRAM cells (write disturbance), which is even more serious on large size arrays due to the switching voltage variation and the influence of interconnect resistance.

Thermal crosstalk is another critical issue that must be considered in a 3D cross-point array, when the feature size shrinks down to the nanometer scale. It is widely accepted that the resistive switching behavior is due to the formation and rupture of conductive filament, and the Joule heat generated during the operation, especially the reset operation, plays an important role in this process [37]. As demonstrated by Sun et al. [38] in a 3D cross-point array the thermal transfer is fast along the WL/BLs and conductive filaments (CFs) of RRAMs in both horizontal and vertical directions. It means that, in a cross-point array, the Joule heat generated inside a RRAM device not only determines its own switching behavior, but it also influences the properties of neighboring devices through the thermal transfer. Thus, a larger cross-point array with more stack layers suffers from this thermal transfer issue more severely, and it is more likely to generate a high enough temperature in disturbed RRAM cells during cycling, thus leading to the thermal dissolution of the

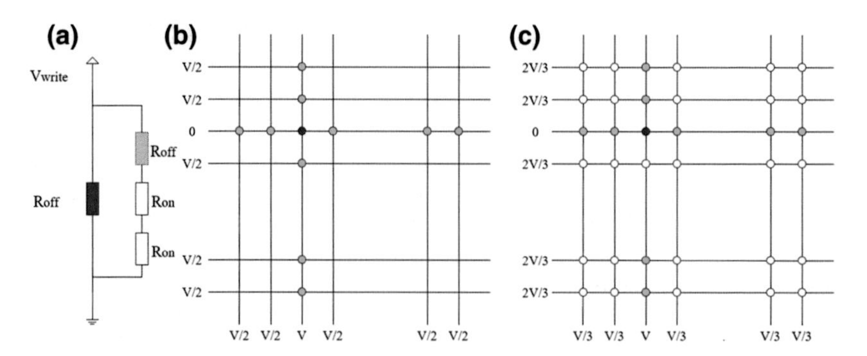

Fig. 8.15 **a** The equivalent circuit of a single 2×2 cross-point array, **b** schematic of V/2 write scheme, **c** schematic of V/3 write scheme

filament and premature loss of the resistance state of RRAM. Moreover, the thermal crosstalk can limit the scaling of the integrated array due to its growing severity with a reduced feature size F.

To alleviate thermal crosstalk effects, several schemes can be adopted. One of them is decreasing the reset current, which could effectively suppress thermal crosstalk [38]. A simple cycle-rehabilitate technique can also be used, i.e. erasing and reprogramming the LRS of RRAM cells in the array after a certain number of operation cycles. Of course, we need to make sure that the deteriorated LRS can still be used to distinguish HRS and LRS [38].

8.2.2.3 Solutions

To handle the sneak-path issues, especially misreading, and increase memory density, one of the most attractive approaches is concatenating a memory element with a two-terminal selector device. The function of the selector device is weakening the leakage through sneak paths according to two important features of sneak paths: (1) the current flows through at least one of the unselected devices in the reverse direction; (2) more than three unselected devices share the whole applied voltage [33]. It means there are at least two categories of selector devices to address the sneak paths. Ideally, if there is a selector device which permits the forward current and prohibits the reverse current, or there is a selector device which turns on at high voltage and turns off at low voltage, the sneak leakage could be avoided. The former is an asymmetric device, like rectifying diodes, and the latter is a nonlinear device, like mixed-ionic electron conduction (MIEC).

Now, a simple quantitative estimation can help us analyzing the influence of the selector in a single layer cross-point array. The approach is similar to the analysis done in Fig. 8.14, but here we simplify it by having m equal to n, as shown in Fig. 8.16. For the cross-point array integrated with asymmetric selectors, the reverse parasitic resistance is much larger than the forward parasitic resistance, which means that the driving resistance is R_2. Instead, for the cross-point array integrated with nonlinear selectors, R_1 and R_3 share the majority of the read voltage due to much fewer parallel devices in the equivalent region.

Based on above analysis, one of the most critical characteristics for selectors is the nonlinearity factor (corresponding to the on/off ratio) α, which is defined for asymmetric and nonlinear devices as shown in Fig. 8.17. Similar to the worst case analysis done for misreading, let's search the limitation for a single cross-point array integrated with selectors. For asymmetric selectors, there must be:

$$R_{off}@V_{read}//\frac{R_{on}@-V_{read}}{(n-1)^2} > R_{on}@V_{read}//\frac{R_{off}@-V_{read}}{(n-1)^2}$$

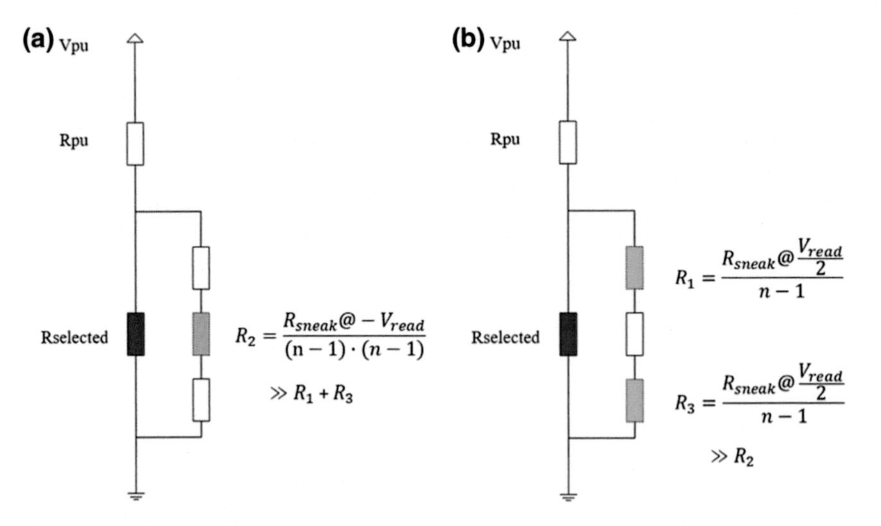

Fig. 8.16 Equivalent circuits of a n × n single layer cross-point array integrated with asymmetric selectors (**a**) and nonlinear selectors (**b**)

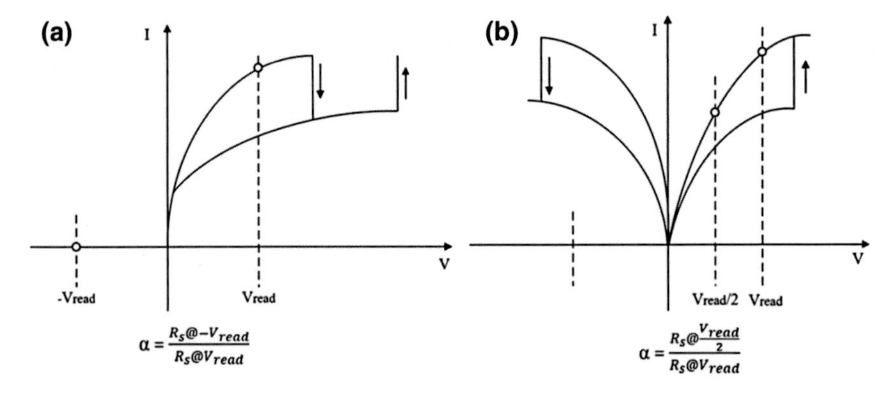

Fig. 8.17 Definition of the nonlinearity factor α for asymmetric devices (**a**) and nonlinear devices (**b**). R_s stands for the resistance of the integrated memory cell in a certain state

Then,

$$\left(1 - \frac{(n-1)^2}{\alpha}\right) \cdot (\beta - 1) > 0$$

Obviously, the R_{off}/R_{on}-ratio β is larger than 1, so,

$$\alpha > (n-1)^2 \propto n^2$$

Likely, for nonlinear selectors, there must be:

$$R_{off}@V_{read}//\frac{2\cdot R_{on}@\frac{V_{read}}{2}}{n-1} > R_{on}@V_{read}//\frac{2\cdot R_{off}@\frac{V_{read}}{2}}{n-1}$$

Then,

$$\left(1-\frac{n-1}{2\cdot\alpha}\right)\cdot(\beta-1)>0$$

So,

$$\alpha > \frac{n-1}{2} \propto n$$

That is to say that the limitation to the array size, defined as $n \times n$, is proportional to α for asymmetric selectors, but to α^2 for nonlinear selectors, which means a huge superiority of nonlinear selectors for high-density application [36].

In addition to selector devices, a single RRAM cell with self-rectifying [27, 30, 32, 53–55] has a similar effect, while complementary resistive switches (CRSs) [56–58] create a novel way to suppress sneak path issues.

8.2.3 Selector Devices

A select transistor is the best solution to fight against the sneak path issue. For this 1T1R (1 Transistor—1 RRAM) cell, the RRAM cell is connected to the drain of the transistor. Turn on voltage is applied on the gate of the transistor during write and read processes. Due to the excellent on/off characteristics of transistors, the sneak path is entirely eliminated. However, the conventional planar 1T1R cell usually takes more than $8F^2$ area, and this is not very well suited for 3D integration. To solve this problem, Wang et al. [39] proposed a 3D vertical transistor based 1T1R cell. RRAM cells were stacked onto the tip of the vertical nano-pillar transistors array, which enabled a $4F^2$ density. Besides using transistors, Wang et al. [40] proposed a 3D 1BJT1R cell with $4F^2$ density. A BJT supplies much higher drive current than a MOSFET transistor, and thus it is a better solution for a process shrink.

Two-terminal selectors are preferred for high density 3D RRAM array. Generally, selectors have very simple device structures but complex I-V behaviors. Although the selector based 1S1R (1 Selector—1 RRAM) cells cannot entirely eliminate the sneak path effect, compared with 1T1R cells, the fabrication cost can be significantly reduced. Therefore, different kinds of selectors have been developed to replace transistors.

The most critical challenge for selectors is their current drive capability. Since RRAM has a filamentary conduction mechanism, the LRS current remains constant or slightly changes as the device size shrinks. In the previous report [6], the RESET currents of RRAM are most between 20–200 μA for different sizes. In contrast, the currents of most kinds of selectors are linearly dependent on the device area. If the selector cannot supply enough current density, the 1S1R cell cannot work when the size scales down. Besides current density, on/off ratio, threshold voltage, endurance, and switching speed are also key parameters for selectors.

8.2.3.1 Diodes

A diode is a kind of selector, with asymmetric I-V characteristics. Si-based diodes can supply larger current density but cannot be easily adopted for 3D integration due to the unbearable thermal budget and other process issues [17]. Therefore, oxide-based diodes are proposed to be fabricated at room temperatures and over metal substrates. The combination of the p-type and n-type materials should be carefully designed by considering the bandgap width and Fermi level. Baek et al. [17] developed a p-NiO/n-TiO_2 diode, and Lee et al. [18] developed a p-CuO/n-IZO diode with >10^3 on/off ratio.

Since the diode can only work on the forward direction, bipolar RRAM is not compatible with such a selector. However, compared with unipolar RRAM, bipolar RRAM has better endurance, uniformity, and controllability. To be compatible with bipolar RRAM, other kinds of diodes, such as Zener diodes, Reverse-conduction diodes, and oxide-based Schottky diodes were proposed. With these diodes, large reverse current can be supplied when reverse voltage is larger than a threshold. The write voltage should be larger than this threshold voltage, while the read voltage should be lower than the threshold voltage.

The major drawback of the oxide diode is the limited current density. The CuO/IZO diode can only supply 10^4 A/cm^2 current density [18], which cannot drive a RRAM device when its size scales below 0.5 μm. The current density can be improved by using oxide Schottky diodes [41], but it is still not acceptable for 100 nm applications. The low current density of the oxide diode is attributed to the poor conductivity of the oxide materials and the current restriction of the junction. To solve these problems, tunneling selectors were proposed, and they seem to be very promising for the 1S1R configuration.

8.2.3.2 Tunneling-Based Nonlinear Selectors

Tunneling selectors usually have a metal/insulator (semiconductor)/metal (MIM or MSM) symmetric structure. As illustrated in Fig. 8.18, under low voltage biasing, the tunneling selector can be seen as a back-to-back Schottky diode. The conduction mechanisms are dominated by the reverse Schottky current or direct tunneling, which is very small. As the applied voltage increases, the width of the

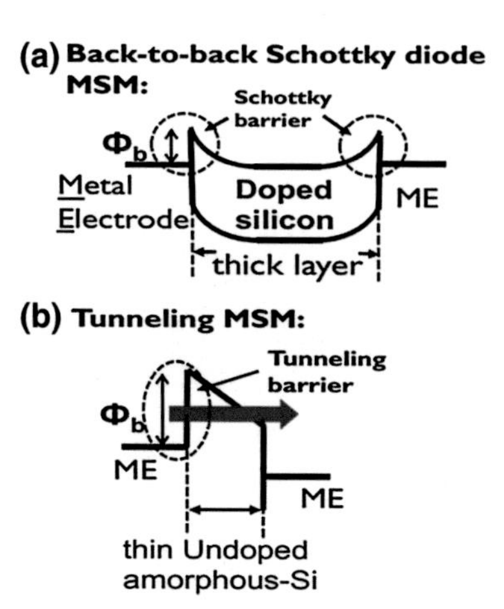

Fig. 8.18 **a** Band diagram of the conventional MSM selector (*at no bias*), with doped Si, and rectifying back-to-back Schottky diode. **b** Band diagram of the tunneling MSM selector (*under bias*), with ultrathin undoped amorphous-Silicon (a-Si), which behaves as a low-bandgap tunnel dielectric [42]

Schottky barrier decreases, and FN tunneling gradually becomes the relevant part [42]. Compared with PN junction and Schottky junction, FN tunneling can supply much higher current density, and the I-V curve is symmetric, being then compatible with bipolar RRAM. The nonlinear I-V is due to FN tunneling and, therefore, barrier's height and width should be designed to achieve the highest nonlinearity. Huang et al. [36] proposed a Ni/TiOx/Ni selector built on a flexible substrate. 1000 times on/off ratio (I(Vr)/I(Vr/2)) was achieved. Zhang et al. [42] proposed a TiN/a-Si/TiN selector with >1 MA/cm^2 current density and $\sim 10^3$ on/off ratio. Similar to the Si P/N type diode, this selector needs high temperature annealing. Kawahara et al. proposed an all-nitride TaN/SiNx/TaN selector, and they built an 8 Mb 3D cross-point array on top of it. Lee et al. [43] proposed a Ta doped TiO_2 varistor with $>3 \times 10^7$ A/cm^2 current density and $\sim 10^4$ on/off ratio. The selector was fabricated by atomic layer deposition (ALD), which is suitable for 3D vertical RRAM array architecture. Later, Woo et al. demonstrated a fully CMOS compatible stack (W/Ta_2O_5/TaO_x/TiO_2/TiN) with only <10 nm oxide thickness. They achieved a current density higher than 10^7 A/cm^2.

8.2.3.3 Volatile Switching Selectors

Another promising selector utilizes the threshold switching mechanism of some dielectric materials. Threshold switching is a kind of volatile resistive switching caused by Insulator-Metal-Transition (IMT). Under zero or low voltage bias, the material is an insulator; as the voltage increases beyond a threshold value, the material becomes conductive because of a phase change, and the current suddenly increases. When the voltage decreases below the threshold, the material changes back to insulator without any memory effect. Such selectors have very high on/off ratio and enough current density for high density and large size array integration. Lee et al. [44] proposed a TiN/TiO_x/Ta/TiN threshold selector. Cha et al. [31]

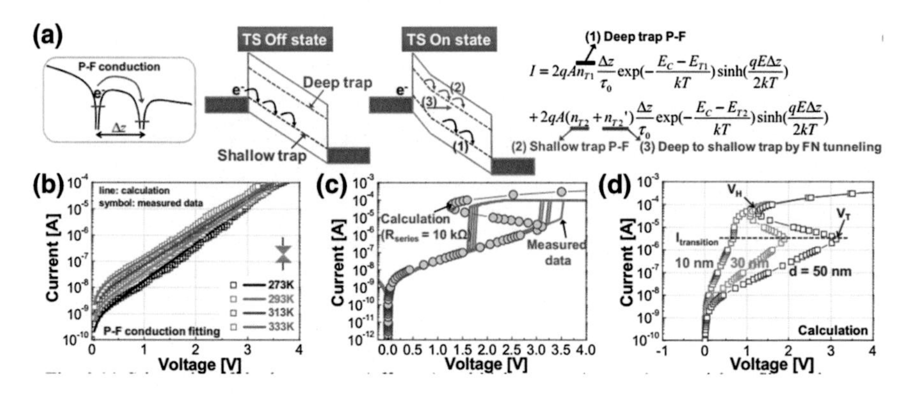

$$I = 2qAn_{T1}\frac{\Delta z}{\tau_0}\exp(-\frac{E_C - E_{T1}}{kT})\sinh(\frac{qE\Delta z}{2kT}) + 2qA(n_{T2} + n_{T2}')\frac{\Delta z}{\tau_0}\exp(-\frac{E_C - E_{T2}}{kT})\sinh(\frac{qE\Delta z}{2kT})$$

Fig. 8.19 **a** low-current (*off state*) and high-current (*on-state*) potential profiles and energy distribution of the electrons in the amorphous chalcogenide. **b** Measured and calculated I–V characteristics at temperatures of 0, 20, 40, and 60 °C. **c** Measured and calculated I–V characteristics of the threshold switching (TS) device. **d** Calculated I-V characteristics of the TS device for different thicknesses [48]

proposed NbO_2-based threshold selector with 3D vertical structure. The IMT was attributed to the thermal induced transition from distorted-rutile phase to rutile phase in NbO_2. They also found that selector device with vertical structure has lower current compared to the planar structure. Jo et al. [45] developed a 3D-stackable Field Assisted Superlinear Threshold (FAST) selector. A record on/off ratio of 10^{10} was demonstrated. The phase change in the FAST selector was electric field driven instead of thermal driven, and thus the off current could be significantly reduced. The current density was higher than 5×10^6 A/cm^2. However, the detailed stack materials were not published. Meanwhile, Lee et al. [46] proposed an AsTeGeSiN-based selector with $>10^7$ A/cm^2 current density. The threshold resistive transition originates from the Mott effect induced by the electronic charge injection [48], as shown in Fig. 8.19. Yang et al. [47] demonstrated a similar doped-chalcogenide based selector with $>10^7$ on/off ratio.

8.2.3.4 Mixed Ionic-Electronic Conductors (MIECs)

IBM proposed another type of selector, based on mixed ionic-electronic conductors (MIECs) [50–52]. MIEC is a material that conducts ions and electronic charge carriers (electrons and/or holes), and its I-V characteristic depends on the relevant defects, such as defects nature, concentration and local neutrality [49]. Figure 8.20a shows a typical structure of a MIEC-based selector, with the MIEC material sandwiched between two electrodes (at least one must be inert or non-Cu-ionizing). These materials have a significant amount of mobile Cu, and under negative bias of the top electrode (TEC), Cu^+ ions are pulled away from the bottom electrode (BEC), leaving vacancies that act as acceptors [51]. The increased acceptor (and hole) concentration near the BEC depends exponentially on the applied bias [49],

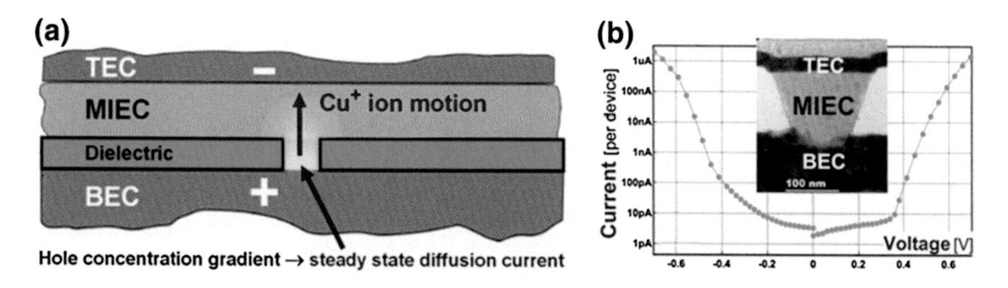

Fig. 8.20 **a** The structure of a MIEC-based selector, and **b** its I–*V* characteristics [50, 51]

and its vertical gradient results in a steady-state hole diffusion current. Thus, negative bias on TEC produces very tight, exponential diode-like I-V characteristics. On the contrary, Cu^+ ions swept from the large TEC into the tiny via form a metallic filament and current abruptly increases [51], as shown in Fig. 8.20b. Moreover, a series of excellent properties, such as high on current densities (due to the large fraction of mobile Cu), ultra-low leakage (<10 pA), and fast operation speeds, were also observed in follow-up studies satisfying the needs for large scale cross-point arrays [50, 52].

It seems that the threshold switching selectors have larger on/off ratio, larger current density, and lower threshold voltage, while the tunneling selectors have better 3D integration process compatibility. Both of them can switch at ~ ns speed. Looking forward, the reliability of the selector should be studied with more attention, because the selector is turned on during both reads and writes and, therefore, the endurance of selector should be much better than the endurance of the RRAM device itself.

8.2.4 Self-rectifying RRAM

Implementing a selector requires the introduction of additional layers of material, which is definitely not helping the 3D integration. Especially with 3D vertical architectures, the selector takes additional area and the $4F^2$ density cannot be achieved when the feature size scales below the thickness of the 1S1R stack. Therefore, it is very important to develop RRAM devices with self-rectifying characteristics.

Tran et al. [53] developed Ni/HfO_x/n^+-Si stacked unipolar RRAM with asymmetric positive/negative I-V curve. The ratio of forward/reverse current was larger than 10^3. The diode-like behavior was attributed to the electron hopping barrier between the conductive filament in the HfO_x layer and conduction band of n^+-Si, as illustrated in Fig. 8.21. Park et al. [27] demonstrated self-rectifying characteristics by inserting barrier layers between electrodes and oxide layers. Detailed materials and processes were not published. They found that, after introducing a thin interface layer, asymmetric positive/negative I-V curve could be realized and the current ratio

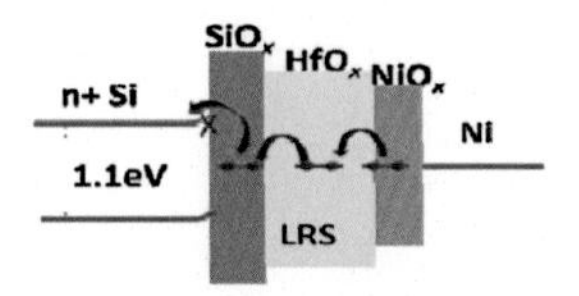

Fig. 8.21 Reverse current transport diagram in n^+-Si/HfO_x/Ni [53]

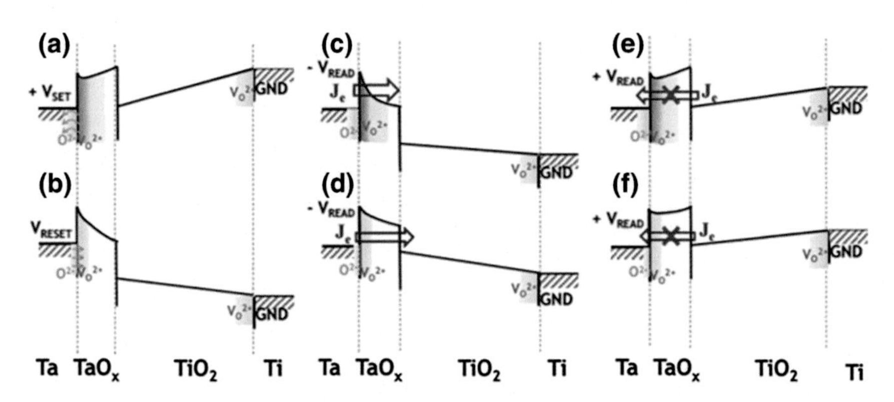

Fig. 8.22 Band diagrams of the Ta/TaO_x/TiO_2/Ti vertical RRAM at **a** set, **b** reset, **c** LRS read at −2 V, **d** HRS read at −2 V, **e** LRS read at +2 V, **f** HRS read at +2 V. Oxygen ions (O^{2-}) migration in TaO_x using Ta as a reservoir is driven by bipolar electric field at set/reset. The oxygen-vacancy (V_o^{2+}) concentration modulates the Schottky barrier at the Ta/TaO_x interface at negative read. Current conduction is dominated by the TaO_x/TiO_2 barrier at positive read [32]

reached ~100. Similarly, Chen et al. found that, by inserting a TiON interfacial layer between the TiN electrode and the HfO_x layer, >10 nonlinearity of the I-V curve can be achieved. The nonlinearity derives from the electron tunneling through the thin interfacial layer. Based on this finding, they developed a TiN/TiON/HfO_x/Pt 3D vertical RRAM array [30].

Hsu et al. [32] proposed a Ta/TaO_x/TiO_2/Ti 3D vertical RRAM with over 10^3 self-rectifying ratio. The TaO_x layer acted as both a switching layer and a tunneling barrier. Similarly, TiN/Al_2O_3/TiO_2/TiN and Ti/HfO_2/TiO_x/Pt RRAM devices with ~100 self-rectifying ratio were demonstrated by Govoreanu et al. and Lee et al. [54, 55]. It was found that TiO_x plays an important role for the current rectifying characteristics. However, the retention of these three stacks was not acceptable for data storage application due to the interfacial switching mechanism (Fig. 8.22).

Compared with 1S1R cell, self-rectifying RRAM devices offer higher integration density and there is no need to consider the endurance degradation, switching delay, and variation issues of the selectors. However, the performances of self-rectifying RRAM devices have insufficient self-rectifying ratio and poor retention issues and cannot meet the requirement of data storage application [32, 54, 55]. Additional research effort is required in this field.

8.2.5 Complementary RRAM

Instead of utilizing the on/off ratio to suppress sneak leakage as mentioned above, Waser et al. invented *Complementary Resistive Switches* (CRSs), consisting of two bipolar memristive elements [56].

As shown in Fig. 8.23, a CRS is made of a memristive element A (Fig. 8.23a) with symmetric bipolar I-V characteristic (Fig. 8.23b), combined with a memristive element B with a reversed material order (Fig. 8.23c) and inverse I-V characteristic (Fig. 8.23d). In the initial state, both memristive elements are in HRS, which is the "OFF" state. An initialization process is needed to bring CRS to its stable switching states by applying a voltage $V < 2 \cdot V_{th,3}$ or $V > 2 \cdot V_{th,1}$. If the initialization voltage $V > 2 \cdot V_{th,1}$, element B switches to LRS, while element A stays in HRS, which is the "0" state. On the contrary, if the initialization voltage $V < 2 \cdot V_{th,3}$, element A switches to LRS, while element B stays in HRS, which is the "1" state. After initialization if, for example, the CRS is in "1" state, almost all the voltage applied to the CRS drops over element B. As the voltage increases to $V_{th,1}$, which is slightly larger than the V_{set} of element B, element B switches to LRS, while element A remains in LRS, which is the "ON" state. At this point, the voltage drop on both elements is similar. When the voltage increases to $V_{th,2}$, i.e. when the voltage drop over element A reaches its V_{reset}, element A switches to HRS. Then, CRS holds the "0" state even if the voltage becomes higher. Similarly, "0" state turns to "ON" state, and then to "1" state, as the voltage increases with opposite polarity.

The "OFF" state exists only before initialization and it doesn't contribute to the storage mechanism. After initialization, the circulating switching between "0" state and "1" state becomes true by the application of a write voltage $V < V_{th,4}$ or $V > V_{th,2}$, respectively. Different from conventional RRAM arrays, where the information is stored as the difference between HRS and LRS, CRS utilizes the switch between "0" state and "1" states. During read operations, a voltage value between $V_{th,1}$ and $V_{th,2}$ is applied. A low current is observed if CRS stays in the "0" state and the state remains the same. In case of the "1" state, the read voltage itself triggers the "ON" state, thus leading to a higher current. As a matter of fact, this operation destroys the information, which has to be restored by an additional re-write operation.

As mentioned above, a CRS, independently from being in "0" state or "1" state, consists of a memristive element in HRS and another one in LRS; its resistance is always high if $V_{th,3} < V < V_{th,1}$. This is equivalent to say that the total resistance of the array becomes independent from the stored information pattern [56]. With a proper voltage scheme, such that the voltage drops over all unselected devices satisfies $V_{th,3} < V < V_{th,1}$, like V/2 or V/3 voltage schemes [35], the CRS-based arrays heavily suppress the sneak path issues. In subsequent studies, Waser et al. demonstrated the vertical integration of CRS cells based on $Cu/SiO_2/Pt$ bipolar resistive switches with nearly perfect symmetry and high resistance ratios, showing

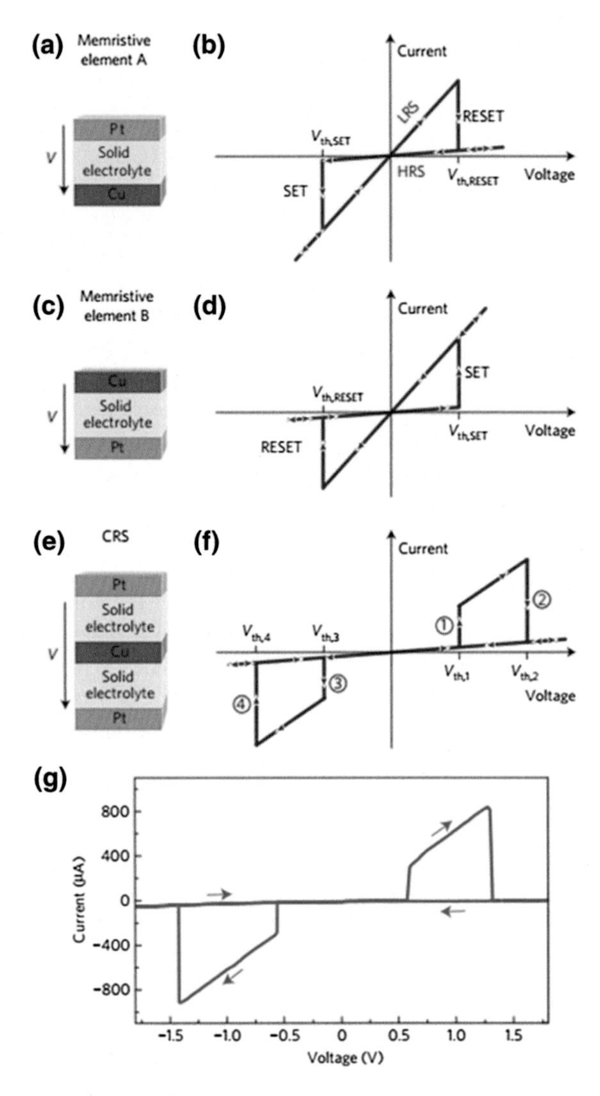

Fig. 8.23 Complementary Resistive Switch (CRS). **a** The structure of memristive element A and **b** its I–V characteristic. **c** The structure of memristive element B and **d** its I–V characteristic. **e** The structure of the whole CRS and **f** its I–V characteristic. **g** Measured I–V curve of two serial $5 \times 5\ \mu m^2$ Pt/SiO_2/GeSe/Cu (30 nm Pt, 3 nm SiO_2, 25 nm GeSe and 70 nm Cu) crossjunction memristive elements: $V_{th,1} = 0.58$ V, $V_{th,3} = -0.56$ V, $V_{th,2} = 1.3$ V, $V_{th,4} = -1.4$ V, $I(V_{th,1}) = 270\ \mu A$, $I(V_{th,3}) = -280\ \mu A$, $(V_{th,2}) = 860\ \mu A$, $I(V_{th,4}) = -910\ \mu A$ (measured done with a 940 Ω series resistor) [56]

the potential for high-density cross-point arrays [57]. Nardi et al. [58] were the first to propose complementary resistive switches consisting of a single RRAM cell with natural asymmetry reset states, thus simplifying the complex multilayer-stack structure, and significantly advancing the progress of CRS technology.

8.3 Analysis of 3D RRAM Array

Compared to 2D arrays, 3D RRAM arrays offer higher density, but they are also more prone to sneak paths and IR drop issues. Yu et al. [59] compared the maximum array size and integration density of 2D and 3D vertical RRAM arrays with different feature sizes. They built a 3D resistor network considering both the resistance of interconnects and a dynamic model of RRAM [69], as illustrated in Fig. 8.24. A virtual node is used to simulate the influence of the plane electrode. It was found that the array size is mainly limited by the degradation of write margin caused by IR drop effects. The read and write margins of both 2D and 3D array degrade as the feature size scales down due to the increased interconnect resistance. For the same feature size, the 2D array has better read/write margin than 3D array but, obviously, the array size and integration density are much smaller. To make a relatively fair comparison, they kept the same array size and same array read/write margin, and found that 3D array has 18 × higher density over 2D array, and it can relax the feature size from 13 nm to 26 nm, which means that the fabrication cost can be significantly reduced.

Furthermore, Deng et al. compared the performance of 3D RRAM arrays with different architectures. They simulated a 3D cross-point array, a vertical array with line shaped horizontal electrode, and a vertical array with plane shaped horizontal electrode. They considered not only the interconnect resistance but also the capacitance of interconnects and RRAM cell in the simulation; in this way, they could evaluate both RC delays and dynamic power. They found that 3D cross-point array has the best array performance, given the same feature size. The plane shape vertical array has better read/write margin and access speed compared to line shape vertical array, because of the lower interconnect resistance of the metal plane, but it shows higher power consumption because there are more sneak paths. This result indicates that 3D cross-point array is better for high performance application, whereas 3D vertical array is better for low cost application. Deng et al. also found that, given the same array size, stacking more layers can offer better array

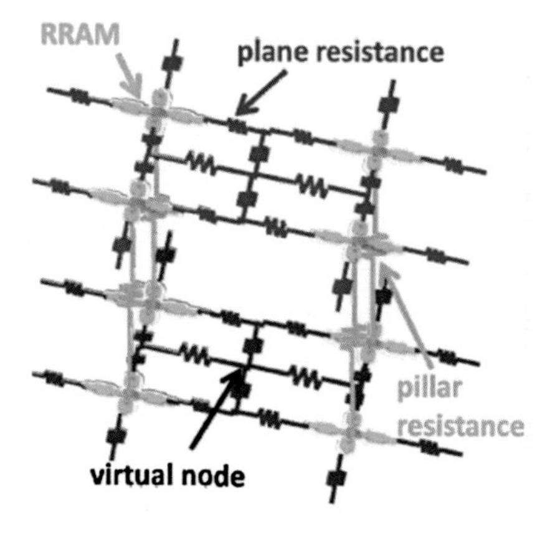

Fig. 8.24 Schematic of a small 3D resistor network which can be extended to build a whole 3D array. A virtual node is added to simulate the plane resistance [59]

Table 8.1 Summary of design tradeoffs

	Total bits	Integration density*	RC delay	Energy
F ↓	↓↓	↑	↓	↓
T_i ↓	↑	↑	↑	↑
T_m ↓	↓	↑↑	↑	↓
R_{RRAM} ↑	↑↑	\	↑	↓↓
*Stacks ↑	\	↑	↓	↓

F is the minimum feature size; T_i and T_m are the thickness of the isolation layer and of the metal plane electrode, respectively; R_{RRAM} is the resistance of a RRAM cell in LRS

*Stacks is the number of stacked layers [29]

performance (read/write margin, speed, and power). They summarized the design guidelines for 3D RRAM array in the Table 8.1 [29].

The scaling limit of 3D RRAM arrays was also investigated. Besides horizontal scaling, which is mainly limited by lithography technology, vertical scaling is crucial for 3D vertical arrays. Vertical scaling refers to the thickness reduction of metal and insulating layers, which translates into a reduction of the vertical interconnect resistance and an increased number of stacked layers. The minimum thickness of insulating layer is limited by the breakdown issue. For example, if we use PECVD SiO_2 as the insulating layer, at least 6 nm thickness is required to prevent its breakdown under the stress of 4 V/10 years. At the same time, the thickness of the metal layer should be carefully designed, since a thinner metal increases the horizontal interconnect resistance. Bai et al. [19] proposed graphene or *Carbon Nano Tube* (CNT) as the electrode to reduce the thickness while keeping the interconnect resistance low. By using this method, the array size and integration density can be significantly improved.

8.4 Progress of 3D RRAM

8.4.1 Intel and Micron 3D XPoint Memory

Recently, Intel and Micron announced a novel 3D XPoint memory. They claimed that 3D XPoint technology is a major breakthrough in memory process technology since the introduction of NAND flash in 1989, with unique material compounds and

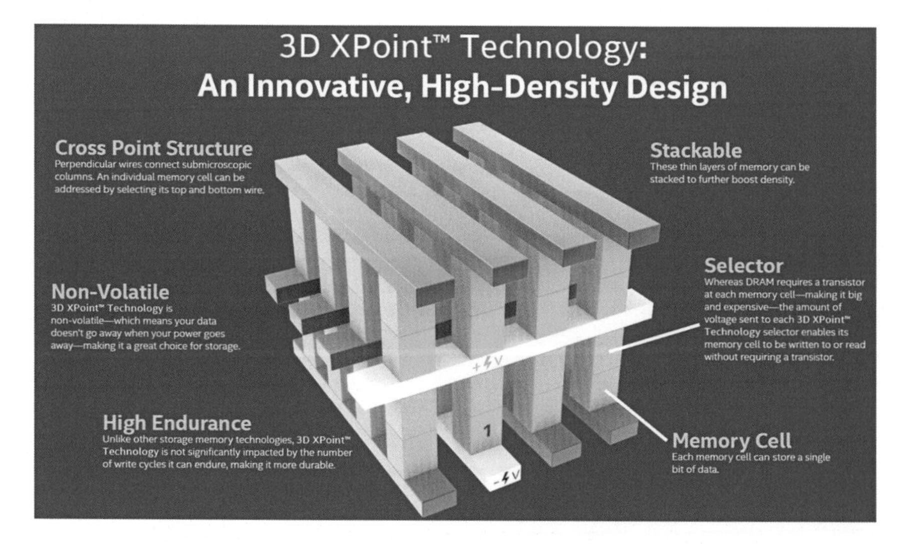

Fig. 8.25 Schematic of Intel/Micron 3D XPoint technology [60]

a cross-point architecture, as shown in Fig. 8.25. The novel 3D XPoint memory can be seen as a storage layer between DRAM and NAND, because it is 10 times denser than DRAM, and speed and endurance can be 1000× higher than NAND Flash memory. Intel and Micron reported that the first product is a two-layer 128Gbit developed on a 20 nm process technology node.

As we speak, there are no public details about this memory cell, but it is well-grounded to presume that 3D XPoint Technology is one type of RRAM technology under extended definition, like ReRAM, PCM, CBRAM and so on. It is worth highlighting that, in recent years, Micron has put a lot of effort into RRAM technology. In 2014, they presented a 16 GB RRAM designed in a 27 nm node, with 200 MB/s write and 1 GB/s read [61].

8.4.2 Sandisk and Toshiba 32 Gbit 3D Cross-Point RRAM

Liu et al. [62, 63] from Sandisk demonstrated a 32Gb 3D RRAM chip in 24-nm process. Cross-point architecture with shared BL and WL between adjacent blocks was adopted to allow multiple memory layers to be stacked above the supporting circuitry, as shown in Fig. 8.26. Each memory block is composed of 4k × 2k cells to allow most of the array supporting circuitry to be placed under the array, thus minimizing the die area overhead. Each memory cell consists of metal oxide as the switching element and a diode as the selector.

Moreover, the chip includes the whole peripheral circuit and many significant design tricks are adopted to improve the performance: a pipelined array control

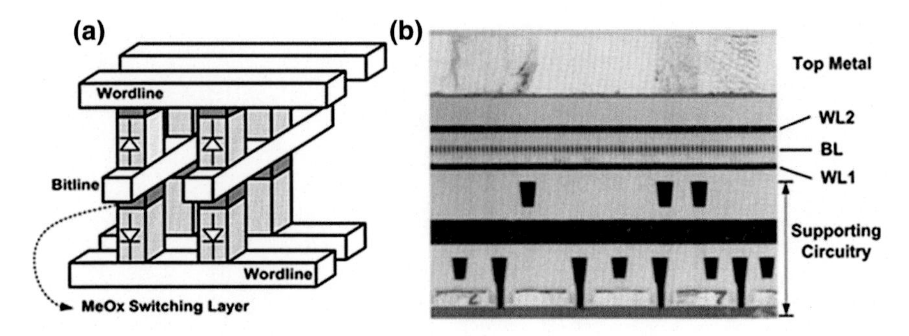

Fig. 8.26 Sandisk 3D RRAM chip: **a** memory cell structure and **b** TEM image of memory array and supporting circuitry [63]

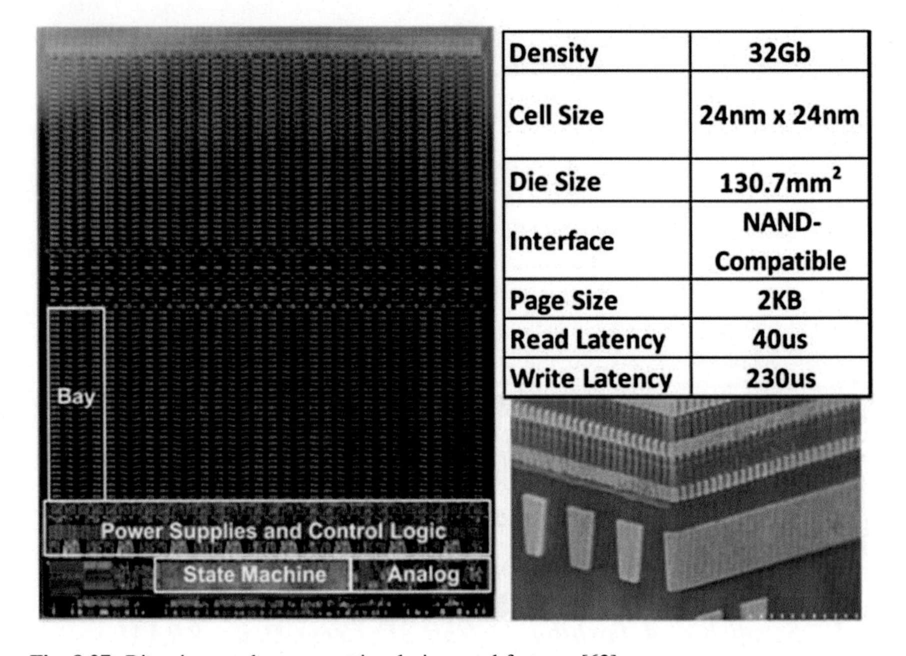

Density	32Gb
Cell Size	24nm x 24nm
Die Size	130.7mm^2
Interface	NAND-Compatible
Page Size	2KB
Read Latency	40us
Write Latency	230us

Fig. 8.27 Die micrograph, cross-sectional view, and features [62]

scheme reduces the performance impact by up to 40 % resulting from shared sense amplifier and array control circuit; Smart Read scheme during sensing allows the memory cells to be read at a more accurate bias level and it improves the sensing capability for weak cells; leakage current compensation scheme during programming mitigates the effect of leakage current caused by the sneak path; a dynamic charge pump control scheme optimizes the charge pump configuration for power consumption based on the operation conditions [63]. The chip photo and its device features are reported in Fig. 8.27.

8.4.3 *Crossbar 3D RRAM*

In addition to the well established semiconductor companies, a startup named Crossbar has made great progresses in the 3D RRAM development. At IEDM 2014, Jo et al. [45] introduced a 3D-stackable Field Assisted Superlinear Threshold (FAST) selector. An on/off ratio of 10^{10} was demonstrated, and the sneak current in the 4Mbit 1S1R cross-point array was suppressed below 0.1 nA, as shown in Fig. 8.28. Later on, they reported a 3D stackable 1S1R passive cross-point RRAM based on FAST selector technology [64]. The selector offered excellent performance characteristics, such as selectivity higher than 10^7, fast turn-on slope (<5 mV/decade), the ability to tune the threshold voltage, and an endurance greater than 10^{11} cycles. This 1S1R integration demonstration shows that the selector-subthreshold current is <10 pA while offering >10^2 memory ON/OFF ratio and >10^6 selectivity during cycling. It means that the sneak path challenge has been overcome, and high-density and high-performance memory applications based on RRAM will be available in the near future.

Moreover, Crossbar claimed that it has proven the manufacturability of 3D RRAM with a working array produced in a commercial fab, integrated with CMOS controller. The company is currently completing the characterization and optimization of this device and it plans to bring its first product to market in the embedded SoC market, while continuing the development of high-density storage class memory arrays in advanced process geometries (Fig. 8.29).

8.4.4 *Others*

Hsieh et al. [66] proposed a 3D Via RRAM with 28 nm CMOS process. The RRAM cell with Ta/TaN/TaON structure was formed between Cu Via and the landed Cu metal layer in 28 nm single damascene process. 4 layers with shared BL and WL were realized, as shown in Fig. 8.30. Unipolar switching with <5 V and <500uA RESET current was presented. Later on, a self-rectifying twin-bit RRAM

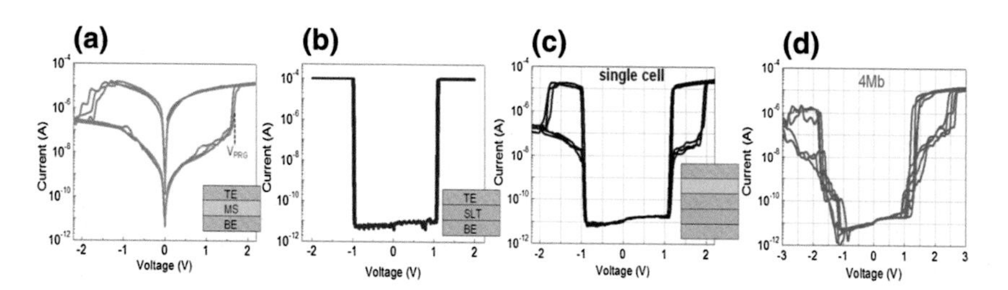

Fig. 8.28 Integration of RRAM devices with FAST selectors done at Crossbar. I-V characteristics of a single cell level **a** RRAM, **b** selector, and **c** integrated 1S1R device. **d** I–V characteristics of a 4Mbit passive cross-point array based on 1S1R [45]

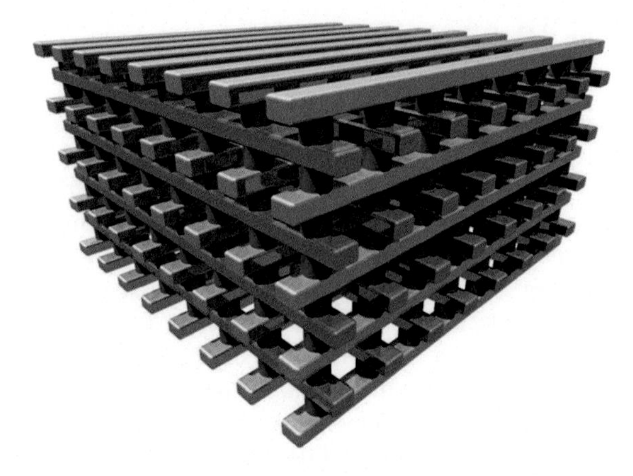

Fig. 8.29 Schematic of Crossbar's 3D RRAM: it is CMOS compatible for easy integration [65]

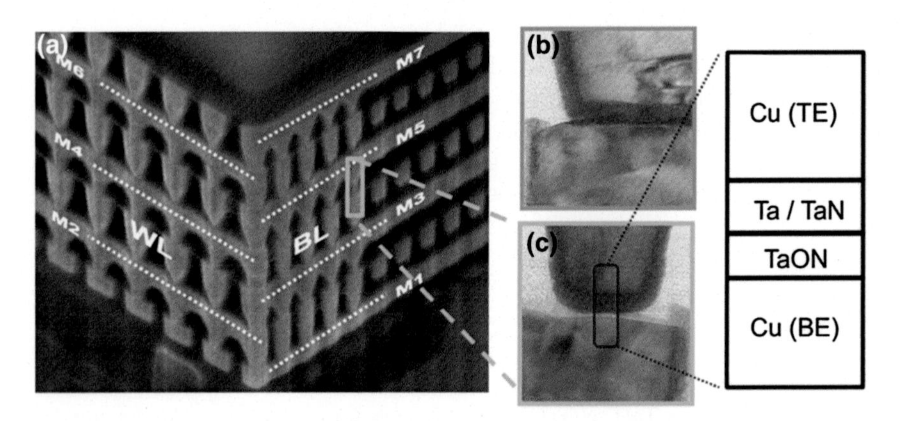

Fig. 8.30 **a** SEM picture of 28 nm TaON based 3D Cross-point Via RRAM **b** TEM picture of a standard Cu Via **c** TEM picture of Via RRAM cell with 30 nm × 30 nm cell size [66]

in a 3D interweaved cross-point array was proposed in 28 nm CMOS BEOL process by the same group [67].

For 3D VRRAM, Chien et al. [21] have demonstrated 2 layers vertical 3D RRAM stack by using a $W/WO_x/W$ stack. The schematic view and corresponding STEM image of 2 layers vertical RRAM are shown in Fig. 8.31a, b, respectively.

In this stack, bottom W serves as plane electrode and top W as pillar electrode, whereas WO_x is used as the storage medium. The pillar electrode is the heart of this structure: it is connected to the drain side of an access transistor which controls and isolates 4 ReRAM elements in the 2-layer structure (Fig. 8.31a). Another advantage of this structure is that we can reduce the device size by controlling the plane electrode thickness. In this case W thickness was $\sim$10 nm which means that this 3D RRAM device can be scaled below the 10 nm node (Fig. 8.31b). The authors

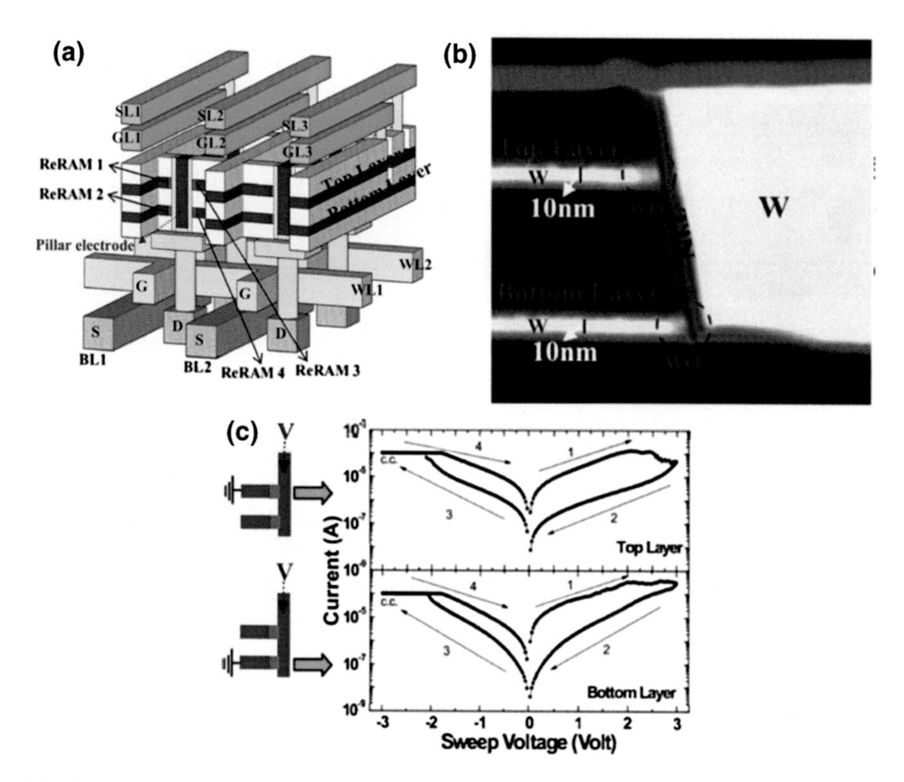

Fig. 8.31 **a** Proposed 3D structure for WO_x based ReRAM and **b** STEM profile for the self-aligned WO_x/SP-$TiNO_x$ cell. W thickness is 10 nrn and it defines one side of the ReRAM cell. **c** DC resistive switching behaviors for top and bottom layer devices, respectively. The device can be RESET by applying a positive pulse voltage to the pillar electrode, and SET by a negative pulse. A SET current of 100 μA is used to prevent oxide breakdown [21]

also reported typical bipolar I-V characteristics, as shown in Fig. 8.31c. A positive voltage applied to pillar electrode can RESET the device while a negative voltage can SET the device. The operating voltage was ± 3 V and current limitation of 100 μA was used to prevent permanent breakdown.

Li et al. [24] reported sidewall electrode technology by introducing ultra-small, functional HfO2-based resistive random access memory (RRAM) device ($<1 \times 3$ nm^2) which is very well suited for high density memory application. In order to increase the number of vertical layers, Bai et al. [19] developed three-layer 3D vertical Pt/AlO_δ/Ta_2O_{5-x}/TaO_y/Pt RRAM cell structure, which is sketched in Fig. 8.32a. The Pt pillar electrode and plane electrode serve as top electrode (TE) and bottom electrode (BE), respectively, while the resistive-switching layer is located along the sidewall of TE. This structure has huge potential for future high-density integration and per-bit lithography cost reduction, because of the high number of stacked layers. Figure 8.32b shows the cross-sectional TEM image of typical 3D vertical AlO_δ/Ta_2O_{5-x}/TaO_y RRAM devices. The magnified TEM image

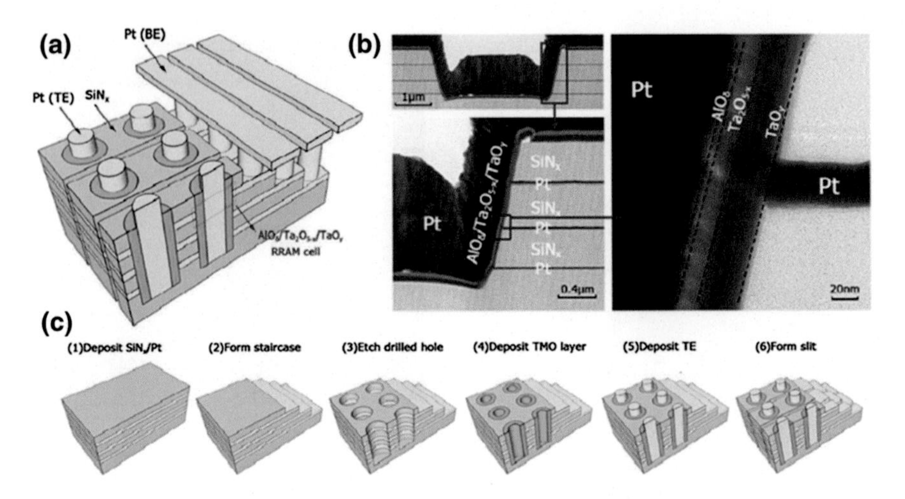

Fig. 8.32 **a** The schematic view of three-layer 3D vertical RRAM; **b** the cross-sectional TEM image of vertical $AlO_\delta/Ta_2O_{5-x}/TaO_y$ RRAM cells; **c** the fabrication process of 3D RRAM [19]

reveals the multiple layer structure of the switching layer. The detailed fabrication process of 3 layers vertical RRAM device is described in Fig. 8.32c. 20 nm Pt BE layer and 300 nm Si_3N_4 dielectric layer are stacked alternately by PVD and PECVD method, at first. Then the plane electrode staircase is formed by dry etching Si_3N_4 layers with C_4F_8/CF_4 plasma and Pt layers with Cl_2/Ar plasma alternately. In the third step, the drilled holes to place vertical RRAM cells are dry etched with Cl_2/Ar plasma. Next, $AlO_\delta/Ta_2O_{5-x}/TaO_y$ TMO resistive-switching layers are deposited on the sidewall of the drilled holes by reactive sputtering. Then 200 nm Pt is deposited by PVD to form the pillar electrodes. Finally, the slits of plane electrodes are dry etched to reduce the interference with neighbor cells. For further scaling along the vertical direction, the concept of edge electrode have been developed by introducing graphene and *Carbon Nano Tube* (CNT) by Prof. Wu's group [25]. In this case, they used graphene and CNT as plane electrode, Pt as pillar electrode, and Ta_2O_{5-x}/TaO_y acted as switching medium. VRRAM shows significant possibilities for future high density and below 10 nm technology node applications.

To check the uniformity, Bai et al. studied cumulative probability distributions of HRS-LRS and SET-RESET voltages of each layer, as reported in Fig. 8.33a, b, respectively. Figure 8.33a shows the resistance distributions of HRS and LRS. The low resistances are about 1 kΩ, while the high resistances are about 1 MΩ. The resistance distributions of cells in all three layers shows good uniformity. It is worth noting that the HRS/LRS resistance window is larger than 1000, which is very suitable for MLC operations. Figure 8.33b shows the distributions of switching voltages during SET and RESET processes. The SET voltages range from 2.08 to 2.15 V, while the RESET voltages are from 0.9 to 2 V. In terms of reliability, it is necessary to perform long P/E endurance and robust data retention. In this research,

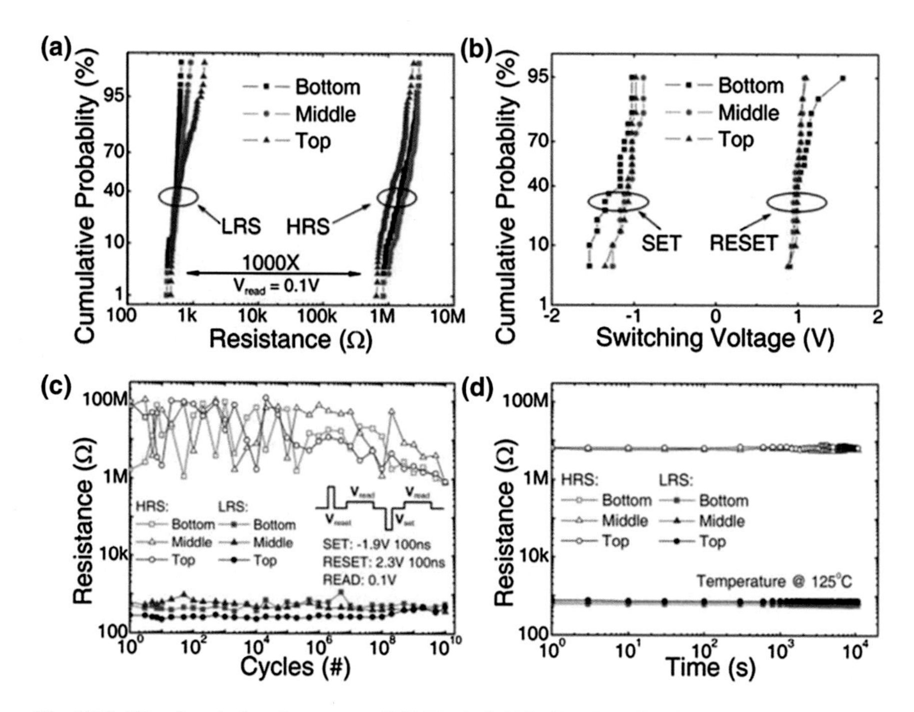

Fig. 8.33 The electrical performance of 3D Vertical $AlO_\delta/Ta_2O_{5-x}/TaO_y$ RRAM. **a** Cumulative probability distribution of HRS and LRS. **b** The switching voltage distributions during SET and RESET. **c** 10^{10} switching cycles are demonstrated. **d** All cells show retention of more than 10^4 s at 125 °C [19]

more than 10^{10} switching endurance cycles were demonstrated, by using a pulse condition of −1.6 V/100 ns for SET, and 1.8 V/100 ns for RESET, as depicted in Fig. 8.33c. Excellent retention was observed too, as shown in Fig. 8.33d.

To enhance storage density, one of the most important ways is multi-level cell (MLC) operation because more logic bits can be stored within a single memory cell. Bai et al. reported two methods to achieve MLC: (1) *Current Controlled Scheme* (CCS) based on controlling the current compliance I_c during SET (e.g. 5 μA, 50 μA, 500 μA); (2) *Voltage Controlled Scheme* (VCS) based on controlling the stop voltage V_{stop} during RESET (e.g. 1.9, 2.4, 2.8 V). Both methods are shown in Fig. 8.34a. The devices in CCS share similar HRS, while, with VCS, the devices have similar LRS. The resistance states of 4 level MLC are defined as following: Level 1 is the LRS with resistance from 1 to 10 kΩ; Level 2 and Level 3 are the intermediate states with resistance of 100 kΩ and 1 MΩ, respectively; and Level 4 is the HRS with resistance of 10 MΩ. As shown in Fig. 8.34b, all MLC levels show good resistance distribution uniformity among three layers in the same pillar. Figure 8.34c shows MLC cycling tests of three vertical RRAM sharing the same pillar electrode by using VCS method. 100 resistive-switching cycles were performed without any degradation. Similar MLC performances were achieved by using CCS.

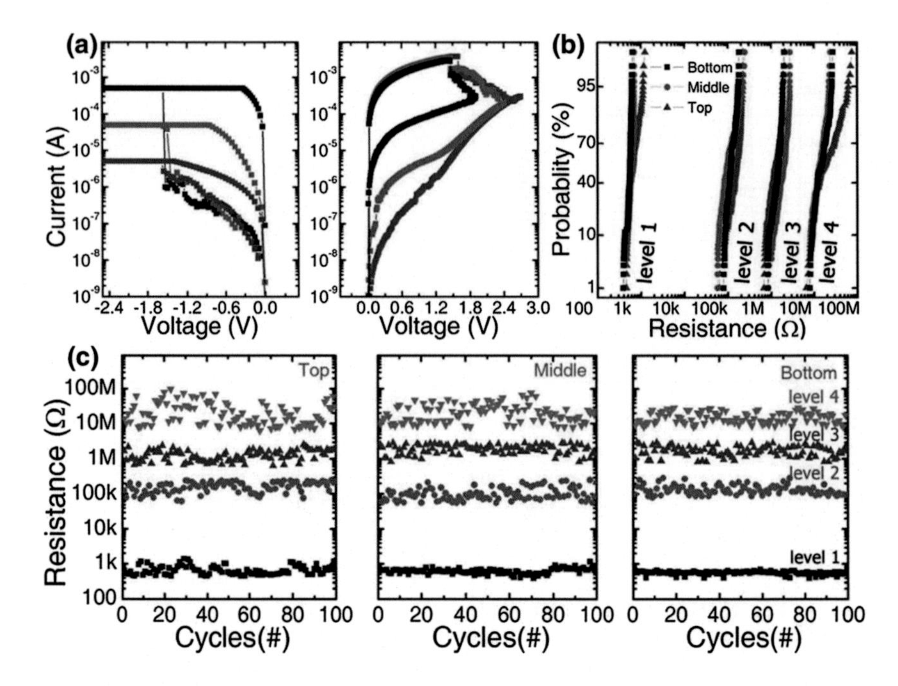

Fig. 8.34 **a** Two methods to achieve MLC operations: CCS and VCS; **b** resistance distributions of cells in three layers with 4 MLC levels: 1 kΩ, 100 kΩ, 1 MΩ and 10 MΩ; **c** stable MLC resistive-switching cycles [19]

8.5 Challenges and Future Outlook for 3D RRAM

With all the development done over the past decades, 3D RRAM technology is making progress towards higher-density and better-performance next generation nonvolatile memories. Moreover, it may also play an important role in nonvolatile logic [6], and it is even seen as a disruptive technology in the semiconductor industry for its potential to overthrow the memory hierarchy and the conventional von Neumann based computer architectures [56]. However, there are still some serious challenges limiting the application of 3D RRAM, some coming from the properties of memory cells and some others from the 3D architecture.

It is a fact that the switching mechanisms of RRAM haven't been thoroughly understood until now. The switching mechanism of the same sandwich structure is usually explained in a different way by different research groups, probably due to the differences in the fabrication process; of course, this is not helping in revealing the underlying switching mechanisms. Moreover, though many papers have reported excellent performances of RRAM devices for endurance, retention or any other parameter, all these good results were based on different physical structures and different materials. Seeking for a single memory cell with all the advantages at the same time is still on the way [15].

In terms of large-scale manufacturing of 3D RRAM, uniformity and density are a concern. There are variations of switching voltages and resistances from cycle-to-cycle and from device-to-device [6]. The origin of the former is widely believed to be the stochastic nature of the conductive filaments (CFs), while the latter may be attributed to the inconsistent fabrication processes. Without uniformity, these variations end up constraining operating voltages and narrowing down the resistance gap between LRS and HRS, thus limiting the bit density.

Theoretically, switching elements can scale below 10 nm with good performances of the conductive filaments [68]. The switching elements can be also easily integrated in 3D structures with a minimum size of $4F^2/n$ (F is the minimum feature size, n is the number of 3D-stacked memory layers). However, the journey towards 3D RRAM is also full of challenges, especially sneak paths, as mentioned in Sect. 8.2.2.

Despite all these challenges, 3D RRAM technology doesn't stagnate, and great progresses have been made by researchers all over the world, as reported in Sect. 8.4. For sure the field of 3D RRAM is, and it will be, full of technical challenges, research activity and, last but not least, excitement.

References

1. M.-J. Lee et al., Two series oxide resistors applicable to high speed and high density nonvolatile memory. Adv. Mater. **19**(22), 3919–3923 (2007)
2. H.J. Sung, N. Hagop, in *Next Generation Nonvolatile Memory, Its Impact on Computer System* (Resources of Crossbar Company, 2013)
3. B.J. Choi et al., in Electrical performance and scalability of Pt dispersed SiO_2 nanometallic resistance switch. Nano Lett. **13**(7), 3213–3217 (2013)
4. T. Yanagida et al., in Scaling effect on unipolar and bipolar resistive switching of metal oxides. Sci. Rep. **3** (2013)
5. C. Ho et al., 9 nm half-pitch functional resistive memory cell with $< 1\mu A$ programming current using thermally oxidized sub-stoichiometric WOx film, in *2010 International Electron Devices Meeting* (2010)
6. H.-S.P. Wong et al., Metal–oxide RRAM. Proc. IEEE **100**(6), 1951–1970 (2012)
7. H. Ohno et al., Magnetic tunnel junction for nonvolatile CMOS logic, in *Electron Devices Meeting (IEDM), 2010 IEEE International*. IEEE (2010)
8. Y. Watanabe et al., Current-driven insulator–conductor transition and nonvolatile memory in chromium-doped SrTiO3 single crystals. Appl. Phys. Lett. **78**(23), 3738–3740 (2001)
9. C. Rohde et al., Identification of a determining parameter for resistive switching of TiO_2 thin films. Appl. Phys. Lett. **86**(26), 262907 (2005)
10. L.P. Ma, J. Liu, Y. Yang, Organic electrical bistable devices and rewritable memory cells. Appl. Phys. Lett. **80**(16), 2997–2999 (2002)
11. I.G. Baek et al., Highly scalable nonvolatile resistive memory using simple binary oxide driven by asymmetric unipolar voltage pulses, in *Electron Devices Meeting, 2004. IEDM Technical Digest. IEEE International* (2004)
12. H.Y. Lee et al., Evidence and solution of over-RESET problem for HfO x based resistive memory with sub-ns switching speed and high endurance, in *Electron Devices Meeting (IEDM), 2010 IEEE International*. IEEE (2010)

13. B. Govoreanu et al., 10 × 10 nm 2 Hf/HfO x crossbar resistive RAM with excellent performance, reliability and low-energy operation, in *Electron Devices Meeting (IEDM), 2011 IEEE International*. IEEE (2011)
14. M.-J. Lee et al., in A fast, high-endurance and scalable non-volatile memory device made from asymmetric Ta_2O_{5-x}/TaO_{2-x} bilayer structures. Nat. Mater. **10**(8), 625–630 (2011)
15. F. Pan et al., Recent progress in resistive random access memories: Materials, switching mechanisms, and performance. Mater. Sci. Eng. R Rep. **83**(9), 1–59 (2014)
16. H.S. Philip Wong et al., Phase change memory. Proc. IEEE **98**(12), 2201–2227 (2010)
17. I.G. Baek et al., Multi-layer cross-point binary oxide resistive memory (OxRRAM) for post-NAND storage application, in *Electron Devices Meeting, 2005. IEDM Technical Digest. IEEE International*. IEEE (2005)
18. M.-J. Lee et al., 2-stack 1D-1R cross-point structure with oxide diodes as switch elements for high density resistance RAM applications, in *Electron Devices Meeting, 2007. IEDM 2007. IEEE International*. IEEE (2007)
19. Y. Bai et al., in Study of multi-level characteristics for 3D vertical resistive switching memory. Sci. Rep. **4** (2014)
20. S. Yu et al., HfOx-based vertical resistive switching random access memory suitable for bit-cost-effective three-dimensional cross-point architecture. ACS Nano **7**(3), 2320–2325 (2013)
21. W.C. Chien et al., Multi-layer sidewall WOx resistive memory suitable for 3D ReRAM, in *2012 Symposium on VLSI Technology (VLSIT)*. IEEE (2012)
22. J.-Y. Seok et al., A review of three-dimensional resistive switching cross-bar array memories from the integration and materials property points of view. Adv. Funct. Mater. **24**(34), 5316–5339 (2014)
23. C.-W. Hsu et al., Self-rectifying bipolar TaO_x/TiO_2 RRAM with superior endurance over 10 12 cycles for 3D high-density storage-class memory, in *2013 Symposium on VLSI Technology (VLSIT)*. IEEE (2013)
24. K.-S. Li et al., Utilizing Sub-5 nm sidewall electrode technology for atomic-scale resistive memory fabrication, in *2014 Symposium on VLSI Technology (VLSI-Technology): Digest of Technical Papers*. IEEE (2014)
25. Y. Bai et al., in Stacked 3D RRAM array with graphene/CNT as edge electrodes. Sci. Rep. **5** (2015)
26. I.G. Baek et al., Realization of vertical resistive memory (VRRAM) using cost effective 3D process, in *Electron Devices Meeting (IEDM), 2011 IEEE International*. IEEE (2011)
27. S.G. Park et al., A non-linear ReRAM cell with sub-1μA ultralow operating current for high density vertical resistive memory (VRRAM), in *Electron Devices Meeting, 1988. IEDM '88. Technical Digest,* pp. 20.8.1–20.8.4 (2012)
28. H.S. Yoon et al., Vertical cross-point resistance change memory for ultra-high density non-volatile memory applications, in *2009 Symposium on VLSI Technology* (2009)
29. Y. Deng et al., Design and optimization methodology for 3D RRAM arrays, in *Electron Devices Meeting (IEDM), 2013 IEEE International*. IEEE (2013)
30. H.-Y. Chen et al., HfOx based vertical resistive random access memory for cost-effective 3D cross-point architecture without cell selector, in *Electron Devices Meeting (IEDM), 2012 IEEE International*. IEEE (2012)
31. E. Cha et al., Nanoscale (~ 10 nm) 3D vertical ReRAM and NbO 2 threshold selector with TiN electrode, in *Electron Devices Meeting (IEDM), 2013 IEEE International*. IEEE (2013)
32. C.-W. Hsu et al., 3D vertical TaO x/TiO 2 RRAM with over 10 3 self-rectifying ratio and sub-μA operating current, in *2013 IEEE International Electron Devices Meeting* (2013)
33. A. Chen, Nonlinearity and Asymmetry for Device Selection in Cross-Bar Memory Arrays (2015)
34. A. Flocke, G.N. Tobias, Fundamental analysis of resistive nano-crossbars for the use in hybrid Nano/CMOS-memory, in *33rd European Conference on Solid State Circuits (ESSCIRC 2007)*. IEEE (2007)

35. C.-L. Lo et al., in Dependence of read margin on pull-up schemes in high-density one selector–one resistor crossbar array. IEEE Trans. Electr. Devices **60**(1), 420–426 (2013)
36. J.-J. Huang et al., One selector-one resistor (1S1R) crossbar array for high-density flexible memory applications, in *Electron Devices Meeting (IEDM), 2011 IEEE International*. IEEE (2011)
37. P. Sun et al., in Physical model of dynamic Joule heating effect for reset process in conductive-bridge random access memory. J. Comput. Electr. **13**(2), 432–438 (2014)
38. P. Sun et al., in Thermal crosstalk in 3-dimensional RRAM crossbar array. Sci. Rep. **5** (2015)
39. X.P. Wang et al., Highly compact 1T-1R architecture (4F 2 footprint) involving fully CMOS compatible vertical GAA nano-pillar transistors and oxide-based RRAM cells exhibiting excellent NVM properties and ultra-low power operation, in *Electron Devices Meeting (IEDM), 2012 IEEE International*. IEEE (2012)
40. C.-H. Wang et al., Three-dimensional 4F 2 ReRAM cell with CMOS logic compatible process, in *Electron Devices Meeting (IEDM), 2010 IEEE International*. IEEE (2010)
41. G. Tallarida et al., Low temperature rectifying junctions for crossbar non-volatile memory devices, in *2009 IEEE International Memory Workshop* (2009)
42. L. Zhang et al. High-drive current (>1MA/cm 2) and highly nonlinear (>10 3) TiN/amorphous-Silicon/TiN scalable bidirectional selector with excellent reliability and its variability impact on the 1S1R array performance, in *Electron Devices Meeting (IEDM), 2014 IEEE International*. IEEE (2014)
43. W. Lee et al., Varistor-type bidirectional switch (J MAX > 10 7 A/cm 2, selectivity ~ 10 4) for 3D bipolar resistive memory arrays, in *2012 Symposium on VLSI Technology (VLSIT)*. IEEE (2012)
44. D. Lee et al., BEOL compatible (300 °C) TiN/TiO x/Ta/TiN 3D nanoscale (~10 nm) IMT selector, in *Electron Devices Meeting (IEDM), 2013 IEEE International*. IEEE (2013)
45. S.H. Jo et al., 3D-stackable crossbar resistive memory based on field assisted superlinear threshold (FAST) selector, in *Electron Devices Meeting (IEDM), 2014 IEEE International*. IEEE (2014)
46. M.-J. Lee et al., Highly-scalable threshold switching select device based on chaclogenide glasses for 3D nanoscaled memory arrays, in *Electron Devices Meeting (IEDM), 2012 IEEE International*. IEEE (2012)
47. H. Yang et al., Novel selector for high density non-volatile memory with ultra-low holding voltage and 10 7 on/off ratio, in *2015 Symposium on VLSI Technology (VLSI Technology)*. IEEE (2015)
48. S. Kim et al., Performance of threshold switching in chalcogenide glass for 3D stackable selector, in *Proceedings of Symposium on VLSIT* (2013)
49. I. Riess, Mixed ionic–electronic conductors—material properties and applications. Solid State Ionics **157**(1), 1–17 (2003)
50. K. Virwani et al., in Sub-30 nm scaling and high-speed operation of fully-confined access-devices for 3D crosspoint memory based on mixed-ionic-electronic-conduction (MIEC) materials. IEDM Tech. Dig. 36–39 (2012)
51. K. Gopalakrishnan et al., Highly-scalable novel access device based on mixed ionic electronic conduction (MIEC) materials for high density phase change memory (PCM) arrays, in *2010 Symposium on VLSI Technology (VLSIT)*. IEEE (2010)
52. G. Burr et al., Recovery dynamics and fast (sub-50 ns) read operation with access devices for 3D crosspoint memory based on mixed-ionic-electronic-conduction (MIEC), in *2013 Symposium on VLSI Technology (VLSIT)*. IEEE (2013)
53. X.A. Tran et al., Self-rectifying and forming-free unipolar HfOx based-high performance RRAM built by fab-avaialbe materials, in *Electron Devices Meeting, 1988. IEDM '88. Technical Digest*, pp. 31.2.1–31.2.4 (2011)
54. B. Govoreanu et al., Vacancy-modulated conductive oxide resistive RAM (VMCO-RRAM): An area-scalable switching current, self-compliant, highly nonlinear and wide on/off-window resistive switching cell, in *Electron Devices Meeting (IEDM), 2013 IEEE International*. IEEE (2013)

55. S. Lee et al., Selector-less ReRAM with an excellent non-linearity and reliability by the band-gap engineered multi-layer titanium oxide and triangular shaped AC pulse, in *Electron Devices Meeting (IEDM), 2013 IEEE International*. IEEE (2013)
56. E. Linn et al., in Complementary resistive switches for passive nanocrossbar memories. Nat. Mater, **9**(5), 403–406 (2010)
57. R. Rosezin et al., Integrated complementary resistive switches for passive high-density nanocrossbar arrays. Elect. Device Lett. **32**(2), 191–193 (2011)
58. F. Nardi et al., Complementary switching in metal oxides: Toward diode-less crossbar RRAMs, in *Electron Devices Meeting (IEDM), 2011 IEEE International*. IEEE (2011)
59. S. Yu et al., 3d vertical rram-scaling limit analysis and demonstration of 3d array operation, in *2013 Symposium on VLSI Technology (VLSIT)*. IEEE (2013)
60. The information Intel pronounced at IDF 15. http://www.eetimes.com/document.asp?doc_id=1327289
61. R. Fackenthal et al., 19.7 A 16 Gb ReRAM with 200 MB/s write and 1 GB/s read in 27 nm technology. *Solid-State Circuits Conference Digest of Technical Papers (ISSCC), 2014 IEEE International*. IEEE (2014)
62. T.-Y. Liu et al., A 130.7 mm 2 2-layer 32 Gb ReRAM memory device in 24 nm technology, in *Digest of Technical Papers—IEEE International Solid-State Circuits Conference*, pp. 210–211 (2013)
63. T.-Y. Liu et al., in A 130.7-2-layer 32-Gb ReRAM memory device in 24-nm technology. IEEE J. Solid-State Circ. **49**(1), 140–153 (2014)
64. S.H. Jo et al., in Cross-point resistive ram based on field-assisted superlinear threshold selector. IEEE Trans. Elect. Devices (2015)
65. S.H. Jo et al., "Sneak path" breakthrough heralds arrival of ultra-high density resistive memory. Chip Des. (Winter 2015)
66. M.-C. Hsieh et al., Ultra high density 3D via RRAM in pure 28 nm CMOS process, in *Electron Devices Meeting (IEDM), 2013 IEEE International*. IEEE (2013)
67. Y.-W. Chin et al., Point twin-bit RRAM in 3D interweaved cross-point array by Cu BEOL process, in *Electron Devices Meeting (IEDM), 2014 IEEE International*. IEEE (2014)
68. M.-J. Lee et al., in Electrical manipulation of nanofilaments in transition-metal oxides for resistance-based memory. Nano Lett. **9**(4), 1476–1481 (2009)
69. H. Li et al., in A SPICE model of resistive random access memory for large-scale memory array simulation. Elect. Device Lett. **35**(2), 211–213 (2014)

Chapter 9
3D Multi-chip Integration and Packaging Technology for NAND Flash Memories

Herb Huang and Rino Micheloni

9.1 3D Multi-chip Integration

Integration of multiple functions, performance enhancement, smaller package and cost reduction are recognized as common but prevailing requirements for advancing IC package technology in modern electronic systems applications such as smart phones, wearable electronics and *Internet of Things*, or IoT. Ideally, electronic package solutions should enable integration of multiple chips with a plurality and variety of functions, as either raw chips or smaller packages of chips, into a geometrically compact format with improved individual and/or overall performances, but at a reduced total cost, which includes wafer processing, chip to system packaging, testing, yield, and all the necessary stages of fabrication of a *System In Package* (SiP).

9.2 Challenges in Nanometer Devices Fabrication

Closely following the Moore's Law, modern CMOS fabrication technology has already surpassed deep submicron regime and now is marching rapidly into nanometer scales, thanks to astonishing innovation and advances in transistor device design, lithography and Silicon process technologies. However, such advance by no means offers fundamental relief to some of the technical complications of the conventional IC fabrication approach, the so-called *System-On-Chip*

H. Huang (✉)
Semiconductor Manufacturing International Corporation (SMIC), Shanghai, China
e-mail: herb_huang@smics.com

R. Micheloni
Performance Storage BU, Microsemi Corporation, Vimercate, Italy
e-mail: rino.micheloni@ieee.org

R. Micheloni (ed.), *3D Flash Memories*, DOI 10.1007/978-94-017-7512-0_9

or "SOC", which means integrating multiple but different functional units on a single silicon wafer through a unified wafer fabrication process.

One of the most noticeable is the technological challenge that the conventional SOC approach continues to face: fabrication process of different functional devices and interconnects at nanometer scales through a unified wafer process. And some of the conventional process technology approaches can't be used anymore when entering in the nanometer scale. For instance, the floating-gate based embedded memory solution becomes not viable when high-K metal gate (HKMG) is adopted for MOS devices, i.e. at 28 nm or below. It is not viable because the complete fabrication solution for incorporating the floating-gate add-on structure with logic baseline process is technically challenging, and economically less attractive because further complicating the fabrication process and reducing the overall process yield as adding extra process layers and steps.

Re-integration of IC system through advanced system-in-package solutions allows segregation of a single wafer fabrication into individual wafer processes, each optimized for one or several but not all of the needed functional blocks. Each individual fabrication process, thus dedicated and simplified, is then easier to be optimized.

9.3 Challenges of On-Chip Interconnections

Integration of multiple functional circuits into a single chip might also face issues with on-chip interconnections. For example, the design rules and pattern density of interconnect layers might differ between the on-chip memory blocks and the logical blocks on a SOC chip. And elongated interconnects would likely worsen latency of data and signal transmission among different functional blocks, on top of the necessity for adding lithographic steps. High density inter-chip connection through SiP is also an important alternative for resolving the speed bottleneck of on-chip interconnects. Such bottleneck arises particularly due to dramatic increase of parasitic resistances of on-chip interconnects in a 2D planar configuration.

First of all, the net cross section areas of on-chip interconnects, still fabricated with basic Cu damascene process, keeps shrinking together with transistor's design rules. There aren't too many options for reducing the sheet resistance of on-chip interconnection material and the dielectric constant of inter-metal dielectric layers. The approach of fabricating air-gaps in inter-metal dielectric layers is one of the most viable solutions for further boosting the K-value beyond ultra low-K dielectric. Even though this approach was recently successfully deployed in a commercial high performance CPU [1], air-gaps would have to be used cautiously in a multi-layer BEOL (Back-End-Of-Line) due to various manufacturing and reliability concerns (both mechanical and thermal).

Secondly, signal latency becomes severer because of the elongated interconnects between different functional blocks in the conventional planar configuration of SOC chips. Moreover, as data rate gets higher, power consumption over such elongated interconnects turns into another serious problem. Therefore, shorter connections and better signal integrity become even more important to further boost system performance.

9.4 Heterogeneous Integration Through SiP

System integration through SiP is a very effective solution for improving performance of many microelectronic systems which incorporate multiple heterogeneous microelectronic units for providing specific functions. Some of these heterogeneous units might contain active semiconductor devices, such as MOS or diode, while others might have passive devices, either integrated in units or just individually packaged. Furthermore, active semiconductor devices might be individually fabricated on different semiconductor substrates through different fabrication processes in separate units, either packaged or as individual die.

A good example is the RF front-end module widely used in modern cellular phones, which provides the most critical transmitting and receiving functions of radio frequency signals between phones and base stations. Such RF front-end modules typically include RF switches, power amplifiers, low noise amplifiers, filters and controllers. Even though RF switches and low noise amplifiers are usually fabricated on SOI substrate, most of the power amplifiers in modern LTE and even in 3G cellular phones are built on GaAs substrate because of its high power amplification efficiency, while controllers are on bulk silicon to reduce substrate and fabrication costs. RF filters, such as *Bulk Acoustic Wave* (BAW) and *Surface Acoustic Wave* (SAW) filters, are passive devices which are fabricated and packaged individually through particular wafer fabrication and packaging processes, substantially different from all the others used in front-end modules. High density, high reliability SiP becomes the most desirable approach for improving the signal integrity and overall performance of cellular RF front-end modules.

Another good example is the *Micro Electro-Mechanical System* (MEMS) IC used in smartphones, and wearable, consumer and auto electronics. Most of the times, MEMS devices are fabricated and packaged through special processes, in general fairly different from CMOS, which is used for the companion ASIC IC; this second device is used for signal readout and processing, system control and data buffering. Although some of silicon based MEMS devices could be fabricated and integrated with their companion CMOS ASIC by using a unified wafer process, there are still many MEMS devices which have to be fabricated and packaged individually through special processes, either because they have to adopt unique fabrication and packaging approaches or their die sizes are different from the companion CMOS ASIC chip.

Another example is silicon photonics, where devices such as waveguide, splitter, modulator and detector need to be coupled with a light-source laser diode in a seamless micro optical configuration, and electronically with high speed CMOS controller: all of this through SiP. In particular, laser diode is a unique heterogeneous device based on compound semiconductors such as InP.

9.5 Solutions for Size and Cost Reduction

As portable, mobile and wearable electronics become main stream system applications, aggressive shrinkage of all their constituent ICs and other components becomes critical. At the same time, further reduction in costs to end-users for massive applications mandates continuous cost reduction in fabricating and packaging these IC devices, as well as system assembly and testing. Advanced SiP solutions are considered as one of the most effective ways for achieving reduction in both the physical size and overall manufacturing cost of IC component and system packages.

As the first and most obvious example, rapid adoption of interactive touch sensing on flat panel display screens has driven fundamental change in spatial configuration of mobile phone designs: their physical designs are centered with high resolution, enlarged display and touch-sensing screens, typically in rectangular shapes. While their screens keep getting bigger for higher resolutions and better visual benefits, phones have to be thinner and lighter for being more portable. Accordingly, their system boards have to be single layer and packaged in a low profile, requiring not only thinner multilayer PCBs but thinner IC packages, often mounted on both sides. This is where chip stacking or 3D SiP can substantially contribute to reducing the vertical profile of IC packages. Furthermore, to enable higher battery capacity, the total area allocated for system boards continues to shrink dramatically and, of course, this limits the lateral dimension of IC packages and other components.

Besides enabling size reduction, SiP solutions are also expected to enable continuous, sustainable reduction in the overall costs of IC packages. More wafer level assembly and testing process in SiP manufacturing is generally recognized as one of the most effective and sustainable approaches, because it can substantially leverage all the advantages of high precision, consistency and efficiency of automated high speed, high volume wafer fabrication technologies. This is where numerous technological innovations emerge and are rapidly adopted for replacing, either partially or completely, conventional die-based manufacturing practices.

In the following sections, the core technological elements and frameworks of various popular and emerging 3D SiP solutions are discussed.

9.6 3D Multi-chip SiP Technology Solutions

Main components of 3D multi chip SiP are summarized below.

- *Active components* such as IC chips or dies, typically containing I/O pads either on their top sides or, if *Through-Silicon* Via (TSV) is adopted, on their bottom sides.
- *Passive components*, such as capacitors, resistors, filters and other passives either packaged or partially packaged with contact pads, providing supporting electronic functions to the IC chips and the system.

- *Substrates* as dielectric carriers of IC chips and passive components, made of any or a combination of silicon, glass, or polymeric compounds.
- *In-package electrical interconnects* such as bumps, pillars, RDL (Redistribution Layer), and TSV built into the substrates, and micro metal tabs or flexes and metallic wires not cast into any substrate between any two of core components.
- *Out-package electrical interconnects*, such as lead-frame, metal pins and tabs, conductive flexes, and bumps, for providing the packaged system with electrical interconnections to the external world, either to other packaged systems or to an electronic interconnect board.
- *Auxiliary components* for optical, mechanical, thermal and chemical functions; glass covers used in micro optical systems, metal caps with inlet holes for MEMS pressure sensors and microphones, and metal fins as heat sinks in high power IC systems are just few examples.
- *Package fills*, such as polymeric dielectric molding compounds and epoxies, for filling spatial gaps or voids left among any of the above components, for physically mounting them according to the designed spatial configuration, and for preventing external chemical or mechanical attacks from surrounding environment.

All the above are critical for achieving target performance, package size, fabrication costs and reliability.

3D multi-chip SiP can be categorized in many different ways according to their various characteristics: for example, spatial configurations and integration process of components, as discussed below.

9.6.1 2D, 3D Spatial Configurations of Multi-chip SiP and Derivatives

Spatial arrangement of key components, i.e. IC dies and passive components in particular, is the first aspect to consider when architecting the framework of a multi-chip SiP solution. Let's consider the simplest example of a package with 2 components: they could be either placed side by side in a planar or so-called *2D configuration*, or vertically stacked, one on top of each other, in a so-called *3D configuration*.

In a 2D configuration, metal wire bonds (Fig. 9.1) are the most conventional in-package interconnects used as the horizontal, direct electrical connection between the two components placed side by side, either on a substrate or simply molded together without a substrate. In some cases, different in-package interconnects are used, such as micro metal tabs or flexes placed between the two interconnected components. Another option to facilitate the basic 2D system integration is a redistribution layer RDL built on the components themselves.

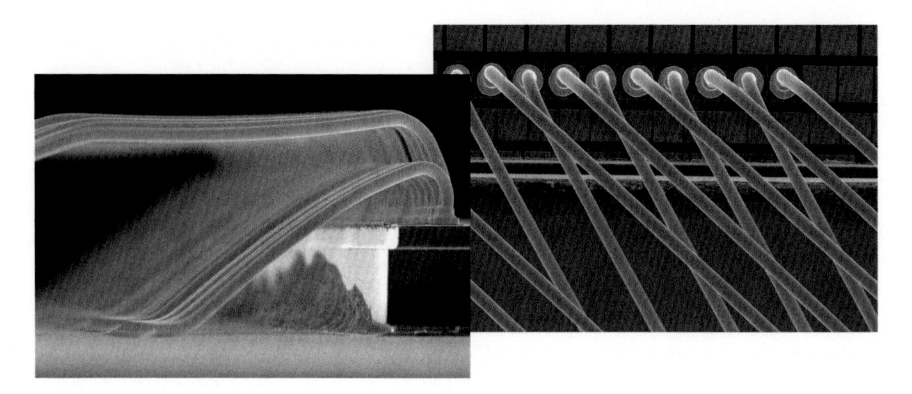

Fig. 9.1 Metal wire bonds

If a substrate is used as the physical carrier of the two components, the in-package interconnects built onto the substrate can facilitate horizontal interconnection between the two carried components. "Bumps" are also commonly used to facilitate vertical connection in the face-to-face physical contact of a component with the carrier substrate and RDL; this is usually referred to as *flip-chip* assembly. Wire bonds can also be used for vertical electrical connection, but they consume extra lateral space.

If one component is carried by another component, instead of a substrate, such a configuration of system assembly is often called 3D stacking. There are two fundamental 3D stacking options for assembling two components: both facing-up and face-to-face. In the face-to-face option, flip-chip assembly is the most viable solution path, employing RDL and bumps as discussed above. In the both facing-up option, either in-package interconnects, such as wire bonds, or a re-routing scheme to the face-to-face option are used. In the case that the top component is an active device or specifically a CMOS chip, integrated or embedded TSV with CMOS must be adopted in such re-routing scheme for later flip-chip assembly.

In general, a 2D configuration of multi-chip SiP occupies a horizontal area larger than a 3D solution. On the other hand, a 3D configuration requires a vertical profile thicker than a 2D configuration, as the components are vertically stacked. Moreover, if wire bonds are adopted for vertical interconnects between stacked components, the over-hang of wire bonds above the top component requires certain clearance and protection, which adds extra height to the system assembly.

From the basic 2D and 3D configurations described above, some interesting derivatives emerge, even for a simple two component system assembly. In case that one active component is vertically stacked onto a substrate which further facilitates re-routing of electrical connection on top surface to its opposite bottom, such substrate is often called an *interposer*. If two or more active or passive components are stacked onto the interposer, even though these components are placed side by side horizontally, the system becomes a hybrid configuration with both 2D and 3D characteristics. The name of 2.5D is widely used in the industry for denoting such configuration of multi-chip SiP. Most recently silicon interposer has become popular because of its capability of providing high density vertical (TSV) and lateral (RDL) interconnects. High performance FPGAs are a good example of commercial success of 2.5D multi-chip SiP [2]. In contrast, multi-layer polymeric substrates, which have been used as interposers for constructing 2.5D multi-chip SiP for years, are cheaper than silicon TSV interposer.

Another interesting derivative is when an active component is stacked vertically with a passive component on its front-face, but has through-via for I/O re-routing to the backside. Widely used in *CMOS Image Sensor* (CIS) and module assembly application, wafer level chip-scale-package is an example where the active CIS chip is capped with a highly transparent glass cover on top, having TSV for re-routing I/O to backside. As an extension from the industry's popular nomenclature of interposer based 2.5D SiP, this is one of the cases called 2.1D SiP [3].

Another alternative is a particular fan-out multi-chip SiP in which two or multiple active or passive components are molded or cast in a planar package and interconnected each other by using a typical 2D multi-chip SiP scheme. If through-vias are also fabricated through the molding compound providing I/O re-routing to backside, such a case is another example of 2.1D SiP: it partially encompasses 2D, 2.5D and 3D multi-chip concepts, but it doesn't exactly adopt the conventional interposer and 2.5D approach.

Modern multi-chip SiP technologies are likely to evolve to versatile *hybrid* spatial configurations incorporating multiple components for different applications, as shown in Fig. 9.2 [3].

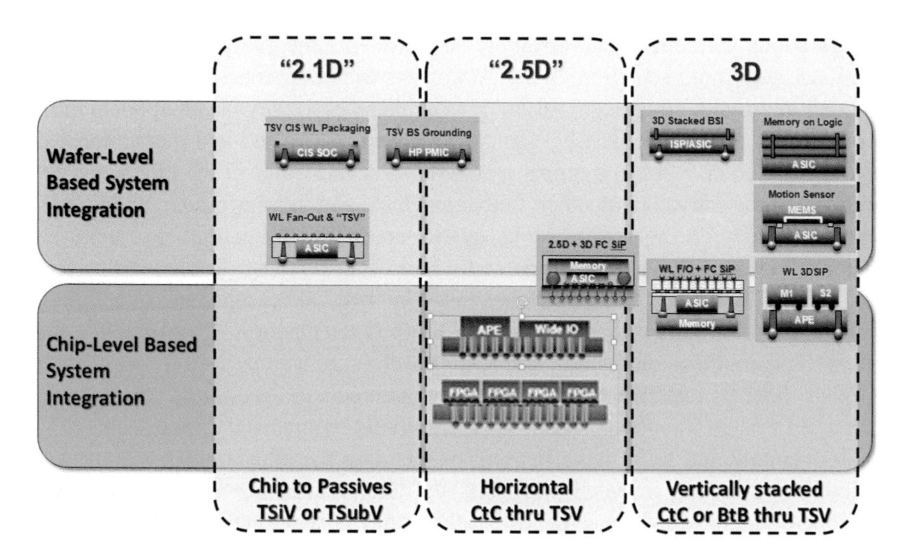

Fig. 9.2 Classification of multi-chip SiP solutions by spatial configurations and system integration [3]

9.6.2 Integration Process of Multi-chip SiP: Die-to-Die, Wafer-to-Wafer and Die-to-Wafer

Multi-chip SiP can also be classified based on manufacturing process schemes for component assembly and system integration.

Considering their fundamental differences in the manufacturing protocol of component assembly, the two fundamental approaches are:

- purely chip or component level based system integration, or *die-to-die* approach;
- pure wafer level based system integration, or *wafer-to-wafer* approach, as shown in Fig. 9.2.

Although the die-to-die approach is less efficient in exploiting advanced automatic batch manufacturing practices, it offers great flexibility in the design of spatial configurations and system integration, as well as in manufacturing. The die-to-die approach can be used with 2D, 3D and any hybrid configurations SiP. There is also a full flexibility in selecting substrates, in-package and out-package interconnects, auxiliary components and filling materials, thus allowing optimal trade-off between system performance and fabrication costs. Another key aspect is the possibility of conducting electrical and other functional tests at the single die level for sorting out quality and performance outliers as early as possible in the integration process; in fact, early discard of bad performers reduces the overall cost of production.

In contrast, the wafer-to-wafer approach is far more efficient under the framework of modern automatic wafer batch manufacturing, thanks to significant advances in certain enabling wafer manufacturing technologies such as precision wafer bonding [4] and TSV. However, it has limited flexibility for custom spatial configurations of component assembly and system integration, as well as for manufacturing implementation. In a pure wafer-to-wafer approach, only the vertical 3D configuration of component assembly is applicable, so that one component built in batch on a first wafer must be vertically stacked, assembled and interconnected with another component in a mirror match on a second wafer of the identical size. Even though electrical and other functional tests can be performed with much higher efficiency during wafer testing, quality and performance outliers cannot be discarded, and they have to be carried over into wafer bonding process and system integration. These outliers or "bad" dies result in "bad" 3D component, thus killing their counterparts. This is the well-known *Known Good Die* or KGD. Moreover, the size of the paired dies on the first and second wafers, as well as the layouts of their I/O pads, have to match, which significantly undercuts the economic efficiency in such a 3D system integration framework and limits its practical application.

To overcome the KGD issue but still maintaining the advantage of wafer-level assembly and testing in batch processes, the intermediate *die-to-wafer* has been developed: silicon dies are vertically stacked onto or assembled with the pairing dies on a "carrier" wafer. Therefore, any outlier "bad" die on the first wafer can be sorted and prevented from being integrated with good dies on the second wafer. Also, neither the size of the paired dies nor the layouts of their I/O pads need to be matched. Furthermore, if their relative sizes match, two or multiple dies can be assembled with one die on the carrier wafer; and to further benefit wafer-level batch processes for post assembly steps, these two or multiple dies being placed side by side horizontally can be molded, with certain dielectric molding compounds, in a wafer-shaped cast onto the carrier wafer. In fact, such a scenario is a hybrid approach which assumes partial aspects of both 2D and 3D spatial configurations of component assembly and system integration.

9.6.3 Challenges of 3D Multi-chip SiP

For a successful commercial deployment a 3D multi-chip SiP must be designed as a viable packaging technology solution for specific system integration and tailored for meeting particular requirements in terms of performance, form factor and cost. Most of such solutions often imply complications beyond the simple 2D and 3D spatial configurations, as well as the pure die-to-die and wafer-to-wafer assembly processes. In other words, appropriate hybrid multi-chip SiP approaches are developed and adopted to specifically support particular application systems and the integration of their components.

Several practical cases, either successfully implemented in commercial applications or under development, are illustrated in Fig. 9.2. *Wafer level chip-scale-package* (WLCSP) of CIS, *2.5D multi-chip SiP* of high performance FPGA or CPU plus memory, and *fan-out multi-chip SiP* have been previously described. Further extension from conventional 2.5D SiP approach is vertical stacking of a high performance memory die onto an ASIC (such as CPU) die, while the ASIC is mounted onto a silicon interposer through flip-chip, hereby named *2.5D + 3D flip-chip SiP*.

Facilitated by precision wafer bonding and TSV technologies, 3D SiP through the straightforward wafer-to-wafer approach has been successfully implemented in few commercial applications: stacked backside illumination (BSI) image sensor SOC, multiple layers of memory stacked on logic, and MEMS sensor stacked onto ASIC. Of course, in all these examples yield is the critical factor for success.

To overcome the size mismatch of stacked dies and to facilitate stacking of two or more dies onto one die on the carrier wafer, a generic *wafer-level 3D SiP* (WL 3D SiP) has been established. In particular, such an approach facilitates high density, low profile and compact SiP of heterogeneous dies, typically fabricated through different wafer processes. With the help of TSV being embedded into the carrier wafer, such an approach can eliminate the need for a carrier substrate underneath the carrier wafer. Such potential benefits are highly desired in modern smart phones, wearable electronics and IoT applications. A typical SiP required by these applications has to incorporate multiple but different sensors together with an ASIC, which serves as a sensor hub and gives “smart” outputs to the user end systems [3].

While no major showstoppers have been identified so far, there are still many challenges to bring 3D multi-chip SiP into scaled commercialization for mainstream applications.

Firstly, in terms of design, new capabilities are required for addressing implications to system-level, 3D floor-planning, implementation, extraction/analysis, test, and IC/package co-design. New design flows have to support unified design intent, abstraction, and convergence with physical and manufacturing data through the whole SiP process to achieve optimal, timely and cost-effective design, which simultaneously integrates three layers of fabrics: ICs and other components, SiP package, and SiP board [5].

Secondly, testing 3D multi-chip SiP poses a number of technical challenges, such as physical access to a die inside a SiP stack and delicate handling of thinned wafers. Like conventional test of many single-die devices, test of 3D multi-chip SiP must also be considered at two levels, wafer sorting for silicon dies and package test after dies are assembled. However, there are far more intermediate steps, such as die stacking and TSV bonding, involved in 3D multi-chip SiP, which provide more testing opportunities before final assembly and packaging. *Design-For-Test* (DFT) is becoming more and more popular as it can provide efficient ways to control and observe individual dies from their I/Os [5].

Thirdly, reliability is a concern, because of the number of components, the pool of materials used is expanded and their spatial configurations are more complicated but compressed into significantly reduced and constrained sizes. For example, TSV has been introduced in many 3D multi-chip SiP cases: its integration with silicon substrate and even with CMOS imposes a number of reliability concerns. In fact, TSV involves the substrate which is usually not the focus for testing. Moreover, metal TSVs, due to the considerable volume of Cu [6], induce thermal mechanical stress to the substrate which not only affects the electronic performance of active devices in silicon substrate, but also results in physical damages and electrical leakage to substrate. In addition, as components with different thermal-mechanical properties are densely packed, both thermal and thermal-mechanical interactions among components become more complicated. Therefore, design for reliability becomes another critical consideration which cannot be ignored in designing and manufacturing a 3D multi-chip SiP solution.

Last but not least, the overall cost for fabricating 3D multi-chip SiP still remains as one of the important obstacles to its scaled commercial deployment into mainstream applications. This is why there is still a lot of R&D activity on a number of already validated 3D multi-chip SiP approaches and particularly. For example, to dramatically reduce the overall fabrication cost of 2.5D SiP, two new alternative approaches are in development, one called “silicon-less integrated module” or “SLIM” and another “silicon wafer integrated fan-out technology” or “SWIFT”, both falling in 2.1D multi-chip SiP category defined in Fig. 9.2. The “SLIM” approach eliminates the costly silicon interposer component and process for 2.5D SiP, while “SWIFT” approach not only eliminates the silicon interposer but also provides wafer-level fan-out [7]. It is foreseeable that similar approaches or further disruptive low cost solutions are yet to emerge and to be deployed in commercial applications in years to come.

9.7 NAND Die Stacking

Let's now take a look at 3D integration of NAND Flash memories from a package perspective.

Small form factors have been one of the main drivers for the success of Flash memory cards (largely in the form of USB stick and Secure Digital, SD), which are shown in Fig. 9.3.

On the other hand, mainly driven by the success of Solid State Drives (SSDs), capacity requirement has grown dramatically to the extent that standard packaging (and design) techniques are no longer able to sustain the pace. In order to solve this issue, two approaches are possible: advanced die stacking and 3D monolithic technologies. This chapter covers the former, while the latter is the main subject of this book.

The standard way to increase capacity is to implement a multi-chip solution, where several dies are stacked together. The advantage of this approach is that it can be applied to existing bare die, as shown in Fig. 9.4: dies are separated by means of a so-called interposer, so that there is enough space for the bonding wires to be connected to the pads. On the other hand, the use of the interposer has the immediate drawback of increasing the height of the multi-chip, and height is one of the most relevant limiting factors for memory cards, and packages in general.

One way to overcome this issue is to exploit both sides of the PCB, as shown in Fig. 9.5: in this way, the PCB acts as interposer, and components are placed on the two sides of the PCB. Height is reduced, but there is an additional constraint on design: in fact, since the lower die is flipped, its pads are no longer matching those of the upper die. The only way to have corresponding pads facing one another is to design the pad section in such a way that pad-to-signal correspondence can be

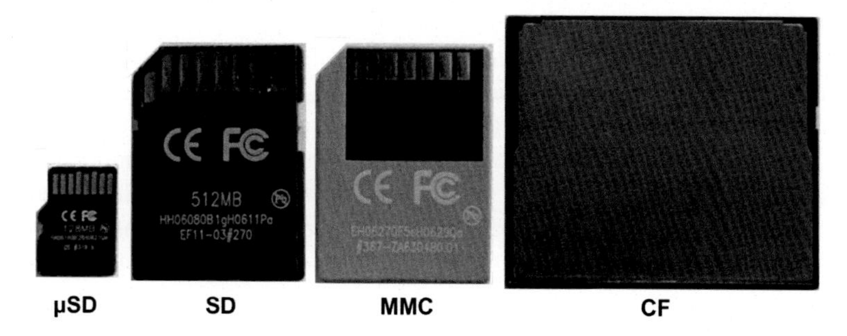

Fig. 9.3 Popular Flash card form factors

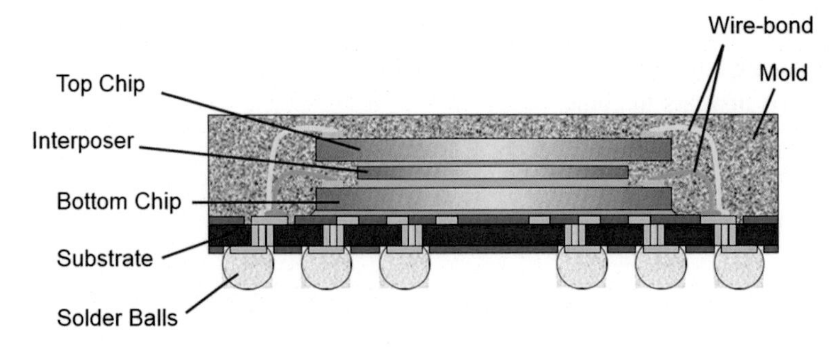

Fig. 9.4 Standard die stacking

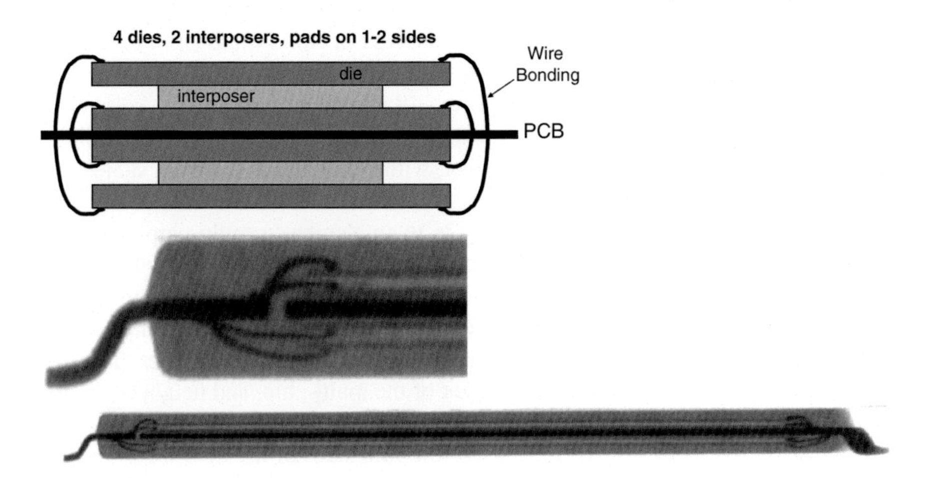

Fig. 9.5 Flipped die stacking

scrambled: that is, when a die is used as the bottom one, it is configured with mirrored-pads. Such a solution is achievable, but chip design is more complex (signals must be multiplexed in order to perform the scramble) and chip area is increased, since it might be necessary to have additional pads to ensure symmetry when flipping.

The real breakthrough is achieved by completely removing the interposer, thus using all the available height for silicon (apart from a minimum overhead due to the die-to-die glue). Figure 9.6 shows an implementation, where a staircase arrangement of the dies is used: any overhead is reduced to the minimum, bonding does not pose any particular issue and chip mechanical reliability is maintained (the dis-overlap between dies is small compared to the die length, and therefore the overall stability is not compromised, since the upmost die does not go beyond the overall center of mass).

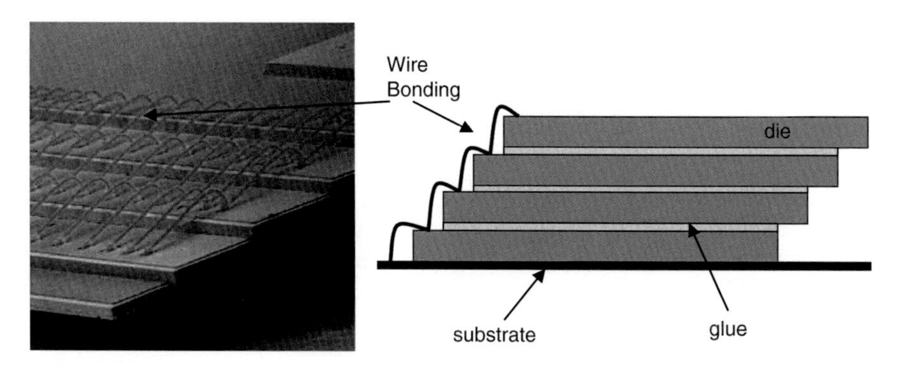

Fig. 9.6 Staircase die stacking: 4 dies, 0 interposers, pads on 1 side

The drawback is that such a solution has a heavy impact on chip design, since all the pads must be located on the same side of the die. In a conventional memory component, pads are arranged along 2 sides of the device: circuitry is then evenly located next to the two pad rows and the array occupies the majority of the central area. Figure 9.7 shows the floorplan of a memory device whose pads lie on 2 opposite sides.

If all pads lie on one side, as shown in Fig. 9.8, chip floorplan is heavily impacted [8]: most of the circuits are moved next to the pads in order to minimize the length of the connection and to optimize circuit placement. But some of the circuits still reside on the opposite side of the die (for instance, part of the decoding logic of the array and part of the page buffers, i.e. the latches where data are stored, either to be written to the memory or when they are read from the memory to be output to the external world).

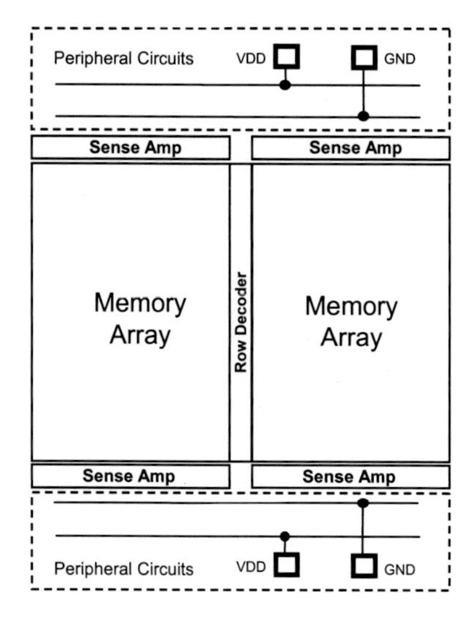

Fig. 9.7 Memory device with pads along opposite sides

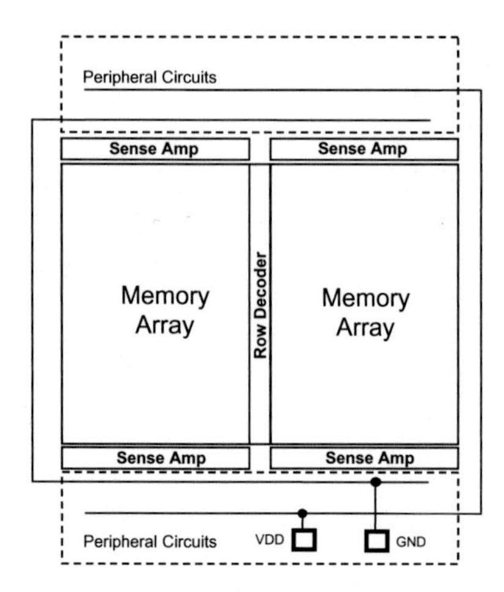

Fig. 9.8 Memory device with pads along one side

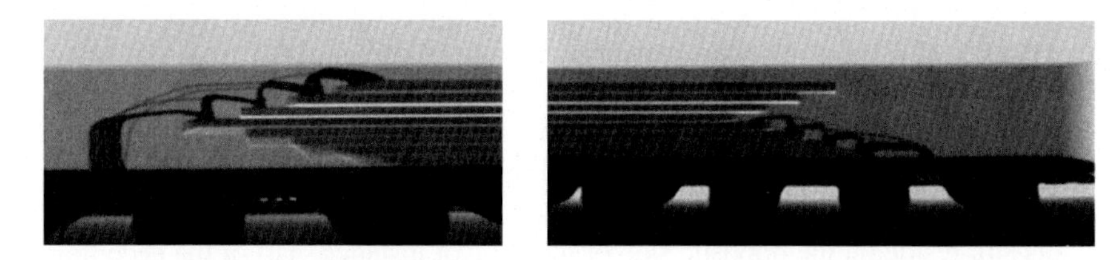

Fig. 9.9 Double-stair die stacking

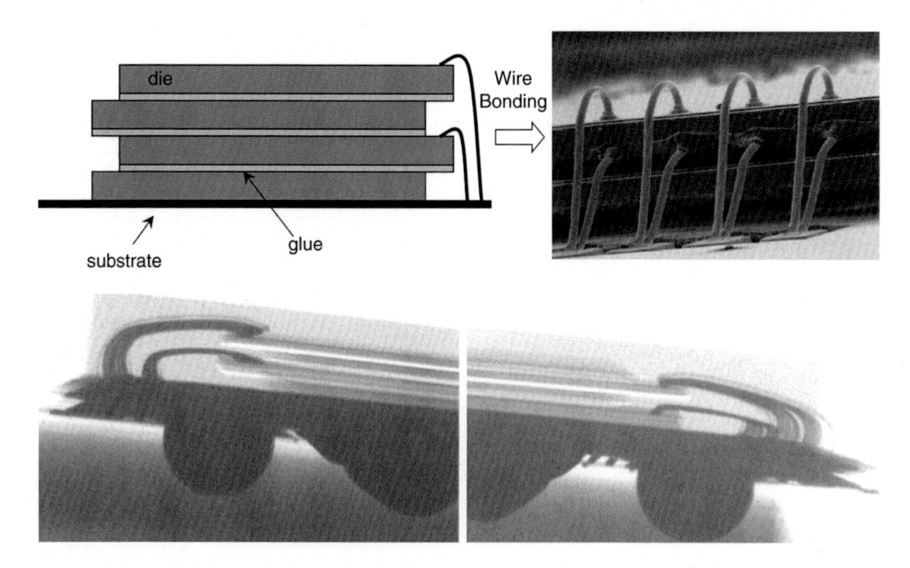

Fig. 9.10 "Zig-zag" die stacking

Of course, those circuits must be connected to the rest of the chip, from both a functional and a power supply point of view. Since all pads are on one side, including power supply, it is necessary to re-design the power rail distribution inside the chip, making sure that size and geometry of the rails are good enough to avoid IR drops issues (in fact, the voltage at the end of the rail is reduced due to the resistive nature of the metal line).

One of the main disadvantages of staircase stacking is the increased size in the direction perpendicular to the pad row: this fact limits the number of dies, given a specific package. When the number of dies grows up, 2 stairs can be combined together: Fig. 9.9 shows 2 stairs of 4 die each.

The "zig-zag" stacking displayed in Fig. 9.10 is another common solution. In this case, thanks to the double side bonding, the overall size can be clearly reduced.

9.8 TSV NAND

As already mentioned above, another stacking option is Through-Silicon-Via (TSV). Application of TSV to NAND Flash memories has been studied for few years [9–18], and Toshiba announced the first NAND product based on TSV technology at Flash Memory Summit 2015 [19]; please refer to Fig. 1.15. With this stacking technology, dies are directly connected without asking for wire bonding, as depicted in Fig. 9.11. A bird's-eye view of one TSV connection is shown in Fig. 9.12. Compared to conventional wire bonding, one of the main advantages of this technology is the reduction of the interconnection length and the associated parasitic RC. As a result, data transfer rate can be definitely improved, as well as power consumption.

DRAM has historically been the main driver for TSV, exactly because of the speed. DDR4 can run up to 2400 MT/s with a clock of 1.2 GHz: at this speed, because of the parasitic capacitance of interconnections and I/O pads, it gets almost impossible to stack more than two dies in the same package with the standard wire

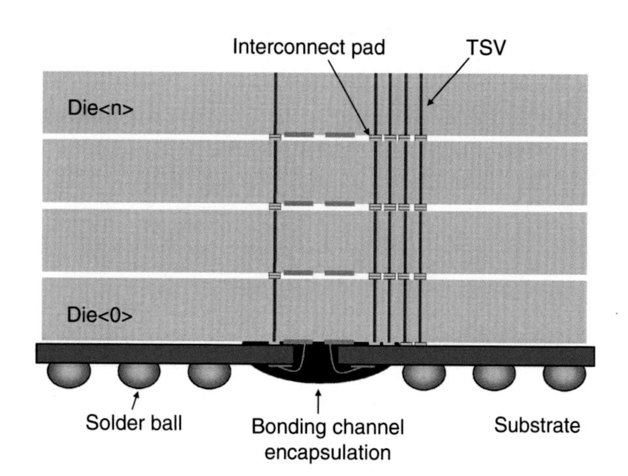

Fig. 9.11 Die stacking with TSV technology

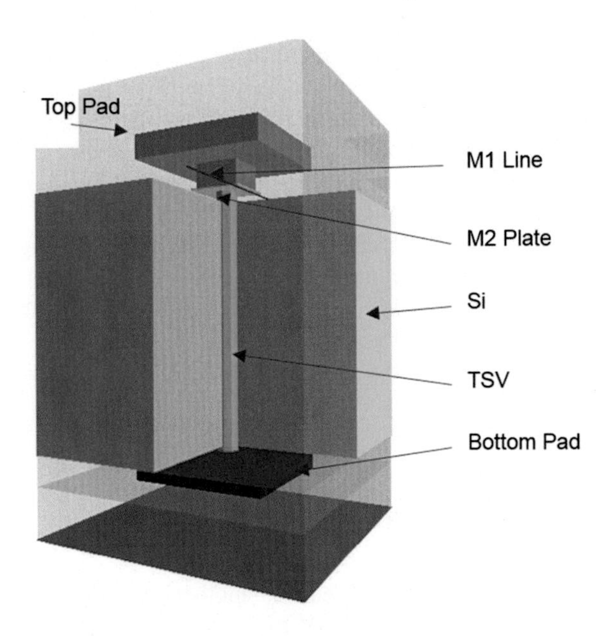

Fig. 9.12 Bird's-eye view of TSV

bonding technology. TSV is an effective way to solve this problem, at least at the package level; buffer chips might be used to increase the overall number of packages at the system level.

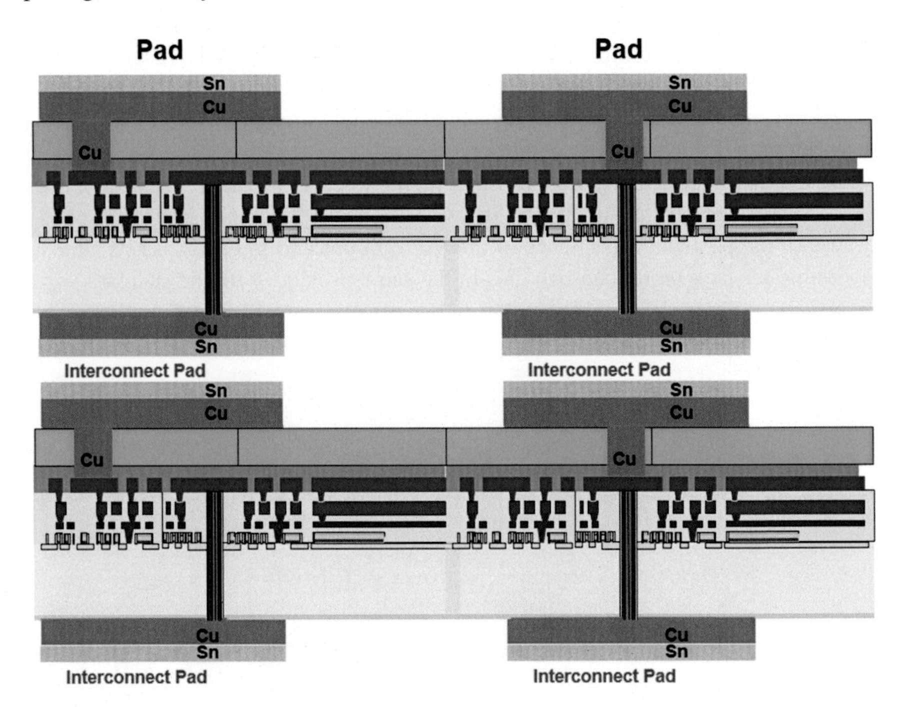

Fig. 9.13 Chip-to-chip connection by using TSVs

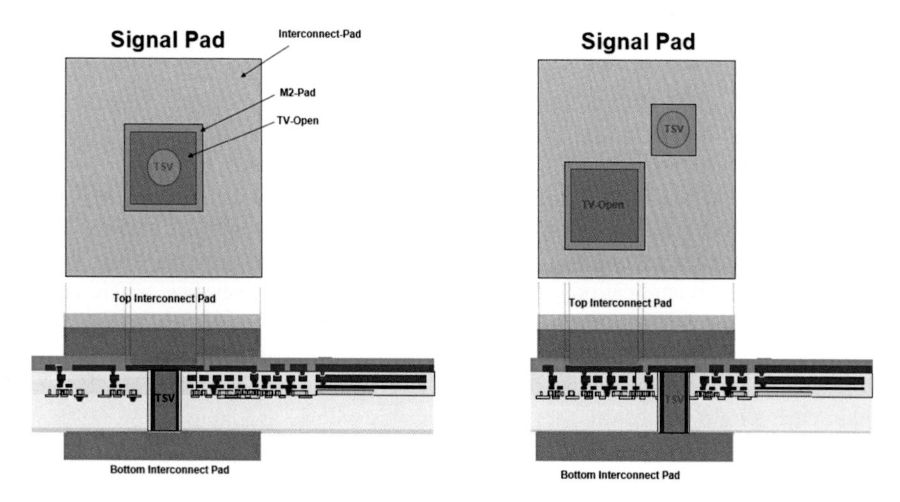

Fig. 9.14 TSVs for signal pads

Figure 9.13 shows few more details about TSV connections between 2 chips. Metal levels are visible in each die (blue rectangles). It is worth highlighting that not always the center of the *interconnect pad* is fully aligned with the center of the TSV itself. This is especially true for control signals, i.e. electrical signals that are

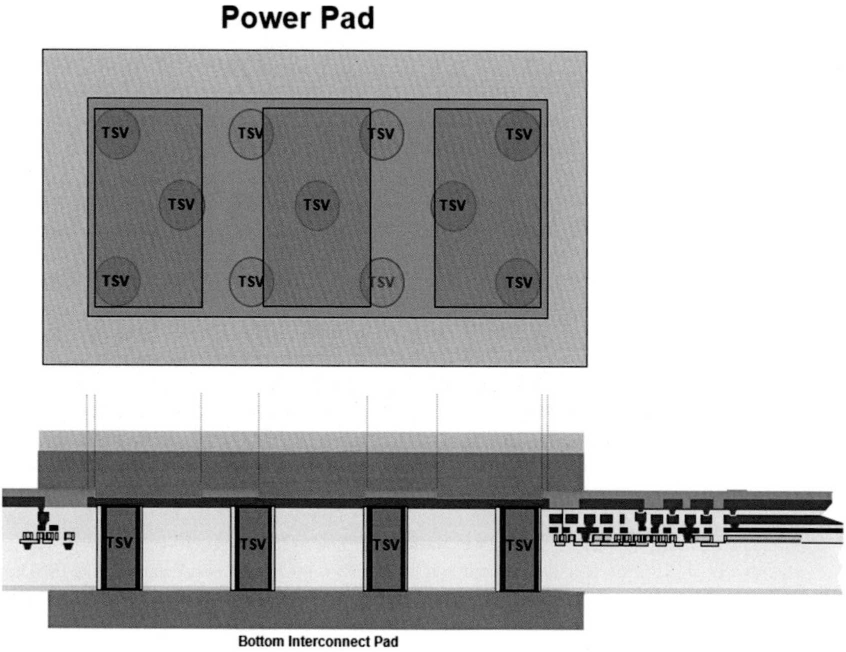

Fig. 9.15 A TSV array is used for power pads

not supposed to drive current into the device. Figure 9.14 includes top views of aligned (left) and misaligned (right) Pad/TSV pairs. Of course, this feature adds great flexibility to the chip design.

Because the current density of each TSV cannot exceed a certain limit, power connections require an array of TSVs, as sketched in Fig. 9.15.

At the end of this chapter, it is worth mentioning that 3D die stacking and 3D monolithic integration (i.e. a single NAND Flash chip with multiple memory layers) can be effectively combined together to serve the storage market, from consumer to enterprise applications. By building a stack of 16 Flash dies, where each single NAND die has a capacity of 768 Gb [20], almost 1.5 TB of storage can already be squeezed in a single package (12 mm × 18 mm)! By leveraging all the effort that Flash vendors are spending in increasing the number of memory layers within each single die, Multi-Tera bytes packages will be available in few years.

References

1. K. Fischer et al., Low-k interconnect stack with multi-layer air gap and tri-metal-insulator-metal capacitors for 14 nm high volume manufacturing, in *2015 IEEE International Interconnect Technology Conference and 2015 IEEE Materials for Advanced Metallization Conference (IITC/MAM)*, 18–21 May 2015, pp. 5–8
2. K. Saban, Xilinx stacked silicon interconnect technology delivers breakthrough FPGA capacity, bandwidth, and power efficiency. Xilinx white paper, WP380 (v1.2), 11 Dec 2012
3. H. Huang, Exploring technology solution paths for expanding 3D devices, interconnects and system integration beyond TSV, in *2014 international symposium on VLSI technology, systems and applications (VLSI-TSA 2014)*, Hsinchu, Taiwan, 28–30 Apr 2014. IEEE Catalog Number: CFP14846-POD. ISBN: 978-1-4799-2218-5
4. M. Zoberbiera, E. Hell, K. Cook, M. Hennemayera, B. Neuberta, Challenges, trends and solutions for 3D interconnects in lithography and wafer level bonding techniques. ECS Trans. **18**(1), 713–719 (2009)
5. Cadence, 3D ICs with TSVs—design challenges and requirements. Cadence white paper, 20 Sep 2010
6. P. Batra et al., Three-dimensional wafer stacking using Cu TSV integrated with 45 nm high performance SOI-CMOS embedded DRAM technology. J. Low Power Electron. Appl. **4**, 77–89 (2014)
7. R. Huemoeller, *Amkor's SLIM & SWIFT Package Technology*. Amkor Technology, 15 May 2015
8. K. Kanda et al., A 120 mm^2 16 Gb 4-MLC NAND flash memory with 43 nm CMOS technology, in *IEEE International Solid-State Circuits Conference Dig. Tech. Papers,* Feb 2008, pp. 430–431
9. C.G. Hwang, New paradigms in the silicon industry, in *International Electron Device Meeting (IEDM)* (2006), pp. 1–8
10. M. Kawano et al., A 3D packaging technology for a 4 Gbit stacked DRAM with 3 Gbps data transfer, in *International Electron Device Meeting (IEDM)* (2006), pp. 1–4
11. S. Hachiya et al., Comprehensive comparison of 3D-TSV integrated solid-state drives (SSDs) with storage class memory and NAND flash memory, in *3D Systems Integration Conference (3DIC), 2015 International*, 31 Aug–2 Sept 2015, pp. TS6.2.1–TS6.2.5

12. C. Sun et al., A high performance and energy-efficient cold data eviction algorithm for 3D-TSV hybrid ReRAM/MLC NAND SSD. IEEE Trans. Circ. Syst. I Regul. Pap. **61**(2), 382–392 (2014)
13. C. Sun et al., Over 10-times high-speed, energy efficient 3D TSV-integrated hybrid ReRAM/MLC NAND SSD by intelligent data fragmentation suppression, in *Design Automation Conference (ASP-DAC), 2013 18th Asia and South Pacific*, 22–25 Jan 2013, pp. 81–82
14. H. Fujii et al., x11 performance increase, x6.9 endurance enhancement, 93 % energy reduction of 3D TSV-integrated hybrid ReRAM/MLC NAND SSDs by data fragmentation suppression, in *2012 symposium on VLSI circuits (VLSIC)*, 13–15 June 2012, pp. 134–135
15. T. Yasufuku et al., Effect of resistance of TSV's on performance of boost converter for low power 3D SSD with NAND flash memories, in *IEEE International Conference on 3D System Integration, 2009. 3DIC 2009*, 28–30 Sept 2009, pp. 1–4
16. X.Q. Shi et al., Development of CMOS-process-compatible interconnect technology for 3D-stacking of NAND flash memory chips, in *Electronic Components and Technology Conference (ECTC), 2010 Proceedings 60th*, 1–4 June 2010, pp.74–78
17. T. Onagi et al., Impact of through-silicon via technology on energy consumption of 3D-integrated solid-state drive systems, in *2015 International Conference on Electronics Packaging and iMAPS All Asia Conference (ICEP-IACC)*, 14–17 Apr 2015, pp. 215–218
18. K. Johguchi et al., Through-silicon via design for a 3-D solid-state drive system with boost converter in a package. IEEE Trans. Compon. Packag. Manuf. Technol. **1**(2), 269–277 (2011)
19. http://toshiba.semicon-storage.com/ap-en/company/news/news-topics/2015/08/memory-20150806-1.html
20. T. Tanaka et al., A 768 Gb 3b/cell 3D-floating-gate NAND flash memory, in *2016 IEEE International Solid-State Circuits Conference (ISSCC), Dig. Tech. Papers*, San Francisco, USA (Feb 2016), pp. 142–143

Chapter 10
BCH and LDPC Error Correction Codes for NAND Flash Memories

Alessia Marelli and Rino Micheloni

Nowadays NAND Flash memories are part of our lives in many ways. The storage world is a completely new world thanks to NAND. As a matter of fact, it wouldn't be possible to have a smartphone without the use of NAND memories as the storage media. After USB keys and digital cameras, Solid State Drives (SSDs) are now the new disruptive application for Flash. Consumer-class ultra-light and ultra-thin laptops require NAND storage, but it is really in the cloud and in enterprise servers that the use of NAND can be a paradigm shift.

Because NAND devices can't be manufactured without defects, the use of Error Correction Codes (ECCs) has always been a common practice. While BCH (Bose-Chaudhuri-Hocquenghem) is a de facto standard for consumer applications, LDPC (Low-Density-Parity-Check) codes are a typical choice in the enterprise world. This is especially true when looking at planar (2D) ultra-scaled (e.g. 15 nm) NAND. Generally speaking, LDPC offers higher correction capabilities, but BCH remains a good solution when bandwidth requirements are very stringent.

As discussed in previous chapters, 3D NAND is becoming a reality in the market. In terms of noise models, 2D and 3D have some commonalities: they are both very complex and they change during the NAND's lifetime!

We do expect 3D NAND to bring new failure models into the game; all the scientists working on ECCs for non-volatile memories will have to put their best effort for getting as close as possible to the Shannon limit.

To set the stage, in this chapter we cover both BCH and LDPC codes. After a brief introduction, we will see the implementation issues when coupling these codes with a "real" NAND communication channel; practical workarounds will also be discussed.

A. Marelli (✉) · R. Micheloni
Performance Storage BU, Microsemi Corporation, Vimercate, Italy
e-mail: alessia.marelli@ieee.org

R. Micheloni
e-mail: rino.micheloni@ieee.org

R. Micheloni (ed.), *3D Flash Memories*, DOI 10.1007/978-94-017-7512-0_10

10.1 Introduction

During life, multiple sources can corrupt the data stored in NAND cells. The most popular way for data recovery, sometimes used in conjunction with other techniques (e.g. signal processing), is the adoption of an error correction code.

ECCs add redundant terms to the message, such that, on the receiver side, it is possible to detect the errors and to recover the message that was "most probably" transmitted. The set of "encoded" data, i.e. data with the added redundant terms, is usually called *codeword*.

In other words, ECC can decrease the native *Raw Bit Error Rate* (RBER) of NAND. Given a RBER as defined in Eq. (10.1)

$$RBER = \frac{Number\ of\ bit\ errors}{Total\ number\ of\ bits} \tag{10.1}$$

and an ECC able to recover t errors, the codeword error rate, sometimes called *Frame Error Rate* (FER) is computed as in Eq. (10.2)

$$\begin{aligned} FER = 1 - \Bigg[& (1 - RBER)^A + \binom{A}{1} RBER(1 - RBER)^{A-1} \\ & + \cdots + \binom{A}{t} RBER^t (1 - RBER)^{A-t} \Bigg] \end{aligned} \tag{10.2}$$

where A is the codeword size.

Figure 10.1 shows the FER for a 1 kB codeword with an ECC able to correct 1, 10, 50 or 100 errors.

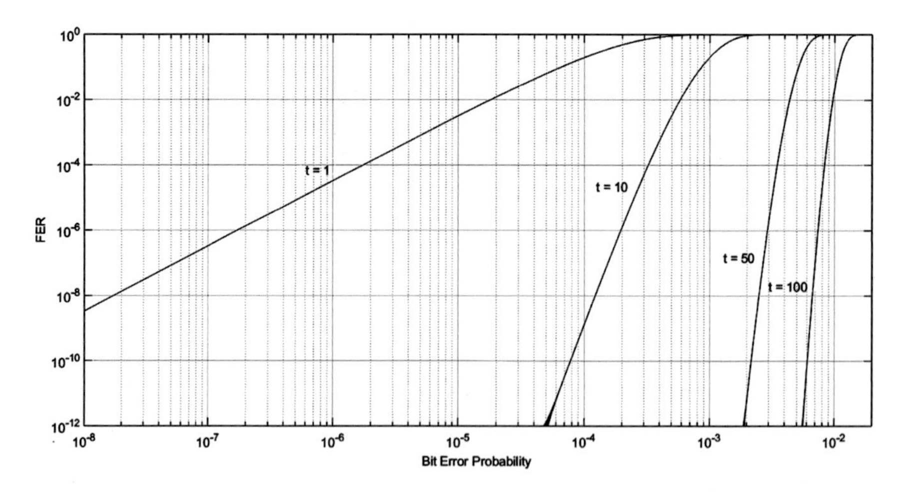

Fig. 10.1 Graph of the frame error probability for a 1 kB codeword with an ECC able to correct 1, 10, 50 and 100 errors

Another quantity used to measure the impact of ECC is the *Uncorrectable Bit Error Rate* (UBER). This is defined as in Eq. (10.3)

$$UBER = \frac{FER}{A} \tag{10.3}$$

A fundamental quantity used to decide which ECC to apply is the Code Rate. The Code Rate is defined as the ratio between the number of protected bits and the total number of transmitted bits (codeword size). If the Code Rate is high, we have few ECC parity bits, i.e. the error correction capability is low. On the other hand, we do not need too much extra space to store them. If the Code Rate is low, we have a higher number of parity bits to protect the data, and the error correction capability is high. In this case, we need more additional space to store the parity bits, and in some cases this is not possible. Even when this is possible, it costs money.

The trade-off between Code Rate and cost ($) is shown in Fig. 10.2. ECC correctability (i.e. the number of correctable bits per codeword) is a function of the Code Rate, as shown in Fig. 10.3. A lower code rate is less efficient in terms of silicon area, but it can recover more errors.

Error correction capability is also influenced by the codeword size (Fig. 10.4). Given the same code rate, the longer the codeword is, the higher the error correction capability is. On the other hand, the longer the codeword is, the more complex the ECC hardware is; a longer latency time to recover the corrupted data is another downside.

In communication theory, *Signal-to-Noise Ratio* (SNR) is usually adopted instead of RBER. Signal-to-noise ratio is a measure that compares the level of a desired signal to the level of the background noise. It is defined as the ratio of the

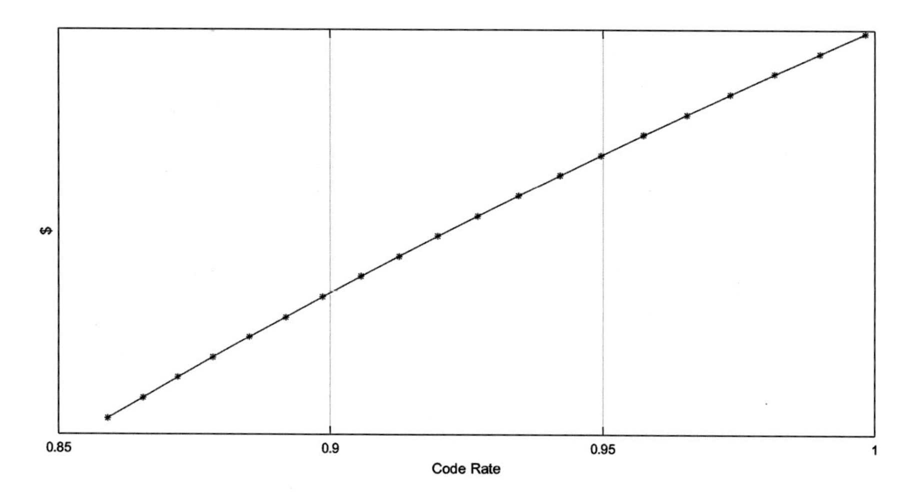

Fig. 10.2 Trade-off between code rate and cost

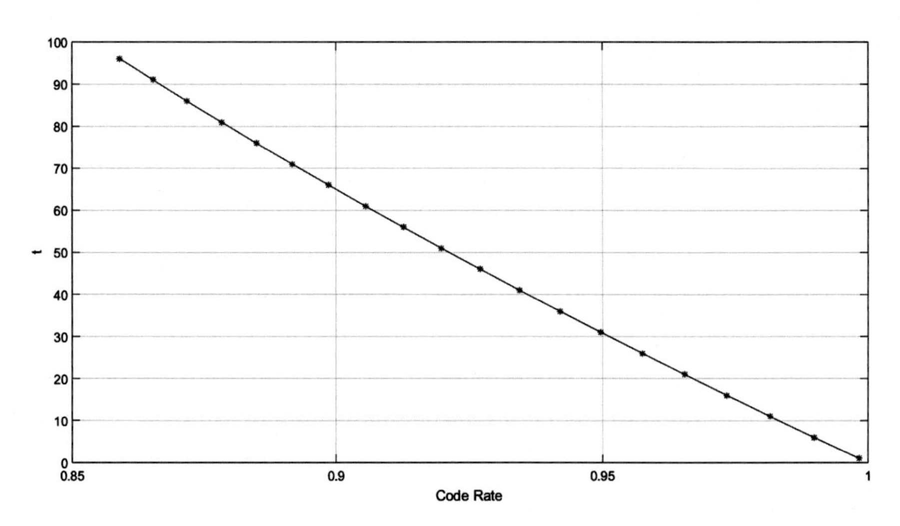

Fig. 10.3 Trade-off between code rate and correctability *t* for a codeword size of 1024 bytes

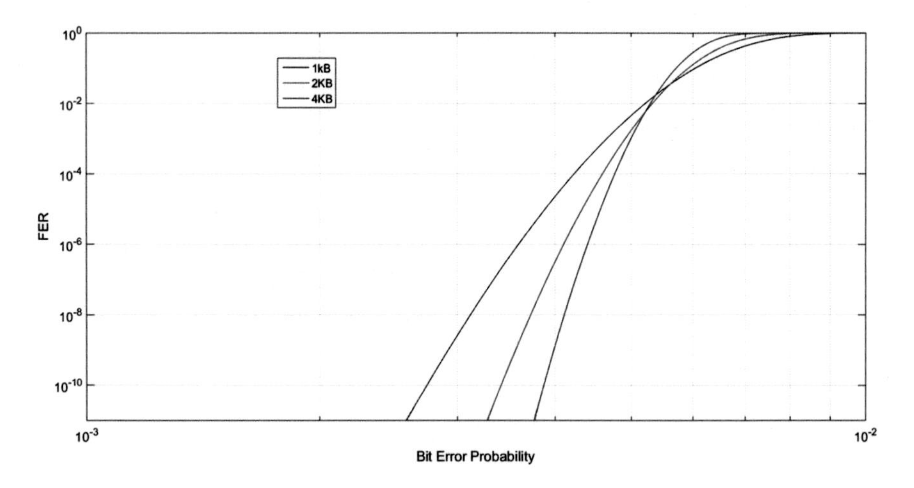

Fig. 10.4 FER versus BER for an ECC with code rate of 0.9 and different codeword sizes

signal power to the noise power, and it is often expressed in decibels. A ratio higher than 1 (i.e. greater than 0 dB) indicates more signal than noise.

Error correction codes belong to the information theory, whose "father" is Shannon; he demonstrated a fundamental theorem known as Shannon Limit [1]. This theorem establishes a limit in terms of achievable signal to noise ratio (SNR) for an error-free communication in a coded system with Code Rate R. The power of Shannon limit is the following: if we can guarantee that SNR does not exceed this limit, then we are sure of the existence of a coded system (with rate R) able to achieve error-free communication. Unfortunately, there isn't any

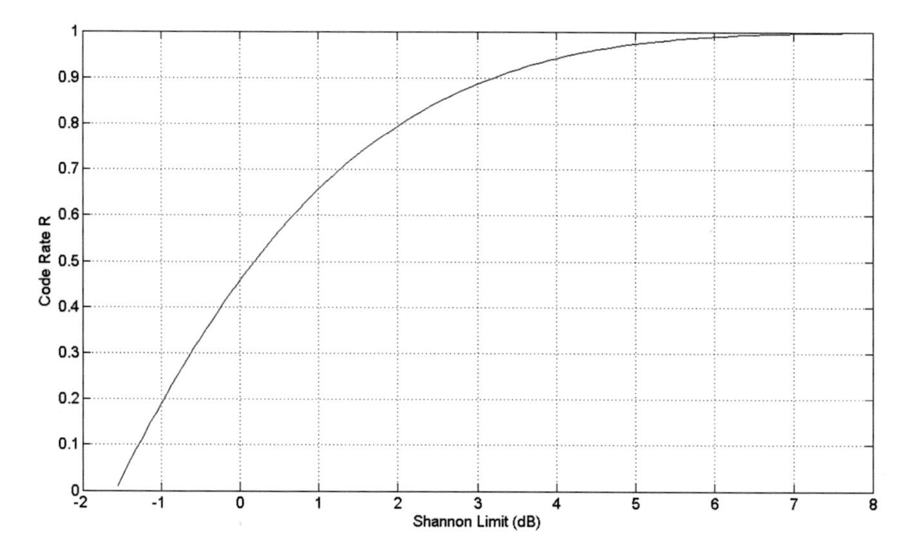

Fig. 10.5 Shannon limit for different code rates

constructive way for building such a coded system; this is why there is an intense research activity for finding codes that get as close as possible to the Shannon limit. The limit can be computed assuming an AWGN (Additive White Gaussian Noise) channel and a BPSK (Binary Phase Shift Keying) modulation (Fig. 10.5).

The achievable SNR can be translated in achievable BER. The Shannon limit is used to evaluate different coded systems: the best code is the one closest to the limit [2, 3].

ECCs can be split in hard decision codes and soft decision codes. This distinction is not based on the structure of the code itself, but on the way the information is treated by the code. A binary hard decision code treats all the data in a digital way, i.e. "0" or "1"; in other words, the analog information is converted into digital format by using one fixed reference level. On the contrary, a soft decision code uses reliability information to take decisions: for example, a "0" is read with a 90 % reliability and a "1" is read with a 10 % reliability. In the following sections, we will see how the soft information applies to NAND Flash, and a comparison between hard and soft codes [3].

Basically, a *Code C* is the set of codewords obtained by associating the q^k messages of length k of the space A to q^k words of length n of the space B in a univocal way. A code is defined as *linear* if, given two codewords, their sum is a codeword. When a code is linear, encoding and decoding can be described with matrix operations.

We define the *generator matrix of a code C* as G. It follows that all the codewords can be obtained as a combination of the rows of G. Therefore, encoding a data message m is equivalent to multiply the message m by the code generator matrix G, according to Eq. (10.4).

$$c = m \cdot G \tag{10.4}$$

G is called in *standard form* or in *systematic form* if $G = (I_k, P)$, where I_k is the identity matrix $k \times k$, and P is a matrix $k \times (n - k)$. If G is in standard form, then the first k symbols of a codeword are called information symbols.

From the matrix G in systematic form, it is straightforward to derive the *parity matrix* $H = (-P^T, \ I_{n-k})$ where P^T is the transpose of P and it is a matrix $(n - k) \times k$, and I_{n-k} is the identity matrix $(n - k) \times (n - k)$ [4, 5].

Systematic codes have the advantage that the data message can clearly be identified in the codeword and, therefore, it can be read before decoding. For codes in non-systematic form the message is no more recognizable in the encoded sequence and it is necessary to have the inverse encoding function to recognize the data sequence.

If C is a linear code with parity matrix H, then $x \cdot H^T$ is called *syndrome* of x. It follows that all the codewords have a syndrome equal to 0.

The syndrome is the key player of decoding. Once a message r is received (i.e. read from the memory), it is necessary to understand if it has been corrupted by calculating:

$$s = x \cdot H^T \tag{10.5}$$

There are two possibilities:

- $s = 0 \Longrightarrow$ the message r is recognized as correct;
- $s \neq 0 \Longrightarrow$ the received message contains some errors.

In the latter case a decoding procedure starts.

In order to understand how many errors a code is able to correct and detect we need a metric. In the coding theory, it is called *minimum distance* or *Hamming distance* d of a code, and it corresponds to the minimum number of different symbols between any two codewords.

A code has *detection capability* v if it is able to recognize all the messages, containing v errors at the most, as corrupted.

The detection capability is related to the minimum distance as described in Eq. (10.6).

$$v = d - 1 \tag{10.6}$$

A code has *correction capability* t if it is able to correct each combination of a number of errors equal to t at the most. The correction capability is calculated from the minimum distance d with Eq. (10.7):

$$t = \left[\frac{d - 1}{2}\right] \tag{10.7}$$

where the square brackets mean the floor function.

Codes can be manipulated or combined depending on applications. The possible operation to increase the minimum distance of a code is the extension: a code $C[n, k]$ is *extended* to a code $C'[n + 1, k]$ by adding one more parity symbol. Generally speaking, for binary codes, the additional parity bit is the total parity of the message. This is calculated as *sum modulo 2* (XOR) of all the bits of the message.

When the "natural" length of the code does not fit the application constraints (e.g. the NAND Flash page), it is possible to change it with the shortening operation: a $C[n, k]$ is *shortened* into a code $C'[n - j, k - j]$ by erasing j columns of the parity matrix. Please note that the columns deleted are the ones corresponding to the user data. With this operation, the Code Rate is decreased.

A similar operation, but with a very different outcome, is the puncturing operation. Puncturing is the process of removing some of the parity bits after encoding. This has the same effect of encoding with an error correction code with a higher rate. The good things is that the same decoder can be used regardless of how many bits have been punctured; therefore, puncturing considerably increases the flexibility of the system without significantly increasing its complexity [6].

10.2 BCH Codes

BCH codes belong to the most important class of cyclic algebraic codes. They were found through independent researches by Hocquenghem in 1959 and by Bose and Ray-Chauduri in 1960 [7, 8].

For BCH codes the minimum distance can be ensured during construction. The definition of the code itself is based on the distance concept and on the Galois field [9, 10].

Let β be an element of the Galois Field $GF(q^m)$. Let b be a non-negative integer. A BCH code with "designed" distance d is generated by the polynomial $g(x)$ of minimal degree that has $d - 1$ consecutive powers of β: $\beta^b, \beta^{b+1}, \ldots, \beta^{b+d-2}$ as roots. Given ψ_i the minimal polynomial of β^{b+i} for $0 \leq i < d - 1$, $g(x)$ is computed as:

$$g(x) = LCM\{\psi_0(x), \psi_1(x), \ldots, \psi_{d-2}(x)\} \tag{10.8}$$

and the protected data size is $k = n - deg(g(x))$.

It is possible to show that the designed d is at least $2t + 1$, hence the code is able to correct t errors.

If we assume $b = 1$, and β a primitive element of $GF(q^m)$, then the code becomes a *narrow-sense* and *primitive* BCH code of length $q^m - 1$ able to correct t errors. We shall now consider narrow-sense primitive BCH codes.

In general, decoding of a BCH code is at least 10 times more complicated than encoding. In this chapter we deal with binary BCH codes only, whose structure is presented in Fig. 10.6.

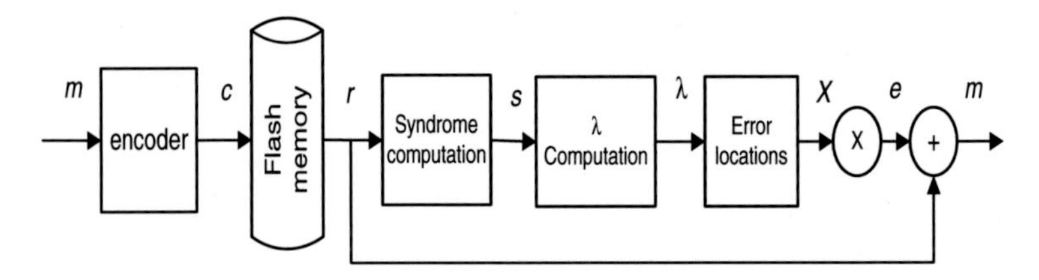

Fig. 10.6 General structure of a binary BCH code

10.2.1 BCH Encoding

Let's assume a BCH code $[n, k]$ with generator polynomial $g(x)$ and a message $m(x)$ to be encoded, which is written as a polynomial of degree $k - 1$.

First of all, the message $m(x)$ is multiplied by x^{n-k} and subsequently divided by $g(x)$, thus obtaining a quotient $q(x)$ and a remainder $r(x)$ in accordance with Eqs. (10.9) and (10.10).

$$\frac{m(x) \cdot x^{n-k}}{g(x)} = q(x) + \frac{r(x)}{g(x)} \tag{10.9}$$

$$m(x) \cdot x^{n-k} + r(x) = q(x) \cdot g(x) \tag{10.10}$$

The multiplication of the message $m(x)$ by x^{n-k} produces, as a result, a polynomial of degree $n - 1$ where the first $n - k$ coefficients, now null, will then be occupied by parity bits.

Therefore, the encoded word $c(x)$ is calculated as:

$$c(x) = m(x) \cdot x^{n-k} + r(x) \tag{10.11}$$

Practical implementation of Eq. (10.11) is depicted in Fig. 10.7. Please note that, since we are considering binary BCH codes, the sum is actually a XOR, while the product is an AND.

The "natural" structure of BCH encoding is sequential; this is not great in high-speed implementations, because it slowly proceeds by byte, word or double word. Figure 10.7b shows the unrolled implementation, assuming a processing of 1 byte at a time [4]. In the figure it is possible to see that the content of each register does not depend on a single input anymore but on a whole byte.

Sequential implementation

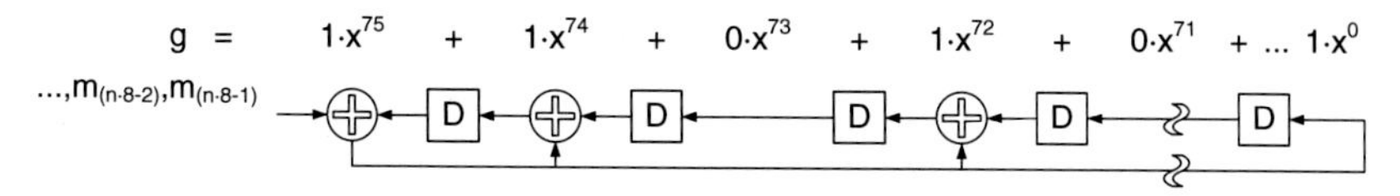

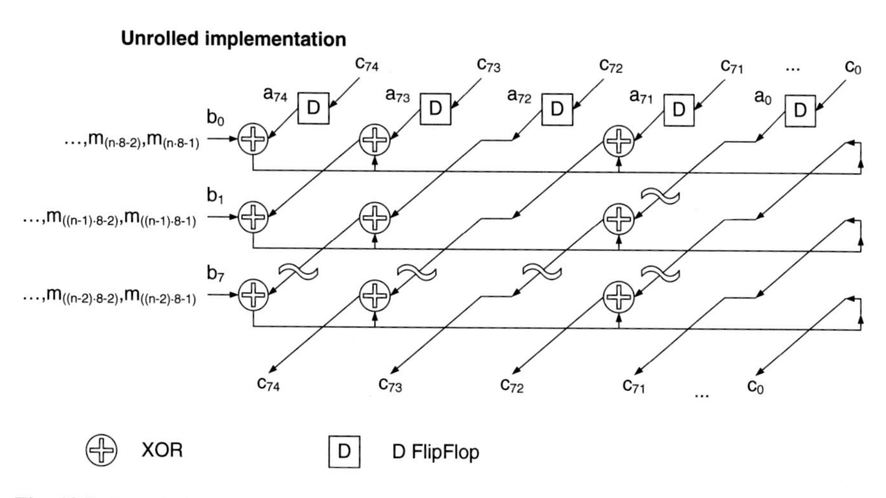

Fig. 10.7 Description of the sequential implementation of the BCH divider: it can be unrolled for a parallel implementation [4]

10.2.2 BCH Decoding

The decoding operation follows three fundamental steps, as shown in Fig. 10.6:

- calculation of the syndromes;
- calculation of the coefficients of the error locator polynomial (usually done with Berlekamp-Massey algorithm [4, 5]);
- calculation of the roots of the error locator polynomial (usually done with Chien algorithm [4, 11]).

During transmission (reading) of the encoded message some errors may occur. Errors can be represented by a polynomial that has coefficient "1" at every error's position:

$$E(x) = E_0 + E_1 x + \cdots + E_{n-1} x^{n-1} \tag{10.12}$$

If ECC can correct t errors, then t non-null coefficients are allowed in Eq. (10.12) at most.

Therefore, the transmitted (read) vector $R(x)$ is:

$$R(x) = c(x) + E(x) \tag{10.13}$$

The first decoding step consists in calculating the $2t$ syndromes for the read message:

$$\frac{R(x)}{\psi_i(x)} = Q_i(x) + \frac{S_i(x)}{\psi_i(x)} \quad \text{with} \quad 1 \leq i \leq 2t \tag{10.14}$$

$$S_i(x) = Q_i(x) \cdot \psi_i(x) + R(x) \quad \text{with} \quad 1 \leq i \leq 2t \tag{10.15}$$

In accordance with Eqs. (10.14) and (10.15), the received vector is divided by each minimal polynomial ψ_i forming the generator polynomial, thus getting a quotient $Q_i(x)$ and a remainder $S_i(x)$ called *syndrome*.

At this point the $2t$ syndromes must be evaluated into the elements β, β^2, β^3,..., β^{2t}, whose ψ_i are the minimal polynomials. According to Eq (10.16), this evaluation is the evaluation of the message received in β, β^2, β^3,..., β^{2t}, since $\psi_i(\beta^i) = 0$ (for $1 \leq i \leq 2t$) because of the definition of minimal polynomial.

$$S_i(\beta^i) = S_i = Q_i(\beta^i) \cdot \psi_i(\beta^i) + R(\beta^i) = R(\beta^i) \tag{10.16}$$

Consequently, the ith syndrome can be calculated either as a remainder of the division between the received message and the minimal polynomial ψ_i, then evaluated in β^i, or as the evaluation in β^i of the received message.

In case there aren't any errors, the received polynomial is a codeword: therefore, the remainder of the division of Eq. (10.14) is null and all the syndromes are identically null. Verifying if the syndromes are identically null is a necessary and sufficient condition to understand if the read message is a codeword (or if some errors occurred).

For binary codes we use the following property:

$$S_{2i} = S_i^2 \tag{10.17}$$

such that we can calculate only t syndromes.

Since the syndromes are computed as the remainder of the division between two polynomials in the Galois field, it is straightforward to understand that the implementation is similar to the one of the encoder.

The Error Locator Polynomial $\Lambda(x)$ is defined as the polynomial whose roots are the inverse of the error positions.

$$\Lambda(x) = \prod_{i=1}^{v} (1 - xX_i) \tag{10.18}$$

The degree of the error locator polynomial gives the number of errors that occurred. As the degree of $\Lambda(x)$ is t at most, in case of more than t errors, $\Lambda(x)$ could erroneously indicate t or less errors.

Coefficients of the error locator polynomial are linked to the syndromes by the following equations

$$\begin{pmatrix} S_{v+1} \\ S_{v+2} \\ S_{v+3} \\ \vdots \\ S_{2v-1} \end{pmatrix} = \begin{pmatrix} S_1 & S_2 & S_3 & \cdots & S_v \\ S_2 & S_3 & S_4 & \cdots & S_{v+1} \\ S_3 & S_4 & \cdots & \cdots & S_{v+2} \\ \vdots & \vdots & \vdots & & \vdots \\ S_v & S_{v+1} & S_{v+2} & \cdots & S_{2v-1} \end{pmatrix} \cdot \begin{pmatrix} \Lambda_v \\ \Lambda_{v-1} \\ \Lambda_{v-2} \\ \vdots \\ \Lambda_1 \end{pmatrix} \tag{10.19}$$

Generally speaking, the method to compute the coefficients of the error locator polynomial is the Berlekamp-Massey algorithm [4, 12].

The philosophy of the Berlekamp algorithm consists in solving the set of Eq. (10.19) in an iterative way by consecutive approximations.

After $2t$ iterations, $\Lambda(x)$ is the error locator polynomial; in the binary case it is possible to perform the Berlekamp algorithm in t iterations. In the following we describe the flow diagram for the inversion-less binary Berlekamp Massey algorithm (Fig. 10.8) [13].

First of all, we define the syndrome polynomial as

$$1 + S = 1 + S_1 z + S_2 z^2 + \cdots + S_{2t-1} z^{2t-1} \tag{10.20}$$

The initial conditions are given as follows:

$$v^{(0)} = 1 \quad k^{(0)} = 1 \quad \text{and} \quad \delta^{(2i)} = 1 \quad \text{if} \quad i<0 \tag{10.21}$$

We define $d^{(2i)}$ as the coefficient of z^{2i+1} in the product $(1 + S(z))v^{(2i)}(z)$.

- If S_{2i+1} is unknown the algorithm is finished;
- otherwise

$$v^{(2i+2)}(z) = \delta^{(2i-2)} v^{(2i)}(z) + d^{(2i)} k^{(2i)}(z) \cdot z \tag{10.22}$$

$$k^{(2i+2)}(z) = \begin{cases} z^2 k^{(2i)}(z) & \text{if} \quad d^{(2i)} = 0 \quad \text{or} \quad \text{if} \quad \deg v^{(2i)}(z) > i \\ z v^{(2i)}(z) & \text{if} \quad d^{(2i)} \neq 0 \quad \text{or} \quad \text{if} \quad \deg v^{(2i)}(z) \leq i \end{cases} \tag{10.23}$$

$$\delta^{(2i)} = \begin{cases} \delta^{(2i-2)} & \text{if} \quad d^{(2i)} = 0 \quad \text{or} \quad \text{if} \quad \deg v^{(2i)}(z) > i \\ d^{(2i)} & \text{if} \quad d^{(2i)} \neq 0 \quad \text{or} \quad \text{if} \quad \deg v^{(2i)}(z) \leq i \end{cases} \tag{10.24}$$

The roots of $v^{(2t)}(z)$ coincide with those of $\Lambda^{(2t)}(z)$.

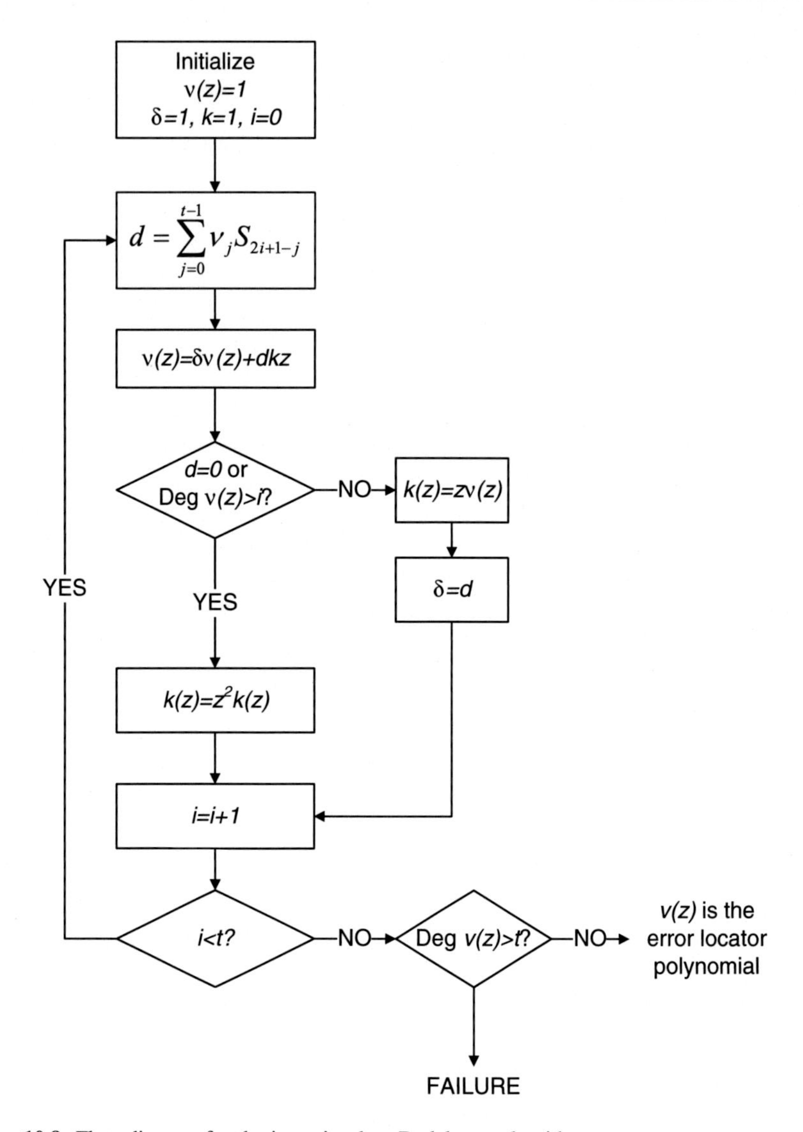

Fig. 10.8 Flow diagram for the inversion-less Berlekamp algorithm

Even if the algorithm is very complex, it usually does not require a parallel implementation, since the size of the memory buffer and the execution latency are acceptable in most of the cases.

The last step of the decoding process consists in searching for the roots of the error locator polynomial, as per Eq. (10.25). If the roots are not coincident and they belong to the Galois field, then it is enough to calculate their inverse to have the error positions. If they are coincident, or they do not belong to the correct field, it

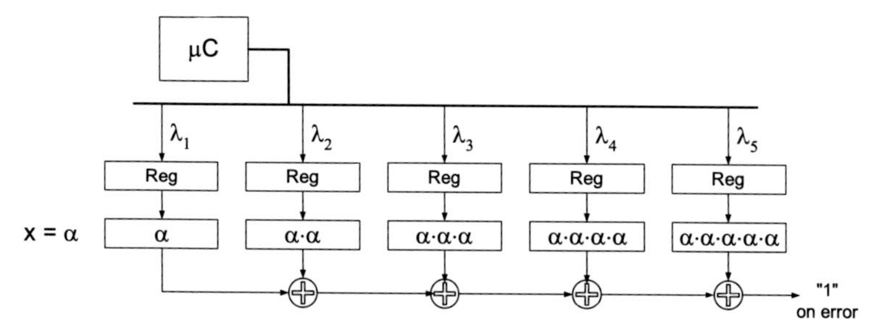

Fig. 10.9 Chien machine for a 5-error BCH: sequential implementation [4]

means that the received message has a distance from a codeword greater than t. In this case an uncorrectable error pattern occurred and the decoding process fails.

$$\Lambda(x) = 1 + \Lambda_1 x + \cdots + \Lambda_t x^t \tag{10.25}$$

To determine the roots of the polynomial, the Chien machine neatly evaluates $\Lambda(x)$ in all the elements of the field $\alpha^0, \alpha^1, \alpha^2, \alpha^3, \ldots \alpha^N$. For each element i of the field such that the polynomial is null, the corresponding position $(2^m - 1 - i)$ is an error position. A possible implementation of the Chien machine is represented in Fig. 10.9.

10.2.3 Multi-channel BCH

When BCH is used in a NAND-based system such as a Solid State Drive, it is necessary to find out a balance between area and bandwidth. In fact, SSDs run several NAND devices in parallel in order to achieve their target performances of bandwidth and IOPS. Usually, NANDs are split in groups called "Flash Channels": channels work in parallel and read/write/erase operations can be interleaved within the same channel (Fig. 10.10). In this multi-channel scenario, multiple encoding and decoding machines are necessary, considering that, especially with ultra-scaled geometries and multi-level storage (Chap. 3), correction is required all the time (because of the high RBER).

In order to keep up with the bandwidth requirements, the most straightforward solution would be one encoder and one decoder per channel. However, this approach is extremely area consuming, especially because of the decoder.

As far as the encoding is concerned, it is very important that the data coming from the host (CPU or Operating System) are dispatched to the various channels without latency. There are three possible approaches, starting from the less area consuming:

- single encoder shared among all Flash channels [14];
- a pool of encoders;
- one encoder per channel.

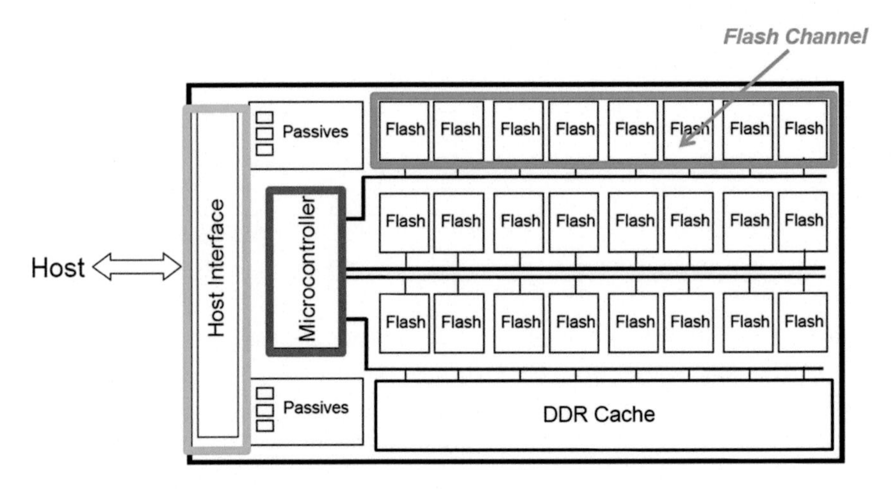

Fig. 10.10 Flash channel inside a solid state drive

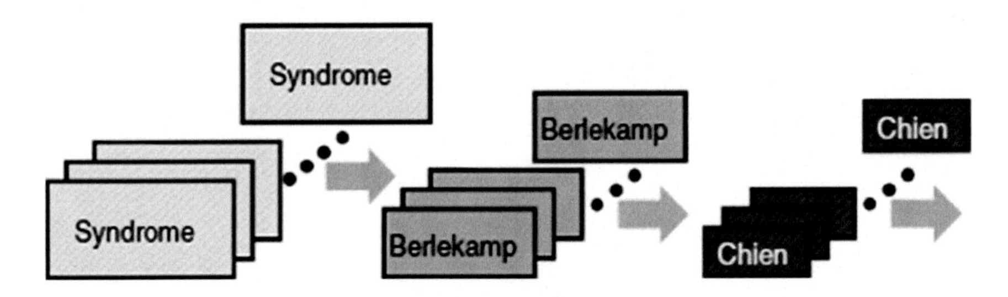

Fig. 10.11 ECC decoding structure for handling multiple channels

As mentioned, the right hardware choice comes from the tradeoff between silicon area and latency.

Let's now move to the decoding phase. The overall structure is shown in Fig. 10.11. In this scenario, number of hardware machines to execute syndrome computation, Berlekamp-Massey algorithm and Chien computation can be different.

Syndrome computation can be treated in the same way as the encoder, since all the read messages require this computation. Execution of the Berlekamp–Massey algorithm is pretty fast because it requires t iterations only.

As described in the previous section, the Chien machine searches for the roots, one at a time. Such operation, carried out for all the bits of the message, results to be very time consuming. Solution is, of course, a parallel architecture. Unlike the parity and the syndromes computation machines, which have to operate with a parallelism equal to the input data parallelism, the Chien machine does not have particular limits other than complexity, area and power consumption. In this parallel

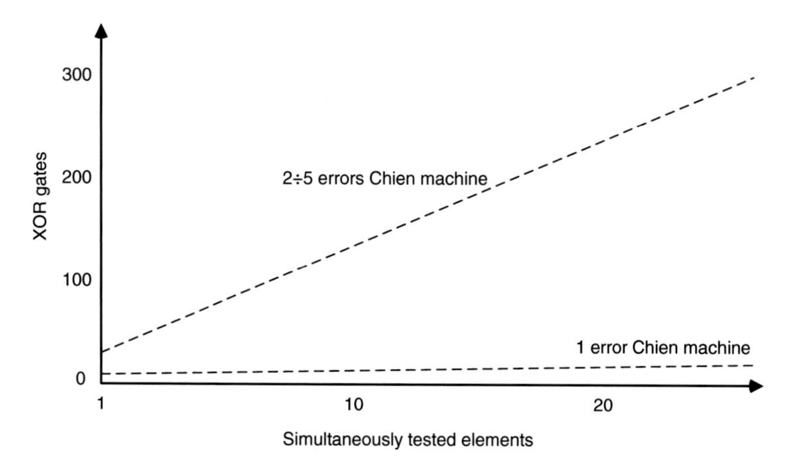

Fig. 10.12 Area impact of the Chien parallelism

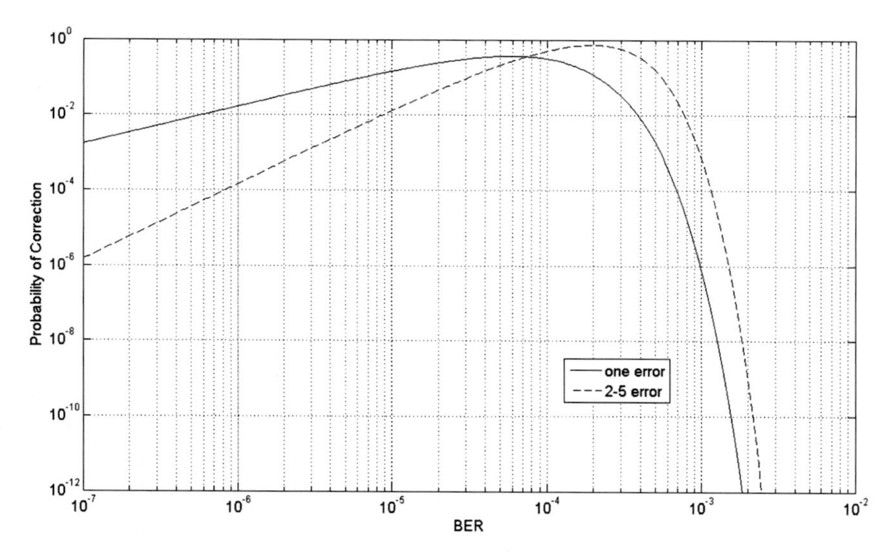

Fig. 10.13 Probability of correction for a 2112-byte page: single error versus 2 ÷ 5 errors

implementation, more error positions are contemporarily evaluated at each computation cycle.

The execution time of the Chien algorithm is usually seen by the system as an additional latency time. If the probability to have one or more errors becomes considerable, this latency can significantly impact the system performance. The downside of the Chien parallelism is the impact on silicon area, as sketched in Fig. 10.12.

Figure 10.13 shows, for a 2112-Byte page, the probability of correcting only one error and the probability of correcting 2 ÷ 5 errors. Assuming a BER of 10^{-6}, we have that the probability of a single error is equal to 1.7×10^{-2} and the probability of 2 ÷ 5 errors is equal to 1.5×10^{-4}, respectively. The probability of a single error is definitely more significant and since the Berlekamp algorithm exactly indicates the number of errors to correct, it may be useful to exploit this information.

The resulting system is, therefore, composed of a couple of Chien machines with different parallelisms, one for the correction of the single error and the second for the correction of 2 ÷ 5 errors (Fig. 10.14).

This solution can be multiplied by any number of machines, especially if the error correction capability t of the BCH we are dealing with is high. In this case, we can compute the frequency of errors t' that is more likely to occur, given the estimated raw bit error rate, and have multiple Chien machine searching for t' roots with a high parallelism. On the contrary, the number of hardware machines to locate the roots of $t^* > t'$ errors can be smaller, and with a smaller parallelism [3, 4].

Of course, the numbers mentioned above are just an example; they might significantly change depending on the NAND technology node and on the number of bits stored within the same physical cell (e.g. MLC or TLC).

10.2.4 Multi-code Rate BCH

As discussed in the introduction of this chapter, it is typical for NAND to deal with noise sources that vary during its lifetime. When NAND is fresh (i.e. few Program/Erase cycles, Chap. 2) and there is no retention, RBER can be pretty low; the situation is totally different at the end of life, i.e. when the device has been read/erased/written multiple times. It follows that it is desirable to have an ECC able to change its correction capability during life.

There are codes for which it is easy to change the code rate, while in other cases it is not that straightforward: BCH is one of them because of its construction. In this section, we present a way for building a multi-code rate BCH with a minimum area overhead.

Encoder is the main issue. As discussed above, parity bits are computed as the remainder of the division between the user data and the generator polynomial, where the latter one is computed as the multiplication among the minimum polynomial of t elements. If we want to adapt a BCH code able to correct t error to correct t' errors, where $t' < t$, the easiest way is to have a second encoder which computes the remainder of the division between the user data and the generator polynomial, where the latter one is computed as the multiplication among the minimum polynomial of t' elements. This approach has a big area overhead, because the encoding area is doubled.

A smarter and a less area consuming way is to derive the parity bits of the t'-code from the parity bits of the t-code. Indeed, for the generator polynomials the equality Eq. (10.26) holds true.

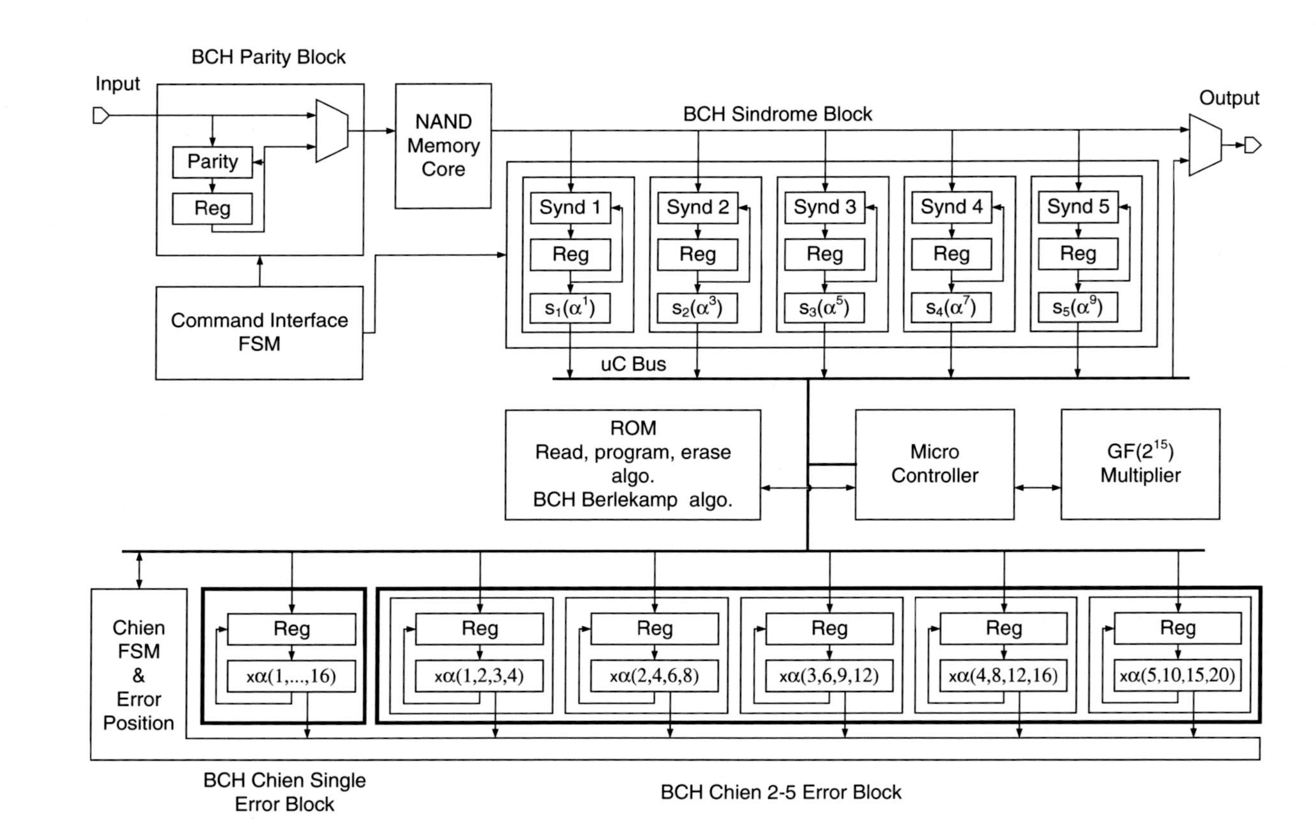

Fig. 10.14 Example of BCH engine including 2 parallel Chien hardware machines [4]

$$g(t,x) = g(t',x) * h(x) \tag{10.26}$$

Parity bits $r(x)$ are computed as remainder of the division between user data $c(x)$ and generator polynomial $g(t, x)$. From the definition of remainder we can write

$$c(x) = q(x) * g(t,x) + r(x) \tag{10.27}$$

where $q(x)$ is the quotient of the division and $deg(r(x)) < deg(g(t, x))$.

The division of $r(x)$ by $g(t', x)$ leads to

$$r(x) = q_1(x) * g(t',x) + r'(x) \tag{10.28}$$

where $q_1(x)$ is the quotient of the division and $deg(r'(x)) < deg(g(t', x))$. By substituting Eqs. (10.26) and (10.28) in Eq. (10.27) we obtain

$$\begin{aligned} c(x) &= q(x) * g(t',x) * h(x) + q_1(x) * g(t',x) + r'(x) \\ &= [q(x) * h(x) + q_1(x)] * g(t',x) + r'(x) \end{aligned} \tag{10.29}$$

It is clear that $r'(x)$ is the remainder of the division between $c(x)$ and $g(t',x)$. The circuit for a multi-code rate BCH encoder is shown in Fig. 10.15.

The overhead for this implementation is a programmable LSFR, which divides the remainder of the first division by a factor of $g(x)$. Of course, we can have more than 2 encoders and multiple programmable LSFRs. Thanks to the LSFR programmability, when NAND is fresh, we can select a BCH code with a small error correction capability, and user data are encoded with two subsequent divisions. When the NAND gets older, we can execute a single division, as the subsequent division is not required anymore.

Decoding is much easier. The syndromes are computed as different divisions by all the factors of $g(t, x)$. If we want to compute the syndromes by using the factors of $g(t', x)$, where $g(t', x)$ is a factor of $g(t, x)$, it is enough to disable the circuits that compute the last $t - t'$ syndromes.

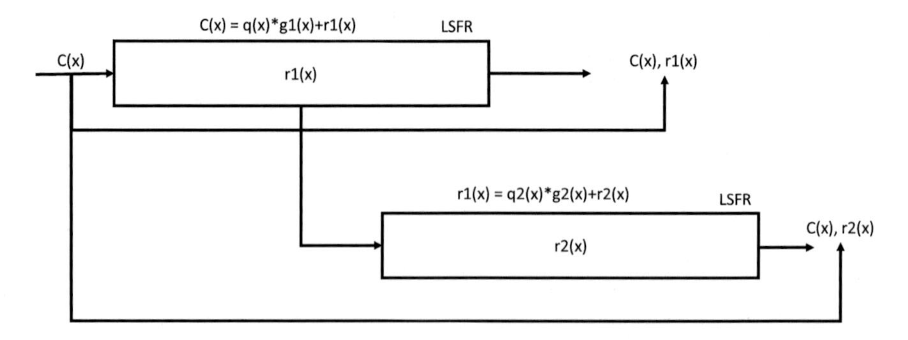

Fig. 10.15 Example of multi-code rate BCH encoder with two different generator polynomials

Berlekamp-Massey algorithm is not impacted by the multi-code rate: it completes in fewer iterations at the beginning of life, with coefficients t' instead of t.

Chien algorithm is not impacted at all. Again, it will stop after finding t' roots instead of t. However, in order to keep up with the SSD's bandwidth, a multi-Chien machine approach (Sect. 10.2.3) is likely to be implemented in a multi-code rate environment.

10.2.5 BCH Detection Properties

BCH codes are not perfect codes: for this reason it is difficult that a codeword with more than t errors moves in the correction sphere of another codeword. The codewords of BCH codes are well separated, and only a number of errors much bigger than t could partially overlap their correction spheres [3]. It follows that the erroneous corrections are made only when the received message is located in a correction sphere different from the original codeword.

Given a binary linear code C able to correct t errors, the probability of mis-correction P_{ME} is defined as the probability that an ideal bounded distance decoder executes erroneous corrections. The weighted probability $P_E(w)$ is the probability of executing erroneous corrections when w errors occur.

Please note that the probability P_{ME} depends on the code C and on the transmission channel.

Theorem 10.2.1 *The weighted probability $P_E(w)$ is computed as:*

$$P_E(w) = \frac{D_w}{\binom{n}{w}} \tag{10.30}$$

where D_w is the number of decodable words and w is in the range $[t + 1, n]$.

The number of decodable words can be computed as

$$D_w = \sum_{i=0}^{n} a_i \sum_{s=0}^{t} N(i, w; s) \tag{10.31}$$

where $N(i, w; s)$ is the number of words with weight w and distance s from a word of weight i. This is computed by Eq. (10.32)

$$N(i, w; s) = \begin{cases} \binom{n-i}{\frac{s+w-i}{2}}\binom{i}{\frac{s-w+1}{2}} & \text{if } |w - i| \le s \\ 0 & \text{if } |w - i| > s \end{cases} \tag{10.32}$$

By substituting Eq. (10.31) in Eq. (10.30) we have:

$$P_E(w) = \frac{\sum_{i=0}^{n} a_i \sum_{s=0}^{t} N(i,w;s)}{\binom{n}{w}} \tag{10.33}$$

P_{ME} is computed based on $P_E(w)$ as described in Eq. (10.34)

$$P_{ME} = \sum_{w=t+1}^{n} P_E(w)\phi(w) \tag{10.34}$$

where $\phi(w)$ is the probability that a word has weight w.

For a Binary Symmetric Channel BSC

$$P_{ME} = \sum_{w=t+1}^{n} D_w p^w (1-p)^{n-w} \tag{10.35}$$

where p is the bit error probability.

D_w can be computed according to Eq. (10.31). Unfortunately, the weights a_i are unknown for BCH codes and must be estimated.

There are a number of different theorems that can help estimating these weights for a BCH code.

Theorem 10.2.2 Peterson Estimation *The weight $a_i$of a primitive BCH code of length n and error correction capability t can be approximated as*

$$a_i \cong \frac{\binom{n}{i}}{(n+1)^t} \tag{10.36}$$

In order to have upper bounds, different correction terms are added to Eq. (10.36).

Figures 10.16 and 10.17 shows P_E and P_{ME} for BCH[16383, 15851, 77], based on the Peterson estimation. Both P_E and P_{ME} exhibit a monotonic behavior.

It follows that the real P_E and P_{ME} behaviour should be increasingly monotonic with a long floor in the middle [11].

When both the code length and the Code Rate are high, this floor can be approximated with

$$Q = 2^{-(n-k)} \sum_{s=0}^{t} \binom{n}{s} \tag{10.37}$$

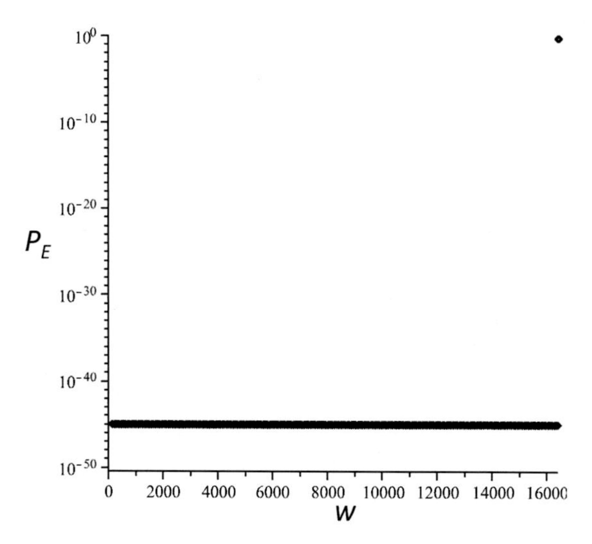

Fig. 10.16 P_E behavior for BCH[16383, 15851, 77] based on the Peterson estimation

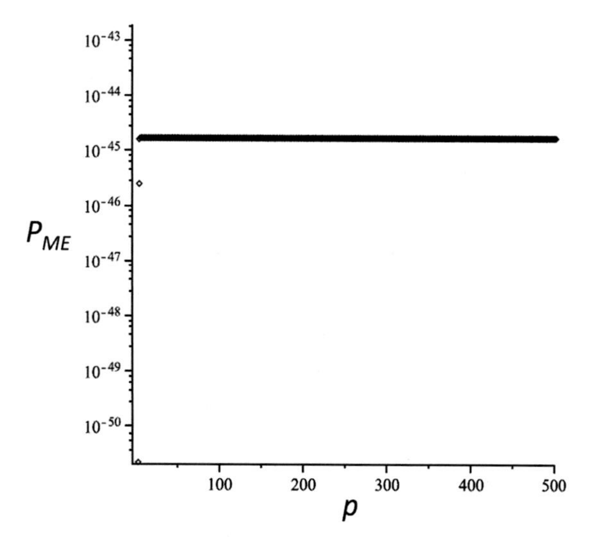

Fig. 10.17 P_{ME} for BCH [16383, 15851, 77] based on the Peterson estimation

To sum up, we can state that the BCH code has a very good detection properties for a long codeword; this feature is well suited for NAND-based systems such as SSDs. In fact, when a catastrophic error occurs or when the error correction capability of the code is overcome, in the vast majority of the cases, BCH signals a decoding failure without attempting erroneous corrections.

Of course, this behaviour becomes really key when BCH is concatenated with another code.

10.3 Low-Density Parity-Check (LDPC) Codes

Since its re-discovery in late 1990s, LDPC code has received a tremendous amount of attentions because of the excellent error-correction capability and experienced a widespread use in many real-life data communication and storage applications. In 1960s Dr. Gallager invented LDPC codes [15], in which two innovative ideas were exploited: iterative decoding and constrained random code construction.

LDPC codes are known as "capacity approaching codes"; in other words, they are a category of codes able to reach a Frame Error Rate very close to the Shannon limit. The main reason is the powerful soft decoding, as shown in Figs. 10.18 and 10.19. Figure 10.18 shows the Shannon limit for 2 BCH codes and 2 hard decoded LDPC codes. In this case, LDPC doesn't show any significant advantage, mainly because of two reasons: the use of hard instead of soft, and the adopted decoding algorithm (i.e. bit-flipping) [5]. LDPC is the clear winner in Fig. 10.19, thanks to the soft decoding. To be fair, the truth is that soft decoding pushes away the Shannon limit; a careful review of the graph reveals that soft LDPC is very close to the Hard Shannon limit, but still far from the Soft Shannon limit.

LDPC are block linear codes defined with a very sparse parity check matrix H. Each matrix can be translated into its corresponding Tanner graph, where there is a number of parity checks equal to the number of the matrix rows called "check nodes"; there is also a number of variable nodes equal to the number of matrix columns. A check node is connected to a variable node if there is a "1" in the corresponding position in the matrix H.

$$H = \begin{pmatrix} 0 & 1 & 0 & 1 & 1 & 0 & 0 & 1 \\ 1 & 1 & \textcircled{1} & 0 & 0 & \textcircled{1} & 0 & 0 \\ 0 & 0 & \textcircled{1} & 0 & 0 & \textcircled{1} & 1 & 1 \\ 1 & 0 & 0 & 1 & 1 & 0 & 1 & 0 \end{pmatrix} \quad (10.38)$$

Figure 10.20 displays the Tanner graph of the matrix described in Eq. (10.38).

Tanner Graph can have cycles; in other words, we can start from a variable node and come back to it by following different paths. The size of the smallest cycle is called *girth* of the LDPC matrix. In Fig. 10.18 the matrix has girth 4 and the cycle is shown with the bold red path; the corresponding 1s in Eq. (10.38) are highlighted with a red circle, and they are the vertices of a rectangle.

Cycles are very dangerous in LDPC decoding because it is there where the decoder can be "trapped", being unable to find a solution.

While, conceptually, the encoder is a multiplication between the transmitted data and the generator matrix G, LDPC codes can be effectively decoded by the iterative *Belief Propagation* (BP) algorithm (also known as *Sum-Product* or *SPA*). BP decoding matches the underlying code bipartite graph: decoding message is computed on each variable node and check node, and iteratively exchanged through the edges between neighboring nodes (Fig. 10.21). At the end of every iteration an estimated codeword is produced; by multiplying this temporary codeword with H,

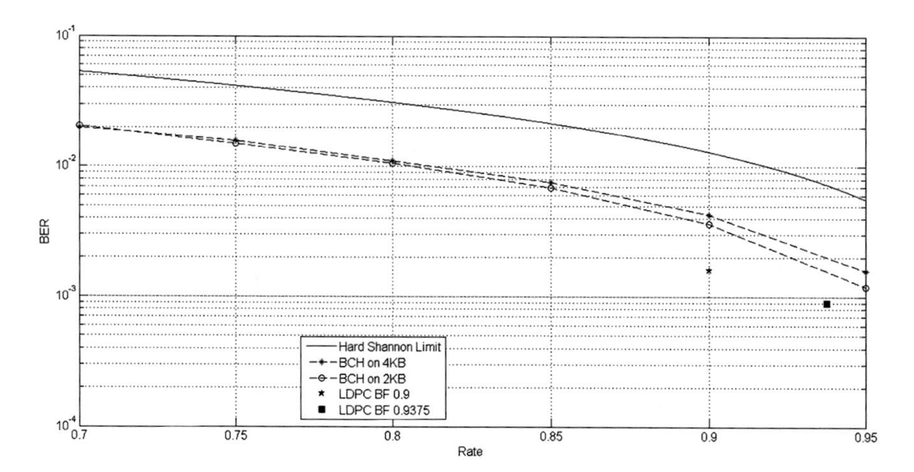

Fig. 10.18 Hard LDPC versus BCH, and hard Shannon limit

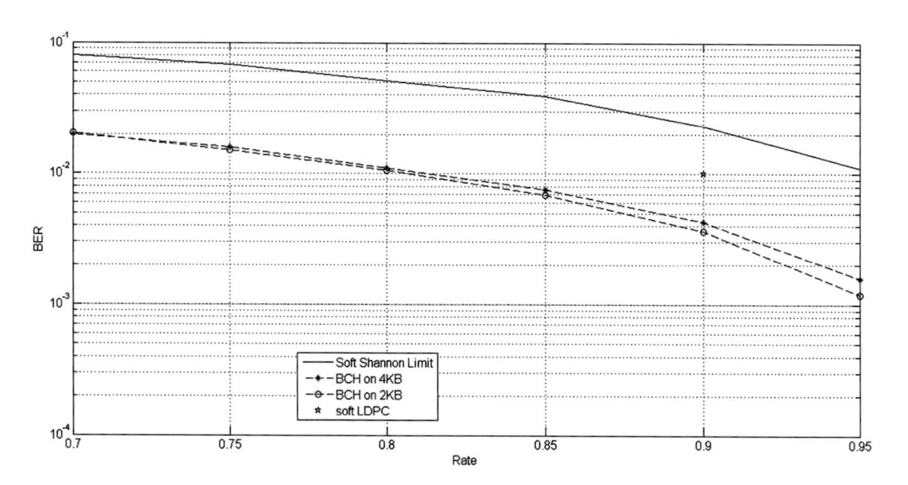

Fig. 10.19 Soft LDPC versus BCH code, and soft Shannon limit

Fig. 10.20 Tanner graph of matrix H of Eq. (10.38)

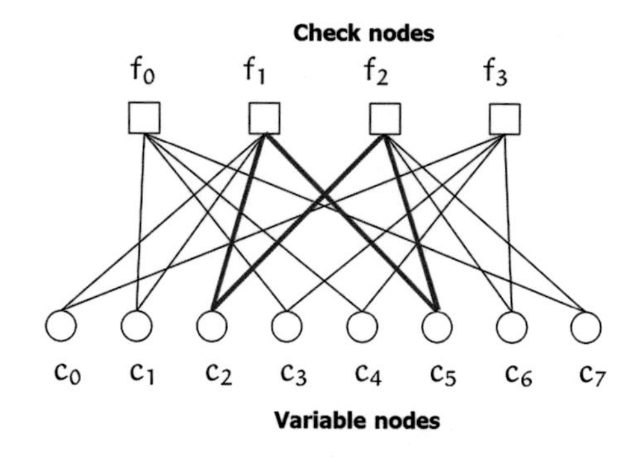

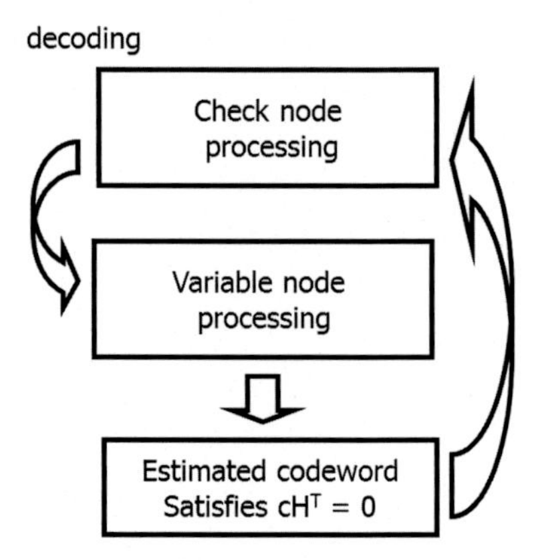

Fig. 10.21 Iterative LDPC decoding

we can check if it is a correct one. If this is the case, then decoding stops, otherwise a new iteration starts. It is well known that BP decoding algorithm works well if the underlying code bipartite graph does not contain too many short cycles. Thus, it is typically required that the graph is 4-cycle free, which is relatively easy to achieve. The construction of graphs with higher order cycle free is definitely not trivial.

There a lot of different LDPC families. The LDPC code is called (j, k)-regular if each variable node has a degree of j and each check node has a degree of k. There are also irregular codes. To be useful for Flash memories, LDPC codes must not only achieve very low decoding error rate with high Code Rates, but also be suitable for high-speed VLSI implementation, with minimal silicon and energy cost. It has been well demonstrated that *Quasi-Cyclic* (QC) LDPC codes are one family of such implementation-oriented LDPC codes. The parity check matrix of a QC-LDPC code consists of arrays of circulants. A circulant is a square matrix in which each row is the cyclic shift of the row above it, and the first row is the cyclic shift of the last row. The parity check matrix H of a QC-LDPC code can be written as

$$H = \begin{bmatrix} H_{1,1} & H_{1,2} & \cdots & H_{1,n} \\ H_{2,1} & H_{2,2} & \cdots & H_{2,n} \\ \vdots & \vdots & \ddots & \vdots \\ H_{m,1} & H_{m,2} & \cdots & H_{m,n} \end{bmatrix} \tag{10.39}$$

where each sub-matrix $H_{i,j}$ is a binary circulant. Data storage systems such as Flash demand very high Code Rates (e.g. 8/9 and higher). It has been proved that LDPCs with best performances are the irregular ones [16–18]. However, with high Code Rates, regular QC-LDPC codes are typically used, because they are easier to implement in hardware. In this case, all the rows have the same number of 1s, all the columns have the same number of 1s, and all the sub-matrices $H_{i,j}$ have the same column weight of 1 or 2. Since LDPC codes are subject to error floor, the code parity check matrix column weight is typically 4, or even higher, in order to ensure a sufficiently low error floor (e.g., error floor only occurs below the decoding failure rate of 10^{-12}) [5]. The regular and cyclic structure of QC-LDPC code parity check matrix can be leveraged to largely improve its encoder and decoder implementation efficiency as described below.

10.3.1 LDPC Codes and NAND Flash Memories

Planar TLC NAND has recently pushed for LDPC codes adoption, mainly because of the very high NAND raw BER. The complexity for tuning LDPC codes to the NAND characteristics is definitely high. The good thing is that the industry already paid the price (in terms of R&D) and today LDPC can be leveraged to foster the 3D evolution (shrink) even more.

A Read operation in the NAND environment is of a hard type by its nature. Sense Amplifiers translate cells threshold voltages into digital values, "0" or "1" (Chap. 3). This is the reason why it is not easy to extract a soft information.

In Fig. 10.22, the two V_{TH} distributions represent the two possible cell states: "0" and "1" (assuming SLC NAND). When distributions overlap, errors pop up. A hard decision decoder reads all the positive values as 0 and the negative ones as 1, so that the overlap area in the figure represents the NAND raw BER. However, A and B are very different errors, because A is a little positive, while B is far away from 0. It's like saying that B is much more likely to be an error than A. By exploiting the exact value of A and B, the decoder can have a better starting point. This is the so called *soft information* and it is measured by the *Log Likelihood Ratio* (LLR).

The LLR for a particular value x is the logarithmic ratio between the probability that the bit x was a 0 given the read value y, and the probability that the bit x was a 1 given the read value y. Given this definition, LLR can be written as:

$$L(u_i) = \log\left[\frac{P(u_i = 0|y)}{P(u_i = 1|y)}\right] \tag{10.40}$$

With NAND it's not possible to know the exact value of the threshold voltage V_{TH}. As an approximation, each overlap area is split in a number of slices, by moving the reference voltages. Figure 10.23 shows a MLC NAND where each overlap area is split in 4 slices, so that each bit (LSB and MSB) is read with 3 soft bits. The higher the number of soft bits, the more accurate the information is. This technique has a cost because each bit has to be read 3 times (in this example). Basically, soft information is asking for read oversampling.

In order to maximize the return on soft information, it is necessary to carefully understand how to move each read reference voltage, and how many times, since each additional read increases the latency.

The interaction between LDPC and NAND Flash is illustrated in Fig. 10.24.

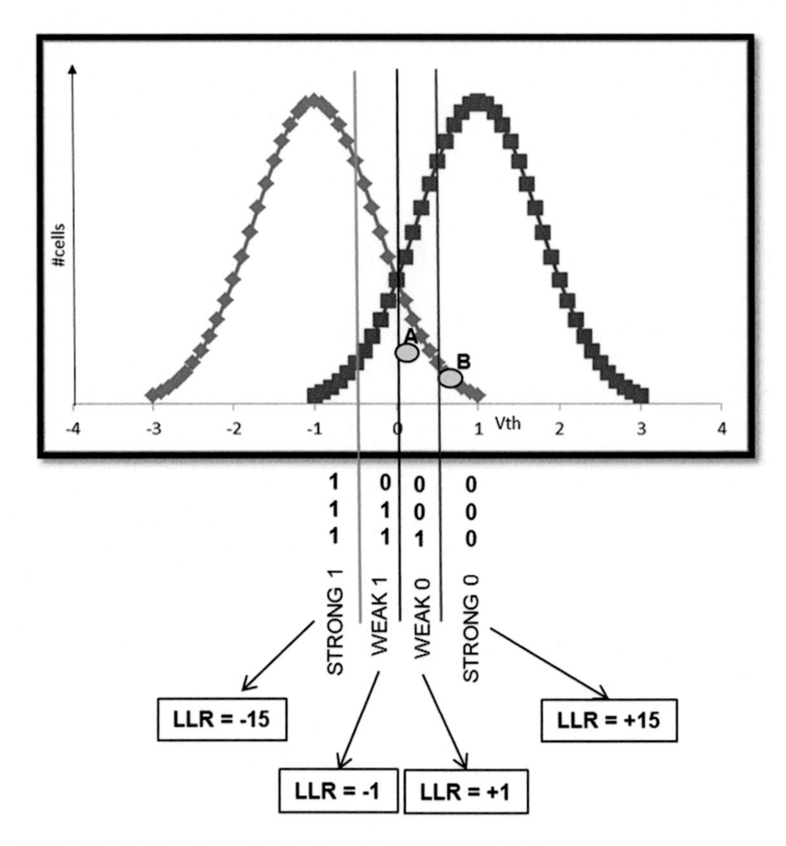

Fig. 10.22 Threshold voltage distributions in SLC NAND flash

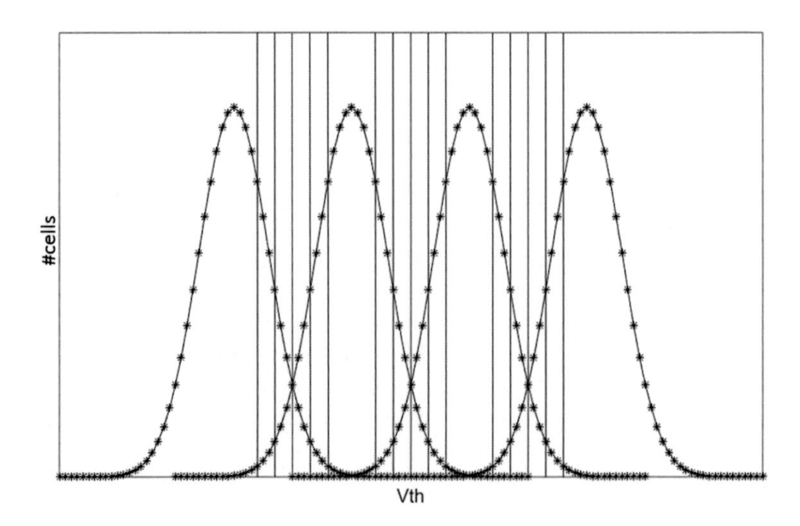

Fig. 10.23 Soft reads in MLC NAND flash

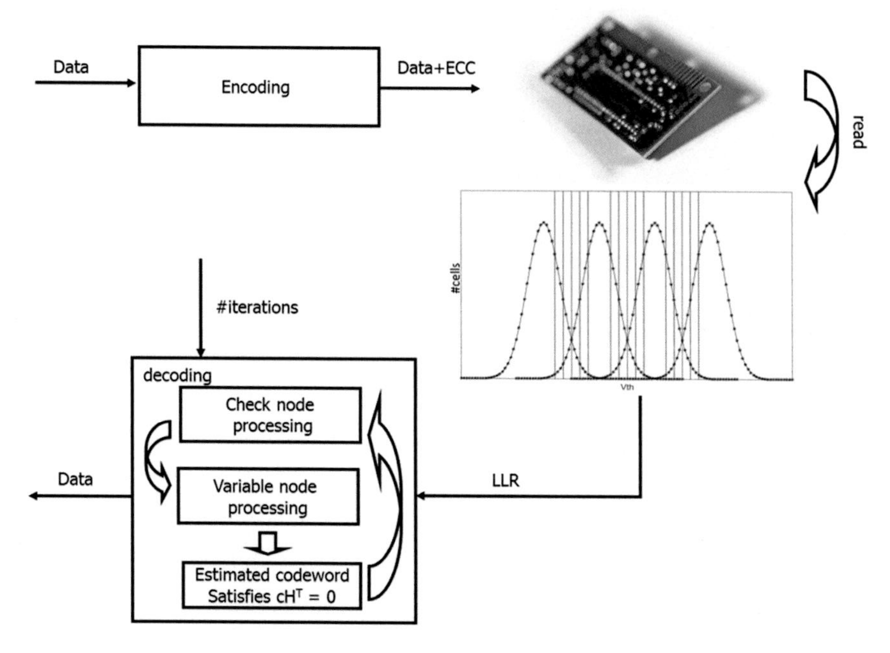

Fig. 10.24 Soft LDPC in the NAND context

10.3.2 LDPC Code Encoding

In the context of LDPC encoder design, the most straightforward approach is to multiply the information bits with the dense generator matrix derived from the sparse parity check matrix. The density of the generator matrix together with a large code length make the parallel implementation of generator matrix-vector multiplication impractical due to very high implementation complexity [19]. Hence, a partially parallel encoder implementation is a must. However, for general non-QC LDPC codes randomly constructed, their dense generator matrices may not have any structural regularity that can be used to develop efficient partially parallel encoder architecture. For QC-LDPC codes, partially parallel encoder design becomes much more affordable. Let's assume that the QC-LDPC code parity check matrix is a $m \times n$ array of circulants, and each circulant is $p \times p$. In the simplest scenario, the matrix has a full rank of $m \cdot p$. We assume that code parity check matrix can be column-wise permuted so that the following sub-array has a full rank of $m \cdot p$:

$$\begin{bmatrix} H_{1,n-m+1} & H_{1,n-m+2} & \cdots & H_{1,n} \\ H_{2,n-m+1} & H_{2,n-m+2} & \cdots & H_{2,n} \\ \vdots & \vdots & \ddots & \vdots \\ H_{m,n-m+1} & H_{m,n-m+2} & \cdots & H_{m,n} \end{bmatrix} \tag{10.41}$$

Let's also consider a systematic encoding, i.e. the first $(n - m) \cdot p$ bits in each codeword are the information bits, and the first $(n - m) \cdot p$ columns of the parity check matrix correspond to the $(n - m) \cdot p$ information bits. Hence, the corresponding generator matrix has the following form:

$$G = \begin{bmatrix} I & O & \cdots & O & G_{1,1} & G_{1,2} & \cdots & G_{1,m} \\ O & I & \cdots & O & G_{2,1} & G_{2,2} & \cdots & G_{2,m} \\ \vdots & \vdots & \ddots & \vdots & \vdots & \vdots & \ddots & \vdots \\ O & O & \cdots & I & G_{n-m,1} & G_{n-m,2} & \cdots & G_{n-m,m} \end{bmatrix} \tag{10.42}$$

where I and O represent identity $p \times p$ matrix and zero $p \times p$ matrix. Being G the generator matrix, it must satisfies $H \cdot G^T = 0$, which clearly suggests that each $G_{i,j}$ should also be a $p \times p$ circulant.

The generator matrix-vector multiplication for QC-LDPC encoding can be carried out in a partially parallel manner by leveraging the inherent cyclic structure of the generator matrix (Fig. 10.25).

If the matrix H is not full rank, the code is semi-systematic. In other words, the matrix G is shown in Eq. (10.43),

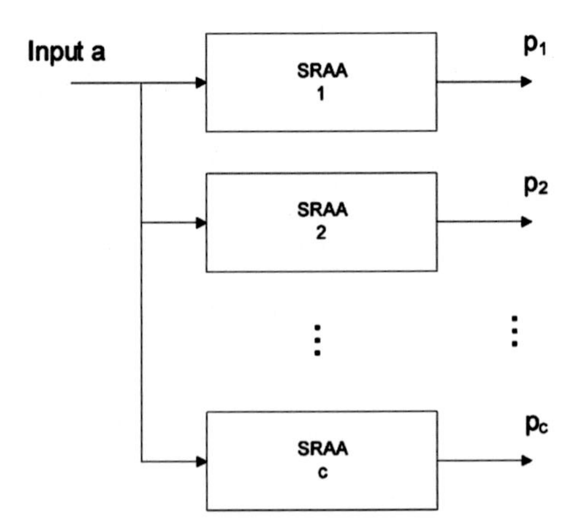

Fig. 10.25 LDPC encoding with a full-rank matrix

$$G = \begin{bmatrix} I & & & & G_{1,1} & \cdots & G_{1,z} \\ & I & & & G_{2,1} & \cdots & G_{2,z} \\ & & \ddots & & \vdots & & \vdots \\ & & & I & G_{n-z,1} & \cdots & G_{n-z,z} \\ 0 & 0 & \cdots & 0 & Q_{1,1} & \cdots & Q_{1,z} \\ \vdots & & & \vdots & \vdots & & \vdots \\ 0 & 0 & \cdots & 0 & Q_{z,1} & \cdots & Q_{z,z} \end{bmatrix} \tag{10.43}$$

where the part represented by Q is neither systematic nor regular (in size).

The hardware structure is represented in Fig. 10.26. The systematic part is equivalent to the one of the full-rank H matrix. The grey part is non-systematic and is not regular since the size of the Qs circulants is not fixed. In addition to that, it is not easy to make it parallel due to its irregularity.

During read, once decoding stops, it is necessary to multiply the non-systematic part by Q^{-1}, in order to recover the original data [20].

As discussed, a semi-systematic implementation is much more complex than a systematic one. When H is not full-rank, a possible workaround is to fix the parity section. Parity-check matrix H on the parity section is composed by a specific circulants. Those circulants can be all-zeros circulants so that matrix H won't be regular anymore. For more detailed discussions on QC-LDPC code encoder design, readers can refer to [19, 21, 22].

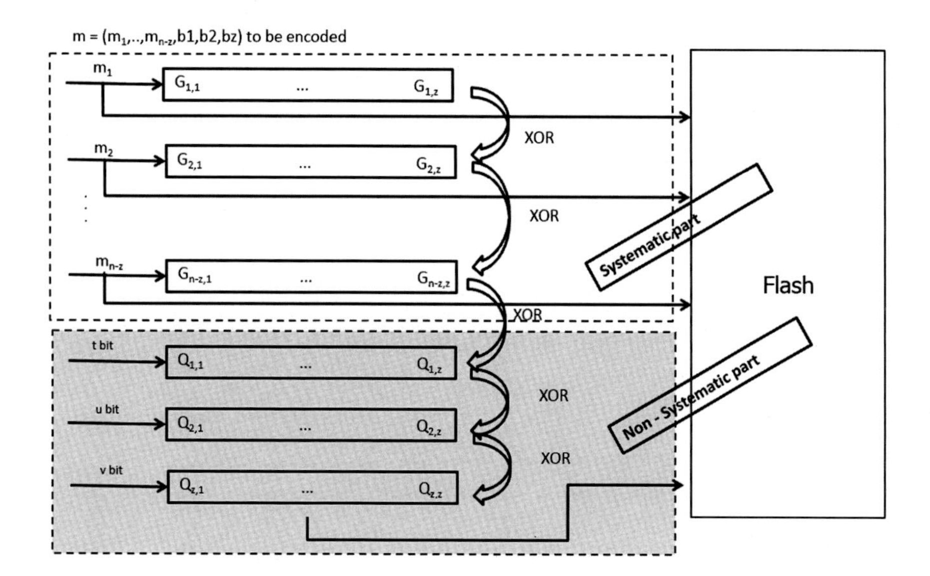

Fig. 10.26 LDPC encoding without a full-rank matrix

10.3.3 LDPC Code Decoding

To understand LDPC decoding, one of the key concepts is the extrinsic information. Here it is explained through an example [5].

We have a troop of 6 soldiers and each soldier wants to know the total number of soldiers in the troop. In Fig. 10.27 we have a linear troop. In this case, each soldier takes the number provided by the neighbour behind, he adds 1, and he transmits the result to the neighbour in front of him. Soldiers at the edges receive a 0 from the side without neighbour. For each soldier, the sum of received and transmitted numbers is equal to the total number of soldiers.

The second troop (Fig. 10.28) is a little more complex, and it requires different rules to pass the information. Each soldier takes all the numbers from his neighbours, he adds 1, and he subtracts the number passed by the neighbour he wants to send the message to. For example, the yellow soldier sends $2 + 3 + 2 + 1 - 2 = 6$ to the green soldier. Soldiers at the edges receive a 0 from the side without neighbour. The sum of the number that a soldier receives from anyone of his neighbours plus the one the soldier passes to that neighbour is equal to the total number of soldiers. This introduces the concept of *extrinsic information*. The idea is that a soldier does not pass to a neighbouring soldier any information that the

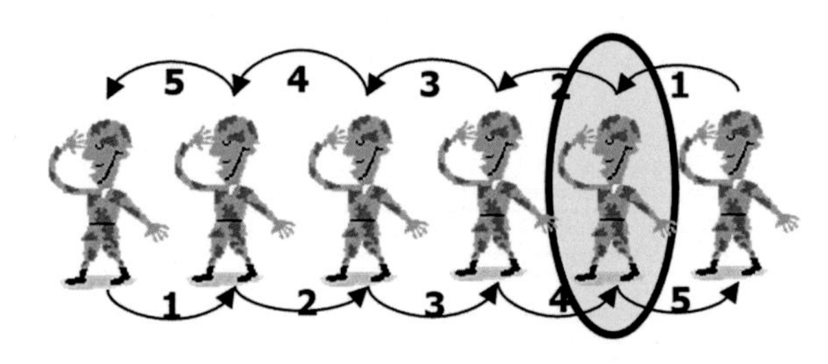

Fig. 10.27 Linear troop

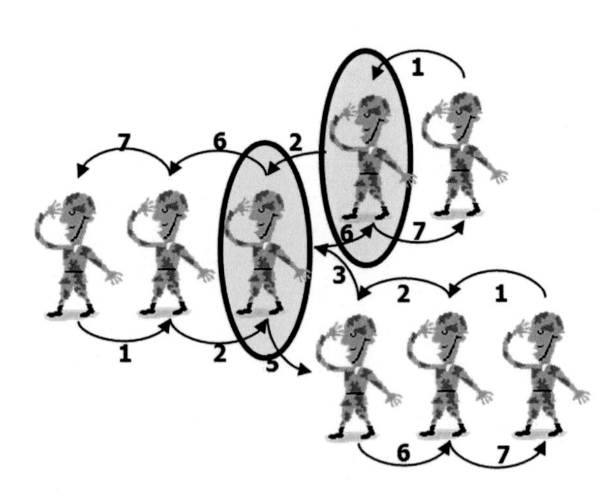

Fig. 10.28 Extrinsic information

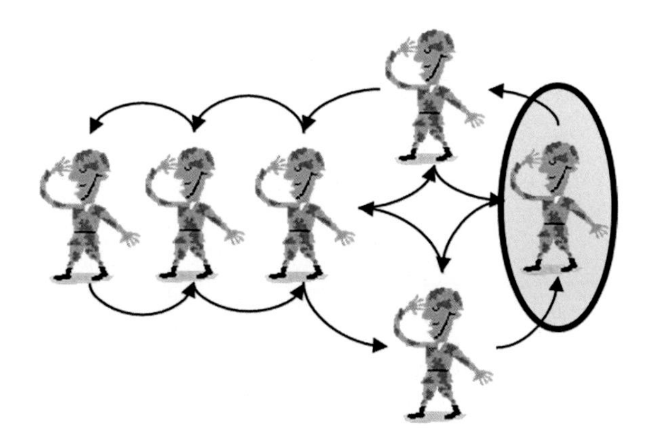

Fig. 10.29 Soldier formation containing a cycle

neighbouring soldier already has; in other words only extrinsic information is passed.

The last troop (Fig. 10.29) contains a cycle. The situation is unsolvable: no matter what counting rule one may devise, the cycle represents a type of positive feedback, both in clockwise and counter-clockwise direction, so that the messages passed within the cycle will increase without bound. This shows that the message passing on a graph cannot be claimed to be optimal if the graph contains one or more cycles. However, while most practical codes contain cycles, it is well known that message-passing decoding performs very well, assuming properly designed codes.

The key innovation behind LDPC codes is the low-density nature of the parity check matrix, which facilitates iterative decoding. Message-passing decoding refers to a collection of low-complexity decoders working in a distributed fashion to decode a received codeword, in a concatenated coded scheme. We can better understand this sentence by using the crossword-puzzle analogy (Fig. 10.30).

Solving a crossword-puzzle proceeds as follows:

- start with all the horizontal words we know → red circles;
- proceed with all the vertical words we know → blue circles;
- re-start to see if we are able to complete more horizontal words given the addition of the vertical words of the previous step → green circles;
- re-start to see if we are able to complete more vertical words → magenta circles;
- keep looping until the crossword-puzzle is completed (or a codeword is found) and stop when either we are not able to solve it (we fell in error floor) or we are too tired (we reached the maximum number of iterations).

Belief Propagation algorithm is the best iterative decoding methods for LDPC. In order to understand it, it might be useful to consider the Tanner Graph of the parity check matrix (Fig. 10.31).

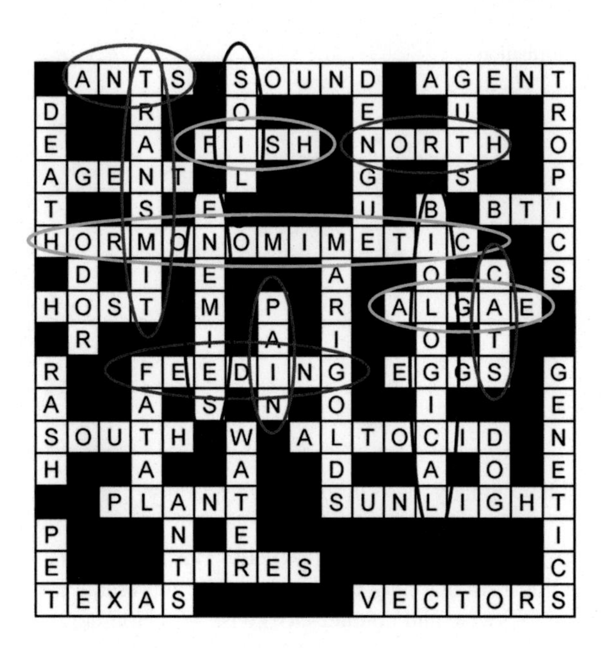

Fig. 10.30 Crossword-puzzle

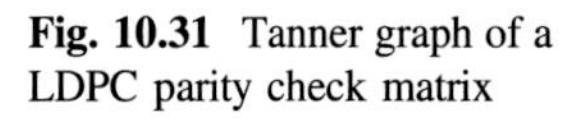

Fig. 10.31 Tanner graph of a LDPC parity check matrix

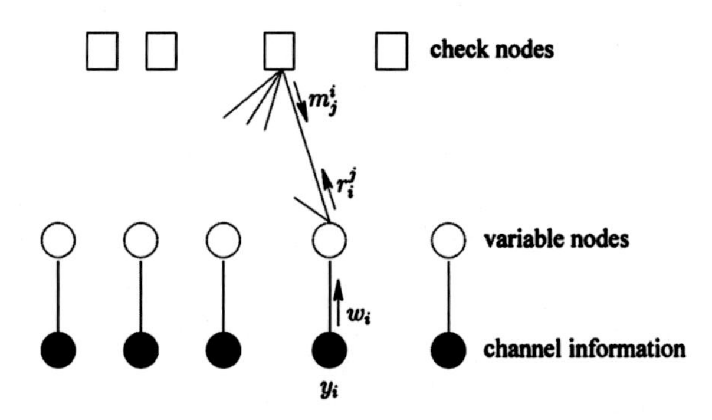

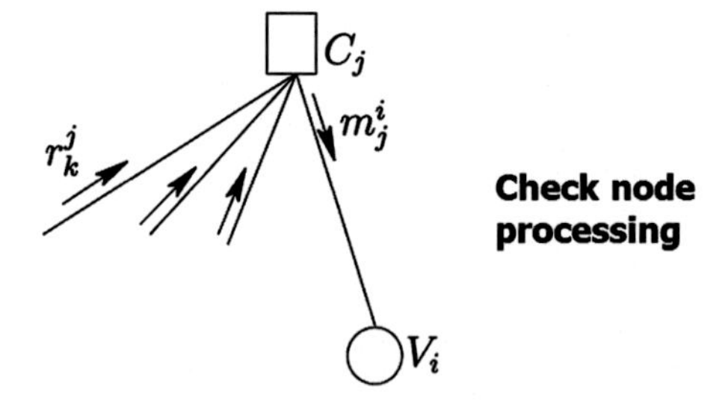

Fig. 10.32 Check node processing of LDPC BP decoding

During the check node processing phase (Fig. 10.32), each check node has to compute the values m it has to send to variable nodes it is connected to. Values are computed according to Eq. (10.44).

$$m_j^i = \prod_{k \in N(j) \backslash \{i\}} sign\left(r_k^j\right) \cdot \phi\left(\sum_{k \in N(j) \backslash \{i\}} \phi\left(\left|r_k^j\right|\right)\right) \tag{10.44}$$

$$\phi(x) = -\log\left(\tanh\left(\frac{x}{2}\right)\right) \tag{10.45}$$

Remembering the soldier example, please note that only the extrinsic information is taken into account: in fact, the value m_i is computed by using all the values sent by the variable nodes connected to that specific check node, except variable node i.

The same idea applies to variable node processing (Fig. 10.33), where value r_j is computed by using all the values sent by the check nodes connected to the variable node, except check node j. Equation (10.46) is used

$$r_i^j = w_i + \sum_{k \in N(i) \backslash \{j\}} m_k^i \tag{10.46}$$

where w are the input LLRs.

Values r represent the estimated codeword. At the end of each iteration this word is multiplied by the transpose of H to check if it is a real codeword. If the result is null, then r is a codeword and the decoding is finished, otherwise a new iteration starts.

The formula used for check node processing is a very complex one and it involves the function tanh, which is sketched in Fig. 10.34.

BP can be approximated with the so-called *min-sum* decoding algorithm: the computational complexity can be largely reduced by paying a small decoding

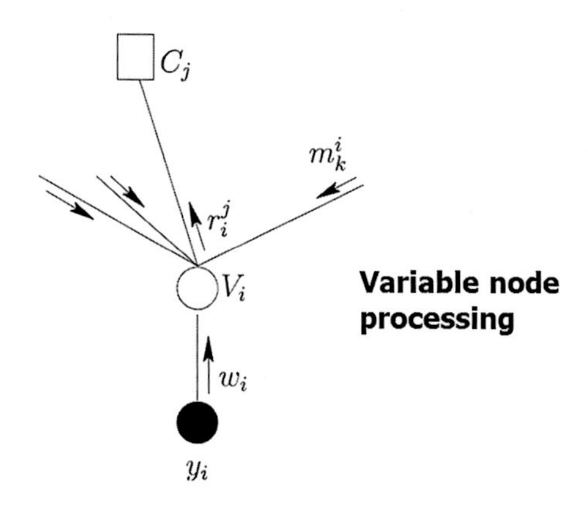

Fig. 10.33 Variable node processing of LDPC BP decoding

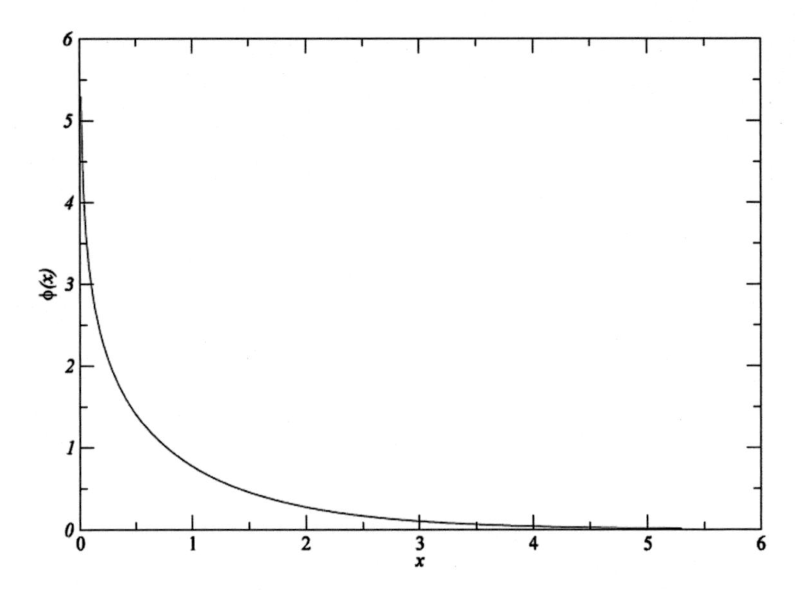

Fig. 10.34 Tanh function

performance degradation. The main difference between BP and min-sum lies in the check node: Eq. (10.44) applies to BP, while the check node processing for min-sum is described by Eq. (10.47).

$$m_j^i = \prod_{k \in N(j) \backslash \{i\}} sign\left(r_k^j\right) \cdot \min_{k \in N(j) \backslash \{i\}} \left|r_k^j\right| \tag{10.47}$$

Therefore, the function $\Phi(x)$ (i.e. tanh), which is typically implemented as LUT, is eliminated in the min-sum decoding algorithm. Min-sum can be further optimized, as described below.

Figure 10.35a shows the comparison between values computed via sum-product (SPA) and values computed via min-sum. Dots on the bisector would mean that min-sum is a great approximation of sum-product, but this is not the case; even the average has a different slope. By introducing an attenuation factor α, the approximation can be much better, as shown in Fig. 10.35b.

In other words, Eq. (10.47) can be computed as

$$m_j^i = \alpha \cdot \prod_{k \in N(j) \backslash \{i\}} sign\left(r_k^j\right) \cdot \min_{k \in N(j) \backslash \{i\}} \left|r_k^j\right| \tag{10.48}$$

The attenuation factors could change at each iteration and they must be properly studied.

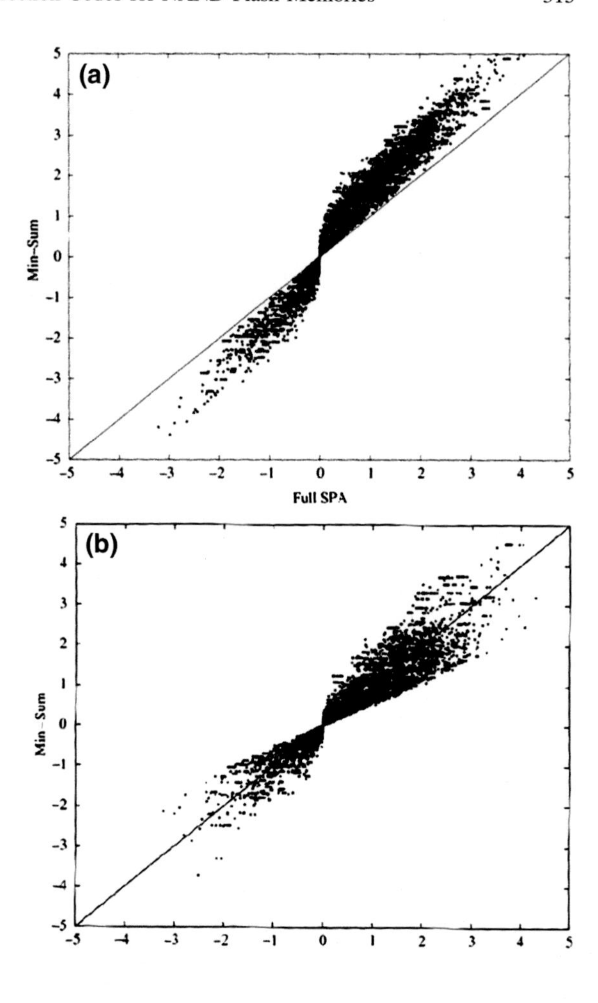

Fig. 10.35 **a** Comparison between check node variables computed with SPA and min-sum. **b** Comparison between check node variables computed with SPA and normalized min-sum [25]

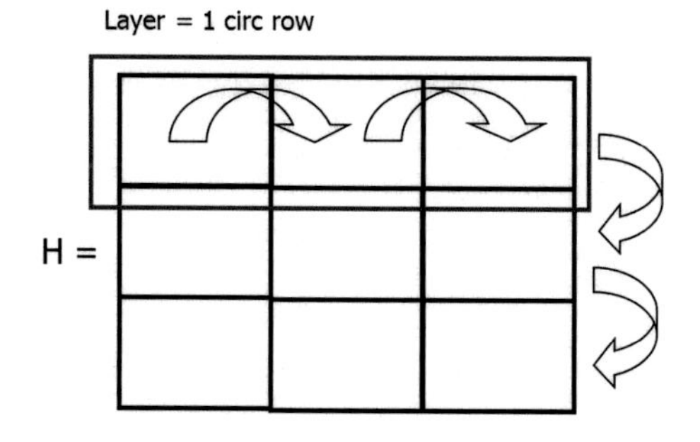

Fig. 10.36 LDPC layered decoding

Regardless of the specific decoding algorithm, the hardware implementation can be parallelized by splitting the circulants processing (for both variable and check nodes), as sketched in Fig. 10.36. This solution is known as “layered decoding”.

Taking again the crossword-puzzle analogy, in the min-sum case we first work on all horizontal words (check nodes) and only then we switch to the vertical words (variable nodes). In the layered case, once we have enough information on the horizontal words (check nodes of one circulant row), we immediately switch to the vertical words (variable nodes). In this way, the computation on the check nodes of the second layer (second circulant row) has a much cleaner input (because it doesn't use the initial variable node value but the one already computed by the first layer). Indeed, layered decoding requires much less iterations than standard min-sum.

10.3.4 QC-LDPC Applied to NAND Flash Memories

For Enterprise SSDs, target UBER is 10^{-16} (Eq. 10.3). Unfortunately, it is not possible to evaluate LDPC performances without simulations, since there aren't any closed formulas like in the BCH case.

In addition to that, LDPC decoding algorithm, because of its iterative nature, has a big drawback known as *error floor* [23, 24, 27].

Figure 10.37 shows how the error floor manifests itself: it is basically a change of the slope at low BER. With BCH it is possible to exactly predict at which BER the resulting UBER will be 10^{-16}; with LDPC we don't know at which BER the error floor will appear and its slope. The only certainty is that it will appear.

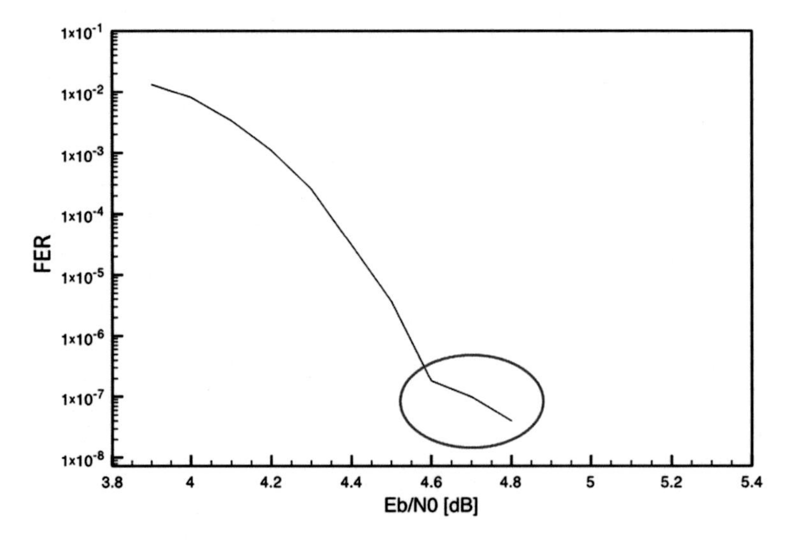

Fig. 10.37 Error floor

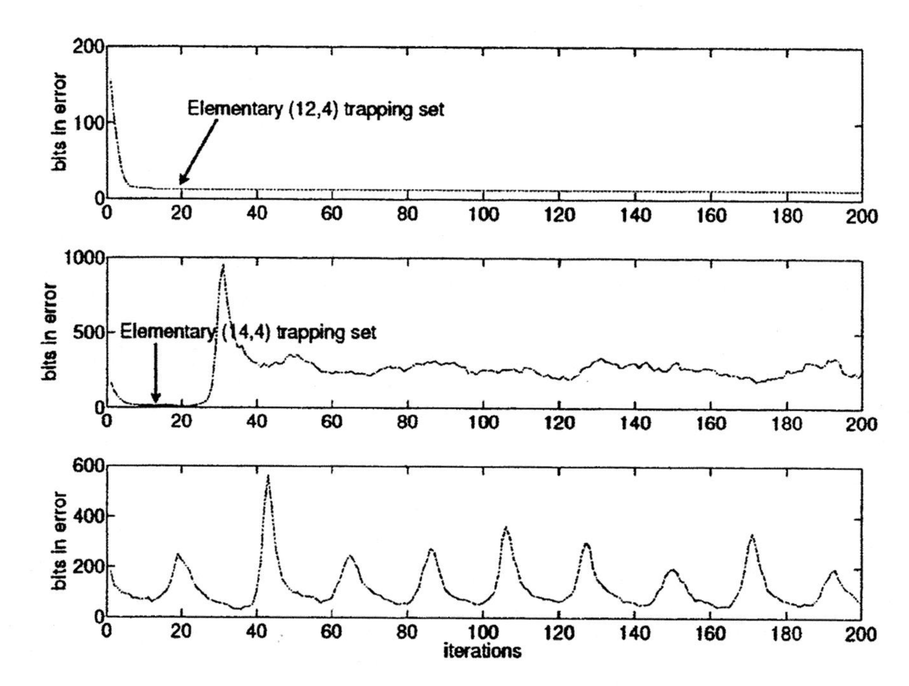

Fig. 10.38 Different kind of trapping sets [26]

It is still a mystery why error floor pops up. Nowadays, mathematicians think that it is due to *trapping sets*. Once the decoder is trapped in a trapping set, values of the variable nodes corresponding to some of the wrong bits become bigger and bigger as decoding proceeds; in other words, at some point, it becomes almost impossible for the decoder to revert its decision. The decoding will reach the maximum number of allowed iterations without finding a codeword.

Because there are 3 different types of trapping sets (Fig. 10.38), the output of the decoder might be:

- a codeword containing few constant errors;
- a codeword containing a random number of errors;
- a codeword that contains a periodical number of errors.

The last one is very dangerous because a codeword with 6 errors can have 200 errors after decoding!

Going back to the simulation topic, software simulations are not a viable solution to reach a UBER of 10^{-16}; hardware co-simulations are a must. A single FPGA can run few hundred million codewords per day, and this acceptable only if the target FER is in the range of 10^{-6}.

On the other hand, because of error floor, it is not possible to approximate the graph below FER of 10^{-6} with a simple straight line. Bottom line, enterprise applications ask for simulations of not less than 10^{13} codewords. One FPGA would

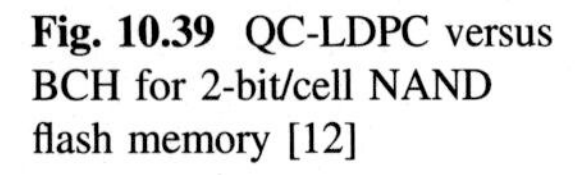

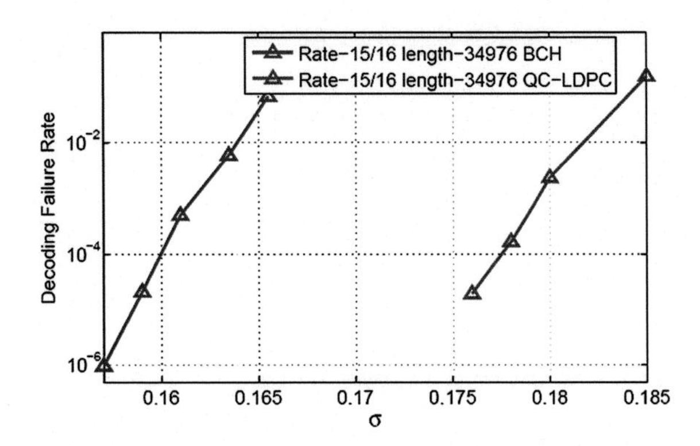

Fig. 10.39 QC-LDPC versus BCH for 2-bit/cell NAND flash memory [12]

need 100,000 days of simulations! This is why networks of FPGAs are the only practical solution to this problem [24].

It is worth highlighting that it is important to run the "correct" simulations. In fact, each parameter change requires a different simulation. For example, it is not possible to extract the soft error floor from the hard error floor. For the same reason, the min-sum decoding error floor can't be used to deduce the floor for the normalized min-sum.

Figure 10.39 shows a comparison between LDPC and BCH on AWGN channel. NAND V_{TH} distributions are modeled as two symmetric Gaussian distributions, whose mean values are $V_{TH} = -1$ and $V_{TH} = +1$, respectively. In this model the NAND raw BER is represented by the variance σ of the distributions.

In order to understand the actual performances of a specific LDPC code, it is fundamental to make simulations based on data extracted from silicon.

Data read from NAND Flash memories are always either a 0 or a 1, as already explained in Sect. 10.3.2. Therefore, the starting point is always hard decoding; if it fails, soft decoding takes over and we need to:

- Re-Read in order to get reliability info for each single bit;
- Map each bit to a LLR value;
- run soft simulations.

Re-Read strategy is described in Sect. 10.3.2: basically, the read reference voltage is shifted, and one or more additional Read operations are performed to understand where bits are located within the voltage distribution.

Table 10.1 Example of LLR values for soft decoding

Value read from NAND flash		
1st read	2nd read (re-read)	LLR
0	0	+7
0	1	+1
1	0	−1
1	1	−7

Each Re-Read operation returns a sequence of 0s and 1s, which can be coupled to the sequence of the previous Read, as shown in Table 10.1.

The LLR sign indicates whether the bit of the 1st Read is more likely to be a 0 or a 1; the magnitude indicates the confidence level associated to the 1st Read. Let's look at a couple of examples: "+1" indicates that we have read a 0 but we are not that confident, while "+7" indicates that we have read a 0 and we are pretty sure about this bit to be correct.

Once each bit of the transmitted message has been mapped to an LLR value, this value is the input for soft decoding simulations, which are used to build curves like the one shown in Fig. 10.19.

To sum up, despite all the challenges related to error floor and soft information, LDPC can successfully be utilized to boost ECC performances, and it is definitely the most promising solution for 3D NAND Flash memories, especially when looking at TLC and QLC storage.

References

1. C.E. Shannon, A mathematical theory of communication. Bell Syst. Tech. J. **27**(379–423), 623–656 (1948)
2. C. Berrou, A. Glavieux, P. Thitimajshima, Near Shannon limit error-correcting coding and decoding: Turbo-codes, in *Proceedings of ICC'93*, Geneve, Switzerland, May 1993, pp. 1064–1070
3. R. Micheloni, A. Marelli, K. Eshghi, *Inside Solid State Drives (SSD)* (Springer, Berlin, 2012)
4. R. Micheloni, A. Marelli, R. Ravasio, *Error Correction Codes for Non-Volatile Memories* (Springer, Berlin, 2008)
5. S. Lin, D.J. Costello, *Error Control Coding* (Prentice Hall, Upper Saddle River, 2004)
6. T.K. Moon, *Error Correcting Coding—Mathematical Methods and Algorithm* (Wiley, NJ, 2005)
7. R.C. Bose, D.K. Ray-Chaudhuri, On a class of error correcting binary group codes. Inf. Control **3**, 68-79 (1960)
8. A. Hocquengheim, Codes Correcteurs d'erreurs. Chiffres 2, Sept 1959
9. M.A. Pellegrini, The (2, 3)-generation of the classical simple groups of dimensions 6 and 7. Bull. Aust. Math. Soc. **93**(1), 61–72 (2016)
10. M.A. Pellegrini, M.C. T. Bellani, The simple classical groups of dimension less than 6 which are (2, 3)-generated. J. Algebra Appl. **14**(10), 1550148 (2015) (15p)
11. M. Kim et al. Decoder error probability of binary linear block codes and its application to binary primitive BCH codes. IEICE Trans. Fundam. (1996)
12. R. Micheloni, L. Crippa, A. Marelli, *Inside NAND Flash Memories* (Springer, Berlin, 2010)
13. E.R. Berlekamp, *Algebraic Coding Theory* (McGraw Hill, New York, 1968)
14. H.O. Burton, Inversionless decoding of binary BCH codes. IEEE Trans. Inf. Theory **17** (1971)
15. Y. Lee, H. Yoo, I. Yoo, I.C. Park, 6.4 Gb/s multi-threaded BCH encoder and decoder for multichannel SSD controllers, in *ISCC Digest of Technical Papers* (2012)
16. R.G. Gallager, Low-density parity-check codes. IRE Trans. Inf. Theory **IT-8**, 21–28 (1962)
17. V. Zyablov, M. Pinsker, Estimation of the error-correction complexity of Gallager low-density codes. Problemy Peredachi Informatsii **11**, 23–26 (1975)
18. R.M. Tanner, A recursive approach to low complexity codes. IEEE Trans. Inf. Theory **IT-27** (5), 533–547 (1981)

19. G.A. Margulis, Explicit constructions of graphs without short cycles and low density codes. Combinatorica **2**(1), 71–78 (1982)
20. Z. Li et al., Efficient encoding of quasi-cyclic low-density parity check codes. IEEE Trans. Commun. (2006)
21. S. Myung et al., Quasi-cyclic LDPC codes for fast encoding. IEEE Inf. Theory June (2005)
22. Z. Li, L. Chen, S. Lin, W. Fong, P.-S. Yeh, Efficient encoding of quasi-cyclic low-density parity-check codes. IEEE Trans. Commun. **54**, 71–81 (2006)
23. T. Richardson, Error floors of LDPC codes, in *Proceedings of the 41st Annual Allerton Conference on Communication*, USA (2003)
24. R. Micheloni et al., Hardware/software co-simulation for error-floor detection in LDPC, in *Proceedings of Flash Memory Summit*, www.flashmemorysummit.com, Santa Clara, CA, USA, 5–7 Aug 2014
25. M. Fossorier et al., Channel Coding: Theory, Algorithms and Application (Wiley, NJ, 2005)
26. S. Landner, O. Milenkovic, Algorithmic and combinatorial analysis of trapping sets in structured LDPC codes, in *International Conference on Wireless Networks* (2005)
27. L. Dolecek et al., Predicting error floors of structured LDPC codes: deterministic bounds and estimates. IEEE J. Sel. Areas Commun. (2009)

Chapter 11
Advanced Algebraic and Graph-Based ECC Schemes for Modern NVMs

Frederic Sala, Clayton Schoeny and Lara Dolecek

In this chapter, we discuss advanced error-correcting code techniques. In particular, we focus on two complementary strategies, asymmetric algebraic codes and non-binary low-density parity-check (LDPC) codes. Both of these techniques are inspired by traditional coding theory; however, in both cases, we depart from classical approaches and develop new concepts specifically designed to take advantage of inherent channel characteristics that describe non-volatile memories.

We focus, in particular, on modern flash devices, including multi-level and 3D flash technology. Flash is a phenomenally popular technology; the attention it has received has led to numerous process innovations. As a result, current implementations of flash, such as 3D flash, contain vast numbers of tightly-packed transistors. Flash cells suffer from a variety of physical issues, including interference/crosstalk (stronger in certain dimensions compared to others due to the packing design parameters in 3D flash), read and write disturbs, charge leakage, and many others. These complex effects are poorly modeled by traditional channels and the resulting errors are not well handled by traditional coding schemes; we must look towards novel approaches. We select two distinct, opposite points of attack. The first is improving on classical algebraic codes, which offer known, efficient encoding and decoding algorithms and are suitable for inexpensive, efficient devices with mild error tolerance requirements. The second is improving on cutting-edge non-binary LDPC codes, which have among the very best error-correcting ability of all known coding schemes, at the cost of more complex encoding and decoding circuitry. Additionally, new algebraic codes are particularly suitable for hard-read channels

F. Sala · C. Schoeny · L. Dolecek (✉)
Electrical Engineering Department, UCLA, Los Angeles, CA 90095, USA
e-mail: dolecek@ee.ucla.edu

F. Sala
e-mail: fredsala@ucla.edu

C. Schoeny
e-mail: cschoeny@ucla.edu

R. Micheloni (ed.), *3D Flash Memories*, DOI 10.1007/978-94-017-7512-0_11

whereas LDPC codes are most beneficial for soft-read channels. The two coding approaches thus target the opposite ends of the flash quality/cost tradeoff curve.

In the case of algebraic codes, we discuss a set of code constructions which rely on traditional symmetric codes, such as BCH codes, as building blocks. The final result is a family of codes specifically tailored towards asymmetric channels, such as the triple-level cell (TLC) flash data storage channel, which can be deployed in both 2D and 3D flash. We introduce a variation of these codes which can handle a type of very specific flash errors, along with codes suitable for the dynamic thresholding scheme, which is effective for non-volatile memories. For a subset of our techniques, we quantify the offered improvements on data sets measured from real flash devices.

In the case of LDPC codes, we present design and optimization techniques that result in non-binary LDPC codes with lowered error floors. The error floor is an effect which reduces the improvement in the output error rate of iteratively-decoded LDPC codes as the input SNR increases; this effect occurs for high SNRs, limiting the applicability of LDPC codes for high-reliability applications, such as flash. In order to resolve this problem, we identify certain subgraph objects, called absorbing sets, which occur in the Tanner graph structure of the LDPC code and contribute to the error floor. We characterize these objects for the non-binary LDPC case and present an algorithm to remove the smallest absorbing sets. Here too, the resulting code construction is tailored for an asymmetric channel. The power of the technique is illustrated for a series of non-binary LDPC codes, including the practical quasi-cyclic (QC-LPDC) codes.

11.1 Asymmetric Algebraic ECCs

One of the most interesting features of real-life memory channels is their asymmetry; that is, the fact that not all errors in such channels occur with equal probability. For example, the channel induced by a multiple-level flash storage device has a vastly higher chance of inducing an error between the erased state and a non-erased state compared to that of errors between two non-erased states.

Traditional coding theory largely does not concern itself with these asymmetries. The binary symmetric (BSC) and binary erasure (BEC) channels are the most frequently studied discrete channels while the additive white Gaussian noise (AWGN) channel is the most used continuous channel. None of these channels model asymmetries beyond the particular channel parameters. As a result, in order to apply tools from traditional coding theory to real-life situations, a symmetric channel is selected based on the worst-case error. This conservative approach allows for a safe margin.

On the other hand, such approaches are also wasteful for asymmetric channels, since a large amount of the strength of the code is then applied towards correcting errors which are rare. This unneeded strength results in a lower-than-necessary code rate, wasting energy or storage capacity. Conversely, if the code rate is kept

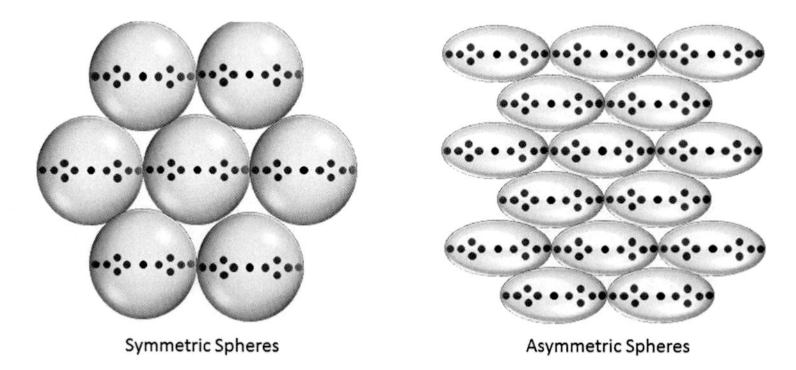

Fig. 11.1 The *left diagram* represents a packing of traditional Hamming spheres that are agnostic to the error distribution. The *right diagram* represents a packing of spheres that are designed with the error asymmetry in mind. The *black dot* at the center of each sphere represents the codeword, and the *red dots* represent the most likely erroneously received words. By targeting the specific error distribution, we can pack more asymmetric spheres than symmetric spheres, which translates into a higher code rate. Note that this is a simplified illustration of n-dimensional spheres

constant, it would be more effective to place the code's power into correcting frequent errors, thus improving the overall error probability of the system. This concept of coding for asymmetries is illustrated in Fig. 11.1.

In the remainder of this section, we discuss asymmetric error-correcting codes. We formalize the intuition presented in the previous discussion. As described earlier, we focus especially on the case of data storage in flash. In fact, data sets collected from production flash devices are available. Since encoders and decoders for algebraic codes are easy to specify and implement, we can test our proposed codes directly on the real data (rather than perform simulations using synthetic data).

11.1.1 Graded-Bit-Error Correcting Codes

We begin by considering the TLC (triple-level cell) flash channel, which, until very recently, was the most advanced and dense flash technology. Despite the name given to these devices, each cell has eight possible charge levels and thus represents three bits of information. The organization of flash devices places each of these three bits on a different page; pages are themselves collected as blocks, which are further organized into planes [1].

This organization allows us to model the TLC flash channel in two natural ways. First, looking at each cell separately, we may view the cell as an 8-ary channel, as each cell has eight possible states. Secondly, we may view each bit separately, since these bits are placed on different pages. In this case, the cell can be modeled as three independent binary channels.

In the case of the 8-ary channel, we may apply non-binary codes. We can use the statistics of the 8-ary channel to estimate the number of errors expected in a block of cells. Using this information and a target error rate, we can select an appropriate code, such, as for example, a code from the 8-ary BCH code family. Similarly, if we view the cell as three independent binary channels, we can select three binary codes, such as three binary BCH codes, based on the error probability of each channel.

It turns out, however, that neither of these approaches is suitable. An $[n,k,t]_8$ BCH code (t-error-correcting 8-ary code of length n and dimension k, e.g., containing 8^k codewords) corrects any t 8-ary errors. For example, an error between states 2 and 6 (a $2 \rightarrow 6$ error) can be corrected just as well as a $1 \rightarrow 2$ error. However, our channel produces vastly more $1 \rightarrow 2$ errors. In particular, most errors are only in one bit of the three-bit binary representation of each 8-ary state.

To illustrate this idea, we present the most frequent errors that occurred on a TLC flash chip over 5000 program/erase (P/E) cycles of operation. Comparing the programmed and errored state for the most frequent errors indeed confirms that most errors occur only in a single bit of the three-bit triplet (Table 11.1).

Table 11.1 Most frequent errors measured in a TLC flash device (3-bit cells)

Programmed state	Errored state	Fraction of errors
000	010	0.2467
000	001	0.2444
111	101	0.0820
111	110	0.0807
000	100	0.0669
011	001	0.0556
100	110	0.0550
011	010	0.0547
100	101	0.0540
111	011	0.0217

The left column gives the intended, programmed state of the cell; the middle column shows the result, which contains an error. The right column gives the fraction of total errors caused. Note that each of these 10 most popular errors contain only one bit in error

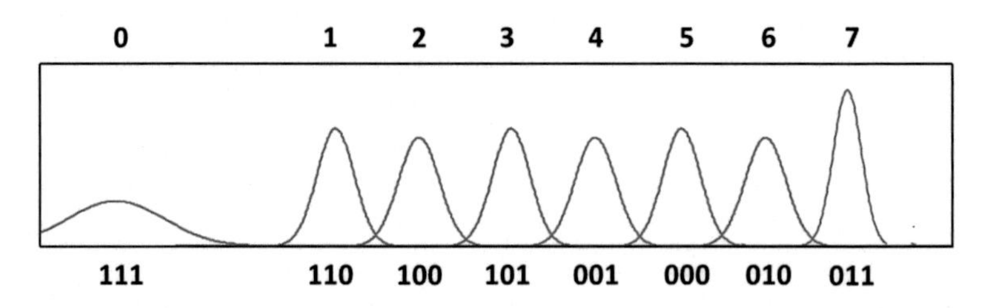

Fig. 11.2 Distributions for different voltage levels in a TLC (3-bit/8-state) cell. The 3-bit representations rely on a Gray code

The reason for this is explained by Fig. 11.2: the 3-bit binary representation of the levels is based on a Gray code, so that going from one consecutive state to the next only changes a single bit. We conclude that the ability to correct many $2 \rightarrow 6$ errors is an inefficiency in the code.

In the binary case, there is a similar problem. As can be predicted from the fact that the three binary channels are really all operating in a single cell, the channels are not independent. In this case, assuming independence underestimates the number of errors where more than one of the bits is in error. That is, errors such as $\mathbf{e} = (1,1,0)$ and $\mathbf{e}' = (1,0,1)$ (here each non-zero value in a triplet represents an error in one of the three bits) are under-represented. In fact, in our TLC device, we measured that the fraction of errors that have 2 bits in error is 0.0314 and the fraction with 3 bits in error is 0.0069. These quantities are far too large to have been produced if the probability of error among each of the pages was independent. Nevertheless, the separate binary code approach is a more accurate model compared to the 8-ary channel.

How can we design a code tailored to specifically deal with such error patterns? First, as shown in the table above, we can profile the channel to discover how many errors are typically single-bit errors and how many are multiple-bit errors. We then seek to introduce a code which corrects errors with precisely these ratios. We detail this notion in the following.

Definition 1 Let $t,v > 0$. Then, a vector $\mathbf{e} = (\mathbf{e}_1, \mathbf{e}_2, \ldots, \mathbf{e}_n)$ over $(GF(2)^m)^n$ is called a [t;v]-**bit error vector** if it satisfies the following two properties:

1. $wt(\mathbf{e}) = |\{i : \mathbf{e}_i \neq 0\}| \leq t$, and
2. for all i, $wt(\mathbf{e}_i) \leq v$.

Definition 2 Let $0 < v_1 < v_2 \leq m$ and $t_1, t_2 > 0$. A vector $\mathbf{e} = (\mathbf{e}_1, \mathbf{e}_2, \ldots, \mathbf{e}_n)$ over $(GF(2)^m)^n$ is a $[t_1, t_2; v_1, v_2]$-**graded bit error vector** if it satisfies the following properties:

1. $wt(\mathbf{e}) = |\{i : \mathbf{e}_i \neq 0\}| \leq t_1 + t_2$,
2. for all i, $wt(\mathbf{e}_i) \leq v_2$, and
3. $|\{i : wt(\mathbf{e}_i) > v_1\}| \leq t_2$.

In the previous definitions, wt() refers to the Hamming weight of the vector (the number of nonzero components in this vector). The basic idea is to introduce a more refined version of the usual definition of error vectors. Rather than simply counting the number of nonzero components, we classify them according to how many bits are in error as well. The first definition is particularly suitable for the case where all errors involve only a small number of bits. The second definition is more flexible: it enables us to profile errors as involving some number of errors in few bits and a (normally smaller) quantity of errors in a larger number of bits. Next, we define codes which are capable of correcting such error patterns:

Definition 3 Let $v, t > 0$. Then, a code C is a [t;v]-**bit error correcting code** if it is capable of correcting every [t;v]-graded bit error vector.

Definition 4 Let $0 < v_1 < v_2 \leq m$ and $t_1, t_2 > 0$. Then, a code C is a $[t_1, t_2; v_1, v_2]$-**graded-bit error correcting code** if it is capable of correcting every $[t_1, t_2; v_1, v_2]$-graded bit error vector.

To see how these definitions work (and how they apply to our asymmetric TLC flash channel), consider the following example. We store vectors of length n, where each element is a 3-bit vector. For the sake of this example, we take $n = 7$. Say we store the vector

$$\mathbf{x} = (000\ \ 110\ \ 010\ \ 101\ \ 000\ \ 111\ \ 000).$$

After some time, we read back the stored data as

$$\mathbf{y} = (111\ \ 110\ \ 110\ \ 101\ \ 010\ \ 111\ \ 010).$$

We can conclude that the error vector was

$$\mathbf{e} = (111\ \ 000\ \ 100\ \ 000\ \ 010\ \ 000\ \ 010).$$

We can classify this error vector as a [3,1; 1,3]-graded-bit-error vector. There are a total of $3 + 1 = 4$ cells in error (4 triplets that are not all 0). Of these four, three have only one bit in error, while the remaining has three bits in error. Based on this, we can take $v_1 = 1$, $v_2 = 3$, and $t_1 = 3$, $t_2 = 1$.

Observe how this classification differs from the much more coarse error definition used by BCH codes. In the case of an 8-ary BCH code, we would simply record that there were 4 errors, not distinguishing between the single-bit and multiple-bit errors.

Next, our goal is to introduce graded bit error-correcting code constructions. As we will see, an operation from linear algebra known as the **tensor product** is a crucial ingredient in these constructions. The tensor product is an operation on matrices defined as follows. Let us say that A is a matrix in $R^{m \times n}$ and B is a matrix in $R^{p \times q}$. Then, the tensor product $A \otimes B$ is defined as

$$A \otimes B = \begin{bmatrix} a_{11}B & \cdots & a_{1n}B \\ \vdots & \ddots & \vdots \\ a_{m1}B & \cdots & a_{mn}B \end{bmatrix}.$$

In other words, $A \otimes B$ is an $mp \times nq$ block matrix where each of the elements of A is (scalar) multiplied by the matrix B. This operation has many important properties in mathematics and physics. In coding theory, it was first used by Wolf to produce a construction of [t;v]-bit error correcting codes [2]:

Construction 1 Let C_A be a code with a parity check matrix given by $H_A = H_2 \otimes H_1$, where H_1 is the parity-check matrix of a binary $[m,k_1,v]_2$ code C_1, and H_2 is the parity-check matrix for a $[n,k2,t]_d$ code C_2, where $d = 2^{m-k1}$. Then, C_A is a [t;v]-bit error correcting code.

We can provide a simple example of such a code construction. For C_1, we use the Hamming code $[3,1,1]_2$, which has parity check matrix

$$H_1 = \begin{bmatrix} 1 & 0 & 1 \\ 0 & 1 & 1 \end{bmatrix}.$$

In other words, we will use as one of our two constituent codes the binary 3-bit repetition code with codewords {000,111}. Next, we can select a different code for C_2. Note that this code, in our case, must be over GF(4), since our final output must be over GF(8), according to our requirements for TLC flash cells. Since we require a code over GF(4), we let α be a primitive element over this finite field. Then, we can take C_2 to be a $[4,2,1]_4$ code, which also corrects one error:

$$H_2 = \begin{bmatrix} 1 & 0 & 1 & 1 \\ 0 & 1 & 1 & \alpha \end{bmatrix}.$$

Then, it is not hard to see that the resulting matrix is

$$H_A = \begin{bmatrix} 1 & \alpha & \alpha^2 & 0 & 0 & 0 & 1 & \alpha & \alpha^2 & 1 & \alpha & \alpha^2 \\ 0 & 0 & 0 & 1 & \alpha & \alpha^2 & 1 & \alpha & \alpha^2 & \alpha & \alpha^2 & 1 \end{bmatrix}.$$

Of course, it is possible to take the binary image of this GF(4) matrix:

$$H_A = \begin{bmatrix} 1 & 0 & 1 & 0 & 0 & 0 & 1 & 0 & 1 & 1 & 0 & 1 \\ 0 & 1 & 1 & 0 & 0 & 0 & 0 & 1 & 1 & 0 & 1 & 1 \\ 0 & 0 & 1 & 1 & 0 & 1 & 1 & 0 & 1 & 0 & 1 & 1 \\ 0 & 0 & 1 & 0 & 1 & 1 & 0 & 1 & 1 & 1 & 1 & 0 \end{bmatrix}.$$

Since H_1 and H_2 are parity-check matrices for single-error correcting codes with the desired properties, we expect C_A (with parity-check matrix H_A) to be a [1;1]-bit error-correcting code. This is indeed the case: we observe that the columns of H_A are all distinct, so that therefore, an error vector with a single bit in error can be corrected. Moreover, if we group the 12-bit long codewords in C_A into 4 groups of 3 bits each, we regain the GF(8) interpretation of the code.

More recently, a construction for the more refined graded-bit-error correcting codes was introduced [3]. This construction relies on the tensor product operation as well; however, the construction is somewhat more sophisticated:

Construction 2 Let C_B be a code with a parity check matrix given by

$$H_B = \begin{bmatrix} H_2 \otimes H_3 \\ H_4 \otimes H_5 \end{bmatrix}.$$

Here, we have C_1 a $[m,k,v_2]_2$ binary code with parity-check matrix H_1. Let $r = m - k$. We take H_1 to be such that the top r_3 rows of H_1 are a parity-check matrix for an $[m,m - r_3,v_1]_2$ code for some $r_3 < r$. This code will be called C_3 (with parity-check matrix H_3). We let H_5 be the submatrix of H_1 including the bottom $r_5 = r - r_3$ rows of H_1. Finally, we let H_2 be the parity-check matrix for a 2^{r3}-ary $[n, k_2,t_1 + t_2]_d$ code C_2 ($d = 2^{r3}$) and H_4 to be the parity-check matrix for a 2^{r5}-ary $[n, k_4,t_2]_f$ code C_4 ($f = 2^{r5}$).

Then, C_B is a $[t_1,t_2;v_1,v_2]$-graded bit error correcting code of length n.

Let us see how the decoding works for this type of code. For the code C_B, we introduce the decoder D_B which takes as an input a vector $\mathbf{y} = \mathbf{c} + \mathbf{e}$, with $\mathbf{c}$ a codeword in C_B and $\mathbf{e}$ a $[t_1,t_2; v_1,v_2]_2^m$-bit error vector. The output here is an estimate $\mathbf{e}'$ of the error vector $\mathbf{e}$ (note that we use a slightly abnormal convention where the output is the error estimate rather than an estimate of the transmitted codeword. The codeword estimate can be computed as $\mathbf{c}' = \mathbf{y} - \mathbf{e}'$). Then, the decoder D_B operates in the following way. Each of the other D_i are the decoders for the corresponding codes C_i.

1. Form the vectors $(\mathbf{s}_1^0,\ldots,\mathbf{s}_n^0)$ from the decoder $D_2(H_2 \cdot (H_1' \cdot \mathbf{y}_1^T,\ldots, H_1' \cdot \mathbf{y}_n^T)^T)$.
2. Set the error $\mathbf{e}^*$ to be $(D_1'(\mathbf{s}_1^0),\ldots,D_1'(\mathbf{s}_n^0))$.
3. Set the codeword $\mathbf{y}'$ to be $\mathbf{y} + \mathbf{e}^*$.
4. Compute $(\mathbf{s}_1',\ldots,\mathbf{s}_n')$ from $D_2(H_2 \cdot (H_1' \cdot \mathbf{y}_1'^T,\ldots,H_1' \cdot \mathbf{y}_n'^T)^T)$.
5. Set $(\mathbf{s}_1'',\ldots,\mathbf{s}_n'')$ to be $D_3(H_3 \cdot (H_1'' \cdot \mathbf{y}_1'^T,\ldots,H_1'' \cdot \mathbf{y}_n'^T)^T)$.
6. Set I to be $\{i : (\mathbf{s}_i',\mathbf{s}_i'') \neq (\mathbf{0},\mathbf{0})\}$.
7. Let $\mathbf{y}''$ satisfy $\mathbf{y}_i'' = \mathbf{y}_i$ if i is in I and $\mathbf{y}_i'' = \mathbf{y}_i'$ if i is not in I.
8. Set $(\mathbf{s}_1^1,\ldots,\mathbf{s}_n^1)$ to be $D_3(H_3 \cdot (H_1'' \cdot \mathbf{y}_1''^T, \ldots, H_1'' \cdot \mathbf{y}_n''^T)^T)$.
9. $\mathbf{e} = (\mathbf{e}_1,\ldots,\mathbf{e}_n)$ where $\mathbf{e}_i = \mathbf{e}_i^*$ if i is not in I and otherwise $\mathbf{e}_i = D_1(\mathbf{s}_i^0,\mathbf{s}_i^1)$.

The basic idea here is to first correct the errors with fewer bits in error. Of course, some errors have too many bits in error, so they will be miscorrected (but only to at most weight $v_1 + v_2$). Next, we detect which errors are the miscorrected ones, and correct them as well.

We also note that all of the non-trivial operations in the decoding procedure are uses of the decoding functions D_1, D_2, D_3. Moreover, each of these operations is performed at most twice. Therefore, the overall decoding complexity is a small constant factor times the complexity of the worst (in terms of complexity) constituent code. So, for example, if we use BCH codes as our constituent codes, our overall decoding algorithm has complexity roughly twice as large as the largest constituent BCH code.

As mentioned, we can test the proposed graded-bit error-correcting codes on actual data collected from TLC flash devices. The data was collected in the following way: random data patterns are written to the device, filling each block. This procedure is repeated for 5000 program/erase (P/E) cycles; each 100 cycles, the data is read back for errors [1].

The comparisons were performed against other BCH codes with the same rate and length. In the three plots in Fig. 11.3, codes had lengths of 4096, 8192, and 16,384, respectively. The purple curve in the bottom figure, for example, indicates a graded bit error correcting code with parameters $[t_1,t_2; v_1,v_2] = [242,8; 1,3]$. The red curves represent 8-ary BCH codes. The blue curves represent (identical) binary BCH codes used to protect each of the 3 bits in the cell separately. The black curves

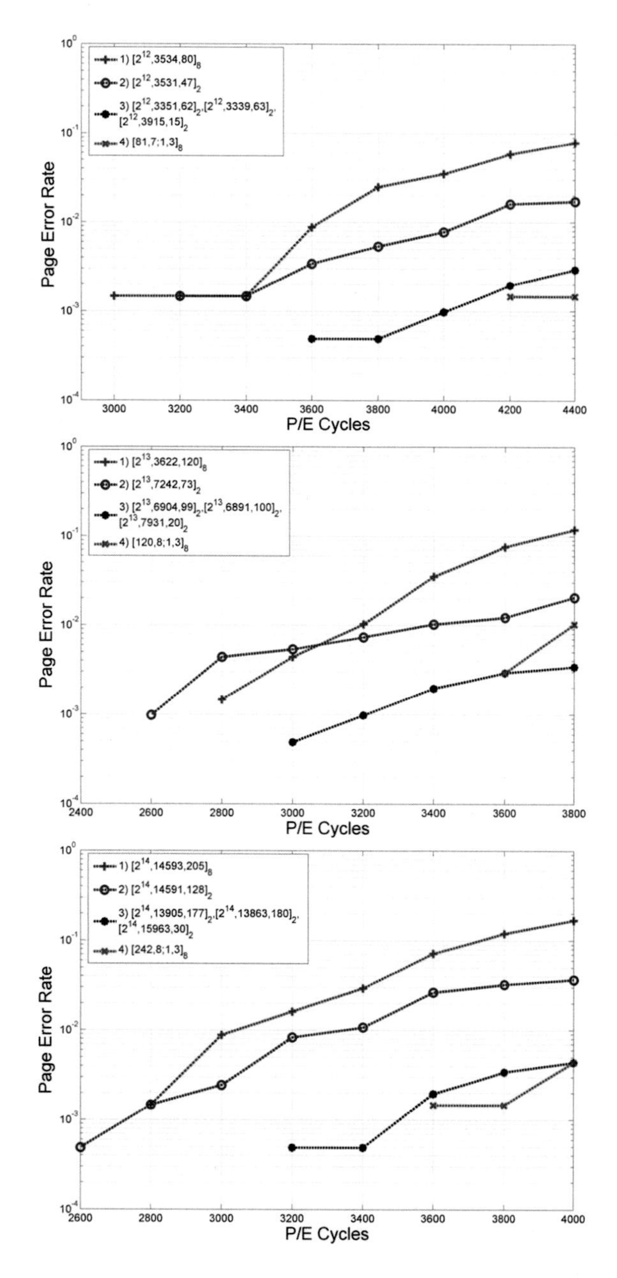

Fig. 11.3 Page error rates (PERs) for codes of with approximately the same length (4096, 8192, and 16,384 bits, respectively) and rate tested using data collected from TLC flash devices after varying numbers of program/erase cycles. The *red*, *blue*, and *black curves* use BCH codes (non-binary, identical binary over the separate pages, and differing binary codes over the separate pages, respectively.) The *purple curves* show our graded-bit error-correcting code construction. As can be seen, our asymmetric construction results in no errors (perfect operation) until much later in the device lifetime compared to traditional codes

represent three binary BCH codes (with different parameters) selected to optimize the error rate on each bit separately. For our graded bit error-correcting code constructions, we also selected BCH codes as our underlying constituent codes, so that the final parity check matrix H_B is produced by stacking the tensor products of parity check matrices of BCH codes.

As can be seen, using graded bit error-correcting codes allows for no errors at all on the measured data until late in the device lifetime (past 3000 P/E cycles). After this point, these codes perform as well or better than the separate binary BCH codes. Meanwhile, the 8-ary symmetric scheme has by far the worst performance.

We must point out that this is not the limit of what can be accomplished by the use of asymmetric codes tailored to handle specific error patterns. We have observed several other types of errors as well [4]. In the case of TLC flash, a careful study of the error patterns indicates that cells can be broadly divided into reliable and unreliable cells, where the unreliable cells are vastly more likely to be in error. In the case of the data set we examined, we noted that a specific set of roughly 65,000 cells (forming approximately 0.05 % of the total number of cells) resulted in more than 50 errors each across the 5000 P/E cycles from the test. In other words, about 10^{-4} of the cells account for more than 10 % of the errors.

What is the behavior of these unreliable cells? We observed that the cells produce these errors specifically when they are programmed to a higher voltage level. TLC cells have 8 possible voltage levels; frequent errors occurred when the unreliable cells were programmed to levels 4–7 (but not levels 0–3). Therefore, it is desirable to introduce a code that has the same features as the graded-bit-error-correcting codes previously discussed while also avoiding programming the unreliable cells to dangerously high levels.

Fortunately, this turns out to be possible. We restrict ourselves to the specific case of TLC flash, which means that we wish to create a code in GF(8), or, equivalently, a binary code of length 3n. Of course, a similar construction can be created for more general cases as well.

Before we proceed, let us introduce an auxiliary code construction, known as "stuck-at" error-correcting codes. First, let the operation $\circ$: $GF(2)^m \times GF(3)^m$ to $GF(2)^m$ be defined so that $\mathbf{b} = \mathbf{a} \circ \mathbf{s}$, where $b_i = s_i$ if $s_i < 2$ and $b_i = a_i$, otherwise. P_j is defined as the set of all vectors $\mathbf{s} = (s_1, \ldots, s_m)$ in P_j so that $|\{i : s_i < 2\}| \le j$. Then, stuck-at error-correcting codes are defined in the following way:

Definition 5 For positive integers m,k,t,j, a $[m,k,t,j]_2$ binary code C is a linear code of length m and dimension k over GF(2) with encoding and decoding maps E_C and D_C such that

1. For all s in P_j and any message h, $E_C(h,\mathbf{s}) \circ \mathbf{s} = E_C(h,\mathbf{s})$, and
2. For any error vector $\mathbf{e}$ in $GF(2)^m$ with $wt(\mathbf{e}) \le t$, $D_C(E_C(h,\mathbf{s}) + \mathbf{e}) = h$.

The idea behind Definition 5 is that even if a particular subset of cells is stuck (the subset has size at most j), the stuck-at error-correcting code C can still recover from errors. As we will see, we can use these types of codes as a building block for our construction; we adapt the stuck-at-error behavior to limit the levels of our target cells.

Next, we introduce our goal codes:

Definition 6 Let n,k,t_1,t_2,j be positive integers where $j,t_1,t_2 < n$. Then, a $[3n,k,t_1,t_2,j]$ dynamic bit-error-correcting code C is a binary linear code of length 3n and dimension k that is capable of correcting any $[t_1,t_2]$-bit-error vector. There is an additional constraint: if we write a codeword in C as $\mathbf{c} = (c_1, c_2, \ldots, c_n)$, where each c_i is an element in GF(8), then, given a set I of size at most j, $c_i \leq 3$ for all i in I and all codewords $\mathbf{c}$ in C.

Definition 6 matches Definition 4 from our previous discussion, but adds the requirement that a particular subset of cells is programmed to low levels only. Therefore, we introduce a construction that builds on Construction 2 and adds a corresponding constraint. Let us also use a simple map from elements in GF(4) to binary vectors of length 2 (again, α is a primitive element in GF(4)):

$$\Gamma(\alpha) = (0,1)^{\mathrm{T}}, \quad \Gamma(\alpha^2) = (1,1)^{\mathrm{T}}, \quad \Gamma(\alpha^3) = (1,0)^{\mathrm{T}}, \quad \text{and} \quad \Gamma(0) = (0,0)^{\mathrm{T}}.$$

Construction 3 Let $H_1 = (\alpha\ \alpha^2\ \alpha^3)$ and $H_2 = (1\ 1\ 1)$, where H_1 is a matrix in $GF(4)^{1\times 3}$ and H_2 is a matrix in $GF(2)^{1\times 3}$. Let H_3 be a parity check matrix for a $[n,k_3, t_1 + t_2]_4$ code C_3. Also, let H_4 be a parity check matrix for a $[n,k_4,t_2,j]_2$ stuck-at-error correcting code (as introduced in Definition 5). Then, a parity check matrix for a $[3n,2k_3 + k_4,t_1,t_2,j]_2$ dynamic bit-error-correcting code of length 3n is given by

$$\mathrm{H} = \begin{bmatrix} \Gamma(H_3 \otimes H_1) \\ H_4 \otimes H_2 \end{bmatrix}.$$

The basic idea here is to slightly modify our previous graded-bit error-correcting construction (Construction 2) by forcing the use of a stuck-at error-correcting code construction. Rather than use it to specifically correct stuck-at errors, we do almost the reverse. The construction allows for mapping a message to one of several possible codewords, in order to deal with the stuck-at behavior. We take advantage of this by selecting the codeword where our unreliable cells are at lower levels.

For the case of our dynamic bit-error correcting codes, too, we can perform simulations to show the advantages of such codes. We perform the comparison against graded bit-error correcting codes (which lack the specific constraint of avoiding high levels in unreliable cells) and other previously mentioned codes, including various BCH codes.

In Fig. 11.4, the page error rates (PERs) are shown for codes of lengths 4096, 8192, and 16,384, respectively. As before, the lengths and rates of the codes compared against are the approximately equal. The same code constructions are shown as before; the green curve shows the new dynamic-bit-error code construction. Exactly 2 unreliable cells were forced to lower levels in the top two plots, while 4 unreliable cells were used in bottom plot. The purple curve shows the graded-bit error-correcting codes. The other curves are the same types of codes used for comparison in the previous sections.

Note that the dynamic-bit error-correcting codes have the best overall PER, while still not exhibiting any errors until very far in the lifetime of the devicc. The additional asymmetry of these codes (compared to the graded-bit error-correcting codes) has granted us an additional half an order of magnitude in PER performance. Therefore, we have the best of both worlds: codes with very good PERs, which do

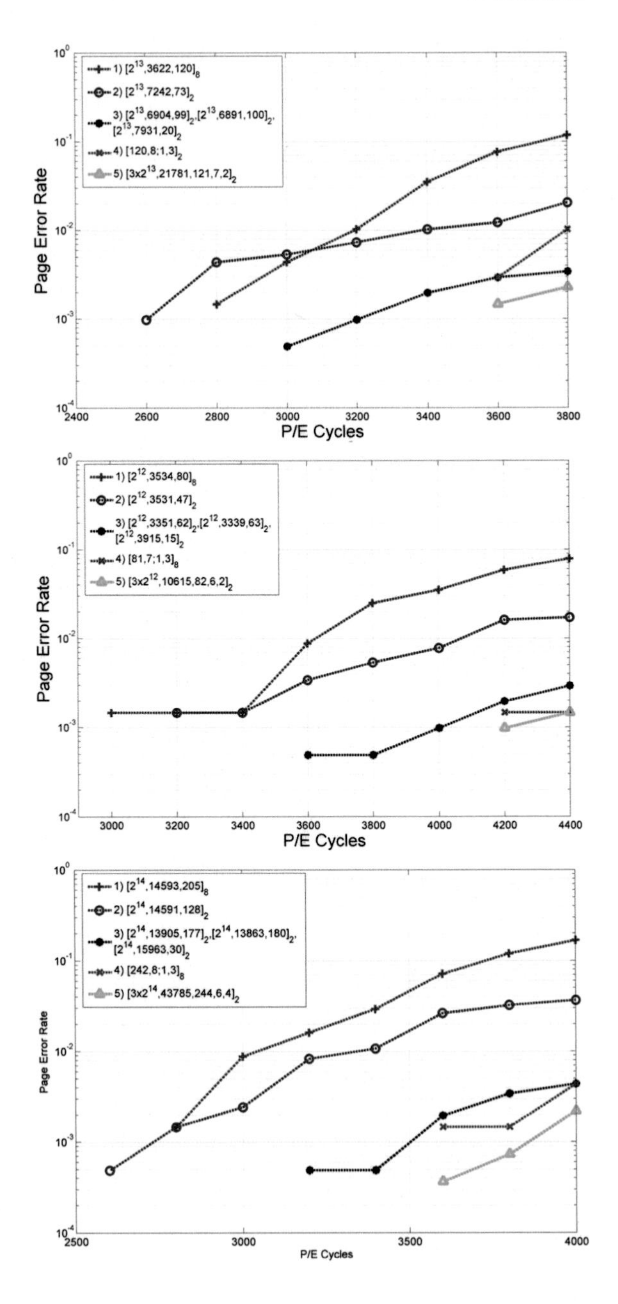

Fig. 11.4 Page error rates (PERs) for codes of with approximately the same length and rate tested with data collected from TLC flash devices after varying numbers of program/erase cycles. As before, the *red*, *blue*, and *black lines* show varying BCH-based code constructions. The *purple curve* shows the graded bit-error correcting code construction. The *green curves* are the new dynamic-bit error-correcting codes, which continue to show no errors until late in the device lifetime, but also offer an additional half-magnitude improvement in overall PER over the graded construction

not allow for any errors at all until late in the operative lifetime of the flash devices. We have successfully taken advantage of asymmetry to produce codes which are a dramatic improvement over traditional, symmetric codes.

11.1.2 Dynamic Thresholds

Another particularly interesting feature of the flash channel is the fact that it is time-varying. The longer the period of time between data write and data access operations, the higher the probability of read error. This property is due to certain physical effects acting on flash transistors. For example, over time, the electrons trapped on the floating gates of flash cells will leak out, escaping these gates. The errors caused by such effects are therefore inherently asymmetric.

In addition to these asymmetries, the time-varying character of the channel is also not considered or exploited by traditional coding techniques. Recall that flash devices work in the following way: the amount of charge on the floating gate is measured and compared to a set of thresholds. The result of this comparison determines the discrete value read out from the device. The thresholds used are traditionally fixed and permanent, ignoring the degradation of the channel over time. Although these fixed thresholds may be suitable for the channel at a particular period of operation, they often prove to be inefficient for differing retention periods.

One solution to this problem is to introduce *dynamic* thresholds, which can be changed over time. Although there are many ways to accomplish this task, there is a particularly simple approach. We set the thresholds in such a way that the distribution of the values in a block of cells is identical when being read as it was upon write [5, 6]. In other words, we use this distribution of values as side information.

Let us formalize this idea. Say that we have a block $\mathbf{x} = (x_1, x_2, \ldots, x_n)$ of n cells that can take on any of the q values $(0,1,\ldots, q-1)$ each. In TLC flash, as in our previous discussion, $q = 8$. Now, some time passes and the written values have become the real values $\mathbf{v} = (v_1, v_2, \ldots, v_n)$. We have the thresholds $\mathbf{t} = (t_1, \ldots, t_{q-1})$, which we use to read $\mathbf{v}$ in the following way: the output $\mathbf{y} = \mathbf{t}(\mathbf{v})$ is given by

$$y_i = a, \quad \text{if} \quad t_a \leq v_i \leq t_{a+1},$$

where we take t_0 to be negative infinity and t_q to be positive infinity.

Now, let us denote by $\mathbf{k} = (k_0, \ldots, k_{q-1})$ the distribution of values in $\mathbf{x}$. That is, $k_a = |\{i \mid x_i = a, \ 1 \leq i \leq n\}|$. Thus, for example, $\mathbf{x} = (1,0,0,3,1,1,1,2)$ has $\mathbf{k}(\mathbf{x}) = (2,4,1,1)$, since $\mathbf{x}$ has 2 values of 0, 4 values of 1, and so on.

Then, we can define dynamic thresholds in the following way:

Definition 7 A threshold vector $\mathbf{t}$ is a *dynamic* threshold if

$$\mathbf{k}(\mathbf{y}) = \mathbf{k}(\mathbf{t}(\mathbf{v})) = \mathbf{k}(\mathbf{x}).$$

For example, say that the vector $\mathbf{x}$ above was written, and the real charge values are given by

$$\mathbf{v} = (1.2, 0.2, 0.6, 2.3, 1.1, 1.0, 1.3, 2.2).$$

Then, if we use the fixed threshold $t^1 = (0.5,1.5,2.5)$, we would read the output

$$\mathbf{y}^1 = \mathbf{t}^1(\mathbf{v}) = (1,0,1,2,1,1,1,2),$$

with errors in the third and fourth positions. However, t^1 is not a dynamic threshold: the distribution of y^1 is $k(y^1) = (1,5,2,0)$, which is not equal to $k(x) = (2,4,1,1)$. Let us instead select a dynamic threshold $t^d = (0.7,2,2.25)$. Then, we correctly read

$$\mathbf{y}^d = \mathbf{t}^d(\mathbf{v}) = (1,0,0,3,1,1,1,2).$$

Of course, dynamic thresholds do not guarantee that the read sequence is error-free. However, they reduce error rates, since to yield an error, two components (with differing initial values) must have their values switched relative to each other. For example, if we have $x_i < x_j$, we must have $v_i > v_j$ to cause an error. This event occurs with lower probability in comparison to simply requiring $x_i < t_{xi}$, which is sufficient for an error in the fixed threshold case.

An illustration of this claim is in Fig. 11.5, where we simulated the degradation of the channel with the passage of time by modeling flash cells as Gaussians with increasing standard deviation over time. We then simulated a block of 10^5 cells by writing random values and reading back for errors using dynamic thresholds versus fixed thresholds. As the standard deviation of the Gaussians modeling the flash channel increases, the dynamic threshold scheme yields a much slower growth in error probability.

This type of simulation offers experimental support to the suggestion that dynamic thresholds outperform fixed thresholds. However, we add a theoretical comparison as well. Let us say that that $N(\mathbf{x},\mathbf{y})$ is the Hamming distance between vectors $\mathbf{x}$ and $\mathbf{y}$. If $\mathbf{y}$ is generated by reading $\mathbf{x}$, that is, $\mathbf{y} = \mathbf{t}(\mathbf{v}(\mathbf{x}))$ is the value read from $\mathbf{v}$ (itself formed by the written values $\mathbf{x}$) using the threshold $\mathbf{t}$, then, we write $N(\mathbf{x},\mathbf{y})$ as $N(\mathbf{t})$. Let us say also that $\mathbf{t}^*$ is the optimal threshold in the sense that $\mathbf{t}^* = \min_t N(\mathbf{x},\mathbf{y})$ for some fixed $\mathbf{x},\mathbf{y}$.

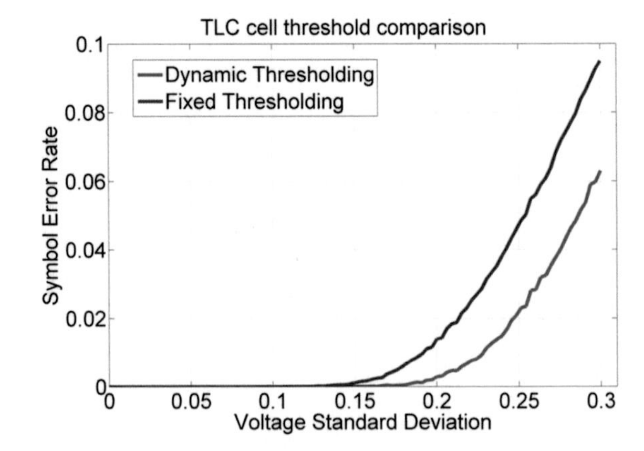

Fig. 11.5 Bit error rates (BERs) in a simple experiment modeling multi-level flash cells as Gaussians with increasing standard deviations over time. Blocks contain 10^5 cells. Dynamic thresholds and fixed thresholds were compared; as can be seen, dynamic thresholds offer improved performance versus traditional fixed thresholds

We also make the assumption that the maximum possible error magnitude is given by r, for some r in {0,..., q − 1}. This is a reasonable assumption for flash: we expect most errors to be of small magnitude, possibly at most 1. With this, we can say that any dynamic threshold $\mathbf{t}^d$ is quite close to the optimal threshold $\mathbf{t}^*$:

$$N(\mathbf{t}^d) \Leftarrow (r+1)N(\mathbf{t}^*).$$

In other words, any dynamic threshold is at most a constant factor (depending on the maximum error magnitude) from the optimal threshold. Of course, this optimal threshold requires knowledge of **x** itself to compute. This knowledge is not available when reading: a reliable estimate of **x** is the goal of the read operation. In other words, the dynamic threshold offers a practical solution that is quite close to an unobtainable optimum.

So far we have not discussed how to generate a set of thresholds. This is, of course, an important practical concern. There are two possible approaches (and a variety of combinations of the two) available [6]. The first is to use the distributions of values in blocks as side information.

This side information is then stored elsewhere. For example, we can store these values in very robust, highly-reliable cells, protected by powerful codes, which can then be read with fixed thresholds with low risk of error.

Another approach is to store data in constant-weight codewords. These codewords have a fixed distribution of values. Since the distribution is fixed, it can be hardcoded into the system from production, bypassing the need to communicate side information at all. The tradeoff here, of course, is the fact that constant-weight codes eliminate certain codewords from being used, yielding a potentially smaller overall rate.

With either approach, we must further protect our system with error-correcting codes. Dynamic thresholds by themselves will not sufficiently reduce the system's error rate to the target rate. This leaves us with the question of what choice of code to select. We could, of course, use an existing, off-the-shelf code, such as BCH codes. However, these schemes ignore the fact that dynamic thresholds yield asymmetric errors, in the same way that asymmetry in 3-bit TLC error vectors are ignored (leading to our improved tensor product-based constructions.) For example, a single component error in a vector cannot occur, since this would change the read codeword distribution, which is not possible with dynamic thresholds by definition. However, traditional codes cannot take advantage of this idea.

Instead, we can propose specialized asymmetric codes that operate specifically on dynamic thresholds.

Definition 8 Let a vector **x** be stored in a system with dynamic thresholds. An error **e** that **x** can experience under dynamic thresholding is called a [t,v]-**DT error** if **e** has at most t non-zero components and if each component has magnitude at most v. A code capable of correcting any [t,v]-DT error is called a [t,v]-**dynamic threshold error correcting (DTEC) code**.

Note that not all [t,v]-error vectors are [t,v]-DT error vectors. For example (1,0,0,0) is a [1,1]-error vector of length 1, but not a DT error vector, since with dynamic thresholds, errors require at least two positions to be non-zero in order to preserve the distribution of values between **x** and **x** + **e**. In other words, there are fewer DT-error vectors than there are error vectors in general. Although a conventional error-correcting code can correct DT errors, it also corrects error vectors which cannot occur with DT errors, which reduces the overall performance of the code by sacrificing rate for unused error-correction strength.

We introduce a type of asymmetric construction that specifically corrects 2 DT errors of any magnitude, thus providing an example of a [2,q − 1]-DTEC code construction:

Construction 4 Let C be a $[n,n-2]_q$ linear block code (of length n and dimension n − 2) over a field F_q with parity-check matrix given by

$$H = \begin{bmatrix} a_1 & a_2 & \dots & a_n \\ a_1^2 & a_2^2 & \dots & a_n^2 \end{bmatrix},$$

where $S = \{a_1, a_2, \dots, a_n\}$ is a subset of distinct elements of F_q. Then, C is a [2, q − 1]-DTEC code if S is a Sidon set (a set with the property that for any four distinct elements a,b,c,d in S, $a + b \neq c + d$).

Note that such a code corrects any 2 errors in dynamic thresholding, while a general 2-error correcting code requires a much larger redundancy. This is the advantage of custom, asymmetric error-correcting codes. It is possible to modify the previous construction to yield codes for other values of limited magnitude r smaller than q − 1.

With, this, we have seen a further example of how to take advantage of asymmetries in order to introduce superior algebraic error-correcting codes.

11.2 Non-binary LDPC Codes

Next, we switch our focus from algebraic codes to graph-based codes. Graph-based codes have a nice advantage over algebraic codes, since it is possible to decode using soft information. In other words, graph-based code decoders can take as inputs fractional (rather than integer) values. Algebraic codes, however, lack this ability. The use of soft information is particularly important for storage devices such as flash, since we can perform multiple reads of the data in order to retrieve more accurate decoder inputs. Soft information thus yields excellent error-correcting performance. We provide more detail on this concept later on in this chapter.

We are particularly interested in one of the most important class of graph-based codes, non-binary low-density parity-check (NB-LDPC) codes. LDPC codes were first introduced in Gallager's seminal doctoral thesis in the 1960s and rediscovered during the 90s. Binary LDPC codes have been extensively studied and have found use in numerous applications.

Non-binary LDPC codes, however, have remained somewhat less well-understood. An early work by Davey and MacKay [7] showed that non-binary

LDPC codes offer better performance compared to their binary counterparts. This performance scales up with the field size parameter. However, this performance gain comes at the cost of decoder complexity. The initial implantation of the LDPC belief-propagation decoder for non-binary codes had a complexity of $O(q^2)$ for a field size of q. However, this complexity can be reduced to a more manageable $O(q \log q)$ by an FFT-based decoder implementation. Other techniques for low-complexity decoding have been proposed, including ones based on linear programming.

In addition to improved decoder complexity, a large number of constructions for non-binary LDPC codes have been proposed over the last ten years. The approaches taken for such constructions vary widely; for example, constructions include quasi-cyclic codes (some based on geometric approaches), protograph-based codes, quantum LDPC codes, and many others [8–10]. The proliferation of such improved works in the non-binary LDPC area of study suggests that such codes are approaching common, practical application. In general, increasing the code length of an LDPC code improves its performance; however, there are diminishing returns. For example, doubling the code length from 1000 bits to 2000 bits typically has a much greater positive effect on performance than doubling the code length from 100,000 bits to 200,000 bits.

However, before common application of non-binary LDPC codes can become reality, there is an additional roadblock to handle. This is the so-called LDPC "error floor". The terminology reflects the appearance of the bit-error or frame-error rate versus SNR for LDPC codes. Initially, as the SNR increases, the BERs/FERs correspondingly improve dramatically; this is the "waterfall regime." However, after a certain point, these curves become increasingly flat, entering the error-floor region. This error floor is a particularly important problem, since many applications for LDPC codes, such as data storage devices, operate at very high SNRs. For example, for Flash memory, the desired output FER often exceeds 10^{-15}; this point lies squarely in the error floor region for many LDPC codes. An illustration of an error floor is shown in Fig. 11.6.

What causes the error floor behavior? This is an important question that has been closely studied in the context of binary LDPC codes [11, 12]. We thus focus our attention on higher performance, non-binary LDPC codes. We begin by explaining the operation of practical LDPC decoders. In the case of storage devices, for example, a small number of probes of the underlying device are allowed. If there is only such probe permitted, we refer to the system as hard-decision. With more than one read, the system is soft-decision, as shown in the bottom of Fig. 11.7. However, only a small number of probes are allowed, due to latency issues. Note that an important problem is setting the reference thresholds (V_{R1},V_{R2},V_{R3} for the single-read case and V_{R1},…,V_{R6} for the two-read case.) A method based on mutual-information optimization was presented in [13].

As a result of the small number of reads, the continuous channel of the storage device has been transformed into a discrete channel. Similarly, in a digital system, the messages are quantized to finite-precision variables. As a result, in practical systems, decoder behavior ends up resembling that of decoders operating over

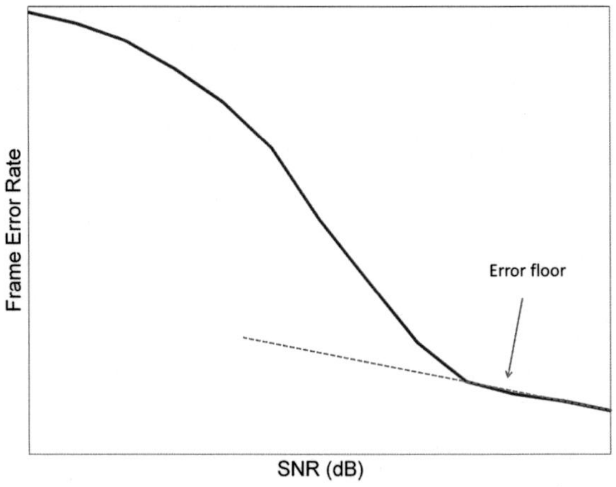

Fig. 11.6 Illustration of the "error floor" behavior of LDPC codes. Initially, as the SNR increases, there is a sharp downward slope as the frame error rate (FER) decreases. However, this slope eventually levels off, leading to much smaller improvement in FER for high SNRs

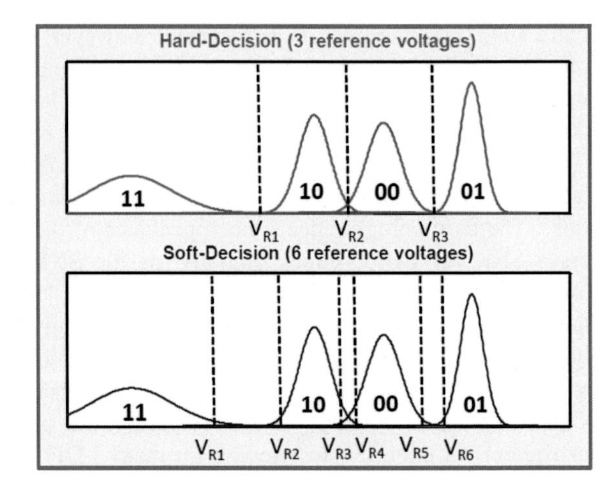

Fig. 11.7 Example of reads in a MLC (4-level) flash device. Hard decision allows only one read, and thus only one output state. For this reason, there is a single distinct threshold separating each state. Soft decision allows for multiple reads; 2 reads are shown on the *bottom figure*. There are many strategies for where to place the thresholds V_{Ri}

much simpler channels, such as discrete memoryless channels. LDPC codes over such channels are very well studied.

The majority of this existing research examines the error floors of binary codes. The error floor is in fact intimately connected to certain objects in the graph structure of the LDPC code. This graph structure is the Tanner graph; the Tanner graph of an LDPC code is a bipartite graph where the two classes of nodes are variable nodes (corresponding to components in LDPC codeword vectors) and check nodes (corresponding to the parity-check equations.) There is an edge between a check node and a variable node if the corresponding component is involved in the corresponding check equation, respectively. An example of a Tanner graph for a Hamming [7,4] binary code is shown below (Fig. 11.8). Of course, this code is not low-density; however, the simple parity-check matrix helps illustrate the idea behind the definition of the Tanner graph.

Since belief-propagation decoders operate on this graph structure, it is not surprising that certain configurations of nodes cause decoding problems. Trapping sets

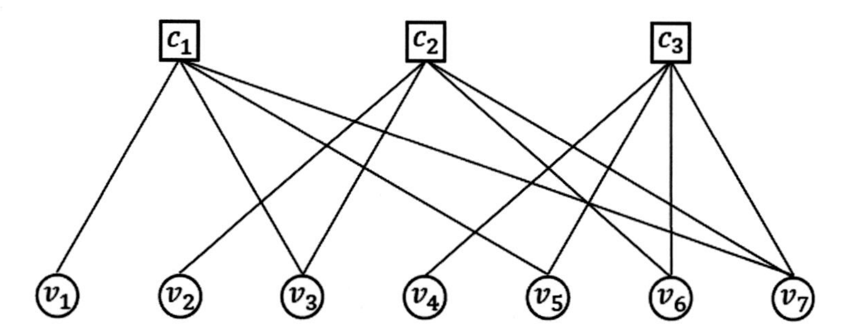

Fig. 11.8 Example of a Tanner graph for the (non-LDPC) Hamming [4, 7] binary code. The *circles* represent the 7 variable nodes corresponding to each bit in the codeword. The *squares* represent the three check equations from the code's parity-check matrix. There is an edge between a check node and a variable node if the corresponding bit is used in the parity-check equation

and absorbing sets are examples of subgraph objects that, when found in the Tanner graph of a particular code, are known to cause errors. These objects have been extensively studied in the case of binary LDPC codes. Many papers have proposed design algorithms for LDPC codes in order to avoid trapping and absorbing sets and thus to remove the error floor behavior [14, 15].

However, this problem is more challenging in the non-binary LDPC case. In this part of the chapter, we explore how to identify, enumerate, and remove absorbing sets for non-binary LDPC codes. We begin with a summary of absorbing sets in the traditional, binary LDPC case.

11.2.1 Binary Trapping/Absorbing Sets

We start with a subgraph of the Tanner graph for a binary LDPC code. The subgraph contains the variable node set V with |V| = a. The variable nodes in V are set to 1 while all other variable nodes are set to 0. The check nodes connected to the vertices in V are divided into sets E and O, where E contains check nodes with an even number of edges to vertices in V, and O has check nodes with an odd number of such edges. Of course, in this configuration, E contains satisfied check nodes while O contains unsatisfied check nodes. Now we can introduce trapping and absorbing sets:

Definition 9 V is an (a,b) trapping set if |O| = b.

Definition 10 V is an (a,b) absorbing set if |O| = b and if each variable node in V has (strictly) more neighbors in E than it does in O.

Definition 11 An elementary absorbing set/trapping set is an absorbing set/trapping set with the added condition that each of the neighboring satisfied check nodes has two edges connected to the set, while each of the neighboring unsatisfied check nodes has exactly one edge connected to the set.

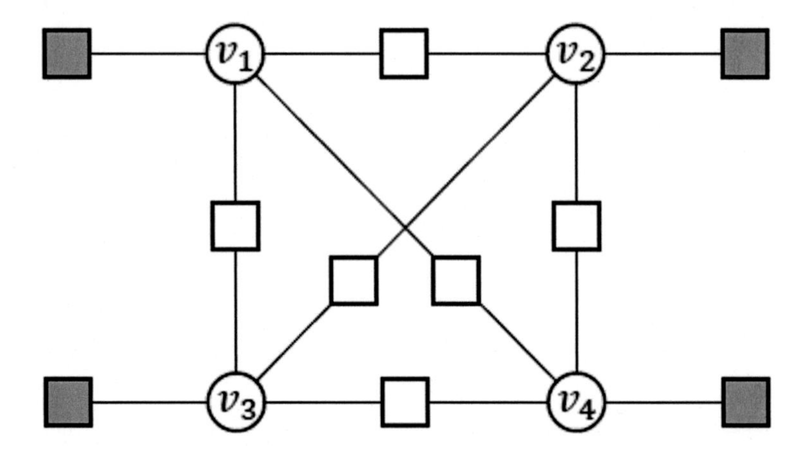

Fig. 11.9 Illustration of a (4,4) binary absorbing set. The *white circles* represent the four variable nodes in the absorbing set. The *white squares* are satisfied check nodes while the *gray squares* are unsatisfied check nodes. Since each of the unsatisfied check nodes has exactly one edge to the variable node set, this is an elementary (4,4) absorbing set

We show an illustration of such a set in Fig. 11.9.

We see that the configuration shown produces a (4,4) absorbing set. We have 4 variable nodes that are connected to 4 unsatisfied check nodes. The unsatisfied check nodes are gray squares, while the satisfied check nodes are white. Note that in addition, this is an elementary (4,4) absorbing set, since each of the unsatisfied check nodes has exactly one edge connecting it to the 4 variable nodes, while each of the satisfied check nodes has exactly two edges to the variable nodes.

Notice the basic idea of the absorbing set: it is a configuration of variable nodes where a majority-logic bit-flipping decoder will make an error (here, we assume that the all zero-codeword was sent) but will be unable to recover from this error. This behavior occurs precisely because of the fact that the majority of the neighboring check nodes are satisfied.

We will also require a few additional graph theory concepts. We can define a vector space on the set of all cycles of an undirected graph. For such a graph G with G = (V,E), the power set of E, 2^E, is a vector space when taking symmetric set difference as the addition operation, the identity function as the negation operation, and the empty set as the additive identity element. Then, the cycle space of the graph G is the subspace of 2^E which has the cycles of G as its elements. Now we apply basic principles from linear algebra:

Definition 12 A set of cycles F in G = (V,E) is the cycle span of G if it forms a basis for the cycle space. Cycles in a cycle span are called fundamental cycles.

We can also introduce a related graph structure, called a variable node (VN) graph. This graph is defined based on the bipartite graph of an elementary absorbing set. The variable node graph contains only variable nodes; these nodes are connected by an edge if they share a degree-two check node as a neighbor.

11.2.2 Non-binary Absorbing Sets

We are now ready to tackle the matter of non-binary absorbing sets. Since we are working in the non-binary regime, the Tanner graph of a code has weights placed on the edges connecting variable and check nodes. This weight is equal to the corresponding non-zero value in the non-binary parity-check matrix of the LDPC code. This adds an additional aspect to the graph structure: there is a *topological* structure (just as is the case with binary codes), but also we now have a weight structure. As a result, non-binary absorbing sets must also satisfy weight conditions.

As before, we seek a configuration where each variable node has more satisfied neighboring check nodes compared to unsatisfied nodes; however, for satisfied/unsatisfied part to be the case, we will require certain relationships between the weights. We show an example of such an absorbing set in Fig. 11.10.

Note that here the edge weights are taken to be nonzero elements from the code's finite field GF(q). This absorbing set has the same topological structure as the previous binary absorbing set example.

In the general case, however, in order for the degree-two check nodes to be satisfied, we need the following relationships to hold over GF(q). Note that the weights are labeled on the earlier diagram.

$$v_1\, w_1 = v_2\, w_2, \quad v_2\, w_3 = v_4\, w_4, \quad v_4\, w_5 = v_3\, w_6,$$
$$v_3\, w_7 = v_1\, w_8, \quad v_2\, w_{11} = v_3\, w_{12}, \quad v_1\, w_9 = v_4\, w_{10}.$$

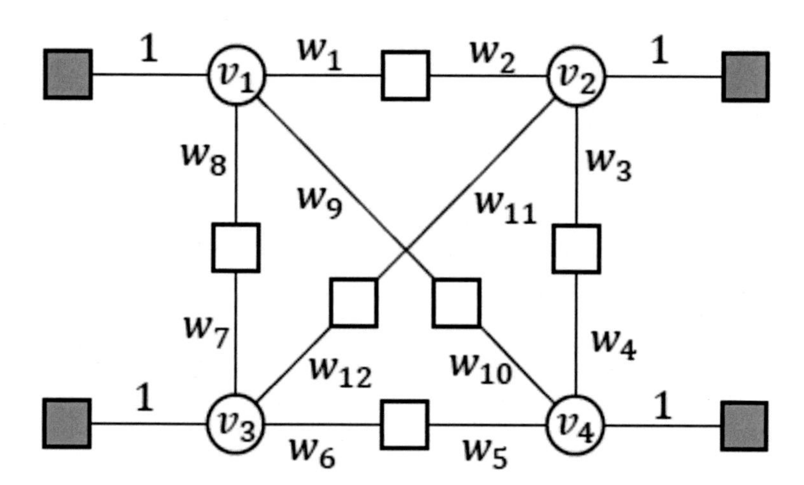

Fig. 11.10 Illustration of a non-binary (4,4) absorbing set. As before, this is an elementary absorbing set. Note that each edge now has a weight; these weights must satisfy certain conditions for the subgraph to form an absorbing set. However, the unlabeled version of the graph forms a binary absorbing set

Each of these equations comes directly from the definition of the Tanner graph. For example, the check node between v_1 and v_2 is only satisfied if (recall that all variable nodes except for $v_1,\ldots,v_4$ are set to 0, while $v_1,\ldots,v_4$ are set to 1) if the corresponding check equation is 0: $v_1w_1 + v_2w_2 = 0$.

If our field size is a power of 2, so that $q = 2^p$, we can eliminate the variable nodes in these equations in order to write a series of conditions exclusively for the weights:

$$w_1\, w_7\, w_{11} = w_2\, w_8\, w_{12}, \quad w_3\, w_5\, w_{12} = w_4\, w_6\, w_{11}, \quad w_2\, w_4\, w_9 = w_1\, w_3\, w_{10},$$

where, as before, the equations are taken over $GF(2^p)$.

The basic idea can be written in a general form to define non-binary absorbing sets:

Definition 13 A set V is an (a,b) absorbing set over GF(q) if there exists an $(l - b)$ x a submatrix B of rank r_B given by elements $b_{j,i}$ for $1 \le j \le l - b$, $1 \le i \le a$ in matrix A satisfying the conditions:

1. If N(B) is the null space of B and $\mathbf{d}_i$, $1 \le i \le b$ is the *i*th row of D where D is given by excluding the matrix B from A, then, there exists $x = [x_1\ x_2\ \ldots\ x_a]^T$ in N(B) such that for x_i is non-zero for all i in {1,…,a} and there is no i such that $\mathbf{d}_i$ $x = 0$.
2. If D contains the elements $d_{j,i}$ for $1 \le j \le b$, $1 \le i \le a$, then, for all i in {1,2,… a}, then

$$\sum_{j=1}^{l-b} S(b_{j,i}) > \sum_{j=1}^{b} S(d_{j,i})$$

Here the function S is an indicator function such that $S(x) = 1$ for x nonzero and 0 for $x = 0$.

We observe that a similar type of adaptation to the non-binary case is possible for trapping sets as well.

We can define non-binary elementary absorbing sets by adding the same condition as we did for the binary absorbing set to elementary binary absorbing set case. In this elementary absorbing set case, we can further manipulate the conditions above to have the following form, which resembles that as shown in our example. Let C_p be a cycle that contains p distinct variable nodes and p distinct check nodes in a graph induced by an (a,b) non-binary absorbing set. Let $C_p = c_1$–v_1–c_2–v_2–⋯–c_p–v_p–c_1. The weight $w2_{i-1}$ is the label on the edge connecting c_i and v_i. Similarly, w_{2i} is the label on the edge connecting v_i and c_{i+1}. Then, we have the following.

Lemma 1 *If the field size parameter* $q = 2^p$, *then, every cycle* C_p *satisfies the following relationship:*

$$\prod_{k=1}^{p} w_{2k-1} = \prod_{k=1}^{p} w_{2k}.$$

In the case of elementary non-binary absorbing sets, we now have a simple breakdown of the definition: the (unweighted) topological structure must be a binary absorbing set, and, in addition, the weights must satisfy the equation given in Lemma 1.

Now that we have set up our definitions and identified just what a non-binary absorbing set is, we are ready to examine how to improve the performance of our non-binary codes.

11.2.3 Performance Analysis and Implications

We begin by identifying how frequently the weight conditions can be satisfied. This is an important question, since if the conditions are not met, we do not have an absorbing set. This concept is described in the following theorem.

Theorem 1 *We have an (a,b) unlabeled (binary) elementary absorbing set with e satisfied check nodes. Then,*

1. *A fraction of $(q-1)^{a-e-1}$ of edge weight assignments (taken over GF(q)) produce non-binary elementary absorbing sets.*
2. *A fraction of $e(q-1)^{a-e-1}(q-2)$ of edge weight assignments (again taken over GF(q)) produce (a,b + 1) non-binary trapping sets.*

The proof of the theorem relies on simple counting arguments based on the graph-theoretic ideas already introduced.

The theorem implies that there is a larger number (by a factor of $e(q-2)$) non-binary trapping sets compared to non-binary absorbing sets, assuming that the code and its weights are generated randomly. In practice, however, simulation results show that error profiles do not involve any errors from trapping sets. On the other hand, error profiles do show errors that are a result of absorbing sets. What explains this behavior? The idea is that quantization used in decoding algorithms results in these belief propagations acting in a way similar to majority-logic bit flipping decoders. Such decoders do not struggle with the more general trapping set errors, but, as we see from the definition of absorbing sets, these decoders will produce errors when faced with absorbing sets.

For this reason, it is more desirable to find a way to remove or reduce absorbing sets from the Tanner graphs of non-binary LDPC codes. First, we note that we must target certain absorbing set parameters over others. In the error-floor regime, which is at high SNR, errors typically only include a small number of variable nodes. Therefore, we look at small absorbing sets. In fact, the performance of the LDPC decoder will be dominated by the smallest absorbing set, which is also typically an elementary absorbing set. The goal thus becomes to maximize the size of the smallest absorbing set.

We are now ready to introduce an algorithm that eliminates problematic absorbing sets from non-binary LDPC code Tanner graphs. As described, the key idea is to manipulate the edge weights in such a way that the subgraphs involved are no longer absorbing sets. The algorithm is given below. We use one additional term: an absorbing set A is a child of an absorbing set B if A is a subgraph of B. We call B a parent of A.

Algorithm 1:

1. **Input**: The Tanner graph G; edge weights taken over GF(q).
2. Find U_j, the set of all (a_j,b_j) absorbing sets in the binary, unlabeled version of G (using the existing techniques in [16] for the general case, or, in specific cases, more sophisticated approaches, such as the technique in [14] for circulant-based codes, including quasi-cyclic codes).
3. Select W, the set of non-binary absorbing sets to be removed.
4. X is the set of non-binary absorbing sets that cannot be removed.
5. Start with X empty.
6. A is the set of absorbing sets in W which have been processed (either eliminated from G or placed into the list X).
7. Start with A empty.
8. For every edge j in T, C_j is defined to be the set of re-weighted cycles which include edge j.
9. For all j in T, start with C_j empty.
10. Find (a_j,b_j), the smallest-size non-binary absorbing set in W\A.
11. If this set is a child of another absorbing set already in A, go to 31.
12. **for** all u in U_j,
13. Find F_u, a set of fundamental cycles of u.
14. Let E_u be the set of edges in u.
15. For an edge k in E_u, set M_k to be the set of cycles in F_u including k.
16. **if** (5) is satisfied for all the cycles in F_u, **then**
17. Find an edge i in E_u with edge weight w_i minimizing $|C_i|$
18. **if** there is exists nonzero w_i' (not equal to w_i) so that all cycles including I do not satisfy (5), **then**
19. Replace w_i with w_i'
20. **else**
21. Set E_u to be E_u\i
22. **if** E_u is empty **then**
23. Set X to be X union U_j and go to 31.
24. **else**
25. Go to 17.
26. **end if**
27. **end if**
28. For every edge e in the cycles of M_i, update C_e to C_e union M_e.
29. **end if**
30. **end for**
31. Add (a,b) absorbing set to the list A.
32. If A is not equal to W, go to 9.
33. If X is empty, all absorbing sets have been removed. Otherwise, it is not possible to remove the sets in X.

The basic concepts in Algorithm 1 follow. First, we select the elementary absorbing sets that we wish to eliminate. These sets must be determined according to the code parameters (for example, column weight, girth, etc.). Next, among this group of sets, we examine the smallest absorbing set. We look for binary versions of this set in the unlabeled (that is, the binary version of the) Tanner graph. If the fundamental cycles of these absorbing sets satisfy the formula in Lemma 1, we modify the weight of one of the edges to some other non-zero element in GF(q). We make this choice in such a way that the previously removed absorbing sets are not brought back. The process continues once the current absorbing set has been removed.

Let us see an error profile (giving errors due to various absorbing sets) for a particular choice of code. We show the effects of the previous algorithm on these errors (Table 11.2).

Each of the error types refers to an (a,b) absorbing set which causes the error. The algorithm removes all of these absorbing sets (up to the (8,2) set), which dramatically reduces the number of errors. Here, the code is a non-binary LDPC code over GF(4). The length was 2904 bits, the SNR was 5.1 dB, the code rate was 0.878, and the column weight was 4.

Next, in Fig. 11.11, we show a performance plot showing the effect of using the algorithm (which is labeled A-method). We also compare with several other algorithms which also attempt to resolve error floors through absorbing/trapping set modification. In particular, we compare against the approach presented in [17], which we refer to as the 'P-method'. This approach attempts to cancel all cycles of length l in the Tanner graph, where l is between the girth g and a l_{max} parameter. This cancellation has the effect of removing certain absorbing sets as well (in particular very small ones.) Another method we compare against, which we refer to as the 'N-method', was proposed in [18].

Here, the figure shows frame error rate (FER) versus SNR for the original codes and the three methods discussed. Note that the previously introduced algorithm produces the best overall improvement in the FER. The code length was approximately 2930 bits, the rate was 0.88, the column weight was 4, and the QSPA-FFT decoder was used for decoding.

In the case of non-binary quasi-cyclic (NB-QC LDPC) codes, which are a very practical class of non-binary LDPC codes, we have the results shown in Fig. 11.12. Here the length is approximately 1400, the rate approximately 0.81, the column weight 4, and, again, the QSPA-FFT decoder was used.

Table 11.2 Error profile for non-binary LDPC code over GF(4)

Error type	(4,4)	(5,0)	(5,2)	(6,2)	(6,4)	(6,6)	(7,4)	(8,2)	others
Original	35	7	9	11	17	21	8	10	10
After Alg.	0	0	0	0	0	0	0	0	9

Code length is 2904 bits and code rate is 0.878. The code has column weight 4 and the SNR is 5.1 dB. Shown are the error profiles due to various absorbing sets before and after the absorbing set removal algorithm

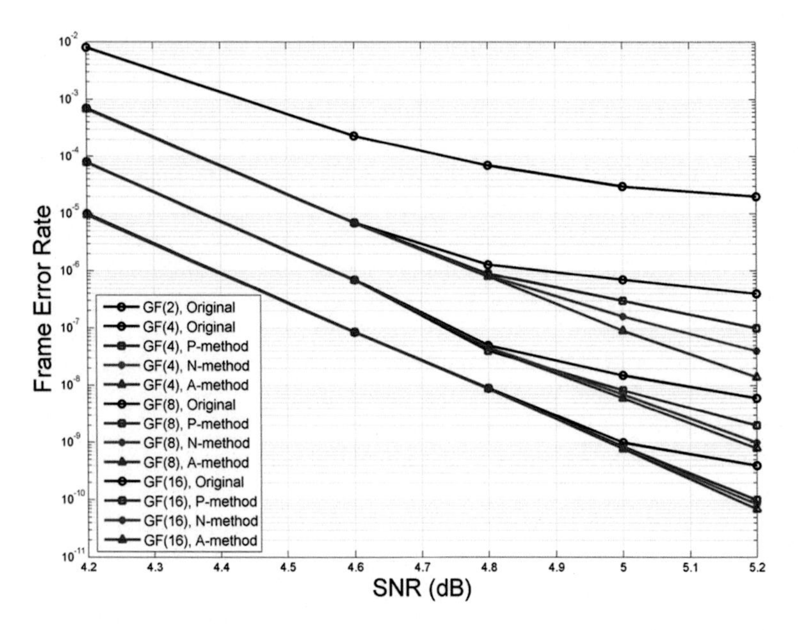

Fig. 11.11 SNR versus frame error rate (FER) for binary and non-binary LDPC codes over several fields. The codes had approximate length 2930 bits, rate 0.88, and column weight 4. The *curves* include the original, unmodified codes along with codes resulting from several methods aimed at improving non-binary LDPC codes. The method described in Algorithm 1 is labeled A-method; this method yields the most significant improvement in FER

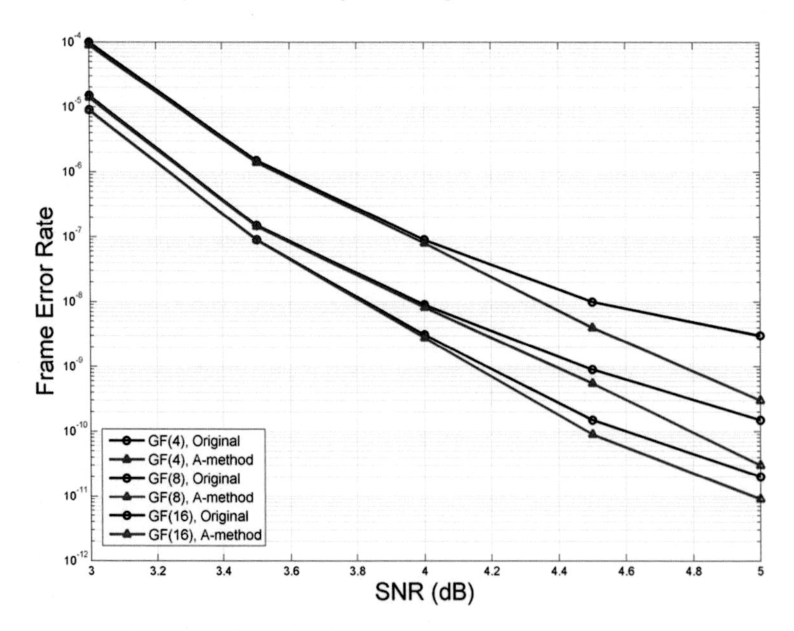

Fig. 11.12 SNR versus FER plot for non-binary QC-LPDC with different field sizes. We compare the original code to the codes improved by using the A-method from Algorithm 1. Note that the improvement is strongest in smaller field sizes

We note the fact that the improvement over the baseline diminishes as the field size q grows. The reason for this is the fact that since there are many more choices of edge weights for larger field sizes, absorbing sets naturally occur with smaller probability, so that there are fewer of them to remove through any of the possible algorithms.

11.3 Summary

In this chapter, two classes of non-standard codes were studied. The first class is composed of algebraic codes, which rely only on hard information and are suitable for applications where simple and efficient decoding is necessary, but error tolerance is more relaxed, such as inexpensive data storage devices. The second class is made up of LDPC codes, which have more complex decoding, but offer extremely good performance. LDPC codes are thus suitable for applications that require extreme reliability. Flash devices occupy the entire spectrum between these two endpoints. Both of the classes of advanced codes we study offer significant improvements over their traditional counterparts, but, at the same time, present more challenges in terms of constructions, design choices, and analyses.

First, we examined asymmetric algebraic codes. We motivated this study by looking at asymmetric channels modeling the physical channels of data storage devices. It was shown that using conventional symmetric codes was wasteful, either in terms of code rate or error-correcting ability. Two types of asymmetric codes were discussed: graded-bit error-correcting codes based on the tensor-product operation, and dynamic threshold-based codes relying on the dynamic thresholding side-information technique.

Afterwards, we looked at non-binary LDPC codes, which offer better performance compared to the frequently-studied binary LPDC codes. We examined the error-floor problem in the non-binary case and defined the underlying non-binary absorbing set objects that result in the error floor. We introduced an algorithm that can efficiently remove the problematic small absorbing sets from the Tanner graph of a non-binary LDPC code. Simulation results showed significant improvement over baseline non-binary LDPC codes.

References

1. E. Yaakobi, L. Grupp, P.H. Siegel, S. Swanson, J.K. Wolf, Characterization and error-correcting codes for TLC flash memories, in *Proceedings on IEEE International Conference on Computing, Networking, and Communications (CCNC)*, Maui, HI, Jan–Feb 2012, pp. 486–491
2. J.K. Wolf, An introduction to tensor product codes and applications to digital storage systems, in *Proceedings on IEEE Information Theory Workshop (ITW)*, Punta del Este, Uruguay, Oct 2006, pp. 6–10

3. R. Gabrys, E. Yaakobi, L. Dolecek, Graded bit-error-correcting codes with applications to flash memory. IEEE Trans. Inf. Theory **59**(4), 2315–2327 (2013)
4. R. Gabrys, F. Sala, L. Dolecek, Coding for unreliable flash memory cells. IEEE Commun. Lett. **18**(9), 1491–1494 (2014)
5. H. Zhou, A. Jiang, J. Bruck, Error-correcting schemes with dynamic thresholds in non-volatile memories, in *Proceedings on IEEE International Symposium Information Theory (ISIT)*, St. Petersburg, Russia, Jul–Aug 2011, pp. 2143–2147
6. F. Sala, R. Gabrys, L. Dolecek, Dynamic threshold schemes for multi-level non-volatile memories. IEEE Trans. Commun. **61**(7), 2624–2634 (2013)
7. M.C. Davey, D. MacKay, Low-density parity check codes over GF(q). IEEE Commun. Lett. **2** (6), 165–167 (1998)
8. A. Bazarsky, N. Presman, S. Litsyn, Design of non-binary quasicyclic LDPC codes by ACE optimization, in *Proceedings on IEEE Information Theory Workshop (ITW)*, Seville, Spain, Sep 2013, pp. 1–5
9. L. Dolecek, D. Divsalar, Y. Sun, B. Amiri, Non-binary protograph-based LDPC codes: enumerators, analysis, and designs. IEEE Trans. Inf. Theory **60**(7), 3913–3941 (2014)
10. I. Andriyanova, D. Maurice, J.-P. Tillich, Quantum LDPC codes obtained by non-binary constructions, in *Proceedings on IEEE International Symposium Information Theory (ISIT)*, Cambridge, MA, Jul 2012, pp. 343–347
11. L. Dolecek, Z. Zhang, V. Anantharam, M.J. Wainwright, B. Nikolic, Analysis of absorbing sets and fully absorbing sets of array-based LDPC codes. IEEE Trans. Inf. Theory **56**(1), 181–201 (2010)
12. T.J. Richardson, Error floors of LDPC codes, in *Proceedings on IEEE Allerton Conference on Communication, Control, and Computing*, Monticello, IL, Oct 2013, pp. 1426–1435
13. J. Wang, K. Vakilinia, T.-Y. Chen, T. Courtade, G. Dong, T. Zhang, H. Shankar, R. Wesel, Enhanced precision through multiple reads for LDPC decoding in Flash memories. IEEE J. Sel. Areas Commun. **32**(5), 880–891 (2014)
14. J. Wang, L. Dolecek, R. Wesel, The cycle consistency matrix approach to absorbing sets in separable circulant-based LDPC codes. IEEE Trans. Inf. Theory **59**(4), 2293–2314 (2013)
15. D.V. Nguyen, S.K. Chilappagari, M.W. Marcellin, B. Vasic, On the construction of structured LDPC codes free of small trapping sets. IEEE Trans. Inf. Theory **58**(4), 2280–2302 (2012)
16. M. Karimi, A.H. Banihashemi, Efficient algorithm for finding dominant trapping sets of LDPC codes. IEEE Trans. Inf. Theory **58**(11), 6942–6958 (2012)
17. C. Poulliat, M. Fossorier, D. Declercq, Design of regular $(2, d_c) -$ LDPC codes over GF(q) using their binary images. IEEE Trans. Commun. **56**(10), 1626–1635 (2008)
18. T. Nozaki, K. Kasai, K. Sakaniwa, Analysis of error floors of non-binary LDPC codes over MBIOS channel, in *Proceedings on IEEE International Conference on Communication (ICC)*, Kyoto, Japan, Jun 2011, pp. 1–5

Chapter 12
System-Level Considerations on Design of 3D NAND Flash Memories

Chao Sun and Ken Takeuchi

Abstract This chapter introduces the design of three-dimensional (3D) NAND flash memory with the implications from the system side. For conventional two-dimensional (2D) scaling, it is facing various limitations such as lithography cost and cell-to-cell coupling interference. To sustain the trend of bit-cost reduction beyond 10 nm technology node, 3D NAND flash memory is considered as the next generation technique. Further, emerging memories called storage-class memories (SCMs) such as resistive RAM (ReRAM), phase change RAM (PRAM) and magnetoresistive RAM (MRAM) will revolutionize the storage system design. By introducing SCM into the solid-state drive (SSD), hybrid SCM/3D-NAND flash SSD and all SCM SSD achieve much higher write performance than all 3D-NAND flash SSD due to SCM's fast speed. In addition, the performance of the SSD is workload dependent. Thus, it is meaningful to obtain the design guidelines of 3D NAND flash for both all 3D-NAND flash SSD and hybrid SCM/3D-NAND flash SSD with representative real-world workloads.

Keywords NAND flash memory · Storage-class memory (SCM) · Solid-state drive · Flash translation layer · Garbage collection

12.1 Introduction

There is a growing demand for NAND flash memory due to its fast speed, low power, and high reliability. NAND flash memory based storage systems are being widely used from consumer electronic products like SD cards to enterprise applications like solid-state drives (SSDs) in servers and data centers. SSDs designed for enterprise applications are discussed in this chapter. As described in Sect. 12.2, the

C. Sun (✉) · K. Takeuchi
Chuo University, Tokyo, Japan
e-mail: sun@takeuchi-lab.org

K. Takeuchi
e-mail: takeuchi@takeuchi-lab.org

R. Micheloni (ed.), *3D Flash Memories*, DOI 10.1007/978-94-017-7512-0_12

bottleneck of NAND flash based SSDs lies in the write performance rather than the read performance due to the inherent characteristics of the NAND flash. Hence, the write performance of the SSD should be improved to meet the increasing requirement for high performance storage in this big data era.

On the other hand, storage class memories (SCMs) such as the resistive RAM (ReRAM), phase change RAM (PRAM) and magnetoresistive RAM (MRAM) are attracting more and more attention due to their faster speed, higher endurance and lower power consumption than the NAND flash memory. SCMs bridge the bandgap between the DRAM and NAND flash memory. According to the speed and capacity, SCM devices are divided into two categories: DRAM-like and NAND-like. DRAM-like SCMs are called memory-type SCM (M-SCM) such as the MRAM while NAND-like SCMs are named storage-type SCM (S-SCM) such as the ReRAM and PRAM. The hybrid M-SCM/3D NAND flash SSD and all S-SCM SSD have been proposed as the next generation SSDs.

In this chapter, techniques to improve the write performance of the SSD are introduced. Three SSDs including all 3D-NAND flash SSD, hybrid M-SCM/3D-NAND flash SSD and all S-SCM SSDs are discussed. Evaluated with representative real-world workloads, it is found that the write performance of the 3D-NAND flash-based SSDs is workload dependent. According to the system-level evaluation results, the design guidelines of the 3D-NAND flash for the SSDs are obtained.

12.2 Background of Solid-State Drive

Figure 12.1 shows the memory hierarchy of the computer system. Top layer's memories have a faster speed but smaller capacity (high bit cost). In contract, bottom layer's memories have a slower speed but larger capacity (low bit cost). SCMs, NAND flash and hard disk drive (HDD) are non-volatile. In the memory

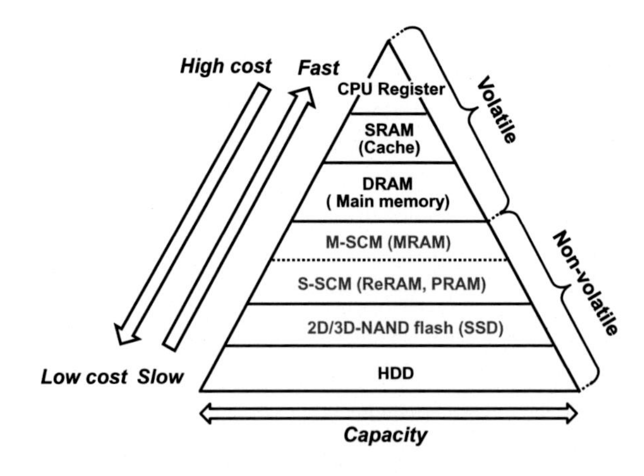

Fig. 12.1 Memory hierarchy

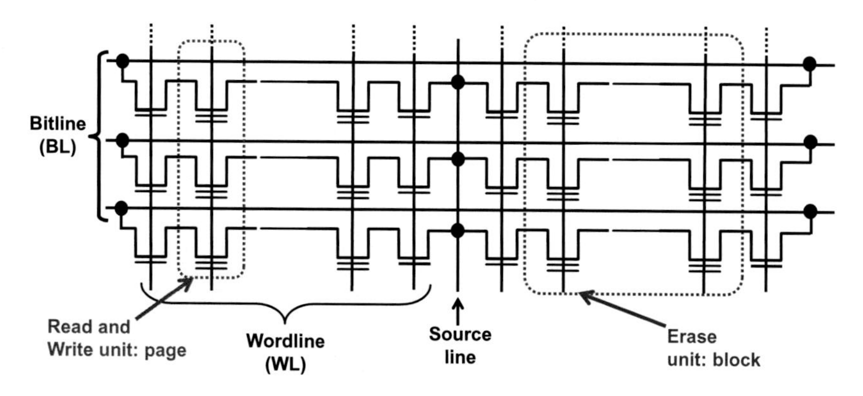

Fig. 12.2 NAND flash organization [1]

hierarchy, NAND flash memory lies between the SCM and HDD. Since the bit cost of the NAND flash memory is continuously reducing by scaling and multi-bit technology, SSD becomes cost-effective as an alternative of HDD.

Figure 12.2 illustrates the organization of the NAND flash memory [1]. The memory cells connected with the same word-line consists of a page, which is the read and write unit of the NAND flash. A block is the erase unit. There are typical 128–256 pages in a block for the multi-cell level (MLC) NAND flash.

The architecture of the all 3D-NAND flash SSD, hybrid M-SCM/3D-NAND flash SSD and all S-SCM SSD are described in Fig. 12.3. The key component of the SSD is the SSD controller that integrates the flash translation layer (FTL) enabling SSD to work as a block device. As shown in Fig. 12.4, the basic but very critical function in the FTL is the logical-to-physical address translation, required due to the prohibited in-place overwrite characteristics of the NAND flash memory. According to the mapping granularity, the address translation can be classified into the page-level mapping, block-level mapping and hybrid mapping. When a page data overwrite

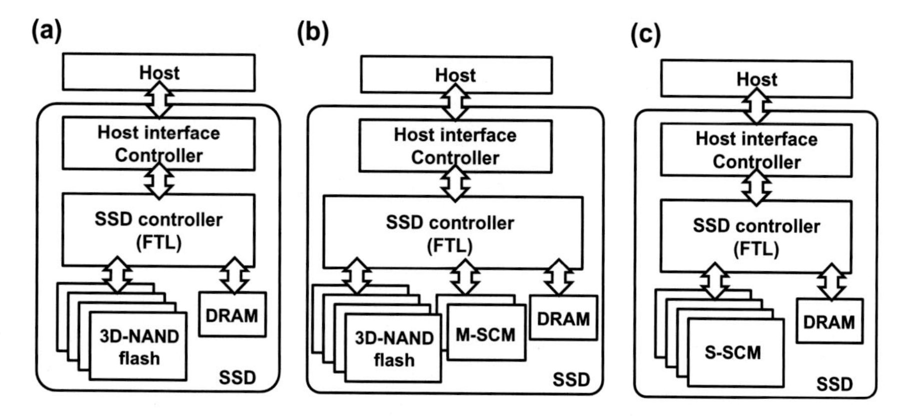

Fig. 12.3 SSD architectures of **a** all 3D-NAND flash SSD, **b** hybrid M-SCM/3D-NAND flash SSD and **c** all S-SCM SSD

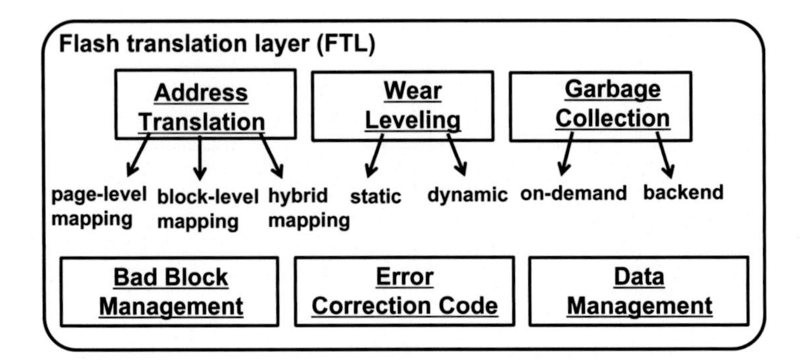

Fig. 12.4 Essential functions in the flash translation layer (FTL)

happens, the old data is read from the old page, merged with the new data, and written to a new page. After that, the old page is invalidated. Hence, there are three page statuses in the SSD: free page, page with the valid data and page with the invalid data. Frequently accessed data (hot data) will create massive number of invalid pages. When the SSD free space reduces to below a threshold (a few free blocks in a plane), an operation, garbage collection (GC), in FTL, is triggered on-demand or in the backend to reclaim one or more old blocks. Before erasing an old block, all the valid pages in the block have to be copied to the free spaces in another block, as shown in Fig. 12.5 [2]. Thus, the latency of the GC increases with the number of valid pages in the recycling block. When such page-copy overhead is large, the GC would become the bottleneck of the SSD write performance. Furthermore, the wear leveling in the FTL guarantees the NAND flash blocks are worn out evenly to maximize the lifetime of the SSD. According to whether the static data in NAND flash blocks are periodically moved around or not, wear leveling can be classified to static and dynamic wear leveling. Other functions like the error correction code (ECC) and bad block management (BBM) are also essential.

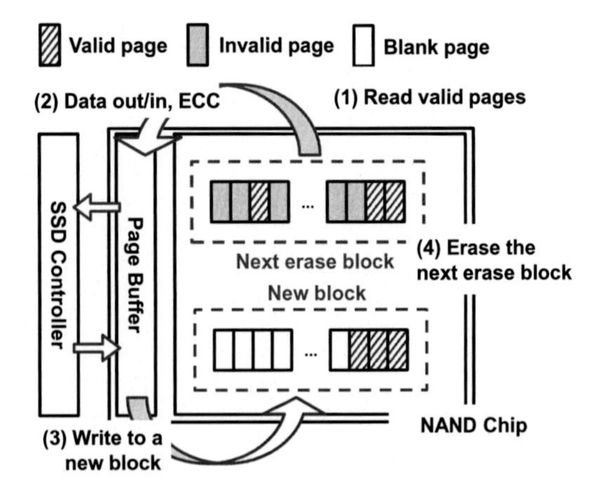

Fig. 12.5 Garbage collection (GC) operation [2]

12.3 SSD Performance Improvement Techniques

This chapter introduces three techniques to improve the write performance of 3D-NAND flash-based SSD: storage engine assisted SSD (SEA-SSD), logical block address (LBA) scrambled SSD and hybrid M-SCM/3D-NAND flash SSD. The first two techniques are based on the SSD controller and middleware co-design. The last technique introduces the M-SCM into the SSD system. Finally, the design of the all S-SCM SSD, as the long-term solution, is presented as well.

12.3.1 Storage Engine Assisted SSD (SEA-SSD)

Database is one of the most widely used applications in enterprise servers. The middleware, storage engine of the database, controls when and where the data should be stored in the storage. Therefore, the first technique, storage engine assisted SSD (SEA-SSD), co-designs the storage engine with the SSD controller to improve the SSD write performance for the database. It is based on the idea that the upper layer of the storage stack holds much richer information than the lower layer. For the current SSD, it only receives the information from the block device layer of the operation system (OS), which includes the data, data size and data address. The information is quite limited.

Figure 12.6 shows the comparisons between the conventional computer system and proposed computer system with SEA-SSD [3]. Due to reasons (i) the storage layers like the file system, block layer etc. are optimized for the conventional HDD but not for SSD, conventional OS is inefficient for SSD storage [4–8], (ii) great engineering effort is required if the hint messages have to go through all the layers in the OS, the OS is simply bypassed in the proposed SEA-SSD. Hints are passed from the SE (Storage Engine) to the SSD controller in order to store data more efficiently.

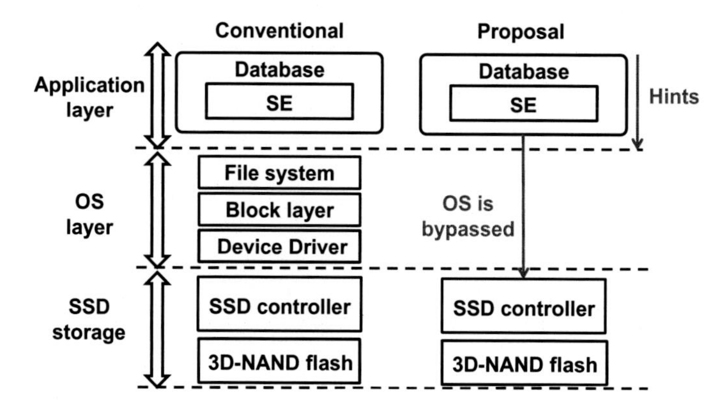

Fig. 12.6 SEA-SSD concept [3]

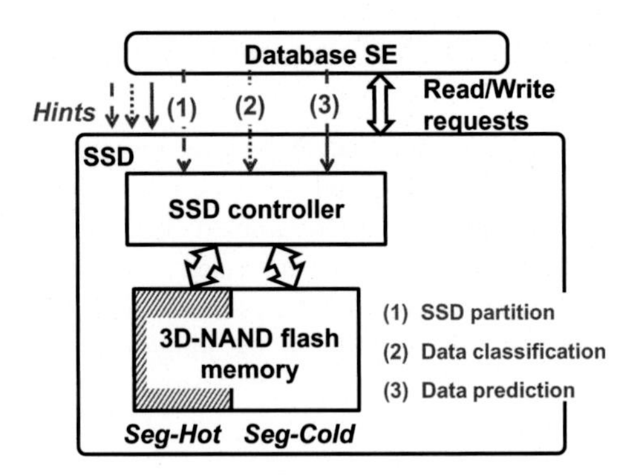

Fig. 12.7 SEA-SSD architecture [3]

Figure 12.7 presents the architecture of the SEA-SSD [3]. Each 3D-NAND flash chip is divided into two logical segments. *Seg-Hot* for the hot data (frequently accessed data) and *Seg-Cold* for the cold data (seldom accessed data). By aggregating data with similar activities in the same block, the GC overhead can be reduced. To determine the size of each segment, the first kind of hint is sent to the SSD controller, which is based on the strong correlation between the SE settings and hot data size. For the Innodb storage engine, the settings include the buffer pool and redo log sizes. The buffer pool caches the frequently accessed data (hot data) while redo log is used for crash recovery that guarantees the durability of the Innodb storage engine. The second hint is for data preliminary classification with a dynamic data model. If the data is judged as hot by the storage engine when it is flushed, logical "1" is sent to the SSD controller indicating that the data is hot and thus stored in the Seg-Hot, as shown in Fig. 12.8 [3]. Otherwise, it is simply stored in Seg-Cold. As the activities of the data stored in the 3D-NAND flash memory change with time, the data is predicted again with the third hint when the GC is triggered in the SSD. The third hint is the logical address of the page data that enters the flush list for the first time, since the data will be flushed to the SSD soon. As shown in Fig. 12.8 [3], such data should be stored in the Seg-Hot while other data are stored in Seg-Cold after the GC.

To evaluate the SEA-SSD, a database and SSD coupled simulator has been developed, which is over 20-times faster than the virtual platform, based on the Synopsys Platform Architect [9]. From the evaluation results, the write performance

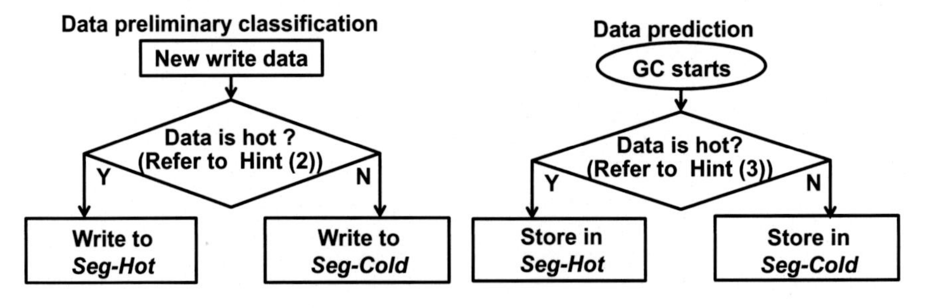

Fig. 12.8 SEA-SSD data management algorithms [3]

is improved by 24 % at maximum. Moreover, maximum 16 % energy consumption and 19 % lifetime enhancement are achieved.

12.3.2 Logical Block Address (LBA) Scrambled SSD

SEA-SSD is a design specially optimized for the database application. As a general solution to improve the write performance of the all 3D-NAND flash SSD for all applications, the logical block address (LBA) scrambled SSD is proposed [10]. A middleware LBA scrambler based on the address remapping technology is added to the existed SSD system.

The concept of the LBA scrambler is to reduce the page-copy overhead of the GC actively, as explained in Fig. 12.9 [10]. There are three kinds of pages in the SSD: valid pages, free pages and invalid pages. Among the valid pages, pages that still own free space are called *fragmented pages*. As mentioned in Sect. 12.2, all the valid pages in the next erase block have to be copied to the free space of another block, which leads to the degradation of the SSD write performance. Thus, LBA scrambler is proposed to write small data to the remaining free space of the fragmented pages in the next erase block actively. Due to the overwrite, all these fragmented pages in the next erase block become invalid and the data in the SSD become less fragmented.

Figure 12.10 illustrates the LBA scrambled SSD based computer system [10]. To achieve the address remapping, the LBA scrambler introduced another logical address called scrambled LBA (SLBA). After LBA scrambling, the data address SLBA is sent to the SSD controller. SSD controller writes data to a physical 3D-NAND flash page by the logical-to-physical table in the FTL (SSD controller).

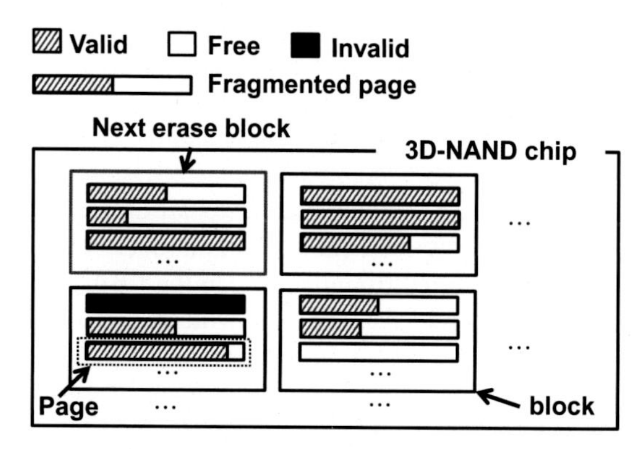

Fig. 12.9 Concept of the LBA scrambler [10]

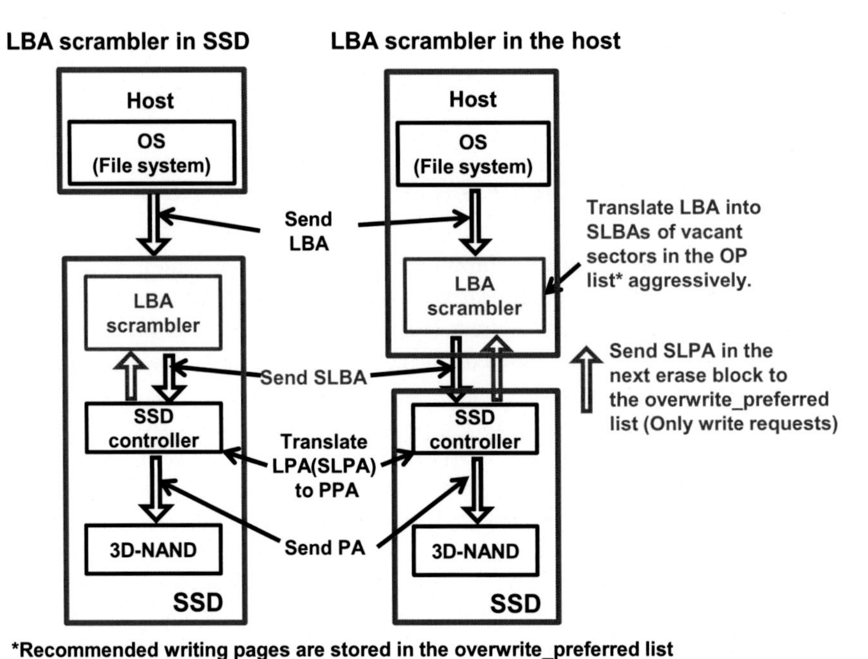

Fig. 12.10 LBA scrambled SSD [10]

To inform the LBA scrambler about the fragmented page addresses, the scrambled logical page address (SLPA) of the fragmented page in the next erase block is sent to the LBA scrambler by the SSD controller. To record the address remapping between the LBA and SLBA, the LBA_to_SLBA table and unused_SLBA table are maintained in DRAM. As shown in Fig. 12.10, the LBA scrambler can locate in either the SSD or host [10]. When it locates in the SSD, a large DRAM capacity will be required for the SSD, but no interface modification is necessary. In contrast, a small DRAM capacity is required if LBA scrambler locates in the host. However, due to the communication between the LBA scrambler and the SSD controller, the interface of the SSD has to be upgraded. The algorithm flowchart of the LBA scrambler is shown in Fig. 12.11. For every *N* write requests, the hint (overwrite_preferred list) is updated with information transferred from the FTL to the LBA scrambler [10]. By referring to the overwrite_preferred list, the new write requests are generated for writing the fragmented pages in the next erase block actively. Moreover, the unaligned writes will create fragmented pages and additional overwrites as shown in Fig. 12.12. With the LBA scrambler's address remapping, the problem of unaligned writes for NAND flash memory can be eliminated.

From the evaluation results, maximum 394 % write performance improvement, 56 % energy consumption reduction and 55 % endurance enhancement are achieved with the LBA scrambler, compared with the SSD system without the LBA scrambler.

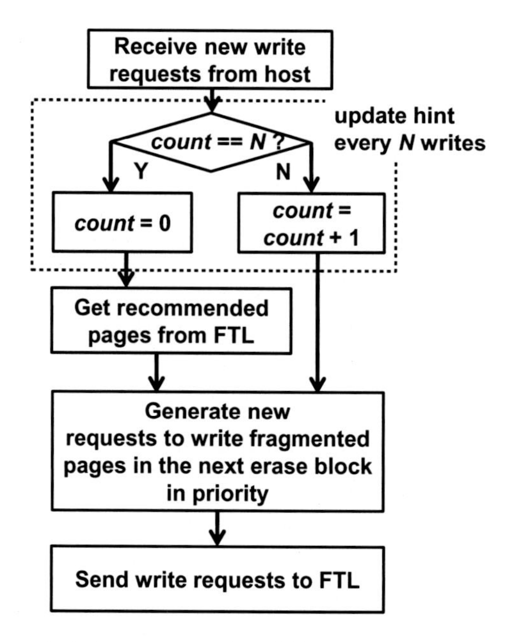

Fig. 12.11 LBA scrambled algorithm flowchart [10]

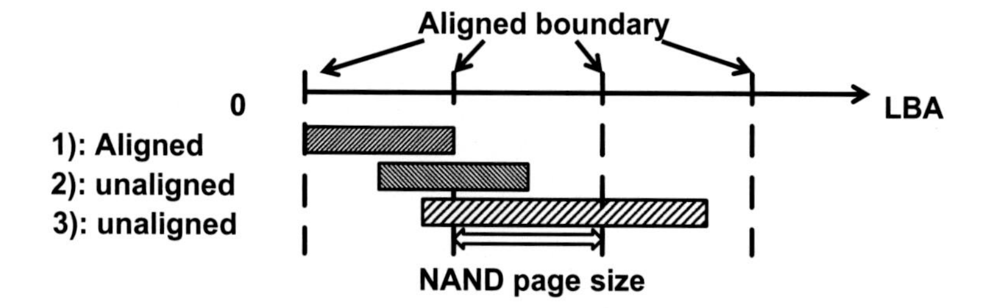

Fig. 12.12 Aligned and unaligned writes for the NAND flash memory

12.3.3 Hybrid M-SCM/3D-NAND Flash SSD

Both the SEA-SSD and LBA scrambled SSD adopts a middleware and SSD controller co-design methodology to improve the write performance of all 3D-NAND flash SSD by reducing SSD GC overhead. However, the write performance improvement of the SSD is limited by the read and write performance of the 3D-NAND flash memory.

On the other hand, SCM is much faster, more energy-efficient and endurable than 3D-NAND flash memory. It is non-volatile and supports in-place overwrite. Thus, both memory and storage systems are under a revolution due to SCM, as shown in Fig. 12.13 [11]. M-SCM is used in both memory and storage systems. In

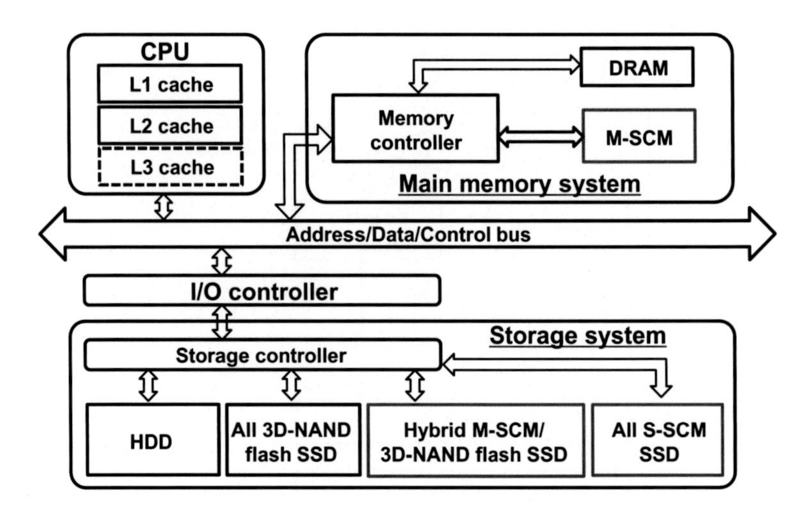

Fig. 12.13 Main memory and storage systems revolution due to SCM [11]

addition, S-SCM is used only in the storage system. By introducing SCM into the SSD system, the hybrid M-SCM/3D-NAND flash SSD is proposed to improve the write performance of the all 3D-NAND flash memory [12]. From the measurement results of the SCM (ReRAM) device, the verify cycles for write success vary with the write/erase (W/E) cycle. Thus, NAND-like interface with ready/busy status is adopted for the SCM. As shown in Fig. 12.3, M-SCM is used as a storage device for the SSD rather than a simple cache [13]. The data fragmentation suppression algorithm [12] and cold data eviction algorithm [13] are developed for the hybrid M-SCM/3D-NAND flash SSD considering both the data activity and data size. In the SSD controller, a least recently used (LRU) table is used to record the page

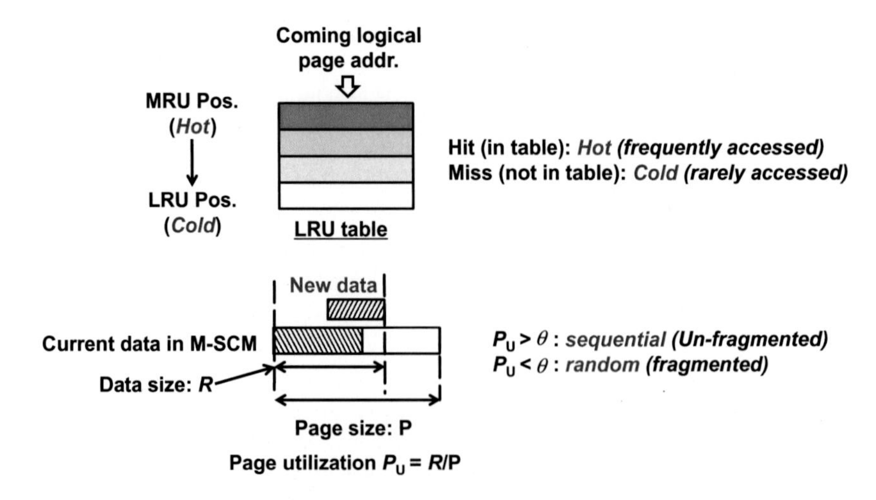

Fig. 12.14 Criteria for data classification [11]

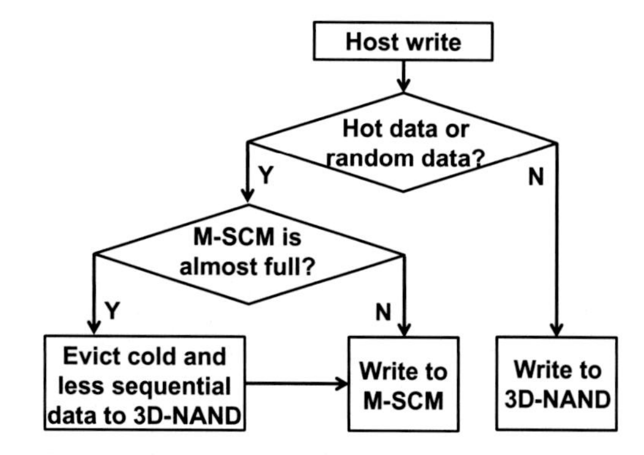

Fig. 12.15 Data management algorithm of the hybrid M-SCM/NAND flash SSD [13]

data's access history. As shown in Fig. 12.14, when the logical page address (LPA) of the page data hits the LRU, the page data is considered as hot (frequently accessed). Otherwise, the page data is judged as cold (seldom accessed). In addition, according to page utilization (data size divided by the page size), the page data are divided into two kinds: random (fragmented) and sequential (un-fragmented). When the page data size is over a threshold (θ), it is considered as the sequential data. The data storage policy of the hybrid SSD is to store hot data or random data in the M-SCM while cold and sequential data in the 3D-NAND flash. Hot data can update in-place and random data can accumulate and become sequential in M-SCM. The algorithm flowchart of the data management algorithm of the hybrid M-SCM/3D-NAND flash SSD is described in Fig. 12.15 [13]. When M-SCM becomes almost full, cold and less fragmented M-SCM data are evicted to the 3D-NAND flash. For the eviction procedure, the threshold to judge the data is sequential or random is a dynamic value, increasing/decreasing according to the data storage status in M-SCM. When it is hard to find the eviction candidates with the current threshold, the threshold is reduced to relax the restriction.

For the hybrid M-SCM/3D-NAND flash SSD, there are two important design considerations. Firstly, understand the M-SCM capacity and latency requirement for representative applications. Secondly, understand the effect of the 3D-NAND organization on the SSD write performance. Before the analyses, the SSD workloads are classified into four categories: hot and random, hot and sequential, cold and random, cold and sequential, as shown in Fig. 12.16 [14]. A large value of the average overwrite, defined as the total write data size divided by the user data size, indicates the workload is hot since it contains many hot data. In addition, the percentage of the random write request determines whether the workload is random or sequential. Here, half of the NAND flash page size is used as the threshold to judge the page is random or sequential.

From the evaluation results, increasing M-SCM capacity is more efficient to boost the write performance of the SSD than increasing the 3D-NAND flash overprovisioning (additional capacity over the user data size) with hot and random workload. Both increasing M-SCM capacity and 3D-NAND flash overprovisioning is capable of improving the SSD write performance. However, neither increasing

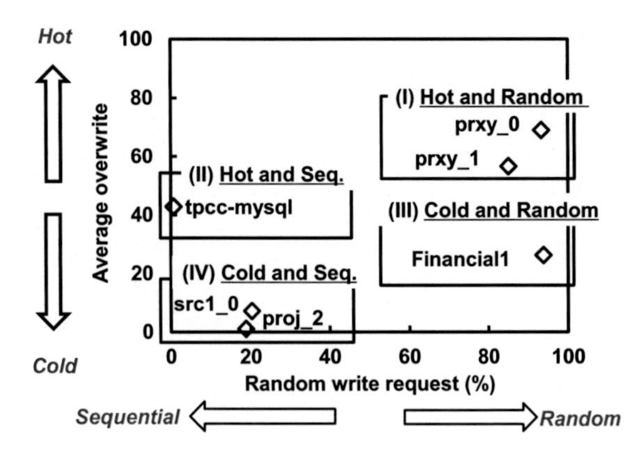

Fig. 12.16 SSD workload classification according to the data activity and size [14]

M-SCM capacity nor 3D-NAND flash overprovisioning is very effective for the write performance boost of the cold and sequential workload. Therefore, introducing M-SCM to the SSD is most suitable for the hot and random workload but not cost-efficient for the cold and sequential workload. Generally, less than 10 % M-SCM/3D-NAND flash capacity ratio is enough for the representative workloads with a fixed M-SCM latency of 100 ns/sector. On the other hand, a faster speed can be achieved by increasing the chip area for memory chip design. For example, write speed can be increased by enlarging the internal write unit and read speed can be improved by adding select devices for the reduction of bit line capacitance. When the speed of M-SCM is increased, the maximum throughput of the hybrid M-SCM/3D-NAND flash SSD is improved. As shown in Fig. 12.17, the write performance of the hybrid SSD saturates when the capacity of the M-SCM is over a threshold for a proxy server application (prxy_0), which is a hot and random intensive [11]. For other workloads such as Financial1, which is from a financial server, there is no trend of saturating when increasing the M-SCM/3D-NAND flash

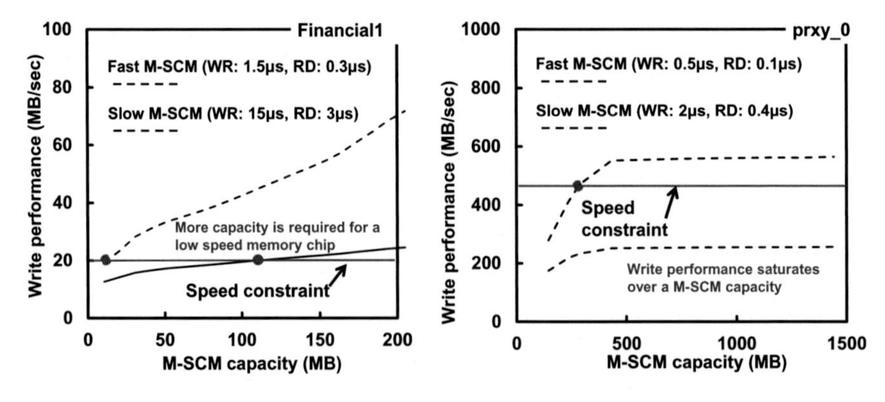

Fig. 12.17 SSD write performance dependency on the M-SCM capacity, latency and application [24]

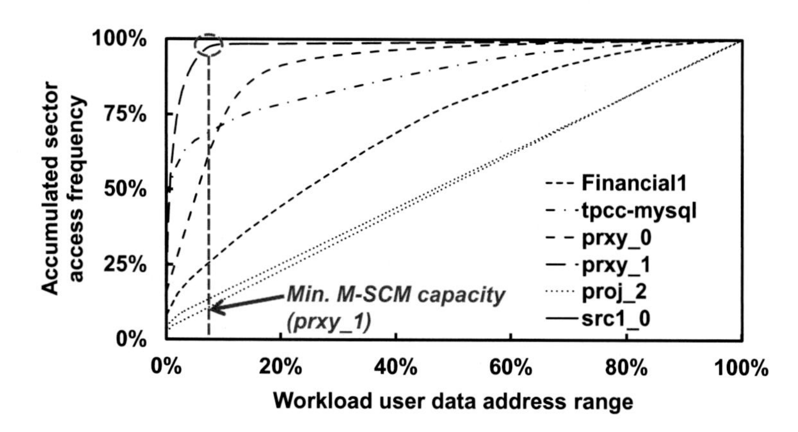

Fig. 12.18 Minimum M-SCM capacity requirement

capacity ratio. From the evaluation results of various workloads, the write performance of the hybrid M-SCM/3D-NAND flash SSD is workload/application dependent. Moreover, less M-SCM capacity is required for a faster speed M-SCM to reach the target application throughput. From the system point of view, there is a tradeoff between the M-SCM capacity requirement and M-SCM speed. Thus, there would be a cost-effective M-SCM chip design for a certain application, which is discussed in [11] by establishing optimistic and pessimistic SCM area cost models.

The minimum M-SCM capacity for the hybrid SSD is shown in Fig. 12.18 by analyzing the SSD workload. It illustrates the relationship between the accumulated sector write frequency and the address range of the write user data. For example, 25 % access frequency at 20 % user data address range means 25 % of the accesses occur at the top 20 % address of the user data. The turning point of the curve indicates the end of the frequently accessed data, usually random data, which requires high input output per second (IOPS). High slope value of the curve shows the most critical data, determining the minimum M-SCM capacity for the hybrid SSD. For Financial1 workload, M-SCM capacity should be over 40 % of the user data size to cover 75 % of the sector accesses. However, due to the temporal and spatial localities, the actual required SCM capacity as a write cache buffer is much smaller than 40 %. The rising trend of the curve is consistent with the results in Fig. 12.17. Increasing M-SCM capacity is effective to improve the hybrid SSD throughput for Financial1 workload. Further, as the slope value of the "Financial1" curve is smaller than that of prxy_0 and prxy_1, increasing SCM capacity is more effective for boosting the performance of the proxy server applications (prxy_0 and prxy_1). From Fig. 12.18, M-SCM capacity of less than 20 % of the user data size is enough for the proxy server application.

Since the conventional planar scaling of the NAND flash memory is facing various limitations making it harder and harder to reduce the fabrication cost and guarantee the memory reliability, 3D technology becomes a viable way to continue the trend of bit cost reduction for the NAND flash memory. Several 3D-NAND

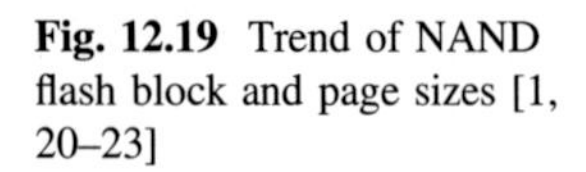

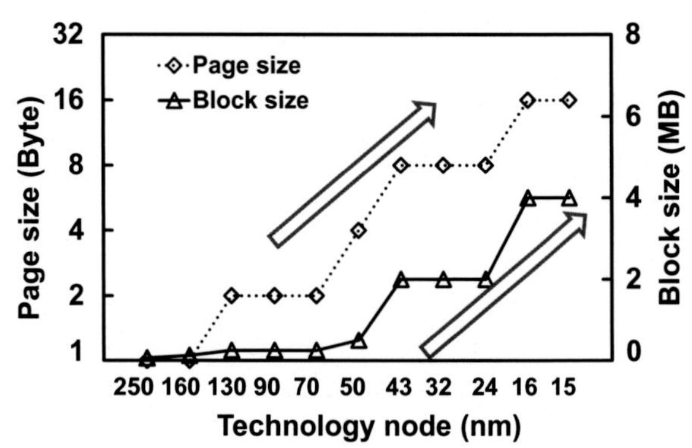

Fig. 12.19 Trend of NAND flash block and page sizes [1, 20–23]

flash architectures have been proposed. For example, terabit cell array transistor (TCAT) [15], Pipe-Shaped bit-cost scalable (P-BiCS) [16], vertical stacked array transistor (VSAT) [17], and dual control gate surrounding floating gate (DC-SF) [18]. There are two types of the 3D array: vertical channel and vertical gate. The current flows vertically and horizontally for the vertical channel type array and vertical gate array, respectively. 3D-NAND increases the bit density in the vertical direction (Z-dimension) in additional to the XY-dimensions. In case of the PiBCS 3D-NAND flash memory, the capacity of the 3D-NAND flash is increased by stacking more layers, which also compensates the problem of the reduced cell density in the XY-dimensions due to the non-scalable BiCS hole's diameter [19].

For design of the 3D-NAND flash memory, the NAND organization is critically important to the performance and cost of the circuits. A group of NAND flash memory cells is connected in series as a NAND flash string. Multiple NAND flash strings sharing the same substrate consists a NAND flash block as the erase unit. By asserting a high voltage on the substrate to eject the electrons from the floating gate of the memory cells in the block, the erase operation is executed. In the block, the memory cells connected with the same word-line is a page. It is the read/write unit. With the scaling, there is an increasing trend for the page and block sizes of the

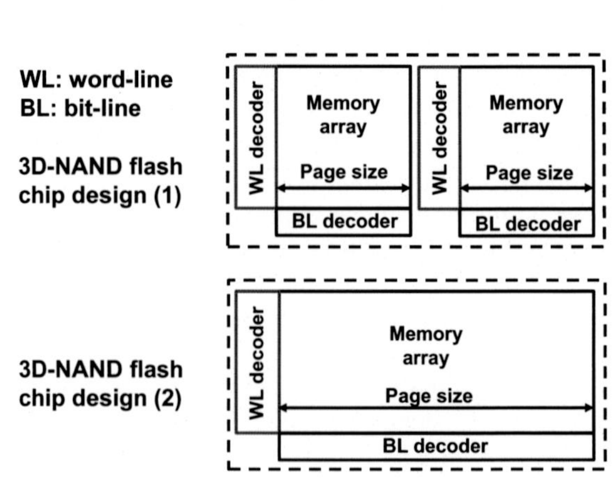

Fig. 12.20 Chip design of the 3D-NAND flash memory [24]

NAND flash memory, as shown in Fig. 12.19 [1, 20–23]. The typical size of the NAND flash page size is 8 kB and block size is 2 MB (256 pages in a block). With the advent of 3D-NAND flash memory, larger page and block sizes can be easily adopted. Take P-BiCS 3D-NAND for example, the block size of the NAND flash is doubled by doubling the stack layers. On the other hand, as shown in Fig. 12.20, adopting a large page size design reduces the word-line decoder area overhead of the NAND flash memory chip compared to the design of adopting a small page size [24]. However, it is not true that larger page or block size is better for the performance of the real-world applications.

Figure 12.21 shows the evaluation results of the block size sensitivity analysis for the all 3D-NAND flash memory [24]. The page size is fixed to 16 kB. Take

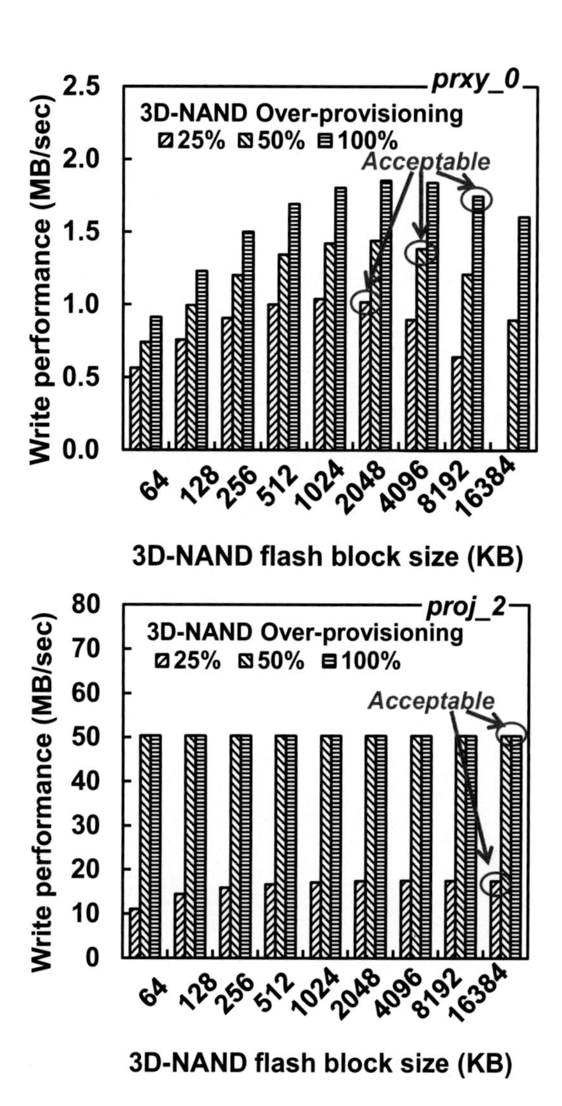

Fig. 12.21 Block size evaluation of the all 3D-NAND flash memory [24]. 2 MB is the typical block size [25]

prxy_0, a firewall/web proxy server workload for example, there is a maximum value for the write performance of the all 3D-NAND flash SSD with a certain capacity. Too small or too large block size decreases the write performance. Large block size may induce long GC latency while small block size will reduce the erase throughput. Assuming 10 % write performance is tolerable for the all 3D-NAND flash SSD, the acceptable block sizes for the all 3D-NAND flash memory with 25, 50 and 100 % over-provisioning are 2, 4 and 8 MB, respectively. Higher all 3D-NAND flash SSD capacity accepts a larger block size. Moreover, for the proj_2, a project directories sever workload (cold and sequential), the write performance saturates when the block size is over a threshold. Even the block size is as large as 16 MB, there is no write performance degradation, which is great for the application of 3D-NAND flash. It is because such workload seldom leads to the triggering of the GC. The cold and sequential data just fill in the block.

The analyses of page size sensitivity of the all 3D-NAND flash SSD are presented in Fig. 12.22 [24]. The block size is fixed to 4 MB. Similar to the block size sensitivity, over large page size or small page size degrades the write performance

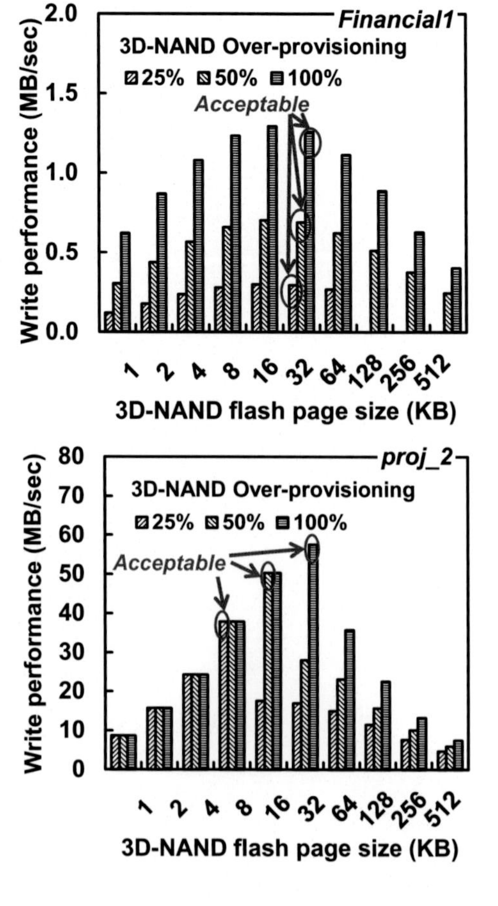

Fig. 12.22 Page size evaluation of the all 3D-NAND flash memory [24]. 8 kB is the typical page size [25]

of the all 3D-NAND flash SSD. Large page size is good for the sequential write throughput but the page overwrite count would be large (more page overwrite overhead). In addition, fewer pages exist in the case of larger page size. Thus, the GC will be triggered more frequently. In contrast, small page size induces less page overwrite overhead but it is not good for sequential writes. Further, more pages may need to be copied during GC. From the experimental results, for workloads like Financial1 a financial online transaction processing (OLTP) sever workload, the acceptable page sizes are the same at 25, 50 and 100 % SSD capacity overprovisioning cases. Only for proj_2 that is cold and sequential, larger page size is acceptable for a larger SSD capacity over-provisioning.

Moreover, larger block and page sizes are acceptable for the M-SCM/3D-NAND flash hybrid SSD, as shown in Figs. 12.23 and 12.24 [24]. Fix the page size as 16 kB

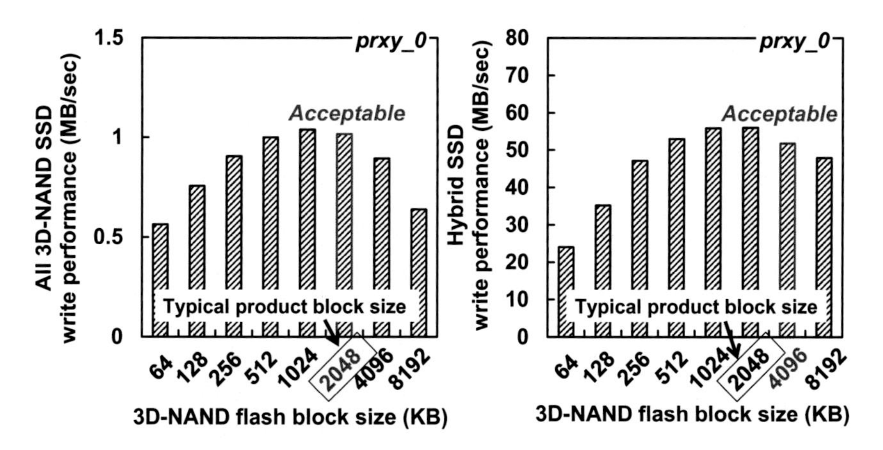

Fig. 12.23 Block size evaluation of the hybrid M-SCM/3D-NAND flash memory [24]

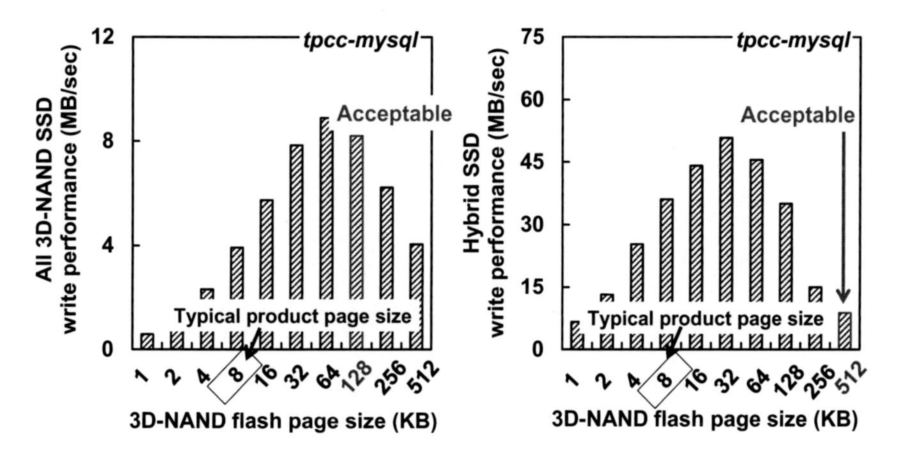

Fig. 12.24 Page size evaluation of the hybrid M-SCM/3D-NAND flash memory [24]

and 3D-NAND flash over-provisioning as 25 % for the prxy_0 workload. The acceptable block size is 2 MB in the case of the all 3D-NAND flash SSD, which is the same size of the current typical NAND flash block size [25]. In the case of the hybrid M-SCM/3D-NAND flash SSD, the M-SCM/3D-NAND flash capacity ratio is set as 8.5 %. Assuming the bottom line of the design of the hybrid SSD is its write performance should be larger than that of the all 3D-NAND flash SSD, the acceptable block size of the hybrid SSD is 4 MB, which is 2 times that of the 3D-NAND flash acceptable block size. It indicates that with M-SCM, the stacking layers of the 3D-NAND flash could be doubled for the prxy_0 workload. As shown in Fig. 12.24, the typical page size of the NAND flash product is 8 kB [25]. The acceptable page sizes for the all 3D-NAND flash SSD and hybrid M-SCM/3D-NAND flash SSD are 128 and 512 kB, respectively, with the tpcc-mysql (a relational database workload).

The comparison results of the 3D-NAND flash design for the all 3D-NAND flash SSD and hybrid M-SCM/3D-NAND flash SSD are summarized in Fig. 12.25 [24]. With M-SCM, the acceptable block and page sizes are enlarged by 4 times and 64 times, respectively. The 3D-NAND flash stacking layers could be quadruple for the hybrid SSD compared with the all 3D-NAND flash SSD, without any write performance degradation.

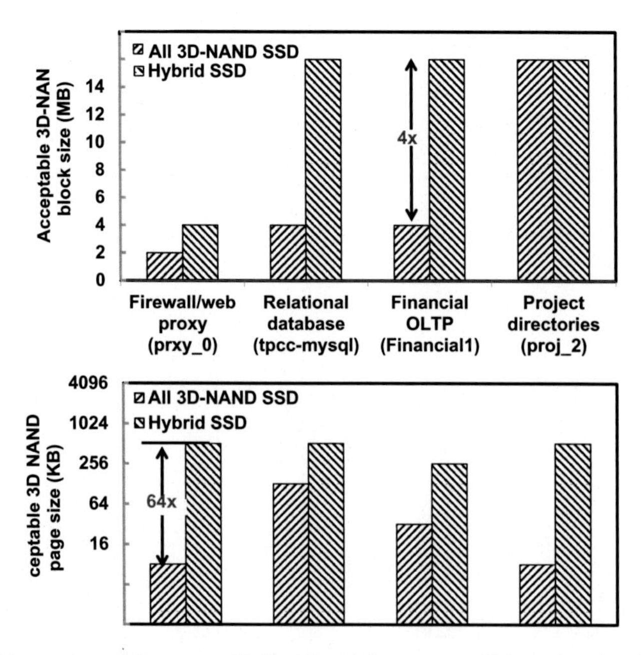

Fig. 12.25 Comparison of the acceptable 3D-NAND flash page and block sizes for all 3D-NAND flash SSD and hybrid M-SCM/3D-NAND SSD [24]

12.3.4 All S-SCM SSD

When the SCM technology matures and the cost is reduced to be competitive to that of the NAND flash memory, the all S-SCM SSD becomes a viable solution to replace the current NAND flash-based SSD. In this section, we present the wear leveling, S-SCM I/O data toggle rate, S-SCM latency design for the all S-SCM SSD.

For S-SCM candidates like ReRAM, the device endurance is limited (10^7 for 50 nm HfO2 ReRAM [26]), although it is high compared with the NAND flash memory. Therefore, wear leveling is required for S-SCM. A simple wear leveling algorithm is shown in Fig. 12.26, which is operated in a page-level. Thus, a wear leveling triggering threshold δ is maintained for each page of the S-SCM. Further, to monitor the endurance of each sector (512 Byte), that is the minimum access unit in the block device, the write/erase (W/E) cycle for each sector is maintained. When the maximum W/E cycle of the sectors in the page i is smaller than $\delta_{page(i)}$, in-place page overwrite is executed. Otherwise, the wear leveling is triggered. During the wear leveling, the data in the old page i is read out, merged with the new data and written to a new page j. After that, the wear leveling triggering threshold of page i is updated with a constant window threshold σ, to raise the bar of the wear leveling triggering. Actually, there are many hot spots (some addresses are frequently

Fig. 12.26 Wear leveling algorithm flowchart for all S-SCM SSD [27]

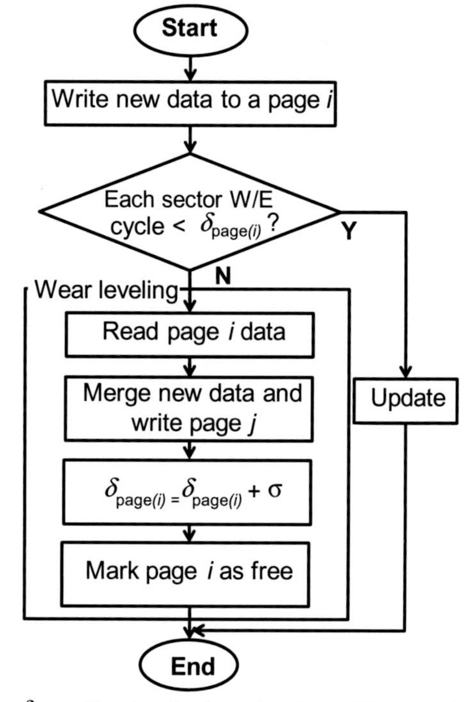

written) in the workload, the endurance of the all S-SCM SSD can be greatly enhanced with the wear leveling. For instance, with σ = 5, the maximum W/E cycle of the sector in the S-SCM without the wear leveling is over 3000 times higher than that of the all S-SCM SSD with wear leveling procedure.

By adjusting the value of σ, the wear leveling triggering interval can be controlled. A small σ triggers the wear leveling easily to make all pages wear out evenly. However, such configuration would degrade the all S-SCM SSD performance due to the additional page-copy operations. Therefore, σ could be adjusted to balance the SSD performance and endurance [27].

Figure 12.27 shows the all S-SCM SSD write performance dependency on the I/O data toggle rate, S-SCM latency (assuming the some read and write latency) and applications [27]. M-SCM latency is set as 100 ns/sector. Different from the 3D-NAND flash based SSD, there is little write performance dependency on the applications. The trend is the same. The slight difference is due to the S-SCM wear leveling and total write data size. When S-SCM speed is faster, a higher data toggle rate should be adopted to fully exploit the write performance of the all S-SCM SSD. By keeping S-SCM latency as 1 μs, the speeds of the all S-SCM SSD and hybrid M-SCM/3D-NAND flash SSD are compared in Fig. 12.28a, which illustrates the breakpoint of the I/O data toggle rate at 25 % fixed M-SCM/3D-NAND flash ratio [27]. Take tpcc-mysql workload for example, over 500 MHz I/O data toggle rate makes the all S-SCM SSD faster than the hybrid M-SCM/3D-NAND flash SSD. On the other hand, at a fixed I/O data toggle rate of 1066 MHz, the breakpoint of the S-SCM latency can be analyzed, as shown in Fig. 12.28b [27]. Faster S-SCM

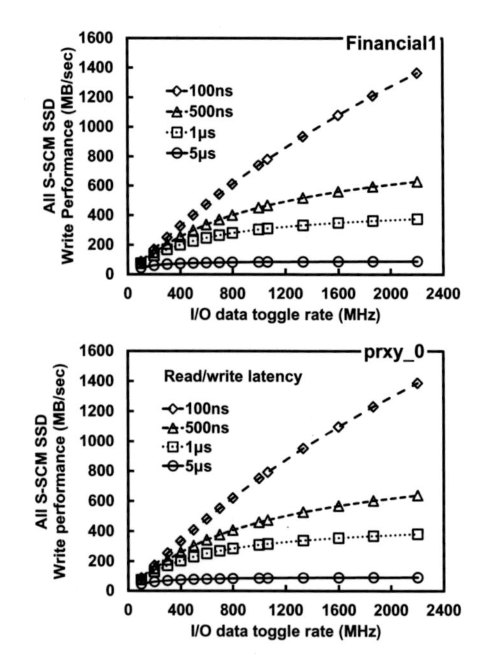

Fig. 12.27 All S-SCM SSD write performance dependency on the I/O data toggle rate, S-SCM latency and applications [27]

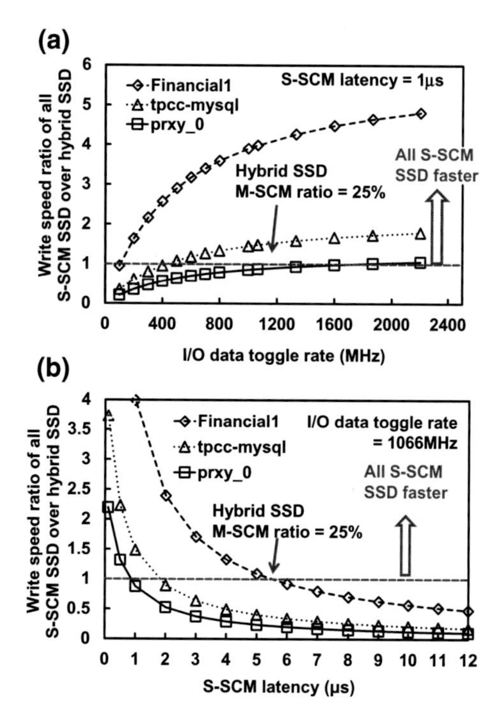

Fig. 12.28 Write speed comparison of the all S-SCM SSD and hybrid M-SCM/3D-NAND flash SSD [27]

device creates a faster all S-SCM SSD. In the case of the Financial1 workload, the hybrid M-SCM/3D-NAND flash SSD owns a faster speed than the all S-SCM SSD if the S-SCM latency is over 5 μs.

Moreover, the cost of each SSD can be compared easily according to the capacity configurations of the memories inside the SSD, by assuming the bit cost of each memory device.

It is interesting to know how speedy storage system can be achieved by SCM, compared with the NAND flash. Expression (12.1) shows the calculation of the latency of a page write operation T_{NAND}, where $T_{\mathrm{CMD\cdot N}}$ is the latency of issuing the program command and programming addresses, $T_{\mathrm{IO\cdot N}}$ is the time of loading the data into the data register of the NAND flash memory, $T_{\mathrm{MEM\cdot N}}$ is the time of storing the data from the data register to the memory array (array programming time).

$$T_{\mathrm{NAND}} = T_{\mathrm{CMD\cdot N}} + T_{\mathrm{IO\cdot N}} + T_{\mathrm{MEM\cdot N}} \quad (12.1)$$

$T_{\mathrm{CMD\cdot N}}$ is only a few clocks long. Compared with $T_{\mathrm{IO\cdot N}}$ and $T_{\mathrm{MEM\cdot N}}$, it is negligible small. $T_{\mathrm{IO\cdot N}}$ is inversely proportional to the data toggle rate $P_{\mathrm{Toggle\cdot N}}$, defined by the NAND flash interface, as shown in formula (12.2), where L_{NAND} is the NAND flash page size and $W_{\mathrm{DAT\cdot N}}$ is the width of the data bus.

$$T_{\mathrm{IO \cdot N}} = \frac{L_{\mathrm{NAND}}}{W_{\mathrm{DAT \cdot N}}} \times \frac{1}{P_{\mathrm{Toggle \cdot N}}} \tag{12.2}$$

When the data toggle rate $P_{\mathrm{Toggle \cdot N}}$ is high, $T_{\mathrm{IO \cdot N}}$ is reduced. Usually, $T_{\mathrm{IO \cdot N}}$ should be much smaller than $T_{\mathrm{MEM \cdot N}}$. If $T_{\mathrm{IO \cdot N}}$ is comparable or even higher than $T_{\mathrm{MEM \cdot N}}$, the interface becomes the memory device performance bottleneck. On the other hand, a page cache program may be supported by the NAND flash memory, which uses the internal cache register to improve the NAND flash program throughput, as shown in Fig. 12.29. During the 1st program data storing from the data register to the memory array, the 2nd program data can be input to the page cache. By the page cache, the $T_{\mathrm{CMD \cdot N}}$ and $T_{\mathrm{IO \cdot N}}$ are hidden. As a result, the latency of writing N NAND flash pages T_{N} can be calculated by (12.3) and (12.4).

$$T_{\mathrm{N}} = T_{\mathrm{CMD \cdot N}} + T_{\mathrm{IO \cdot N}} + N \times T_{\mathrm{MEM \cdot N}} \tag{12.3}$$

$$T_{\mathrm{N}} = T_{\mathrm{CMD \cdot N}} + T_{\mathrm{IO \cdot N}} + \left\lceil \frac{L_{\mathrm{DAT}}}{L_{\mathrm{NAND}}} \right\rceil \times T_{\mathrm{MEM \cdot N}} \tag{12.4}$$

Note that (12.4) is true only when the $T_{\mathrm{MEM \cdot N}}$ is much longer than the $T_{\mathrm{CMD \cdot N}}$ and $T_{\mathrm{IO \cdot N}}$. When the $T_{\mathrm{IO \cdot N}}$ is large, the page cache cannot completed be hidden. Moreover, the page cache does not work in the random write case. Without page cache effect, the sequential write latency of the NAND flash memory is expressed by (12.5):

$$T_{\mathrm{N}} = \left\lceil \frac{L_{\mathrm{DAT}}}{L_{\mathrm{NAND}}} \right\rceil \times \left(T_{\mathrm{CMD \cdot N}} + T_{\mathrm{IO \cdot N}} + T_{\mathrm{MEM \cdot N}}\right) \tag{12.5}$$

Additionally, the sequential write throughput ratio of the SCM device is expressed by (12.6), where the write unit of SCM is L_{SCM}. $T_{\mathrm{CMD \cdot S}}$ is the latency of

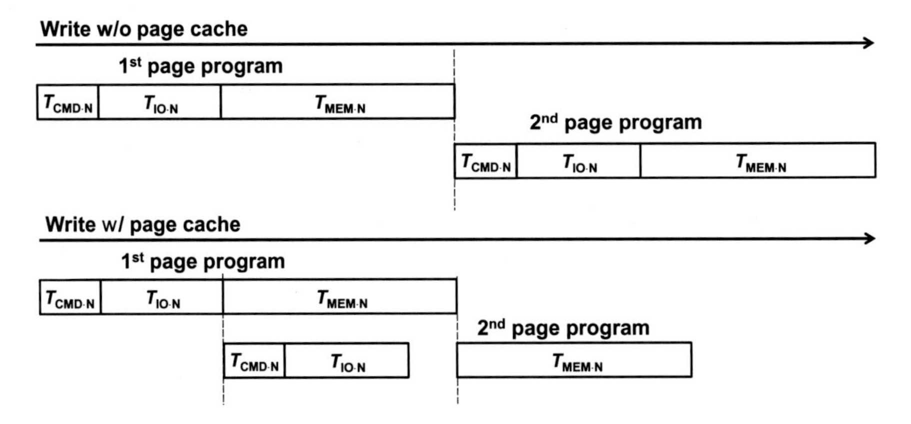

Fig. 12.29 Page cache operation of the NAND flash memory

issuing the program command and programming addresses, $T_{\mathrm{IO\cdot S}}$ is the time of loading the data into the data register of the SCM, $T_{\mathrm{MEM\cdot S}}$ is the time of storing the data from the data register to the memory array (array programming time). Since SCM is fast, page cache is not designed.

$$T_{\mathrm{S}} = \left\lceil \frac{L_{\mathrm{DAT}}}{L_{\mathrm{SCM}}} \right\rceil \times (T_{\mathrm{CMD\cdot S}} + T_{\mathrm{IO\cdot S}} + T_{\mathrm{MEM\cdot S}}) \tag{12.6}$$

According to formulas (12.2), (12.4) and (12.6), the SCM and NAND flash device write performance ratio $R_{\mathrm{S/N}}$ can be *estimated* by expression (12.7) without the considerations of the GC and overwrites in NAND flash only SSD and wear leveling in All SCM SSD. $W_{\mathrm{DAT\cdot S}}$ is the memory bus width of SCM. $P_{\mathrm{Toggle\cdot S}}$ is the data toggle rate of the SCM.

$$\begin{aligned} R_{\mathrm{S/N}} &= \frac{T_{\mathrm{CMD\cdot N}} + T_{\mathrm{IO\cdot N}} + \left\lceil \frac{L_{\mathrm{DAT}}}{L_{\mathrm{NAND}}} \right\rceil \times T_{\mathrm{MEM\cdot N}}}{\left\lceil \frac{L_{\mathrm{DAT}}}{L_{\mathrm{SCM}}} \right\rceil \times (T_{\mathrm{CMD\cdot S}} + T_{\mathrm{IO\cdot S}} + T_{\mathrm{MEM\cdot S}})} \\ &\approx \frac{\frac{L_{\mathrm{NAND}}}{W_{\mathrm{DAT\cdot N}}} \times \frac{1}{P_{\mathrm{Toggle\cdot N}}} + \left\lceil \frac{L_{\mathrm{DAT}}}{L_{\mathrm{NAND}}} \right\rceil \times T_{\mathrm{MEM\cdot N}}}{\left\lceil \frac{L_{\mathrm{DAT}}}{L_{\mathrm{SCM}}} \right\rceil \times \left(\frac{L_{\mathrm{SCM}}}{W_{\mathrm{DAT\cdot S}}} \times \frac{1}{P_{\mathrm{Toggle\cdot S}}} + T_{\mathrm{MEM\cdot S}}\right)} \end{aligned} \tag{12.7}$$

Assuming $T_{\mathrm{MEM\cdot N}}$ = 1.6 ms, L_{NAND} = 16,384 Bytes (32 sectors), L_{SCM} = 512 Bytes (1 sector), $P_{\mathrm{Toggle\cdot N}}$ = 400 Mbps/pin, $P_{\mathrm{Toggle\cdot S}}$ = 1066 Mbps/pin, $W_{\mathrm{DAT\cdot N}}$ = 1 Byte and $W_{\mathrm{DAT\cdot S}}$ = 1 Byte. The relationship of $R_{\mathrm{S/N}}$ and $T_{\mathrm{MEM\cdot S}}$ is shown in Fig. 12.30. From Fig. 12.30, it can be found that single SCM chip can achieve over 1000-times performance gain than the single NAND flash chip if SCM write latency is below 1 μs. For random writes ($L_{\mathrm{DAT}} < L_{\mathrm{NAND}}$), $R_{\mathrm{S/N}}$ is equivalent to the IOPS ratio of the SCM over the NAND flash memory. IOPS is used as the metric of

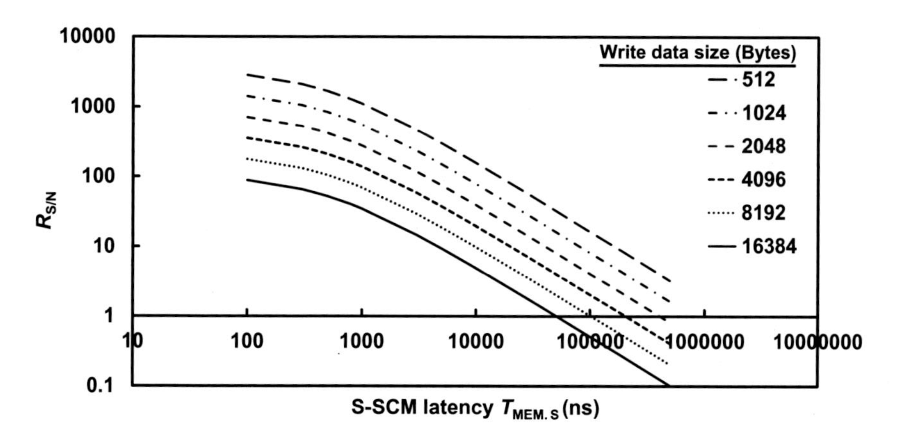

Fig. 12.30 S-SCM and 3D-NAND flash memory throughput ratio

the random write performance. If SCM has a latency of 100 μs, $R_{S/N}$ is less than 100 at any L_{DAT}. Furthermore, $R_{S/N}$ becomes less than 1 when the $L_{DAT} > L_{NAND}$. Thus, such a long latency SCM device is not cost-efficient. The curves almost overlap in Fig. 12.30 at the condition of $L_{DAT} > L_{NAND}$ (sequential write), which shows the lowest $R_{S/N}$ that the SCM can achieve. Below 50 μs, the SCM still has a higher performance than the NAND flash memory. However, the SCM is slower than the NAND flash memory if its latency is over 50 μs. It indicates that the NAND flash is good for the sequential write, thanks to the low write energy of the memory cell. In other words, achieving over 1000-times sequential write performance boost (>1 NAND flash page size) for SCM device is extremely difficult.

A recent ReRAM device [28] and MLC NAND flash memory [29] presented in ISSCC 2014 have the program latencies 10 μs/2048 Bytes and 1185 μs/16 kB, respectively. Comparing the SSDs made of these two devices, the performance gain of the all ReRAM SSD with 512 B random data pattern can be over 100-times (over 50-times at 4 kB random data pattern). To improve the system performance gain, the SCM chip latency should be further reduced or the device program current has to be reduced for increasing the number of parallel program cells (corresponding to L_{SCM}).

For the enterprise applications, the random data access is the bottleneck. Since the real workload is the mixture of random and sequential data, MB/s is still used as the overall performance metric throughout the chapter. The average IOPS $IOPS_{Avg}$ can be calculated by formula (12.8):

$$IOPS_{AVG} = \frac{Throughput_{AVG} \times 1024}{RS_{AVG}} \tag{12.8}$$

where RS_{AVG} is the average request size (kB) and $Throughput_{AVG}$ is the average SSD performance (MB/s). If the SSD average write performance is 20 MB/s with the 512 Byte random write data pattern, the average IOPS is equal to 40,960. When the SSD is tested with 4 kB random data, the average SSD IOPS is 5120.

12.4 Summary and Conclusion

Three techniques are presented in this chapter to improve the write performance of the 3D-NAND flash-based SSD. Table 12.1 summarizes the pros and cons of the techniques in this chapter. The SEA-SSD and LBA scrambled SSD are short-term solutions, which co-design the middleware and SSD controller. To overcome the hurdle of limited write performance of the NAND flash memory, hybrid M-SCM/3D-NAND flash memory SSD is the mid solution which is able to improve the conventional all 3D-NAND flash SSD write performance by over 10 times. In the long run, all S-SCM SSD is promising. With the system proposals like the SEA-SSD and LBA scrambler, the latency design parameters for the 3D-NAND flash memory are relieved. On the other hand, a larger page size and block size can

Table 12.1 3D-NAND flash design with system-level considerations

Solution	Pros	Cons	Implications for 3D-NAND flash design
SEA-SSD [3] (short-term)	• Low cost • Use upper layer information • Low data management complexity	• Only for database application	• Relaxed latency and endurance design constraints for 3D-NAND flash based SSD by integrating these techniques • Larger page and block sizes are acceptable for hybrid M-SCM/3D-NAND flash SSD, compared with the all 3D-NAND flash SSD • Customized 3D-NAND flash design enables write performance optimization for each application
LBA scrambler [10] (short-term)	• Low cost • Eliminate page fragmentation due to unaligned writes • Enhance write speed for all representative SSD workloads	• More DRAM capacity is required for tables • Interface may need to be modified	
Hybrid M-SCM/3D-NAND flash hybrid SSD [13] (mid-term)	• Cost-effective (compared to all M-SCM SSD) • Improve SSD write speed, power and reliability greatly	• Complicated tiered storage/cache algorithm is required	
All S-SCM SSD [27] (long-term)	• Little write performance dependency on application • In-place overwrite • No garbage collection is required • >100 times SSD write speed boost is possible	• High cost (currently) • S-SCMs like ReRAM are still in early development stage	

be accepted for the NAND flash memory by introducing M-SCM into the SSD, which is good for the design of the 3D-NAND flash memory. Due to the write performance dependency on the application, custom 3D-NAND flash design enables the performance optimization for each application.

References

1. K. Takeuchi, Novel co-design of NAND flash memory and NAND flash controller circuits for sub-30 nm low-power high-speed solid-state drives (SSD). IEEE J. Solid-State Circ. **44**(4), 1227–1234 (2009)
2. K. Takeuchi et al., A 56 nm CMOS 99 mm^2 8 Gb multi-level NAND flash memory with 10-MB/s program throughput. IEEE J. Solid-State Circ. **42**(1), 219–232 (2007)
3. C. Sun et al., SEA-SSD: a storage engine assisted SSD with application-coupled simulation platform. IEEE Trans. Circ. Syst. I **62**(1), 120–129 (2015)
4. FusionIO, http://www.fusionio.com/press-releases/fusion-io-software-development-kit-enables-native-flash-memory-access
5. A.M. Caulfield et al., Moneta: a high-performance storage array architecture for next-generation, non-volatile memories, in *Proceedings of the Annual IEEE/ACM International Symposium on Microarchitecture (MICRO)* (2010), pp. 385–395
6. A.M. Caulfield et al., Providing safe, user space access to fast, solid state disks, in *Proceedings of the International Conference on Architectural Support for Programming Languages and Operating Systems (ASPLOS)* (2012), pp. 387–400
7. A. Trivedi et al., Unified high-performance I/O: one stack to rule them all, in *Proceedings of the USENIX Workshop on Hot Topics in Operating Systems (HotOS)* (2013)
8. S. Peter et al., Towards high-performance application-level storage management, in *Proceedings of the Workshop on Hot Topics in Storage and File System (HotStorage)* (2014)
9. Synopsys Platform Architect, http://www.synopsys.com/Systems/ArchitectureDesign/Pages/PlatformArchitect.aspx
10. C. Sun et al., LBA scrambler: a NAND flash aware data management scheme for high-performance solid-state drives. IEEE Trans. Very Large Scale Integr. (VLSI) Syst. (2015 in press)
11. C. Sun et al., Cost, capacity and performance analyses for hybrid SCM/NAND flash SSD. IEEE Trans. Circ. Syst. I **61**(8), 2360–2369 (2014)
12. H. Fujii et al., x11 performance increase, x6.9 endurance enhancement, 93 % energy reduction of 3D TSV-integrated hybrid ReRAM/MLC NAND SSDs by data fragmentation suppression, in *IEEE Symposium on VLSI Circuits* (2012), pp. 134–135
13. C. Sun et al., A high performance and energy-efficient cold data eviction algorithm for 3D-TSV hybrid ReRAM/MLC NAND SSD. IEEE Trans. Circ. Syst. I **61**(2), 382–392 (2014)
14. C. Sun et al., SCM capacity and NAND over-provisioning requirements for SCM/NAND flash hybrid enterprise SSD, in *Proceedings on International Memory Workshop (IMW)* (2013), pp. 64–67
15. J. Jang et al., Vertical cell array using TCAT (Terabit Cell Array Transistor) technology for ultra high density NAND flash memory, in *IEEE Symposium on VLSI Technology* (2009), pp. 192–193
16. R. Katsumata et al., Pipe-shaped BiCS flash memory with 16 stacked layers and multi-level-cell operation for ultra high density storage devices, in *IEEE Symposium on VLSI Technology* (2009), pp. 136–137
17. J. Kim et al., Novel Vertical-Stacked-Array-Transistor (VSAT) for ultra-high-density and cost-effective NAND flash memory devices and SSD (Solid State Drive), in *IEEE Symposium on VLSI Technology* (2009), pp. 186–187
18. S.J. Whang et al., Novel 3-dimentional dual control-gate with surrounding floating-gate (DC-SF) NAND flash cell for 1 Tb file storage application, in *IEEE International Electron Devices Meeting (IEDM)* (2010), pp. 668–671
19. K. Miyaji et al., Control gate length, spacing, channel hole diameter, and stacked layer number design for bit-cost scalable-type three-dimensional stackable NAND flash memory. Jpn. J. Appl. Phys. **53**, 024201 (2014)

20. C. Lee et al., A 32 Gb MLC NAND-flash memory with Vth-endurance-enhancing schemes in 32 nm CMOS, in *Proceedings IEEE International Solid-State Circuits Conference (ISSCC)*, Feb 2010, pp. 446–447
21. K. Fukuda et al., A 151 mm^2 64 Gb MLC flash memory in 24 nm cmos technology, in *Proceedings on IEEE International Solid-State Circuits Conference (ISSCC)*, Feb 2011, pp. 198–199
22. S. Choi et al., A 93.4 mm^2 64 Gb MLC NAND-flash memory with 16 nm cmos technology, in *Proceedings on IEEE International Solid-State Circuits Conference (ISSCC)*, Feb 2014, pp. 328–329
23. M. Sako et al., A low-power 64 Gb MLC NAND-flash memory in 15 nm cmos technology, in *Proceedings on IEEE International Solid-State Circuits Conference (ISSCC)*, Feb 2014, pp. 128–129
24. C. Sun et al., A workload-aware-design of 3D-NAND flash memory for enterprise SSDs, in *International Symposium on Quality Electronic Design (ISQED)*, March 2014, pp. 554–561
25. Micron Technology Inc., MT29512G08CUCAB data sheet, Nov 2009, http://micron.com
26. K. Higuchi et al., Evaluation of voltage vs. pulse width modulation and feedback during set/reset verify-programming to achieve 10 million cycles for 50 nm HfO2 ReRAM. Solid-State Electron. **91**, 67–73 (2014)
27. T. Onagi et al., Design guidelines of storage class memory based solid-state drives to balance performance, power, endurance and cost. Jpn. J. Appl. Phys. (JJAP) (2015 in press)
28. R. Fackenthal et al., A 16 Gb ReRAM with 200 MB/s write and 1 GB/s read in 27 nm technology, in *Proceedings on IEEE International Solid-State Circuits Conference (ISSCC)*, Feb 2014, pp. 338–339
29. M. Helm et al., A 128 Gb MLC NAND-flash device using 16 nm planar cell, in *Proceedings on IEEE International Solid-State Circuits Conference (ISSCC)*, Feb 2014, pp. 326–327

Index

R. Micheloni (ed.), *3D Flash Memories*, DOI 10.1007/978-94-017-7512-0